T0314220

**Introduction to System Science
with MATLAB**

Introduction to System Science with MATLAB

Second Edition

Gary Marlin Sandquist

and

Zakary Robert Wilde

Registered Offices
John Wiley & Sons, Inc., 111 River Street, Hoboken, NJ 07030, USA
John Wiley & Sons Ltd, The Atrium, Southern Gate, Chichester, West Sussex, PO19 8SQ, UK

Editorial Office
111 River Street, Hoboken, NJ 07030, USA

For details of our global editorial offices, customer services, and more information about Wiley products visit us at www.wiley.com.

Wiley also publishes its books in a variety of electronic formats and by print-on-demand. Some content that appears in standard print versions of this book may not be available in other formats.

Library of Congress Cataloging-in-Publication Data applied for
Hardback ISBN: 9781119213963

Cover Design: Wiley
Cover Image: © BEST-BACKGROUNDS/Shutterstock

Set in 9.5/12.5pt STIXTwoText by Straive, Pondicherry, India
Printed and bound by CPI Group (UK) Ltd, Croydon, CR0 4YY

C9781119213963_221022

Contents

Preface

Purpose. The purpose of this text in system science is to provide the reader with a comprehensive, rigorous introduction to the recognition, definition, quantitative modeling, approximation, analysis, and evaluation of rational systems. The novel approach to system science developed herein is a fundamental one, based upon the principle of causality. This concept is pursued across all disciplines that are amenable to rational, deterministic logic. The goal of the textbook is to sufficiently develop those essential perceptions and skills within the reader so that the development and analysis of any rational system may be approached with confidence and competence. However, the reader is cautioned that no textbook alone in system science can provide the full resources and level of competence required to master the subject and contribute to the knowledge and understanding of systems in a significant manner.

Prerequisites. The emphasis of this textbook is on quantitative system science, and thus a basic background and skill with mathematics is essential for success. For full mathematical comprehension of all text material, the reader should be familiar with ordinary and partial differential equations and basic definitions and operations with matrices. However, with a basic understanding of ordinary differential equations and perhaps the assistance of a good instructor, the reader should be able to comprehend most material in the book.

Content and Arrangement. Chapter 1 introduces the principle of causality, or cause and effect, which is the basis for understanding quantitative modeling of rational systems. Chapter 2 provides the basic concepts and terms employed in system science. The mathematical formulation of the basic system equations governing any quantitative system is presented in Chapter 3. In Chapter 4 the mathematical treatment of single-input systems is explored for both continuous and discrete system models. The mathematical methods pertinent to multiple-input systems are developed in Chapter 5. Chapter 6 presents some of the essential methods required to quantitatively establish and model systems. In Chapter 7 the development and analysis of systems with linear kernels is addressed. Chapter 8 provides some of the important methods for processing and analyzing systems. Specific system science applications are demonstrated for a broad variety of system topics in Chapter 9. Chapter 10 describes paradigms for performing system modeling.

Suggested Course Outlines. The entire text taken in chapter sequence is suitable for two semesters or three quarter courses meeting three to five hours per week. Other reasonable chapter sequences are shown in the following diagram:

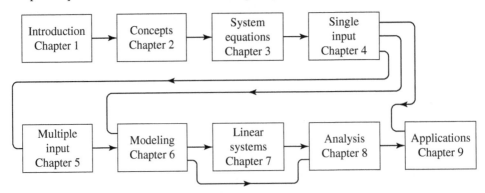

A detailed study schedule, appropriate for various disciplines, is given in Section 1.3.2 of Chapter 1. Both reading and applicable problems assignments are provided.

Examples and Exercises. Generally examples are provided to demonstrate each major system concept and method presented. Furthermore, over 750 system problems and exercises of graded difficulty are provided for developing system skills.

Computers and System Science. It is apparent that the digital computer, with its combined capabilities of high speed, great precision, massive data storage, and near universal availability, will have a profound impact upon system science. Even now a system investigator no longer needs to possess nor exercise broad and profound analytical and numerical skills to pursue sophisticated system studies. For the most part, even micro or personal computers have the capacity, speed, and software required to perform surprisingly sophisticated system analysis functions. And in the future, the system investigator will be able to devote major efforts toward accurate modeling and comprehensive evaluation of systems and leave the burdensome numerical tasks to the computer. Chapter 8 presents numerical algorithms in BASIC and FORTRAN for solving multiple-output system equations.

References. References, pertinent to the material in each chapter, are given at the end of the respective chapters. A rather complete bibliography for each of the major fields of application of system science is given at the end of Chapter 9.

Acknowledgments. The development and production of this book has been realized through the talents and efforts of many people. Former teachers, colleagues, and students have contributed in various ways. Professor R.F. Boehm and the staff in the Department of Mechanical and Industrial Engineering provided the resources and personnel to create the draft text. The staff of John Wiley & Sons have provided the professional effort and support to produce the finished textbook. Finally, I acknowledge my appreciation and love to my wife Kristine and children – Titia, Julia, Taunia, Cynthia, and Carl for permitting me the uninterrupted time necessary.

1

Introduction

In this introductory chapter, we will endeavor to define and describe system science as a rigorous, quantitative, scientific discipline for gaining and extending knowledge about ourselves and our environment and rational universe. We first introduce the principle of causality, which provides the essential basis for the effective practice of system science. The causality principle allows us to employ system science with reasonable confidence that our efforts are rational and useful. General applications of the causality principle will be examined, together with a brief background on the discipline of system science and its relation and contributions to the general pursuit of knowledge. The allied fields that system science embraces are briefly outlined to provide an overview of the great breadth and utility of system science. Finally, the chapter closes with a statement of the goals and an outline of this text with alternative study schedules for specific fields of application in system science.

1.1 System Science

Epistemology, which encompasses system science, is that branch of science that embraces the acquisition of knowledge by humans. Epistemology is derived from the Greek root "episteme" meaning to know or have knowledge. Epistemology attempts to address the various ways in which we seek to comprehend and validate our observations of and experiences with our environment and ourselves. Our observational abilities include our mental, physical, and emotional perceptions of our surroundings as well as our five-sensatory perceptions (sight, sound, touch, smell, and taste). Epistemology also addresses how we distinguish and identify accurate and useful knowledge from inaccurate or irrational knowledge as we seek to understand our environment and ourselves. Humans are constantly engaged in improving their understanding and knowledge of their environment. The rational, scientific, and quantitative corner stone in epistemology for gathering data, modeling, and quantifying knowledge about our environment and all the entities therein constitutes the field of System Science.

In practice, knowledge and understanding of our world are gained through system science, primarily by building rational models. These models are usually quantitative models, which attempt to represent our observations of "reality" as we perceive and measure these

Introduction to System Science with MATLAB, Second Edition.
Gary Marlin Sandquist and Zakary Robert Wilde.
© 2023 John Wiley & Sons Ltd. Published 2023 by John Wiley & Sons Ltd.

observations. Then we test these models for their value and accuracy in predicting and validating these observations. The critical test for the utility of these models is usually based upon experimental evidence from future observations of the actual system and comparison with our model for the system. We seek to expand and improve these models to provide a better description of our environment. However, it should be evident that no set of models can ever capture and assess all information regarding a particular subject of study. We can never completely describe and absolutely model any entity in our environment. Even if such a model or set of models could approach fully describing the entity, the models would probably be too complicated and unyielding to be of practical value for human comprehension and use.

The entities or subject matter in our environment that we seek to study and understand through system science can generally be addressed with a variety of different models that describe or simulate the subject in a given manner. The models chosen and developed will depend upon the subject addressed. The model's value is judged on the model's ability to describe actual observed behavior and yield correct predictions of future behavior of the given subject. Ideally, these models should be as simple as possible but as accurate as necessary to provide the understanding and level of knowledge we desire concerning the subject under investigation.

System science is a scientific domain that embraces virtually all scientific and technical disciplines. Using quantitative, mathematical methods for quantification of these models, system science addresses the physical sciences, engineering and technology, life sciences and medicine, the social and behavioral sciences, and has even been applied to philosophy and religion. This very broad venue for system science renders it a difficult subject to confine to a single academic discipline. Consequently, few formal academic programs exist that train students and grant degrees in these broad areas of system science. Nevertheless, it is through the effective and diverse applications of system science that we continually expand and improve our knowledge of our environment. To be a successful system science investigator or practitioner, the candidate should possess strong quantitative mathematical skills, computer competence, and basic knowledge in the areas to be studied and modeled.

This text attempts to provide the mathematical background and modeling tools and skills required to study any rational system with the assistance of MATLAB. Rational systems are those systems that satisfy the principle of causality, which will be explained in Section 1.3 of this chapter. The overall goal of this text is to introduce the reader to the recognition, definition, quantitative modeling, approximation, analysis, and evaluation of rational systems.

1.1.1 Definition of System Science

By the term "system science," we denote the total collection of knowledge, methods, data, and skills available for the identification, abstraction, modeling, quantification, analysis, synthesis, evaluation, and observation and control of rational systems and their behavior. An elementary definition of system science that is practical and comprehensive is given below.

System Science System science is that scientific discipline that employs quantitative, analytical, and numerical methods for analyzing, modeling, assessing, and predicting the nature and behavior of rational systems.

1.2 Principle of Causality

1.2.1 Definition

A basic principle of all science, engineering, technology, and rational knowledge, in general, is the implicit assumption that in any rational system, a measurable cause produces a measurable effect. This basic assumption is often referred to as the "principle of causality" or simply the "principle of cause and effect." The principle of causality can be briefly epitomized as follows.

Causality Principle The causality principle asserts that for every effect or output response exhibited by any rational system, there exists a definite set of causes or input stimuli that influenced or produced that effect.

Furthermore, the principle generally implies that identical causes imposed upon identical rational systems produce identical effects. For systems classified as stochastic systems, this equality for identical causes and effects must be interpreted on a statistical basis since uncertainty may exist either in achieving identical, repeatable tests of the system or because of the intrinsic stochastic behavior of the system. Quantum Mechanics recognizes the inherent uncertainty in the causation and measurement and prediction of possible events in the behavior of systems described by quantum mechanics. Quantum Mechanics asserts that any attempt to measure physically the response or behavior of a system affects its response. However, if the effects of a given set of inputs to a quantum mechanical system are properly interpreted within the methodology of quantum mechanics (e.g., quantum electrodynamics [QED] and quantum chromodynamics [QCD]) then these systems also satisfy the principal of causality. The standard model of particle physics (SMPP) provides the current quantum mechanical basis to address and model this field of physics.

For classical mechanics using Newton's second law of motion, if identical external forces (the causes) are imposed on identical material particles (the system), then the particles will cause each exhibit identical accelerations (the effects). This assumption of identical causes producing identical effects in identical systems is essential if the output response of an identical system is to be rational and predictable. Furthermore, it is the ability to predict the behavior of a given rational system before the actual response of the system occurs that makes the study of rational systems so important and useful. Thus, in principle, the behavior of any rational system can be analyzed and predicted for any set of causes or inputs using the methods and tools provided by system science. However, accounting for all or even the most significant controlling inputs or causes imposed on the system and knowing quantitatively how these inputs influence the system's response or outputs are generally very demanding tasks requiring training, skills, imagination, experience, and good judgment. Nevertheless, in spite of possible difficulties in analyzing and understanding complex systems, it is through the study of systems and their behavior that we develop a rational knowledge of our universe and ourselves. Furthermore, it is because of the principle of causality that the development of such knowledge is possible and useful.

1.2.2 Common Examples

The principle of causality is often associated with the concept of "determinism." Unfortunately, the scientific concept of determinism has been altered and redefined by some social scientists and philosophers. Certain implications of "social and philosophical determinism" are not appropriate or useful for the operational needs of system science. Determinism appropriate for system science is the concept that any event or condition that occurs or exists anywhere in the universe is or was caused or influenced by some set of preceding events or conditions, which acted on the universe to produce the event or condition that is observed or exists.

Again, for certain systems called "stochastic systems," this equality of effects must be considered on a statistical and probabilistic basis since it may not be practical or even possible to achieve an identical, completely repeatable test of these systems or the system is inherently stochastic. For example, identical coins randomly tossed may exhibit various results; but will, after many tosses, show the same statistical outcome. Furthermore, the spontaneous time for decay of a single radioactive nucleus cannot be predicted, but the decay rate of a very large number of the same radioactive nuclei can be predicted with great accuracy. Thus, as long as the methods of probability and statistics are properly applied to stochastic systems, these systems also satisfy the principle of causality if interpreted on a consistent statistical basis.

Quantum mechanics exhibits an important principle called the "Heisenberg Uncertainty Principle" that certain sets of physical measurements cannot be performed with arbitrary accuracy. For example, the precise, simultaneous measurement of an object's position (Δx) and corresponding momentum (Δp) or energy (ΔE) and event time (Δt) is limited to a very small error parameter determined as follows

$$\Delta p\, \Delta x \geq \frac{h}{2\pi} \quad \text{and} \quad \Delta E\, \Delta t \geq \frac{h}{2\pi}$$

where h is called Planck's constant ($h \cong 6.63 \times 10^{-34}$ Joule-seconds). However, the numerical value of this fundamental constant is so small in the world of normal human senses that the uncertainty in a typical physical measurement is unobservable. For example, the quantum uncertainty in determining the position of an automobile traveling at 60 miles per hour on the highway is less than the size of a single electron. Therefore, the uncertainty in position or speed is indiscernible with human sensory perceptions. However, the discovery and existence of the Uncertainty Principle provide important insight into the fundamental properties of nature. In effect, absolute physical measurements are not possible and do not even exist in SMPP for certain physical observations. These measurements must be interpreted within the domain of the applicable rules established for quantum mechanics.

In general, determinism assumes that quantum indeterminacy associated with quantum mechanics models can be ignored for macroscopic system models. Random quantum events equalize in the limit as large numbers of microscopic inputs enter the system. So, the laws of quantum mechanics asymptotically approach the laws for macroscopic system models and result in systems that are deterministic. Quantum effects rarely alter predictions of macroscopic system models that are accurate at all practical scales. Even objects as small as DNA molecules, biological cells, nerve impulses, and other microscopic entities generally behave as deterministic systems and can be modeled ignoring quantum indeterminacy.

Returning to practical applications of determinism as used here in system science, we find that the use of determinism is common and essential for ordinary human experience and observations. Consider, for example, the following statements that express deterministic observations or conclusions as given in Table 1-1.

The purpose of giving the statements in Table 1-1 is not to assert the statement's validity or truth, but only to indicate that each statement assumes the existence of a causal system in which inputs or causes imposed on that system result in given outputs or effects from the system that are consequences of these inputs. Thus, a cause in the form of input, condition, change, event, or other stimulus defined as the variable "**c**" for cause(s) when imposed upon a rational system produces or determines a new effect, situation, condition, change, event or response by the system yields the output variable "**e**." We define the variable **e** for effect(s). (Note that bold fonts for mathematical quantities indicate multiple input and output effects.) Formally, we can express such a cause and effect dependency or sequence for any rational system as follows:

$$\mathbf{c}\{\mathbf{cause(s)}\} \rightarrow \boxed{\textbf{SYSTEM}} \rightarrow \mathbf{e}\{\mathbf{effect(s)}\}$$

Table 1-1 Common Statements with Cause and Effect Consequences

Statement	Cause (C) and Effect (E)
1. Human overpopulation is the major cause of poverty and starvation in a country	(C) Overpopulation (E) Poverty and starvation
2. You only get out of life what you put into it	(C) Input to life (E) Output from life
3. Wars are the result of bad economic conditions	(C) Bad economics (E) Wars
4. Money is the root of all evil	(C) Money (E) Evil
5. Cigarette smoking causes lung cancer	(C) Smoking (E) Lung cancer
6. The sudden movement of adjacent tectonic earthen plates causes earthquakes	(C) Tectonic plate movement (E) Earthquakes
7. Adolf Hitler was the primary cause for initiating World War II	(C) Adolf Hitler (E) World War II
8. Human destiny is determined by the stars	(C) Stars (E) Human destiny
9. Inflation occurs when too much money chases too few goods	(C) More money than goods (E) Inflation
10. God's blessings are predicated on obedience to His commandments	(C) Obedience (E) Blessings
11. The Fatwas of Osama bin Laden provided an initial cause for world terrorism	(C) Fatwas (E) Terrorism
12. Gravitational force of attraction between mass particles is proportional to their mass and inversely proportional to separation distance squared	(C) Mass separation distance (E) Gravitation attraction force

It is implicit and imperative, therefore, that if the causes are absent or changed, then the effects produced or determined by the system will be absent or changed. Mathematically, we can establish a functional dependency or mapping between the causes (c) and effects (e) for a given system. We can express this dependence through the following "system equation of state" or simply the "system equation" as follows:

$$\mathbf{e} = g\,(\mathbf{c})$$

This mathematical relationship is simply an abstract, concise way of expressing the cause and effect principle or relationship. The $\mathbf{e}\{\text{effect(s)}\}$ issuing from a given system is dependent on how the $\mathbf{c}\{\text{causes(s)}\}$ acts upon on the given system. This response of the system due to the causes is described by the function $g(\mathbf{c})$. If the inputs (i.e., $\mathbf{c}$'s) to the system vary or change, the system's outputs (i.e., $\mathbf{e}$'s) or response will change accordingly as determined by the function or system kernel $g(\mathbf{c})$. We will show in subsequent chapters that the function $g(\mathbf{c})$ may also depend upon the effect $\mathbf{e}$ through feedback, so we have for the general system equation that

$$\mathbf{e} = g\,(\mathbf{e}, \mathbf{c})$$

where $g(\mathbf{e}, \mathbf{c})$ is now defined as the "**system kernel.**" The system kernel is the defining mathematical function that determines the quantitative behavior of the output variables $\mathbf{e}$ arising from the consequences of the system kernel $g(\mathbf{e}, \mathbf{c})$ acting upon the system.

The principle of causality also has rather interesting and profound philosophical implications that influence us and our awareness and understanding of the natural universe. The principle of causality is, in reality, not a proven law or even a self-evident truth and is a consequence of the "Second Law of Thermodynamics" or the assumption that causes precede effects. It is an extension of faith and belief that the universe, or more correctly, our perception and observations of the universe reveal a rational, comprehensible, objective universe. The implicit trust is that the universe is not inherently capricious and unpredictable but can be logically perceived and understood by the human mind through the application of system science as conceived by humans. Indeed, the principle of causality assumes that all real events occurring in our universe follow a rational, sequential pattern of cause and effect that may be perceived through human senses and understood by rational minds through logical interpretation of these perceptions.

It is paradoxical that many humans believe that science and the "scientific method" of inquiry do not assume certain basic, unproven premises and do not require faith in the validity of these premises. In reality, the belief (with great success) that human logic and mathematics contrived in the human mind can be successfully employed to predict and understand any event in the external universe is truly a profound act of faith and a miracle of the highest level. That this scientific act of faith is justified and successful is confirmed by the scientific and technical achievements and progress of the human family.

1.2.3 Relationship to System Science

In general, the formal mathematical analysis of rational systems has been limited for the most part to the physical sciences, engineering, and technology. The principle of causality and the rigorous analysis of scientific and technological-based systems resulting from the principle have proven very useful and productive. Indeed, we have developed very accurate and detailed system models for such complex and diverse physical and technical systems as our universe as formed by the "big bang," the solar system, complex chemical reactions, weapon systems, digital computers, electronic devices, and DNA the complete human genome. However, the effective application of the principle of causality to other rational systems and rigorous mathematical modeling and analysis of social (e.g., political, economic, cultural, ethnic, historic, religious) and esthetic (art, music, sculpture) systems has not occurred. It is claimed by some that mathematical modeling and quantitative analysis are not applicable or even possible for such systems. Admittedly, developing suitably acceptable quantitative models for such nontechnical-based systems is novel and challenging, and devising suitable means for measuring inputs and outputs can be difficult; but certainly, more can be done. Furthermore, even though these systems are generally complex with numerous inputs and outputs, matrix methods are available for formulating and manipulating large sets of data and equations. In addition, with the ubiquitous availability of powerful computers and their capacity for rapid and accurate processing of the extensive data files associated with large, complex systems, former unexamined rational systems can now be modeled and analyzed. The behavior of these relatively unexplored systems in the social, political, economic, cultural, ethnic, historic, religious, and esthetic fields can and should be addressed and studied using quantitative system science tools. The growing use of computer-based artificial intelligence (AI) for modeling and analysis is now ubiquitous in human endeavors.

1.3 Overview of System Science

1.3.1 Historical Background

The concept of cause and effect is ancient, and the operation of the principle was undoubtedly observed by early human societies as they witnessed the consequences of cause and effect for natural processes and their own acts. On the other hand, many of humanity's superstitions and objects of fear and worship were directed at phenomena and events that seemed inexplicable and without apparent causation.

The abstract modeling of systems and the mathematical analysis and evaluation of causal behavior is a relatively recent development in human history. A particularly significant development, which historically contributed to our present confidence in system science, was Sir Isaac Newton's brilliant applications of the "laws of motion" that he discovered in 1650. Newton successfully applied his discoveries to the motion of celestial bodies. This astonishingly accurate system modeling effort allowed Newton and others to predict the motion and location of terrestrial and celestial bodies with an accuracy limited only by measurement errors in experimental observations. Because of this success, Newton became an object of worldwide fame, admiration, and near deification by his contemporary society.

The technological impacts of system science became practically apparent with the advent of the Industrial Age or Industrial Revolution, as society began to design and build machines to accomplish and industrialize tasks previously performed with human or animal labor or natural forces. The energy required to power these machines was derived principally from fossil fuels, which were discovered and exploited in an evolutionary fashion to support the expanding energy requirements of the Industrial Age.

Since the industrial revolution, the growth in our inventory and breadth of technological devices and machines (which are simply rational physical systems fabricated by humans) and their impact on civilization is increasing at an exponential rate. The extent of this tremendous growth and environmental impact is humorously but cynically reflected by some detractors of science and technology. These critics bemoan that our advanced science and technology are responsible for empty aluminum Coca-Cola cans now found in remote, terrestrial jungles, and the manufactured radioactive element; plutonium-238 has been placed by the Apollo Space Mission on the moon as a remote power source. However, despite some anxiety and discomfort in adjusting to and correcting for rapid technological changes and a nostalgic desire for "the good old days" and "a return to simpler times," the essential dependency of the escalating human population on technology is remarkably evident. In reality, few people or nations seek to eliminate or reduce technological developments in sanitation, health care, food production, housing, electrical power, industry, transportation, internet, and communications. The global acceptance and ubiquitous presence of the personal cell phone is indisputable evidence of humanity's acceptance and addiction to modern technology.

Although most of the essential mathematical theory and analysis methods presently used in system science were developed much earlier, the first significantly directed efforts generally associated with fashioning a distinguishable discipline of system science as practiced today apparently began with the advent of World War II. Scientists, mathematicians, and engineers were employed by both the allied and axis countries to support each country's war efforts. Operations research, which is the application of rigorous mathematical methods to the modeling and analysis of military fields in system science such as submarine search, code-making and breaking, gun tracking, amphibious assaults, and ship movements, became an active, well-financed discipline. In retrospect, one of the most significant technical developments that occurred because of activities in system science applied to war efforts was the successful development of the electronic-based digital computer-based upon information theory and contributions of Claude Shannon, John von Neumann, and many others. With the subsequent development and universal deployment of these electronic data processors, the accurate and rapid processing of large databases and analysis of quantitative system models is now possible, practical, and omnipresent.

Shortly after the end of World War II, a strong development trend in system science arose based upon the use of transform methods from operational mathematics. In particular, the use of Laplace and Fourier transforms to develop transfer functions for the temporal description of systems became widespread. Many of the methods of analysis and synthesis in control theory, such as root locus methods, phase space and frequency domain analysis, gain and phase shift, Bode diagrams, signal flow graphs, and stability analysis, are traceable to prolific applications from operational mathematics and complex variable theory.

Current system science developments are directed at the analysis of multiple input and output systems using matrix methods for system formulation and analysis. Since operational methods such as Laplace and Fourier transform methods are essentially limited to linear systems with constant coefficients or parameters, much of the current education and advanced research in system science is directed at many variables, nonlinear systems often intensified with feedback. These systems generally require the solution of large sets of differential or difference equations using a vast array of numerical methods. These numerical methods require massive data storage capacity and extremely fast digital computers for practical analysis and evaluation.

It is apparent that the modern computer, with its uniquely combined capabilities of high speed, extreme accuracy, massive data storage, and near-universal availability, will profoundly affect essentially all areas of human development and enterprise and especially the field of system science.

1.3.2 Major System Science Achievements in the Twentieth Century

Although early system science featured great minds such Aristotle, daVinci, Euclid, Euler, Faraday, Galileo, Gauss, Leibniz, and course Isaac Newton throughout the past, the twentieth century has also posited significant system science agents who have brought the field of System Science into its current level of remarkable achievement.

Twentieth-century Systems Scientists can be sorted into approximate decade periods. From 1900 to 1950, system founders included Ludwig von Bertalanffy, Kenneth Boulding, Ralph Gerard, George Klir, James Miller, Anatol Rapoport, and John von Neumann. These scientists initiated and stimulated the scholarly basis and breadth that system science now addresses across diverse disciplines of natural and social science. From 1950 and later, scientists like Per Bak, Peter Checkland, Hugh Everett, Andrew Sage, and Stephen Wolfram have continued to make significant contributions to system science.

Table 1-2 provides a partial list of prominent system scientists given in alphabetical order who are listed in Wikipedia, the free encyclopedia. The System Scientist's "name, birth date and death date, and system science contribution" are provided with brief career details. Significantly, Table 1-2 includes three Nobel Laureates and the foremost mathematician, John von Neumann, of the twentieth century. A detailed description of these individuals can be obtained by entering the name from Table 1-2 into the Wikipedia search block where an expanded, detailed description for that person can be found.

1.3.3 Measurable Systems and Quantitative Modeling

The major system science emphasis of this text is on the quantitative modeling and analysis of rational systems. By this, we mean that, for us, a major goal in the successful investigation of any system is the establishment of a rational plan for assigning and measuring numerical values for all the system inputs and outputs to be modeled. The assignment of these values can be accomplished by the creation and application of explicit mathematical functions that express the system's change of output values in terms of inputs, outputs, and the changes in inputs that produce the output changes. This coupling between input changes and the resulting output changes can be expressed mathematically by the system kernel, to be defined in Chapter 3.

The statement of Lord Kelvin (William Thompson) on the role of measurement succinctly expresses our justification for this marked emphasis on quantitative modeling and measuring of system behavior as follows:

Measurements When you cannot measure it, when you cannot express it in numbers, your knowledge is of a meager and unsatisfactory kind. It may be the beginning of *knowledge, but you have scarcely, in your thoughts, advanced to the stage of science.*
Lord Kelvin Lecture on "Electrical Units of Measurement," 3 May 1883

Table 1-2 Prominent System Scientists in the Twentieth Century

---

Major System Science Scientists: Name, (birth - death year), and Contribution

---

Russell Ackoff (1919–2009) American scientist, management science, operations research

Genrich Altshuller (1926–1998) Russian engineer, scientist, Theory of Inventive Problem Solving

Pyotr Anokhin (1898–1974) Russian biologist, physiologist, cybernetics and functional systems

Leo Apostel (1925–1995) Belgian philosopher, research between exact science and humanities

W. Brian Arthur (1945–) Irish economist, complexity theory in technology and financial markets

Per Bak (1948–2002), Danish theoretical physicist, the concept of self-organized criticality

Bela H. Banathy (1919–2003) Hungarian systems scientist, systems research conferences

Yaneer Bar-Yam (1959–) American physicist, founder of New England Complex Systems Institute

Stafford Beer (1926–2002) British management scientist, operational research, organizations

Ludwig von Bertalanffy (1901–1972) Austrian Canadian biologist, Society for Systems Science

Maamar Bettayeb (1953–) Algerian control theorist, system scientist, singular value decomposition

Harold Stephen Black (1898–1983) American engineer, invented a negative feedback amplifier

Kenneth E. Boulding (1910–1993) British economist, educator, religious mystic, systems scientist

Valentino Braitenberg (1926–2011) German neurologist, cybernetics, embodied cognitive science

Richard Peirce Brent (1946–) Australian mathematician, computer science, Brent's method

Walter F. Buckley (1922–2006) American sociologist, applied general systems theory in sociology

Peter Checkland (1930–) British management scientist, developed soft systems methodology

C. West Churchman (1913–2004) American philosopher, educator, management science

James J. Collins (1965–) American bioengineer, stochastic resonance, biological dynamics

Vladimir Damgov (1947–2006) Bulgarian physicist, mathematician, nonlinear space dynamics

George Dantzig (1914–2005) American mathematician, founder of linear programming

David Easton (1917–2014) Canadian political scientist, systems theory for political science

Joshua M. Epstein (1956–) American epidemiologist, social and economic dynamics

Hugh Everett (1930–1982) American physicist, "Many Worlds Quantum Mechanic model"

J. Doyne Farmer (1952–) American physicist, founder of chaos theory, complexity economics

Mitchell Feigenbaum (1944–2019) American physicist, math, chaos, Feigenbaum constants

Peter C. Fishburn (1936–2021) American system scientist, pioneer in decision making processes

Irmgard Flügge-Lotz (1903–1974) German-American mathematician, control theory developer

Heinz von Foerster (1911–2002) Austrian-American scientist, physics, philosophy, cybernetics

Jay Forrester (1918–2016) American computer engineer, System Dynamics founder

Charles François (1922–2019) Belgian cybernetician, Encyclopedia of Systems Cybernetics

Ralph Gerard (1900–1974) American behavioral scientist, Nervous system pioneer

Murray Gell-Mann (1929–2019) American physicist, Nobel Prize in elementary particle theory

Harry H. Goode (1909–1960) American computer and systems engineer, management science

Arthur David Hall III (1925–2006) American electrical engineer, Bell Lab founder of IEEE

Friedrich Hayek (1899–1992) German economist, Nobel prize economics, complex phenomena

Francis Heylighen (1960–) Belgian cybernetician, self-organization, evolution of complex systems

John Henry Holland (1929–2015) American complex systems, founder genetic algorithms

Michael C. Jackson (1951–) British systems scientist, Systems Thinking, Cybernetics

J. A. Scott Kelso (1947–) Irish neuroscientist, founder Center for Complex Systems, Brain Science

Table 1-2   (Continued)

Major System Science Scientists: Name, (birth - death year), and Contribution

---

Faina Mihajlovna Kirillova (1931–2020) Belarusian mathematician, optimal control theory

George Klir (1932–2016) Czech-American computer scientist, intelligent and fuzzy systems

Klaus Krippendorff (1932–) German cyberneticist, mathematical foundations of cybernetics

Christopher Langton (1949–) American biologist and computer science, artificial life research

Edward Norton Lorenz (1917–2008) American mathematician, meteorologist, weather prediction

Niklas Luhmann (1927–1998) German sociologist, systems theory for political science

Donella Meadows (1941–2001) American environmental scientist, coauthor Limits to Growth

John von Neumann (1903–1957) Hungarian-American, foremost 20th century mathematician

Eugene Odum (1913–2002) American scientist, ecosystems, pioneer of ecosystem ecology

Ilya Prigogine (1917–2003) Russian physical chemist, Nobel prize irreversible thermodynamics

Anatol Rapoport (1911–2007) Russian psychologist, International Society for Systems Science

Eberhardt Rechtin (1926–2006) American systems engineer, aerospace systems and architecture

Andrew P. Sage (1933–2014) American systems engineer, information technology and engineering

Claude Shannon (1916–2001) American mathematician, cryptographer, information theory

Thomas B. Sheridan (1931–) American mechanical engineer, robotics and remote control theory

Georgiy Starostin (1976–) Russian linguistic research, Santa Fe Institute Human Languages

Len Troncale (1943–) American biologist, systems theorist, Cellular and Molecular Biology

John Warfield (1925–2009) American electrical engineer, Systems, Man, Cybernetics Society

Brian Wilson (1933–) British systems scientist, developer of Soft Systems methodology

Stephen Wolfram (1959–) British-American physicist, Computer scientist, business man

Warren Weaver (1894–1978) American mathematician, communication scientist.

Norbert Wiener (1894–1964) American mathematician, founder of cybernetics, Tauberian theorems

Lotfi Asker Zadeh (1916–2017) American-Iranian mathematician, founder of fuzzy concepts

Erik Christopher Zeeman (1925–2016) Japanese-British mathematician, topology /singularity theory

---

Qualitative models of systems that consist of verbal or sensual descriptions and qualifiers are useful in system science and generally serve as the initial stage in recognizing and establishing a causality situation for a given system. However, the quantitative system model with its measurable inputs and outputs can be unambiguously examined, quantitatively measured, and used to predict the behavior of the system that provides real understanding and knowledge about the system. Furthermore, and most significantly, a quantitative model may be subjected to experimental verification and critical peer review. Generally, the quantitative system model will stand or fall on how accurately the model reflects experimental observations and measurements in the real world.

Let us consider and compare a qualitative system associated with a cause and effect observation and a quantitative model for the same system that describes the system's input and output changes in an explicit, measurable, mathematical relationship.

First, the qualitative model statement: "Overeating will result in weight gain." Most of us would accept this statement as valid. However, upon consideration, we see that important information about the system (the human body), the input (excess food), and the output (excess body weight) is lacking. What constitutes overeating? Are all foods equally weight-producing, and is there some measurable unit associated with food consumption that relates to weight gain? What change in weight will result from overeating by 10% for one month? What are the consequences for undereating? We see that many questions about this system and its relationship between cause and effect remain unanswered.

Now let us consider a specific quantitative model for the same system. Our proposed system equation is as follows for the human body and dieting.

$$\Delta W = a\, W(t)[C(t) - C_0]\Delta t$$

where

ΔW = change in body weight or, equivalently, body mass measured in kilograms
a = individual weight gain factor measured in units of kilograms of body weight change per calorie of food intake
$W(t)$ = individual body weight at time t measured in kilograms
$C(t)$ = total caloric intake of the food consumed at time t per kilogram of body weight at time t measured in units of calories per kilogram of body weight per day
C_0 = mean value of caloric intake required to maintain constant body weight measured in units of calories per kilogram of body weight per day
Δt = passage of time measured in days

Now we observe that for our quantitative model, the associated mathematical statement to be compared with the initial qualitative model statement is..."Overeating, $[C(t) - C_0]\Delta t$, is proportional to a fractional change in body weight, $[\Delta W/W(t)]$." We observe that the input to the system, overeating, is $[\{C(t) - C_0\}\Delta t]$ and the output, change in weight, is $[\Delta W/W(t)]$, and the system is, of course, the human body. Our quantitative system model has the form of a difference equation. However, we will consider the model to be continuous and solve the resulting differential equation by separation of variables and integration to obtain as a solution for the quantitative system equation the following:

$$W(t) = W_0\, \exp\left\{ a\int_0^t [C(t) - C_0]\, dt \right\}$$

where we assume as an initial condition that the human body (the system) weighs or more correct as mass, W_0 (kilograms) at time $t = 0$.

This quantitative system solution can be used to predict the behavior of the system (weight or mass change), identify significant system parameters (e.g., a and C_0), and determine the range of application of the dietary model. To provide an example, suppose we assume by caloric analysis the following values for an average adult male.

$C_0 = 35\ \text{calories}/(\text{kg-day})$
$a = 0.001\ \text{kg/calorie (cal)}$

Furthermore, we want to predict long-term, average trends in weight change, so we let

$$\int_0^t [C(t) - C_0]\, dt = [\langle C\rangle - C_0]\, t$$

where $\langle C \rangle$ is the mean value of caloric intake during the time interval 0 to t. Then for these values and assumptions, our quantitative system solution becomes

$$W(t) = W_0 \exp \{0.001 \left[\langle C \rangle - 35 \right] t \}$$

Now we can ask the question, "What body weight (viz., body mass) increase for an adult male can we expect from overeating 10% for 30 days?" We find from our equation, for $\langle C \rangle = 110\% \times C_0 = 38.5$, that

$$\frac{W}{W_0} = \exp \left[0.001 \text{ kg/cal } (38.5 - 35.0) \text{ cal/(kg-day)} (30 \text{ days}) \right]$$

and

$$W \left(30 \text{ days} \right) = 1.11 \, W_0$$

Therefore, our model predicts an 11% increase in body weight due to overeating by 10% for 30 days. Observe that the model can even be used to predict the weight (as mass) loss sustained from a true "crash diet" of no food for two weeks. Of course, we must recognize the model's limitations and realize that the human body might collapse in this period. However, for this example, assume that adequate water, which has no caloric content, is taken to sustain life for the two weeks. So, for $C(t) = 0$ and $t = 14$ days, we have

$$\frac{W}{W_0} = \exp \left[0.001 \text{ kg/cal} \{ -35 \text{ cal/(kg-day)} \} 14 \text{ days} \right] = 0.61$$

and a weight loss of nearly 40% could be achieved, probably with great risk to human health and even life.

Observe that the quantitative model indicates how other factors that influence weight change might affect the system. For example, a given person's activity level and body metabolism would probably influence the parameters, a and C_0.

It should be readily apparent that a quantitative model for weight gain is a superior description of the human body and its metabolic behavior than a qualitative model. Lord Kelvin's assertion that our knowledge of systems is much greater when we can quantitatively measure the behavior of the system is surely true.

1.3.4 Application of Computers to System Science

Undoubtedly, we have arrived at the age of the electronic digital computer, with its capacity for processing information and data on a scale of volume and speed unknown to our ancestors. Central processor units, which serve as the essential digital system for control of the computer and processing of all numerical operations, can exceed many petaflops (10^{15} flops or floating-point operations per second) in the most advanced computers. This astonishing performance, which continually improves with time, appears to be ultimately limited only by natural laws, such as the speed of light for electrical signal transmission and the packing density of distinguishable physical or electrical states in the matter, such as single atoms.

It is interesting to speculate on the possible applications and capabilities that the computing power available in current high-performance computers (HPCs) can provide. For example, we can crudely assume that the entire written human record is equivalent to about a million (10^6) words (i.e., 2000 pages at 500 words per page) for each person on the earth (with a population estimated to be over 7 billion people). This written record could be processed by an advanced computer in less than an hour, if we ignore the obvious problems of data collection and data input. For a scientific application, three-dimensional space and time-dependent mathematical equation could be numerically evaluated at one million space points for one billion-time steps in a few hours. However, the greatest surprise occurs when we contrast the HPC's arithmetic processing capability with a human processing the same data. Crude estimates reveal that the calculation output of one HPC operating for one hour would exceed the manual calculation output of the earth's entire human population for over a year to provide the same accuracy and reliability for the data processed. Finally, from an economic point of view, about 1 US dollar of computer time would purchase the equivalent of almost ~1 billion US dollars of human computational effort.

It is apparent then that the computer offers accurate data processing capabilities that cannot be approached by humans in a practical way and that, in the area of data processing, computers will be used more and more to satisfy our computational demands as they arise.

Admittedly, access to very fast, advanced computers is limited and costly. However, of greatest significance to the potential applications of computers in system science is present development of personal computers (PCs) that have moderate but generally adequate speed and storage capability and are inexpensive and widely available. It is apparent that personal computers are as common and accessible as cell telephones and television are.

Personal computers offer the system investigator adequate speed and storage to accurately analyze and evaluate most quantitative system models that the investigator may wish to study, including systems with many inputs, outputs, and model parameters.

It is important to stress here that quantitative system models are essential for digital computer processing since the computer uses numbers based upon the two-state binary numbering system for all data processing. Thus, all data must be converted to numerical data for processing by the computer.

1.3.5 Utilization of Computer Software in System Science

A vast array of software programs exists for performing the mathematical operations associated with modeling and evaluating quantitative systems. These programs provide the system analyst with extraordinary power in rapidly and accurately performing the onerous task of manipulating and evaluating the quantitative expressions and complex algebra that frequently accompanies the assessment of system models.

Significantly, computer software has been developed to process and evaluate symbolic mathematical expressions, such as algebraic forms, differential and integral equations, and matrices. For the reader with limited experience and time constraints in manipulating and solving ordinary and partial differential equations, symbolic computer software is a significant and useful resource for modeling. We will demonstrate that capability in this text, particularly in Chapter 9, which addresses specific system science applications for a broad range of system topics. Unfortunately, these symbolic algorithms are somewhat limited for treating those classes of equations (system kernels) that are classified as multi-variable, nonlinear, and exhibit extensive feedback. Nevertheless, the potential for analytical processing of these complex mathematical expressions is promising and improving with time.

A partial list of some of the symbolic software available for system analysis studies is provided in the following list. Because of its availability and useful application to graphics, matrix manipulations, and symbolic applications, we shall focus on MATLAB, a proprietary, multi-paradigm programming language and numeric computing environment developed by MathWorks.

- Macsyma
- Magma
- Maple (Maple provides the symbolic kernel for MATLAB and Mathcad)
- Mathcad
- Mathematica
- MathXpert
- MATLAB (Including the Symbolic Math Toolbox. See Maple above)
- Sage
- Wolfram Alpha
- Vensim Personal Learning Edition (PLE)

Another software of special interest here is Vensim's Personal Learning Edition (PLE) available from Ventana Systems. Ventana Systems Inc. is a Harvard, Massachusetts-based computer firm founded in 1985 for large-scale business simulation models. Ventana Systems hosts the simulation language called Vensim (PLE) that, significantly, is available without charge for educational use and system science modeling. Vensim is a valuable modeling resource for the reader.

Prudently, lest we be overly impressed with the present and future capabilities and applications that are offered by computers, it is important to recognize what even advanced computers cannot provide or perform. Computers cannot provide imagination and creativity in model conception and abstraction. Furthermore, the ability to evaluate, compare, and make judgments about real system behavior and predicted model behavior and then to remodel and innovate to produce improved and more accurate system models are human enterprises. Finally, and perhaps most important, the essential purpose of system science is for humans to gain a better understanding and knowledge of themselves and the universe in which they live. This purpose can be realized only by the human investigator.

Undoubtedly, the best resource for effective system modeling, analysis, and evaluation exists when human investigators effectively couple their unique, homo sapien capabilities with the speed, accuracy, and massive data processing and storage offered by the computer. Such an interactive combination of the computer and the human operator can be viewed as a basic system for the processing and evaluation of system models, as shown in Figure 1-1.

Fig. 1-1 Model for investigating systems.

The computer (the system) receives input information in the form of model data from the computer's external environment. The computer acts upon these data as programmed by the human operator to produce output data descriptive of the model system. The human operator (who functions as a feedback system) analyzes and evaluates these output data from the computer as an independent system and returns modified data (as extrinsic feedback) to the computer system for further computer processing and refinement.

This overall cyclic process can continue until the human operator is satisfied with the final system model that has been created and evaluated. As we shall learn in subsequent chapters, such systems are referred to as feedback-controlled, closed-loop systems and constitute an important field of study in system science.

1.3.6   General Applications of System Science

The successful application of the methods of system science to broad areas throughout the physical and biological sciences is readily apparent. Microbiology, solid-state physics, electronics, organic chemistry, and meteorology are obvious examples where system conception, modeling, analysis, and evaluation have dramatically improved our knowledge and understanding in these sciences. Furthermore, the engineering by-products of system synthesis are evident in the profound impact of technology on human life. Undoubtedly, the extensive utilization of system sciences in all scientific and technical fields will continue. Furthermore, these same fields provide the background and basis for many of the best examples of system modeling and investigation that can be employed and studied in textbooks in system science.

However, a very important but relatively untapped field for the methods and applications of system science lies in areas outside the physical sciences, where historically, quantitative, mathematical based investigations have not been applied to the myriad of systems that exist in the social sciences, economics, humanities, and esthetics. The principle of causality, which applies to all rational systems, not just those in the "hard sciences," can be applied to all rational systems. However, significant progress will not be made in applied system science to these "nonstandard" fields unless investigators interested in these fields acquire an adequate working knowledge of both the methods of system science and understanding and competence in these respective fields. It is interesting that many of the most important and unresolved issues confronting human societies today are found in the nonscientific fields of human knowledge and endeavor where system science has had little impact. These social, moral, and economic issues include inflation, unemployment, famine, crime, divorce, drugs, welfare, taxation, military expenditures, international arms buildup, the threat of major outbreaks of war, and finally, terrorism and the risks from weapons of mass destruction.

These issues can affect individuals and nations, and the successful resolution of or improvement in any of these issues is vital to a safe, healthy, prosperous world. However, it is paradoxical that our understanding is limited by considerable ignorance of the essential rational systems associated with each of these issues and ignorance of the important causes and effects that exist. Often, irrational, unsound decisions and judgments arrived at without the benefit of quantitative analysis and evaluation results. Admittedly, these important societal issues, as well as many others on which system science has had little influence, represent complex systems that are difficult to identify and isolate. These systems typically possess many diverse inputs, with compound interaction between these inputs due to feedback and system coupling of the output. Nevertheless, all these real systems are rational and are therefore subject to cause and effect. In spite of the difficulties of system identification and modeling, of measurable assessment and observation, these rational systems obey the principle of causality, and if they can be and are effectively studied, modeled, analyzed, and evaluated, they can hopefully be successfully controlled.

1.4 Outline and Utilization of Text

1.4.1 Outline of Text

In **Chapter 1**, we describe the discipline of system science and introduce the principle of causality, which provides the rational basis for system science. General applications of this principle are given, and cause and effect relationships for various systems are demonstrated. The vital role that the computer plays in support of applications in system science is presented. A brief historical review and future prospective of system science closes Chapter 1.

Chapter 2 introduces the basic terms and concepts employed in system science. These are defined, explained, and demonstrated for the various system models that are encountered in system science.

Chapter 3 The essential task of translating the principle of causality into a quantitative mathematical model for systems is undertaken in Chapter 3. The generalized system equations for both continuous and discrete systems with and without feedback and for any number of inputs and outputs are derived.

In **Chapter 4**, some important mathematical methods for the analytical treatment of system equations for single-input systems are addressed. The general single-input, single-output system equation is examined in considerable detail for appropriate solution methods and schemes. Both discrete and continuous system equations are considered.

Chapter 5 Mathematical methods for the analytical treatment of multiple-input system equations are examined in Chapter 5. The general, discrete, multiple-input system equation is examined, together with a variety of continuous, multiple-input system equations.

Chapter 6 undertakes the tasks of identifying and modeling systems and quantitatively formulating their kernels. Block diagrams, signal-flow graphs, and organizational diagrams are introduced as graphical modeling tools. Varieties of techniques for modeling physical and nonphysical systems, using experimental and heuristic methods, are addressed.

Chapter 7 is devoted to the description and analysis of system equations that are linear or can be acceptably approximated by linear system equations. The methods and consequences of approximating system equations by linear expressions are introduced, and the general solution forms are investigated.

In **Chapter 8**, some of the important methods for processing and analyzing systems are considered, including reduction and modification of inputs and outputs, normalization and transformation of system variables, parameter reduction and minimization, treatment of systems with feedback, and computer analysis of system models.

In **Chapter 9**, a broad variety of systems from many fields are identified, modeled, analyzed, and evaluated. Most of the modeling and analysis concepts, methods, and schemes developed in this text are demonstrated in this chapter that also provides an extensive bibliography of selected texts relevant to system science.

In **Chapter 10**, a basic, generalized, quantitative algorithm or paradigm is developed and demonstrated to undertake the general study and analysis of a typical system. Obviously, there is no single, comprehensive approach for modeling and analyzing all systems, but this algorithm should provide insight into practical and effective steps for approaching a system designated for study and achieving satisfactory results. The chapter examines many tools utilizing MATLAB, particularly the symbolic toolbox for assisting in both addressing, manipulating, developing, and evaluating system equations as they are created, modified, and completed for addressing a given system study and analysis.

1.4.2 Study Schedules by Discipline for this Text

Table 1-3 provides a suggested study schedule outline by discipline for this text. Appropriate chapter sections for reading and applicable problem assignments are provided for several disciplines. These disciplines include applied mathematics, the physical sciences and engineering, business and management, the biological sciences and medicine, social sciences and psychology, history and fine arts, and philosophy and religion.

Suggested use of the table by an instructor or student in a given discipline is to examine the chapter sections outlined in the appropriate discipline and to select a representative set of problems from each chapter. Successful execution of these problem assignments is essential to understand and test comprehension of the chapter material.

1.5 Summary

In this very important introductory chapter, the discipline of system science and the principle of causality are introduced and defined. The basic sequential train of essential cause and consequent effect for all rational systems is examined. Common examples of rational systems are considered, and the concept of determinism is introduced in which chains of cause and effect give rise to all the rational observations and information experienced by humanity. The vital linkage between system science and the principle of causality is examined, and a brief historical overview of the evolution of system science is traced to the present time.

The importance of quantitative system modeling and numerical analysis and evaluation in system science is stressed. The application of computers by systems investigators as powerful tools for system analysis and evaluation is examined. Finally, the application of system science to the needs and interests of the system investigator is examined, and the many opportunities for further observation and study are reviewed.

Table 1-3 Suggested Study Sections and Problem Assignments by Discipline

Discipline	Chpt 1	Chpt 2	Chpt 3	Chpt 4	Chpt 5	Chpt 6	Chpt 7	Chpt 8	Chpt 9	Chpt 10
Applied Mathematics	All	All	All	All	All	All	All	All	9.1, 9.2	All
Problems	1, 4, 6, 8, 10	1, 2, 5, 6, 7	All	All	All	All	All	All	Study	
Biological & Medical Science	All	All	3.1, 3.2.1, 3.2.2	All	Omit	All	All	All	9.1, 9.4, 9.5	All
Problems	1, 3, 5a, 9	1, 2, 4, 8, 9	1a-m, r, t-v, x, z, 4, 6, 12	All	Omit	All	7.1, 7.2, 7.3	All	Study	
Business & Management	All	All	All	All	Omit	All	All	All	9.1, 9.6	All
Problems	1, 4, 8, 9	1, 2, 5, 8, 10	1, 2, 3, 4, 6, 11, 12	All	Omit	All	1, 2, 3	1, 2, 3a-c, 4, 5, 6, 8, 9	Study	
History & Fine Arts	All	All	All	All	Omit	All	All	All	9.1, 9.7, 9.10	All
Problems	1, 2, 3, 9, 10	1, 2, 3, 10	1a, b, e, f, h, I, t.v, x, z, 4, 6, 12	06-Jan	Omit	6.1.1, 6.	All	All	Study	
Philosophy & Religion	All	All	All	All	Omit	All	All	All	9.1, 9.9	All
Problems	1, 2, 5, 7, 10	1, 2, 5, 6, 7	All	All	Omit	All	All	All	Study	
Physical Sciences & Engineering	All	All	All	All	All	All	All	All	9.1, 9.2, 9.3, 9.8	All
Problems	1, 4, 6, 8, 10	1, 2, 5, 6, 7	All	All	All	All	All	All	Study	
Social Science & Psychology	All	All	All	All	Omit	All	All	All	9.1, 9.6	All
Problems	1, 2, 8, 9, 10	1, 2, 3, 4	1a, b, e, f, h, I, t, v, x, z, 4, 6, 12	All	Omit	1a-j, 3, 5, 10, 11	All	All	Study	

Bibliography

BATESON, G., *Mind & Nature: A Necessary Unity*, Hampton Press, 2002. Classic thinking on patterns connecting living beings the environment.

BENDER, E. A., *Introduction to Mathematical Modeling*, Dover, New York, 2000. Basic methods for mathematical modeling of systems.

BERTALANFFY, VON L., *General System Theory Foundations, Development, Applications*, George Braziller, New York, 1968. Established General System Theory (GST) and developed a bridge for interdisciplinary study of systems in social sciences. BERTALANFFY is often considered as the "founding father" of system science.

BORN, M., *Natural Philosophy of Cause and Chance*, Oxford University Press, New York, 1949. Probably the best single volume treatment on the role of causality and determinism in the physical sciences. Born, a deceased Nobel Laureate, related causality to quantum mechanics.

BRIDGEMAN, P. W. *The Nature of Physical Theory*, Dover Publications, New York, 1936. An excellent insight by Bridgeman, another deceased Nobel Laureate, into the mathematical modeling of nature.

BUEDE, D. M., *The Engineering Design of Systems: Models and Methods*, John Wiley & Sons, 2000. Application of system analysis to engineering design.

BUNGE, M., *Causality & Modern Science*, Transaction Publishers, 2011. Concepts and principles concerning causes and effects versus determinism.

CAPRA, F., *Web of Life-A New Scientific Understanding of Living Systems*, Anchor Press, 1997. Applications of system science to living systems.

CHECKLAND, P., *Systems Thinking, Systems Practice*, John Wiley, 1999. 30-year retrospective on Systems thinking and practice.

EDDINGTON, A. S., *The Philosophy of Physical Science*, Cambridge Press, New York, 1949. A stimulating discussion of the author's thoughts and speculations in science.

EDWARDS, P., and A. PAP, eds., *A Modern Introduction to Philosophy*, The Free Press, New York, 1965. A critical discussion of determinism with comments and observations by distinguished philosophers.

FRIGG, R., and S. HARTMANN, *Models in Science, Stanford Encyclopedia of Philosophy*, Spring 2006 Edition. Useful source of modeling in science.

LUHMANN, N., *Social Systems*, Stanford University Press, Palo Alto, CA, 1996. Applications of system science to social systems.

MEADOWS, D. H., *Meadows Thinking in Systems, A Primer Paperback*, 2008. Bestseller in system theory.

MELLA, P., *Systems Thinking: Intelligence in Action*, Addison Wesley, Reading, MA, 2001. A system dynamics modeling approach to problem solving.

MILES, Jr, R. F., ed., *System Concepts*, John Wiley & Sons, New York, 1973. A series of lectures by experts from various fields in systems science.

PRIGOGINE, I., *The End of Certainty*, Simon & Schuster, 1997. Pioneer of chaos and self-organization theory. Nobel Prize in 1977 for work on self-organization reconciling systems theory concepts with system thermodynamics.

RUTHERFORD, A., *Mathematical Modelling Techniques*, Dover, New York, 1994. Classical methods for mathematical modeling of systems.

SAGE, A. P., and J. E. ARMSTRONG, *Introduction to Systems Engineering*, John Wiley & Sons, 2000. Practical introduction of applications of system science in engineering.

SCHROEDINGER, E., *Science Theory and Man*, Dover Publications, New York, 1975. Human perspectives and the relationship to scientific thought by the deceased Nobel Laureate who established the basic mathematical model (Schroedinger wave equation) for quantum mechanics.

SENGE, P., *Fifth Discipline. Art and Practice of Learning Organization*, Doubleday, New York,. 1990. Applications of educational organizational modeling.

WIENER, N., *The Human Use of Human Beings. Cybernetics and Society*, Avon, New York, 1950. Established the field of cybernetics for relationship between computers and humans

Wikipedia, the free encyclopedia, https://en.wikipedia.org/wiki

WOLFRAM, S., *A New Kind of Science*, Wolfram Media, LLC, 2002. A novel and fundamental approach to analysis of entire field of rational systems. Author outlines these concepts and methods using Mathematica in this formidable and massive ~1200-page text.

ZEIGLER, B. P., et al., *Theory of Modeling and Simulation*, Academic Press, New York, 2000. Provides essential needs for system and control engineers for modeling and simulation.

Problems

1 Consider each statement as a system in Table 1-1 in this chapter, together with the respective causes and effects for their validity. Can you identify the system associated with each cause and effect? How would you attempt to establish and confirm each statement? Could some cause and effect factors be interchanged, and the associated statement still be accurate? What other causes or effects might be associated with the systems?

2 For the following sacred, historical, and political writings, identify at least five cause and effect statements in each writing. Reference the location of each statement and identify the system, the cause(s), and the effect(s) for each statement cited. Is a quantitative measurement of the system and its causes and effects possible?

a) Book of Proverbs (Holy Bible)
b) The New Testament (Christianity)
c) The Koran (Islam)
d) The Bhagavad-Gita (Hindu)
e) The Writings of Confucius
f) Theravada Scriptures (Buddhist)
g) Taoist literature (Taoism)
h) US Declaration of Independence
i) U.S. Bill of Rights
j) Mein Kampf (Adolf Hitler)
k) Das Capital (Karl Marx)
1) Communist Manifesto (Marx/ Engels)
m) On Liberty (John Stuart Mill)

n) Magna Carta (English)
o) NATO Alliance Treaty
p) Treaty of Versailles (World War I)
q) Gettysburg Address (A. Lincoln)
r) Salt I and II Treaties
s) United Nations Charter
t) War and Peace (Leo Tolstoy)
u) Macbeth (William Shakespeare)
v) Les Misérables (Victor Hugo)
w) Tale of Two Cities (Charles Dickens)
x) Gulag Archipelago (Alexander Solzhenitsyn)
y) The Inferno (Dante Alighieri)
z) Moby Dick (Herman Melville)

3 Verify the results and correct, if necessary, the data used in Section 1.2.2 for the weight gain model. Nutrition data can be found in standard medical science sources.

4 Assume that an ultimate limit for computer storage density and speed of memory interrogation is based on the properties of pure crystalline silicon, with the further assumption that a single data bit can be stored on and retrieved from each atom of silicon. If naturally occurring silicon's atomic mass number is 28.1, density is 2.33 grams per cubic centimeter, and Avogadro's number is about 6.02×10^{23} atoms in a gram molecular weight of substance, what is the storage capacity for 64-bit words or bytes within a 1 cubic centimeter sphere of silicon? What is the approximate data transfer time within the silicon sphere (assume light speed is $\sim 3.0 \times 10^{10}$ cm/s for signal movement)? If the memory transit time were equivalent to the time to access a storage word, how long would be required to access the entire storage content of the memory?

5 a) Examine and critique the social or philosophical view of "determinism." Write a brief statement of the referenced view and contrast that view with the system science concept of determinism.

    b) Examine the biological view of natural evolution or "Darwinian natural selection." Write a brief statement of the alternative view of "special creation or intelligent design" and contrast the two viewpoints with regard to the system science concept of determinism.

6 Examine Sir Isaac Newton's development of the laws of motion. For each of the three laws, identify the system, the cause(s), and the effect(s).

7 Respond to the claim that "assigning quantitative measurement to human-based qualities such as character, wisdom, and knowledge is impossible, and if it were possible, it would be demeaning and disrespectful of our humanity."

8 Speculate on the advanced capability of computers and their potential applications in the year 2050. Do you believe that it will be possible for computers to be programmed to model, evaluate, and make ethical, moral, legal, and social judgments and thus replace human judgments? Could such a computer program be used to impart unbiased decisions and judgments in legal matters, such as court actions, and would these computer-generated assessments be accepted by society?

9 Suppose that rational, very accurate models for our major social and political problems could be developed. Would the public accept the results and predictions made by such models? Consider a model for the "prevention of nuclear war" as a reference case. How would countries with nuclear weapons respond to such model results and conclusions if the model dictated a prescription for the elimination or international control of a given country's nuclear weapons? How do you respond to denuclearization?

10 Respond to the following assertion: "Since all human, rational knowledge is based upon absolute cause and effect, then the past and future for any system can be fully understood and completely known in principal using system science coupled with complete and accurate interpretation of the causality principle." Consider the assertion by the eminent French scientist and mathematician, Pierre Simone de Laplace, that all future events must rigidly follow from their initial, primal event.

2

Fundamental System Concepts

Before the general study of systems as conceived in system science can be effectively approached, basic terms and concepts must be defined and explained so that mutual understanding and agreement can exist among investigators. The purpose of this chapter is to provide these important definitions and explanations and to generalize the nature and variety of rational systems.

2.1 Definitions of System Concepts and Terms

System science is similar to other scholarly endeavors such as physics, biology, mathematics, and philosophy in that there exists a nomenclature or basic set of words and concepts within the field that have precise definition and special meaning. A common understanding of the nomenclature employed in system science and an agreement on the description of these concepts and definitions of these words and terms are necessary if communication between those working in the same field is to be accurate and effective.

2.1.1 Concept and Definition of a System

The most important concept in the study of system science is that associated with a system itself. Although the word "system" is commonly used in casual conversation and writing, system is often used indiscriminately and incorrectly. When the term "system" is used casually, it is apparent that the concept can be hazy and lacking in precision and rigor. Such casual use of the system when applied incorrectly to the field of system science can lead to misunderstanding and even significant error.

A precise definition of a system with general agreement in system science and will be employed throughout this text is as follows:

> **System** Any collection, grouping, arrangement or set of elements, objects, or entities that may be material or immaterial, tangible or intangible, real or abstract to which a measurable relationship of cause and effect for a given system exists or can be rationally assigned and measured.

Introduction to System Science with MATLAB, Second Edition.
Gary Marlin Sandquist and Zakary Robert Wilde.
© 2023 John Wiley & Sons Ltd. Published 2023 by John Wiley & Sons Ltd.

This is a formal definition, but it is necessary if the full meaning and significance of the system concept is to be appreciated and exploited. We will expand on the concept of a system and related concepts and meanings later in this chapter.

2.1.2 System Causes

A system is a collection of elements that can exhibit cause and effect phenomena consistent with the basic principle of causality. We also need to define precisely and formally the concepts of cause and effect. For the concept of cause, we provide the following definition for a system cause or input to the system:

> **Cause or System Input** Any change, disturbance, perturbation, input, or other external stimulus that exists or is produced within the system's external environment and is or can be applied to or imposed upon the system.

Causes, or system inputs as they are more formally called, are stimuli to the system that arise outside the system from the system's external environment. These stimuli when imposed upon the system cause the system to respond with one or more measurable effects or system outputs. These outputs are related to and caused by the initiating input in accordance with the causality principle.

2.1.3 System Effects

The principle of causality for systems asserts that in every rational system, any and every output response or effect produced by the system is a direct consequence of one or more causes or inputs imposed upon the system. Our definition for a system effect or output is as follows:

> **Effect or System Output** Any change, disturbance, perturbation, output, or response that is produced by or a consequence of the system's response to its inputs and is perceived and measurable in the external environment of the system.

2.1.4 Measurability of System Causes and Effects

We have used the term "measurable" in the definition of several system terms. To specify our intent for this concept as applied to systems, we define measurability as follows:

> **Measurability** Those inputs and outputs associated with the stimulation and response of a system that can be rationally perceived, interpreted, and satisfactorily measured by the common human senses (e.g., sight, hearing, touch, smell, taste, heat, cold), the rational communication media (e.g., language, mathematics, print, music, art), or other rational, logical, or value judgments.

Although it is not always necessary or possible to assign a numerical measure to each of the inputs and outputs of a rational system, it is highly desirable and useful if this can be done. In general, only by defining, comparing, and evaluating numerical values can we effectively perform quantitative analysis and evaluation of a system and its behavior subject to inputs and outputs. Furthermore, computers used for system studies must be supplied with quantitative data for most processing operations.

2.1.5 Isolation of a System from Its External Environment

To make system analysis practical and manageable, it is necessary that the prescribed system under study and its components or elements be identified and effectively isolated from the remaining universe of all other systems and influences that might affect the subject system. Thus, we must define the concept of a system boundary, which provides separation and isolation of the system from its surrounding universe. The system boundary effectively separates the system for any interaction with its environment except for the movement of inputs into the system and outputs from the system. A formal definition of the system boundary is provided as follows:

> **System Boundary** A physical or conceptual boundary that contains all the system's essential elements and effectively and completely isolates the system from its external environment except for those inputs and outputs that are allowed to move across the system boundary.

The system boundary constitutes a barrier to interaction and communication between the system and its surrounding environment and external universe. The only interaction or communication the system is allowed to have with its external environment is through identifiable inputs delivered to the system from the system's environment and identifiable outputs transmitted from the system into the system's environment. Thus, the system's only interaction or coupling with the system's environment is through measurable inputs and outputs crossing the system boundary.

Observe the importance of recognizing and identifying these inputs and outputs that can cross the system boundary. If these inputs and outputs are not identified and properly assessed, the behavior of the system may not be properly understood and not accurately modeled. A simple example here is the fundamental law for conservation of total energy for a closed system (a well-established system model or accepted law of physics). This conservation law applies to a given system only if all inputs and outputs of mass and energy crossing the system boundary are known and correctly measured. If inputs or outputs with significant mass or energy content influence a given system without correct accounting, then the fundamental law for conservation of total energy cannot be accurately employed for that system.

The system boundary separates the system and its contents from the universe surrounding the system and provides a passageway through which inputs and outputs can be identified and measured as they move in and out from the external environment. We carefully define this external environment for the system as follows:

System's External Environment The system's external environment is composed of the remaining universe of everything not contained within the system's boundary. Communication and interaction between the system and its environment occurs only through inputs and outputs crossing the system boundary.

2.1.6 Intrinsic and Extrinsic System Feedback

Another important concept, which requires a formal definition, is that of system feedback. Systems that exhibit feedback are often referred to in systems science as closed-looped systems, since an output is coupled or "closed" with one or more system inputs. Systems with no feedback of any of the outputs are, in this same vein, referred to as open-loop systems, since the input is uncoupled or unaffected by the system's outputs. We will find that systems that exhibit feedback often display unusual or unanticipated behavior, since feedback of an output can markedly affect the successive input, and this cycle may continuously repeat in the presence of feedback.

Two classes of system feedback must be distinguished and defined. The first class of system feedback is intrinsic or internal system feedback, defined as follows:

Intrinsic Feedback The internal feedback of a system output that is modified and recycled within the system to alter inputs delivered to the system. Internal feedback is generally not evident or measurable in the external environment of the system.

The other class of system feedback is extrinsic or external system feedback, which is defined as follows:

Extrinsic Feedback The external feedback of a system output that is modified and recycled within the external environment of the system and alters the inputs to the system in addition to other inputs that are delivered to the system.

Additional system concepts could also be defined and explained here, but we have provided the basic set of concepts for us to begin the study of system science. These additional system concepts will be defined as they are encountered in the text.

2.2 Discussion of System Concepts

With the formal definitions of the major concepts in system science now given, it is appropriate to expand on these concepts.

2.2.1 Concept of a System

The number of possibilities for different rational systems that can be conceived and formulated is unlimited. The only practical constraints on the conception and formulation process are that the resulting system model be rational and consistent with the basic assumptions and definitions of system science. The essential constraint and requirement for analysis is that the system model and all its elements must satisfy the principle of causality. This principle will be translated into formal mathematical form in Chapter 3.

Rational system models are, in practice, only limited by our imagination and modeling skills. System models may range from a model for a single biological organism or cell that we isolate in the laboratory to investigate its behavior and output responses to various inputs, to such intangible and esoteric systems as "man's inherent desire to recognize and worship a Supreme Being." Identifying the significant inputs and outputs for this latter system offers an interesting challenge for theologians or philosophers.

In Chapter 9, we will attempt to provide a generalized classification scheme for major system categories. However, the classification is not exhaustive or complete. Perhaps, as we will come to recognize in our study of system science, the single most important trait or skill that an effective system investigator requires is an active imagination and the ability to innovate and draw correct conclusions as various modeling needs become apparent.

It is also important to realize at the onset of our study of systems that, in general, it is not possible or necessary to model any given system completely and exactly. Being mortal and limited in our understanding of the operation and interaction of systems within the universe, we can only identify and quantify in an approximate way all those factors that define and influence a given system. Furthermore, it is not necessary that various models of a given system be identical. Which and how many elements are assigned to a given system are often somewhat arbitrary and, generally, a matter of considered judgment and experience. How the elements that constitute the system are grouped or arranged into subsystems and how these subsystems of the system are assigned a functional dependence between respective input and output is again a matter of judgment and experience. Often, the beginning system investigator attempts to account for too much detail and unimportant factors in the system. The result is a model that is difficult or impossible to analyze or even formulate. The result is often that the model is abandoned.

One experienced investigator of a given system may formulate and evaluate the model in a much different fashion than another experienced investigator. Yet both investigators may derive useful and accurate information and understanding about the same system and its behavior. Of course, some models of a given system are better representations than other models in that they are more descriptive and accurate. To acquire the training and skills required to conceive, recognize, model, analyze, and evaluate systems with acceptable accuracy and completeness is one of the basic goals of systems science. This text is directed at contributing to that goal.

2.2.2 Isolation of a System from the Environment

To distinguish between the system and its environment and to permit the clear identification of the inputs and outputs of the system, it is necessary to construct in a real or abstract sense a system boundary or enclosure envelope. This boundary or envelope isolates and confines the system and all its essential elements under study from the remainder of the universe and all other systems. All inputs and outputs passing across the system boundary are then determined by their direction of flow, as shown diagrammatically in Figure 2-1. Inputs are those stimuli or signals (causes) that flow from the external environment across

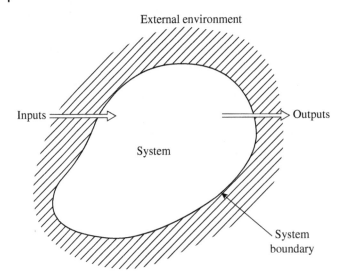

Fig. 2-1 Isolation of a system from its environment.

the system boundary and into the system. Outputs are those responses or signals (effects) that flow from the system across the boundary into the external environment. No other communication or interaction exists or is allowed to occur between the system and its external environment, except through the medium of input–output exchange across the system boundary.

Actually, the creation of boundaries or enclosures for controlling and inventorying certain common systems is a familiar and useful concept used throughout civilization. Sovereign nations (the systems) usually establish national territorial boundaries (viz., land, water, and airspace), control, and monitor the movement of materials and people across these boundaries. In the study of thermodynamics, systems under investigation are isolated and enclosed with a boundary to account for the movement of mass and energy into and out of the system. Even the proverbial family budget involves enclosing a boundary around the wallet, purse, or checking account (the system) and ideally accounting for all income (inputs or credits) and expenses (outputs or debits), with an extrinsic feedback control system generally imposed by a family member to control the system by influencing both the inputs and outputs.

Although, in theory, our system boundary should completely isolate the system from its outside environment so that we can account for all movement of inputs into and outputs out of the system, in practice this is often not possible or practical or generally necessary. What is essential is that we have adequately assessed the various possible inputs and outputs, and we have accounted for the significant inputs that are primarily responsible for the outputs sought. Furthermore, no significant outputs are overlooked or ignored that could seriously affect our analysis and evaluation of the system.

Provocatively, modern cosmology now asserts that about 95% of the energy (~72%) and matter (~23% mass) in our observable universe is "dark" or unknown. However, the observable effects upon cosmic accelerating expansion and galactic behavior are ascribed to these dark entities in a rational, systematic manner.

2.2.3 Identifying and Distinguishing Between Causes and Effects

After the system under study has been defined and the active elements identified, the system must be isolated from its environment by the creation of a boundary or enclosure. It is across this system boundary that all inputs (causes) must enter the system and all outputs (effects) must exit. Now the important task of identifying and quantifying the significant inputs to and outputs from the system begins. Sometimes there is uncertainty as to what are the significant inputs and outputs and which respective signals are inputs and which are outputs.

Generally, distinguishing between inputs and outputs is obvious. For those systems subject to causal time passage, the inputs causing a given output effect must precede that effect in time. This causal time flow is associated with the second law of thermodynamics that requires in any closed system the entropy or randomness and disorder must increase with time. For example, the production of a fertilized egg must be preceded by the fertilization event. The breakage of a glass object is not observed to reassemble itself later.

However, other system inputs and outputs are not so easily individually distinguished and related. For example, does an impoverished nation's famine result in part from widespread disease or does the famine initially contribute to the disease. The conclusion here would depend on the particular nation (i.e., the system), the level of disease and the state of famine at some initial time, and how these factors interacted in time for the nation.

Often, the choice of declaring which variables are inputs and which are outputs for a system is rather arbitrary and can be specified by the system investigator. Consider a gas contained within a closed volume (the system) whose boundary is the control or containment volume for the gas. Whether we consider changes in the volume and confining pressure at the system boundary as inputs and the resulting temperature change and heat flow from the gas across the boundary as the output, or the addition of thermal energy as an input and gas volume and pressure changes as outputs, is arbitrary. The inputs and outputs only need to be specified by the investigator in a consistent manner in either case.

Presumably, what must be guarded against is assuming for the system model a sequence of inputs and outputs that violates a natural restriction or law relating to such cause and effect sequences. Some sample violations include the following:

1) The reduction of total entropy with increasing time in a closed (isolated) system.
2) Time reversal in a macroscopic system, such as reverse aging (note, however, that time reversal may not be a constraint for certain few body microscopic systems such as the elastic collision of atomic particles).
3) The flow of heat from a body at lower temperature to a body at higher temperature without the input of energy or work to produce this heat flow (a violation of the second law of thermodynamics).
4) Negative friction or the action of friction to increase rather than decrease directed motion of an object subjected to friction.
5) General event sequences that are regarded as irreversible or not statistically probable in nature (such as the simultaneous collection and compaction of all the gas molecules in a container to create a substantial void volume within the container).

Actually, examples (2), (3), (4), and (5) are special cases of example (1) (i.e., conditions that result in a reduction of entropy with time for a closed system). Indeed, the increase of entropy in a closed or isolated system is an alternative measure of time change for a natural physical system.

The concept of entropy developed for the physical sciences also plays an interesting role in many other areas of human knowledge. Those systems outside of the physical sciences also demonstrate a tendency to move towards their most probable state of existence. Originally conceived for macroscopic states of matter and their observable properties such as pressure and temperature, entropy is the tendency for any closed collection of matter and energy to move towards that state for the system, which has the highest probability of occurring. This conclusion may appear obvious, but if all of the possible states for any system can be identified and the probability of occurrence of each state determined, then the proper conclusion derived from the concept of entropy is that any system, physical or nonphysical, over time will move to its most probable state of existence. Interestingly, if the most probable state for formal, human civilizations and empires (e.g., Roman, Aztec, Third Reich, USSR, etc.) to attain is eventual collapse, then entropy dictates that such empires will ultimately collapse.

Generally, the experienced system investigator is aware of such natural restrictions and entropy principle associated with physical systems. However, the existence and operation of similar restrictions for intangible and abstract systems is often more difficult to determine and assess. For example, the natural constraints or restrictions that might exist in a social or philosophical system may be difficult to identify, and unambiguous sorting of system inputs (causes) and outputs (effects) may be a challenge. Perhaps the only advice possible here is that the investigator must exercise critical review and judgment, as well as a thorough understanding, in modeling and evaluating all systems.

It is obvious that a common input for many systems, both physical and nonphysical, is the passage of time. Only for an unusual system model would we consider time change as an output. In fact, time change is the only essential input considered for many systems treated in standard systems analysis courses and textbooks. However, our perspective of systems and system science is completely generalized, and we will allow any inputs or outputs that can be associated with a rational system.

After the process of identifying and categorizing the inputs and outputs for a given system is completed, the investigator may find that the number of inputs and/or outputs is too large for the level of effort of analysis and evaluation available. The next step then is to rank the inputs and outputs individually based on their contribution to the system response of interest. A compromise must often be made here between system modeling accuracy and analysis effort, since model complexity can increase rapidly with an increase in the number of inputs or outputs. The system investigator should develop some awareness of and skill for achieving a satisfactory compromise between system accuracy and analysis results as we pursue our study of system science.

2.3 Classification of Systems by Type

An important skill we should develop in system science is the ability to determine accurately the type and characteristics of the system under investigation so that subsequent efforts in modeling, analysis, and evaluation will be most effective. The first, obvious issue to be resolved is that the system under consideration is rational and the principle of causality actually applies. The system and its elements then must be clearly recognized and defined, and a suitable system boundary established. Then the significant and essential inputs and outputs to the system are identified, and a consistent set of measurements is assigned to each input and output.

It is during this process of modeling the inputs and outputs and establishing the quantitative relationship between them that the following important issues arise.

1) Are the changes in inputs into and outputs from the system discrete, that is, do they occur as incremental, discrete step changes? Alternatively, are these changes reasonably considered as continuous? Must these changes be discrete and serially independent?
2) Are the system inputs the result of macroscopic (i.e., contribution of many individual responses) averaging processes, or should a single input or output change be interpreted as a microscopic (i.e., only a few responses) or random result that must be interpreted statistically?
3) Does the system exhibit intrinsic feedback; and, if the system is to be externally controlled, what form of extrinsic, external feedback can and should be imposed for satisfactory system behavior?

Let us now consider each of these system types in detail to ensure that the systems examined are rational and measurable and we can classify the system for modeling and analysis.

2.3.1 Irrational and Immeasurable Systems

By an irrational system, we mean a system that does not obey the principle of causality or, at least, does not appear by rational means and judgment to do so. Violation of causality can occur when effects or outputs issue from the system without apparent causation or the inputs to the system have no rational bearing on the observed system response. Of course, we must be careful in declaring a given system to be irrational simply because of complexity or our failure to adequately identify and monitor valid inputs and outputs to the system. The transmission of inaudible signals through the atmosphere by radio signals (electromagnetic, EM, waves) seemed irrational and mystical because they could not be seen by humans. However, this was demonstrated first by Marconi, and now we are engulfed by EM emission from both human and natural sources. Curiously, astrology has been a major driver for many of these irrational beliefs and fears. Astrology is an irrational, unreasonable, and occult practice. How the presence and movement of extraterrestrial objects in the cosmos can affect common human events and the future is unreasonable. However, current cosmology does entertain the existence of dark matter and dark energy, and even parallel universes, which to the general public may seem irrational.

Let us consider some systems that are deemed to be irrational and without scientific basis or reality.

Irrational Explanation of Existing Events or Prediction of Future Events

1) Astrology, fortune-telling, occult, supernatural, séances, etc.
2) Good and bad luck omens
3) Walking under ladders, breaking mirrors, etc.
4) Burning incense, incantations, séances, etc.

5) Possession of a rabbit's foot or lucky charm
6) Divination, I-ching, crystallomancy (reading of a crystal ball)
7) Using playing cards or tarot or dice to foretell the future
8) Reading tea leaf distributions in a teacup, palmistry, phrenology
9) Presence and distribution of stars or other objects in the cosmos

Null Systems (systems without correlation of cause and effect)

1) Null or zero output systems: system produces no output for any input.
2) Null or zero input systems: system outputs arise without inputs.

It is apparent that in each of these systems the inputs to the respective system bear no rational, quantitative influence on the future and our fortunes or other outputs. How card shuffling, dice throwing, the stirring of tea leaves, or the astronomical distribution of stars could foretell the future is not credible or apparent. Furthermore, passing under a ladder, breaking a mirror, or burning incense (i.e., assuming we are not bodily injured by these acts) bears no influence on our fortune. Moreover, the possession of a rabbit's foot for good luck was certainly not a lucky event for the rabbit that lost the foot. On the other hand, we should be careful in declaring a system as irrational before careful investigation and thoughtful judgment. For example, although gambling may seem capricious, unwise, or even immoral to some, games of chance are generally fully governed by rational models with measurable inputs and outputs, and the correlation between these is usually a model based on probability and statistics. Given random tosses of an unbiased coin (the input), the probability for getting heads (an output) or tails (another output) is one-half and is well modeled by the binomial distribution. In conclusion, our studies in this text are not applicable to irrational systems, and we will not consider such systems any further other than to identify and dismiss them if they arise.

Immeasurable systems are those systems for which a quantitative measure cannot be rationally assigned to the system's inputs or outputs. The neophyte investigator in system science often tends to assess or declare many important system inputs and outputs as immeasurable. Usually, the problem here is not the system, but the lack of skill and imagination on the part of the investigator to define or invent a consistent basis for quantitative measurement and assessment of the input and output variables. Admittedly, this can be a challenging task, particularly for the inexperienced or unskilled investigator, but it generally can be accomplished. What is essential to recognize is that the system model for assigning and scaling a given input or output is rarely unique or does it need to be. All that is required is that measurements be reasonable and consistent. For example, for a system that has cash flow as an input or output, we can simply measure that variable by the number of dollars or other currency involved. But for an output or input such as loyalty, love, or kindness, we must establish some range of value, say −1 to +1, where, for example, −1 represents maximum disloyalty or unkindness and +1 represents maximum loyalty or kindness, and then attempt to interpret what intermediate values within this range signify and how they may be consistently measured. We will see that, with imagination and boldness, we can generally define consistent quantitative measures or scales for most system inputs and outputs. The failure to establish a consistent basis for measuring numerically the inputs and outputs of any rational system should generally be regarded as a weakness in the system model and not a limitation of the actual system. Measurements and quantitative descriptions for most systems are generally best accomplished with numbers.

2.3.2 Continuous and Discrete Systems

Mathematically, a continuous system is a system whose inputs can be continuously varied by arbitrarily small changes, and the system responds to these continuous inputs with an output that is continuously variable. On the other hand, a discrete system is a system in which the inputs change by discrete amounts (like building a brick wall with each successive input being the addition of another brick to the wall). For discrete systems, there are generally certain minimum input step changes that cannot be reduced in size; thus, the output of the system changes in a related, discrete, step change fashion. Another example of a common discrete system is the population of humans in a given region, which obviously changes by integer values.

Although both continuous and discrete systems will be considered in this text, because of simplicity and convenience in modeling and analysis, continuous systems will be emphasized, as is generally the case in system science. Since continuous systems are assumed to exhibit continuously variable inputs (causes, c's) and outputs (effects, e's), the differential representation of changes in these quantities (i.e., dc's and de's) is valid and meaningful. Furthermore, the developments and resources of mathematical sciences associated with continuous variables are available to us for the modeling and analysis of such systems.

However, for actual systems, it is significant to observe that, in reality, physical variables such as energy, matter, momentum, electric charge, and even space and time apparently cannot be indefinitely reduced in magnitude, but that minimum finite quantities (or quanta) appear to exist for each of these quantities. Modern chemistry is based on the assumption that the atom of an element is the smallest basic unit of that element that is of principal interest, and further reduction produces a remnant structure that no longer exhibits the full properties of the atom. Thus, one can continue to reduce the quantity of mass of a given element only to the point to which a final single atom of the element remains. Further reduction destroys the general properties and characteristics associated with the element. This general observation is true for any basic mass particle, whether it be a molecule, atom, nucleus, electron, proton, or neutron. The Standard Model of Particle Physics (or SMPP) considers the electron and quark as the fundamental particles of matter. The recent theories of "String Theory and M-Theory" assume that all entities comprising the universe are composed of very small, open and closed vibrating strings of energy and have properties measured by Planck units. For example, the Planck units for length, time, mass, and energy in SI units are shown in the following table. It is obvious that the magnitude of most of these Planck units is far outside human perception. Contrast the Planck volume at about 10^{-105} cubic meters with the Planck density at about 10^{96} kilogram/cubic meter. This numerical span is greater than 10^{200}. Interestingly, it is estimated that the total mass in our observable universe is about 10^{53} kg or equivalently about 10^{80} hydrogen atoms, and this value is considerably less than 10^{200}.

Table of Planck Units in M-Theory

Planck Unit	Magnitude	SI Units (Abbreviation)
Planck length	1.6×10^{-35}	meter (m)
Planck area	2.1×10^{-70}	square meter (m²)
Planck volume	4.2×10^{-105}	cubic meter (m³)
Planck mass	2.2×10^{-8}	kilogram (kg)
Planck mass	22.0	micrograms (nano kg)
Planck density	5.2×10^{96}	kilogram/cubic meter (kg/m³)
Planck energy	2.0×10^{9}	Joule (J)
Planck temperature	1.4×10^{32}	Kelvin (K)
Planck time	5.4×10^{-44}	second (s)
Planck momentum	1.0×10^{-34}	kilogram-meter/second (kg-m/s)
Planck electric charge	1.9×10^{-18}	Coulomb (C)

Continued arbitrary reduction to an infinitesimal quantity as is implied by a differential mass variable dm is, strictly speaking, not defined; and the smallest mass change that can be imposed on or issued from a system is the minimum mass or energy particle associated with the mass description. Thus, for example, for a system in which we are varying the electron density, the smallest definable change is the addition or removal of a single electron to or from the system.

This same observation on the quantization of mass properties is true for energy, momentum, electric charge, and apparently for space and time as suggested in the preceding table of Planck units. The magnitude of the minimum quantities for these variables entails concepts from advanced topics in atomic and nuclear physics, such as quantum mechanics, and would be an unwarranted diversion of study here. Suffice it to state that the proper and accurate mathematical description of physical systems is apparently most correctly described by discrete variable changes (i.e., Δc's and Δe's), where these quantities represent minimum changes (e.g., minimum energy, electrical charge, space or time quanta, or mass particles) in the cause and effect variables. Usually, however, the size of these minimum definable quantities is sufficiently small compared to the magnitude of the practical changes in the system inputs and outputs considered that the assumption of continuous variation of these quantities is a reasonable and useful assumption.

For nonphysical science systems such as social and philosophical systems, where the causes and effects are not associated with such fundamental physical properties as mass, energy, and space, the question of continuity of system changes in cause and effect variables arises. In many such systems, the nature of the cause and effect changes is obviously discrete, while in other systems the behavior of the system's cause and effect may appear continuously variable. The choice of whether to describe the system by discrete, continuous, or even mixed cause and effect variables must remain with the analyst and is a matter of convenience, training, insight, and analytic skill.

2.3.3 Deterministic and Stochastic Systems

Although both deterministic and stochastic systems obey the principle of causality, essential differences between these two system types must be recognized and properly considered in the modeling and analysis of systems.

Deterministic systems are those systems whose inputs and outputs may be meaningfully measured and interpreted on an effective single event basis. By this, we mean that a given input set will interact with deterministic systems in the same way repeatedly and result in the same statistical output set for each identical input set. Thus, the measurement or sampling of the system's specific response for a given output is uniquely determined by the input, and repeated trials with identical inputs will yield the same outputs.

Stochastic systems are primarily those systems for which it is not practical or even possible to impose identical inputs or sustain these identical input distributions over a sufficient duration to achieve identical outputs for the inputs distributions. For some stochastic systems, the elements composing the system cannot be maintained in or returned to an identical initial state. The concept of "identical inputs resulting in identical outputs" is valid for stochastic systems only in the limit as the set of identical inputs is repeated sufficiently so that distribution function for the possible outputs becomes dominant. The concept of identical conditions cannot be implemented for stochastic systems in any practical way since the inputs and the system cannot or will not return to the identical conditions existing previously. For example, consider the toss of a coin (our system) to a surface, which results in heads showing. If the coin could be tossed again and all factors influencing the behavior of the coin were the same as previously existed, a heads toss would undoubtedly occur again. However, for many stochastic systems, it is not possible to duplicate exactly all factors influencing the system; and because of this uncertainty or statistical variation in both the system and the inputs influencing the system, we assume that a statistical interpretation is appropriate and, indeed, essential to model the system accurately.

Interestingly, the concept of simultaneous measurement of certain physical properties for exact reproducibility of a given state for a system of atomic particles is not permitted in atomic and nuclear physics (i.e., quantum mechanics). The uncertainty principle states that the very act of measuring each particle's dynamic state (e.g., linear and momentum and position, or energy and time) induces an uncertainty in that state, and so the actual state can only be known to within a small, but nonzero, fundamental limit. Therefore, in stochastic systems where statistical variations in the inputs exist or are allowed to occur, what we seek is a practical statistical output that is the "expected" value for the output when the sampling is performed repeatedly under identical conditions. The determination of these expectation values (crudely, the mean or average values) for the system's outputs is an important analysis goal for stochastic systems. The success of modeling efforts for many stochastic systems attests to the feasibility and accuracy of stochastic analysis that are possible.

2.3.4 Feedback Systems

System feedback or output recycling occurs whenever any of a system's outputs is returned or recycled (i.e., feedback) into the system through one or more of the inputs to the system. This modified input then produces a modified output compared to the output that would occur in the absence of feedback. It is essential to recognize two different types or classes of feedback that can possibly exist for a system. These types of feedback are:

1) Intrinsic or internal feedback
2) Extrinsic or external feedback

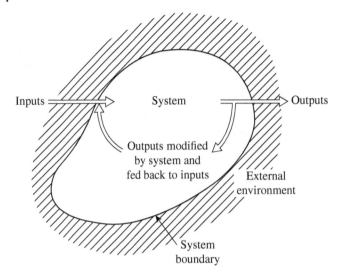

Fig. 2-2 Intrinsic or internal system feedback.

The essential difference between these two types of feedback, as we will observe, is in what region (i.e., within or without the system's boundary) the feedback occurs.

Intrinsic or Internal Feedback

When an output produced for emission from the system is internally modified within the system's boundaries by elements belonging to the system and alters one or more inputs coming into the system, we have intrinsic or internal system feedback. Figure 2-2 displays schematically this process of intrinsic or internal system feedback. Observe that no part of the output being modified by the system crosses the boundaries into the external environment. The only observable effects in the external environment are the actual inputs and outputs that cross the system boundary. Thus, for intrinsic system feedback, the system is responsive to and controls the feedback occurring internally within the system boundary. Therefore, the presence of the intrinsic feedback within the system is often not apparent or even detectable from the system's external environment.

Extrinsic or External Feedback

When an output produced by the system crosses the system boundary and is then modified within the external environment and fed back or returned to alter one or more inputs entering the system, we have extrinsic or external system feedback. Figure 2-3 displays schematically this process of extrinsic or external system feedback. Observe that all modification of the system output occurs in the external environment of the system. Thus, the system is unaware that the inputs it is receiving have been modified because of the behavior of the system's outputs. If the same inputs are duplicated by the system's external environment

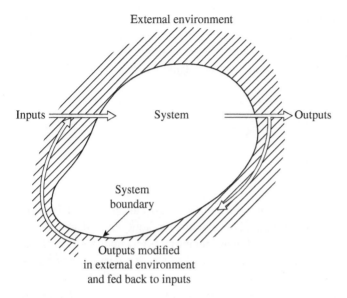

External environment

Inputs — System — Outputs

System
boundary

Outputs modified
in external environment
and fed back to inputs

Fig. 2-3 Extrinsic or external system feedback.

for transmission to the system, whether external feedback of system output occurs or not, the system's response is the same.

A natural observation that arises here is, assuming that the system investigator is not part of the system but observes the behavior of the system from the external environment through the inputs and outputs crossing the system boundary, how is the presence of intrinsic or internal feedback determined? In general, the answer is that intrinsic system feedback cannot be detected unless, upon redefinition of the system and/or its boundary, we can isolate and convert the source of feedback into an extrinsic source for actual identification.

Thus, as an observer situated in the external environment, we are usually limited to observing only the inputs and outputs that cross the system boundary. How the system actually processes these inputs to produce its output response is generally not apparent, and the same output may be accomplished with or without intrinsic feedback. For example, the basic system equation governing the change in total energy of a system with the change in system mass is Einstein's celebrated special law of relativity, $\Delta E = \Delta m \; c^2$, where c^2 is the speed of light squared and ΔE and Δm are incremental changes in energy and mass, respectively. Now, whether "nature" accomplishes this significant process with or without intrinsic feedback cannot be detected simply by examining the system equation. The same system equation can be formulated by assuming a system model that includes or excludes intrinsic feedback. We will increase our understanding and appreciation of this fact as we continue our study of systems. In summary, we will find that most systems exhibit intrinsic feedback and are subject to extrinsic feedback as the systems respond to inputs imposed upon them. This condition makes system science both challenging to learn and execute but rich in understanding and appreciation.

It should be recognized that almost all systems could possess or exhibit both intrinsic and extrinsic feedback. In practice, it is usually possible to control most systems at least for limited ranges of behavior using external feedback. However, nature has a capricious capacity to thwart human's efforts to control natural systems and demonstrate behavior that is both unanticipated and undesirable.

2.3.5 Controllable Systems

A major subdiscipline of system science is system control, which is the study of how to control the output response of certain systems once their basic operation is understood. In fact, historically, much of the motivation and progress in the early study of systems arose as an overt attempt to control or influence the output of certain systems to achieve some desired behavior or response. The analysis, design, and development of a mechanical governor by James Watt in the 1790s to control the speed of steam engines was an important early development in system control; and the success of this simple, but effective governor made a significant contribution to the advent of the Industrial Age. In nature, many natural and inherent factors exist to control the response of systems through feedback. The growth or population change of a plant or animal species is controlled by food supplies, predators, disease, weather, and other natural factors. When population increases, for example, become too large for the ecosystem to support, these control factors through natural feedback processes control and limit increases in population. Indeed, nature with its many species of flora (plant life) and fauna (animal life) is a marvelously complex system demonstrating intrinsic feedback control to provide harmony and balance. Sometimes, unfortunately, when humans attempt to control or influence natural systems, the results can be surprising and occasionally undesirable because of failure to understand fully the consequence of how these modified inputs will affect the system (nature) and its outputs. The worldwide utilization of certain persistent chemicals such as DDT, polychlorinated biphenyls (PCBS), chlordane, dioxin, lead, and asbestos has produced some undesirable effects and long-term problems.

Large cities are good examples of major systems with intrinsic and extrinsic feedback. Usually, the land area and natural resources encompassed by the city are inadequate to support such a large concentration of humans in a single area, and a city with high population is unstable and would collapse without modification of the natural system. So, humans modify the natural system (land, water, agricultural capability, waste disposal, etc.) by modifying inputs and outputs delivered to the city system. Water projects bring water into the city. Food, materials, and supplies also are transported into the city on a nearly continuous basis to support the population. Furthermore, byproducts such as labor and processed materials are produced to exchange for the outside support the city receives in the way of food, water, power, and materials. Garbage, sewage, and other refuse must be removed from the city to prevent the spread of disease and blight, which is a natural tendency with such population concentrations. Of course, these inputs, outputs, and feedback loops are subject to interruption and disturbance, and major changes or breakdowns in supplies can be perilous for the city and its inhabitants.

2.4   System Analysis and Evaluation Using a Computer

2.4.1   Computer Applications to System Analysis

The extremely high speed and accuracy with which computers can process quantitative data was addressed in Chapter 1. The great impact that these ubiquitous resources now have on system science for modeling, analyzing, and data processing is that the system investigator no longer needs to possess broad and profound analytical and numerical skills to analyze large and complex system models. Now, complex models composed of many inputs and outputs and highly coupled nonlinear kernels can usually be analyzed with existing computer software (i.e., computer algorithms written to solve sets of algebraic and differential equations both numerically and symbolically). With the performance, speed, and storage of even desktop and portable notebook computers, the capacity for a system investigator to analyze any reasonable system model now exists.

Thus, the system investigator seeking to model, analyze, and evaluate a system in any field can devote primary effort to scoping and modeling the system. The previously onerous task of manipulating large arrays of equations and iterating mathematical operations can now be committed to the computer and appropriate software packages. The major work of the system investigator is in correctly modeling the system and in accurately evaluating, correcting, and improving the system model as the computer performs the data processing for the system model.

2.4.2   Symbolic Computer Applications to System Analysis

The Symbolic Math Toolbox in MATLAB permits symbolic or algebraic representations of quantitative variables and quantities without immediate reliance on numerical representation. Symbolic entities support the calculus for all mathematical manipulations and computations in system science. The computation capability for these symbolic operations is provided primarily by MAPLE software developed by Waterloo Maple, Inc. A partial list of those symbolic operations that are of particular interest for modeling of systems and their analysis is shown in Table 2-1. It is important to recognize that MATLAB is a dynamically evolving software system and the parameters, definitions, and algorithms provided in this

Table 2-1   MATLAB Symbolic Operations for System Modeling

Subject	Symbolic operations
Calculus	Differentiation, integration, limits, summation, Taylor series
Linear algebra	Inverses, determinants, eigenvalues, singular value decomposition, and canonical forms of symbolic matrices
Simplification	Simplification of algebraic expressions
Solutions	Symbolic solution of algebraic and differential equations
Special functions	Special functions in applied mathematics
Transforms	Laplace, Fourier, and z-transforms and their inverses

text are subject to change with time. The MATLAB nomenclature provided here is illustrative for pedagogical utilization.

These symbolic tools provide the system investigator with significant capabilities in manipulating and solving the system equations encountered in system analysis. The results of these symbolic operation tools are of great value since the results are those that would be obtained by analytical processing by the system investigator. Indeed, much of the labor and skills normally required by the investigator are available using these symbolic tools. Of course, there are limitations to these symbolic tools, and they cannot analytically process or solve any system equation that may be approached by the tools. Nevertheless, these tools greatly enhance the system investigator's capabilities and generally provide independent verification of the investigator's analytical results.

We restrict our use of these symbolic tools primarily to those operations of principal interest for system analysis as follows:

- Algebraic manipulation of equations and matrices
- Solution of systems of differential equations composing system equations
- Graphical presentation of data

It is assumed that the system investigator has or will acquire sufficient knowledge and skill in using MATLAB and only the essential results obtained from MATLAB toolboxes will be presented here. A background on MATLAB is readily obtained from the extensive literature in the field and widespread application of this software package. The reader is again advised that these symbolic tools and other nomenclature may or will change with subsequent releases of MATLAB software. The user of MATLAB should consult the respective specifications for the version implemented.

Symbolic tools are very useful in manipulating and examining the system equations under investigation. For example, finding maximum and minimum values for the symbolic system equation u is easily accomplished using the toolbox function "diff" to perform symbolic differentiation:

$$v = \text{diff}\,(u)$$

where u is the function $u(x)$ to be differentiated with respect to the variable x. The resulting function v is set to zero to obtain the maximum, minimum, and points of infection for $u(x)$. This is easily accomplished using the toolbox function "solve" as follows:

$$v = 0$$

and

$$w = \text{solve}\,(v)$$

Of course, the extremum found, i.e., the minimum, maximum, or point of inflection, can be assessed by finding the value of the derivative of v at the point identified for each extremum value.

MATLAB also provides functions for working with difference equations, which are encountered in discrete model equations. The function "filter" exemplified next:

$$y = \text{filter}\,(a, b, x)$$

also processes data for the vector-dependent variable $x(n)$ and discrete vector parameters a (n) and $b(n)$. For example, consider the following difference equation:

$$a(n)y(n) = a(1)y(1) + a(2)y(2) + \cdots + a(n-1)y(n-1)$$
$$+ b(1)x(1) + b(2)x(2) + \cdots + b(n-1)x(n-1)$$

The system investigator often encounters analytical expressions that required integration. If $u(x)$ is a symbolic expression, then

$$v = \mathrm{int}\,(u,x)$$

attempts to find the related symbolic expression $v(x)$ that is the indefinite integral or anti-derivative of u for the variable of integration x provided that $v(x)$ exists in closed form and MATLAB can determine the indefinite integral. The constant of integration is implied and added by the investigator. If MATLAB is unable to integrate the expression $u(x)$, then the initial command is simply returned as the default function.

Undoubtedly, a very important symbolic operator is that for analytically solving differential equations. The symbolic function "dsolve" computes the symbolic solution for a system of differential equations if available with its solutions algorithms. For example, the system of first order differential equations for the dependent variables fi and independent variable x:

$$Df1 = g1(fi,x), \quad Df2 = g2(fi,x), \quad Df3 = g3(fi,x), \dots \text{ for } i = 1 \text{ to } n$$

where the symbol D denotes the differential operator $D = d/dx$ is solved using the symbolic operator for "dsolve" as follows:

$$S = \mathrm{dsolve}\,['Df1 = g1(fi,x)', 'Df2 = g2(fi,x)', 'Df3 = g3(fi,x)', \dots]$$

The computed solutions are returned in the symbolic structure S and the values for the $f1$, $f2$, $f3$,... are returned by entering each function:

$$f1 = S.fi$$

MATLAB responds with the solution as:

$$fi = \text{solution for } fi$$

and repeating this command for the remaining solutions for each subsequent fn. It is also possible to impose initial conditions on the solutions as shown in the following example:

$$[f1, f2, \dots] = \mathrm{dsolve}\,('Df1 = g1(fi,x)', 'Df2 = g2(fi,x)', \dots$$
$$f1(0) = f10, f2(0) = f20, \dots')$$

Symbolic matrix algebraic operations such as addition, subtraction, and multiplication are provided for matrices that satisfy the proper rules for conformance of rows and columns. Furthermore, transpose and inverse matrices (for non-singular matrices) can also be determined symbolically. Symbolic matrix operations are particularly useful when manipulating multi-input and or multi-output system equations since the results preserve the symbolic expressions for the original system equations. Thus, the symbolic parameters are preserved for easy identification and assessment.

Other symbolic toolbox operations that are very useful and will be employed in examples within this text are the transform operators for Laplace and Fourier transforms. For example, the Laplace transform, $L(s)$, of the function $F(t)$ with respect to the default variable t is found using the symbolic operator "laplace" as follows:

$$L(s) = \text{laplace}\,(F)$$

and to invert the Laplace transform $L(s)$, the symbolic operator "ilaplace" returns the original function $F(t)$ as follows:

$$F(t) = \text{ilaplace}\,(L)$$

Many other useful symbolic operators are available in both the MATLAB toolbox and MAPLE's symbolic operator menu. For the full complement of these operators, the reader is referred to the appropriate literature for these symbolic packages. We shall demonstrate a few of these symbolic operators in examples and problems presented throughout this text.

2.5 Summary

In this chapter, the concept of a rational system with inputs or causes and outputs or effects was defined as any set of objects to which a relationship of cause and effect can be rationally assigned. The importance of quantification or assigning numerical values to inputs and outputs was established with the concept of system measurability. Then, to realize system isolation from the universe of interacting systems, the concept of a system boundary was defined in which only input causes and output effects are able to cross the system boundary.

The concept of system feedback arises when the output of a system is used to influence or affect an input to the system. This feedback modification of the system input can be accomplished by the system itself, in which case we refer to the process as intrinsic or internal feedback. The other case or external feedback is that in which the system's environment receives the output and through extrinsic feedback externally modifies the input to the system.

Then the use of system feedback to control the output response of the system to satisfy some control policy or natural constraints imposed by the system's environment was introduced and exemplified. Such controllable systems constitute a major portion of modern system science.

Irrational systems, which are systems that do not satisfy the principle of causality, are ignored in system science because rational methods of modeling and analysis cannot be consistently applied to such systems.

Deterministic systems are those systems that always exhibit a unique output set for each given set of inputs, regardless of the number of times the same input set is imposed on the system under identical conditions. For stochastic systems, either it is not possible to realize an identical input set for repeated trials on a given system or the output of the system is inherently stochastic and must be interpreted stochastically. Generally, such stochastic systems will exhibit output sets that can be related to the causative input sets through the concepts and methods of statistics and probability. We will classify systems governed by quantum mechanics and uncertainty principles as stochastic systems for our purposes.

Bibliography

BODE, H., *Network Analysis and Feedback Amplifier Design*. Van Nostrand Co., 1945. An early document that was basic to development of electrical engineering feedback control.

BORGHESANI, C., *Quantitative Feedback Theory Toolbox Users Guide*, The Math Works Inc., Natick, MA, 1994. Math Works development of the Toolbox for feedback analysis.

BRIDGEMAN, P., *The Nature of Some of Our Physical Concepts*, Philosophical Library, New York, 1952. Provides B insight into the basis and meaning of our physical models.

COX, D., and H. MILLER, *The Theory of Stochastic Processes*, John Wiley & Sons, New York, 1965. An introduction to stochastic systems and their analysis.

ELLIS, D., and F. LUDWIG, *Systems Philosophy*, Prentice-Hall, Englewood Cliffs, NJ, 1962. Conceptions and models in system science.

FREEMAN, H., *Discrete-Time Systems*, John Wiley & Sons, New York, 1965. An introduction to discrete time systems and general discrete system development.

HOROWITZ, I., *Synthesis of Feedback Systems*, Academic Press, New York, 1963. Standard development for feedback control modelling.

HOUPIS, C., et. al., *Quantitative Feedback Theory: Fundamentals and Applications*. 2ª Edition. CRC Press, Marcel Dekker, NY, 2005

KLIR, G., *An Approach to General Systems Theory*, Van Nostrand Reinhold Co., New York, 1969. A general introduction to system science.

MACIEJOWSKI, J., *Multivariable Feedback Design*, Addison Wesley, Harlow, United Kingdom., 1989. Feedback for multivariate systems.

MARLIN, E. *Process Control: Designing Processes and Control Systems for Dynamic Performance*. McGraw-Hill, New York, 1995. Process feedback control for optimal system behavior.

MSAROVIC, M., *Views on General Systems Theory*, R. E. Krieger Publishing Co., Huntington, NY, 1964. A highly axiomatic treatment of concepts in system science.

ROTHMAN, S., and C. MOSMAN, *Computers and Society, Science Research Associates*, Chicago, 1972. A survey of the impact of computers on human society.

SAGE, A. P., and J. L. MELSA, *System Identification*, Academic Press, New York, 1971. A useful resource for the modeling of systems.

Symbolic Math Toolbox, *User's Guide, Various Versions*, The MathWorks Inc., Natick, MA. An essential reference for this Systems Science Text. The reader should acquire this software to effectively peruse this Text.

VON BERTALANFFY, L., *General System Theory*, George Braziller, New York, 1968. A standard work in system science.

WEYL, H., *Philosophy of Mathematics and Natural Sciences*, Princeton University Press, Princeton, NJ, 1949. An especially readable account written with balance and wisdom about the role of mathematics in the natural sciences.

WILSON, I. G., and M. E. WILSON, *Information, Computers and System Design*, John Wiley & Sons, New York, 1965. Application of computers to system science.

YANIV, O., *Quantitative Feedback Design of Linear and Non-linear Control Systems*. Kluwer Academic Pub., 1999. Addresses both linear and nonlinear control applications.

Problems

1   For the following list of systems and their related input causes and output effects, supply for the empty entry block one or two credible inputs or outputs or identify the system as appropriate. Also, perform the following:

   i) Define a reasonable boundary for containing each system by specifying the essential elements or members of the system.
   ii) Suggest methods for numerically measuring inputs and outputs.
   iii) Decide if system is deterministic or stochastic, continuous or discrete.
   iv) Identify possible sources of internal or intrinsic feedback that might exist within the system.
   v) Suggest a policy for controlling the system by external or extrinsic feedback. What outputs would you feedback to which inputs?

System	Cause(s)	Effect(s)
(a) Nations	-----------------	International wars
(b) British government	Monarchy	--------------------.
(c) People	-----------------	Hunger and poverty
(d) Moon	Oxygen deficiency	-----------------
(e) -----------------	Electronic bonding	Molecules
(f) Food supplies	Poor harvests	-----------------
(g) Germany	-----------------	World War II
(h) U.S. Confederate States	-----------------	Slavery
(i) -----------------	Electrical power input	Refrigeration
(j) Automobile	Driver	-----------------
(k) Books	Authors	-----------------
(1) -----------------	Researchers	Research
(m) Water cycle	Evaporation	-----------------
(n) Electrical power plant	Fuel and operation	-----------------
(o) U.S. Space Program	Funds and personnel	-----------------
(p) Family budget	Income	-----------------
(q) Human digestive system	-----------------	Cell growth
(r) Human kidneys	-----------------	Urine
(s) Computer	Software	-----------------
(t) Universe	-----------------	Closed strings
(u) -----------------	Electrical charge	Voltage
(v) -----------------	The Pope	Catholic faith
(w) Material particles	Motion	-----------------
(x) -----------------	Evolution	Human life
(y) Electrical phenomena	Maxwell's equations	-----------------
(z) Human mind	-----------------	Mathematics

2   Discuss the consequences of an irrational system that exhibits varied responses or outputs without any input stimulation to the system. Is it possible to make quantitative measurements?

3   If all systems composing the universe were completely understood and quantitatively describable, and if all inputs to these systems were known over an appropriate period

of time, would the future of the universe and all future events of its components be completely predictable? Would the future of the universe then be determined or predestined? Would humans still have free will? Discuss and defend your opinion.

4 It is often claimed that quantitative measurements and value assignments cannot be made for human characteristics and qualities. For each of the following human characteristics, determine a means of measurement (define any finite standard of quantitative measure you wish) and a range of values for that characteristic.

(a) body mass	(j) cleanliness	(s) serenity
(b) Effect(s)	(k) hair style	(t) ambition
(c) height	(l) complexion	(u) kindness
(d) eyesight	(m) beauty	(v) loyalty
(e) hearing	(n) charm	(w) honesty
(f) physical strength	(o) sociability	(x) integrity
(g) response time to stimulus	(p) hospitality	(y) sincerity
(h) intelligence	(q) grace	(z) love
(i) speaking skills	(r) gentleness	

5 In this chapter, several examples of irrational systems were discussed in which effects were produced by systems without apparent or creditable inputs existing. See Section 2.3.1. Discuss the system that produces no output response regardless of the inputs delivered to the system. Such systems do have theoretical interest and are generally referred to as null systems, which zero or annihilate all inputs delivered to them so that no response or output from the system is apparent. Can you identify examples and suggest possible applications for such null systems?

6 Speculate on and attempt to identify several systems that have real physical existence and exhibit truly continuous inputs and outputs. What is the physical significance of making the inputs and outputs arbitrarily small?

7 Respond to the following assertion: In reality, all systems are stochastic and so-called laws of science are only tentative temporal trends observed in repeated trials of a given test. In reality, all outcomes are possible for a given experiment. Even important concepts such as conservation of energy and electrical charge are only temporal trends in human experience to be properly interpreted only in a statistical or stochastic sense.

8 What are the consequences if the boundary defining a system changes in time so that the number or nature of elements constituting the system changes in time? What might happen if the system elements move undetected back and forth across the system boundary?

9 Identify five distinct intrinsic feedback systems found in the human body. Discuss how these systems control the system's output. Identify five distinct extrinsic feedback systems that externally control human behavior. Discuss how these extrinsic feedback systems control behavior.

10 Identify possible sources of intrinsic and extrinsic feedback that may or does exist for the following systems. Indicate a basis for quantitative measurement of the feedback.

a) U.S. Supreme Court and U.S. lower federal courts
b) Sea life in oceans and oceanic pollution
c) Terrorist camps throughout world and opposing military actions
d) United Nations Security Council and UN General Assembly
e) Civil wars in Middle East and Islamic extremism
f) Semitic terrorism and Hebrew Zionism
g) Dow Jones stock market averages and stock investors
h) U.S. Pentagon and U.S. defense industry
i) Federal Reserve Board and U.S. Congress
j) U.S. Congress and congressional lobbyists
k) Tooth decay and glucose (sugar)
l) DeBeers Diamond Limited (cartel) and world diamond markets
m) University libraries and university student patrons
n) NATO (North Atlantic Treaty Organization) and Europe
o) Dark matter and dark energy in our observable universe
p) American Stock Exchange and commodity markets
q) Symphony orchestra and musical score
r) Roman Catholic Church and Greek Orthodox Church
s) Pure and applied mathematics
t) European Common Market and Pacific Rim Countries
u) Organized crime and Russian Federation and government
v) London Commodities Market and world trade
w) Covalent molecular bonding and ionic molecular bonding
x) COVID-19 pandemic and impact on human populations
y) OPEC (Organization of Petroleum Exporting Countries) and Europe
z) Military operations by Arab nations against ISIS/ISIL

3

Basic System Equations

In this chapter, we will undertake the essential task of converting the principle of causality into a quantitative mathematical statement for systems with any number of inputs and outputs. We will develop this generalized, fundamental mathematical framework for both discrete and continuous systems and for systems with and without feedback.

3.1 Functional Dependence of System Causes and Effects

A primary objective of system analysis is the determination and evaluation of the relationship between the inputs or causes imposed on a system and the outputs or effects produced by the system because of these inputs. Knowledge and understanding of this relationship are valuable for several reasons. First, accurate knowledge of the relationship between cause and effect for a given system leads to some understanding of the behavior and inner operations or internal mechanics of that system and is a vital step in fully understanding the operation and nature of the system. Second, with the cause and effect relationship adequately known, it is possible to explain the past performance of the system and to predict the output or response of the system for any permissible future input. Third, with the output response of the system known for any input, it is then possible to develop means for controlling or influencing the system in some desirable or optimal manner.

If a complete and accurate relationship between the system output response and any possible input is known for a given system, then the system may be considered to be adequately understood and predictable from a system standpoint. However, the true relationship between system output and input is never completely or exactly known for any system. It is always necessary, therefore, at least for real systems, to make simplifying assumptions and approximations about the system and its interaction with its environment. However, this approximation procedure is usually justified and satisfactory if the response of the system can be reasonably well predicted for a practical range of variation of inputs. Thus, although the complete response of a system is generally not known under all conditions, considerable knowledge and understanding of the system can still be gained and practical means devised for predicting and possibly controlling the system behavior over a particular range of interest. Even qualitative knowledge about some systems is useful

Introduction to System Science with MATLAB, Second Edition.
Gary Marlin Sandquist and Zakary Robert Wilde.
© 2023 John Wiley & Sons Ltd. Published 2023 by John Wiley & Sons Ltd.

and can provide general observations and allow conclusions to be inferred about the system. However, it is not until a quantitative evaluation (i.e., a numerical assignment and measurement of values) can be made about the output response for a given quantitative input that significant progress is made in understanding the system and its operational basis. Thus, our main goal throughout our study of systems is the determination and evaluation of acceptably accurate and quantitative descriptions of rational systems. We can contribute to this goal by investigating, modeling, analyzing, and evaluating a broad variety of quantitative mathematical statements between cause and effect relationships for various systems.

3.1.1 Proportionality Relationship Between Cause and Effect

To formulate a consistent mathematical statement of the relationship between the inputs (causes) and outputs (effects) of a system, we begin by defining incremental variables as follows:

> Δc *is the incremental change of the input or external stimuli imposed on a system that arises from outside the system and affects or perturbs the system in a quantitative, measurable way.*

> Δe *is the incremental change of the output or response of a system that results from the incremental input Δc and is observable and measurable outside the system. The output or response is a result of the input imposed on the system. Thus, for any rational system, it is always assumed that all effects (Δe's) observed issuing from the system are a consequence of one or more instigating causes (Δc) that produced the observed effect.*

Now the basic assumption and the fundamental principle of system analysis is that in any rational system, an incremental change in the input, Δc, is proportional to some associated incremental change in the output Δe, from the system. Thus, we can express this proportionality relationship as follows:

$$\Delta e \propto \Delta c \tag{3.1.1}$$

This proportionality relationship implies that the system output response or effect, Δe, is directly proportional to the input stimulus or cause, Δc, and is the mathematical statement for the principle of cause and effect. That is, every effect produced by a system has one or more instigating causes that were imposed on that system to produce that effect.

The size or magnitude of the incremental changes in the input, Δc, and the related incremental change in the output, Δe, associated with a given system is, of course, dependent on the nature of the system and the nature of the inputs and outputs. Generally, it is assumed that these incremental changes in cause and effect, Δc and Δe, can be made arbitrarily small so that Δc and Δe can be replaced by the related differential quantities dc and de, respectively. Then we have for continuous system response, the following proportionality relationship:

$$de \propto dc \tag{3.1.2}$$

Generally, certain advantages and conveniences result from assuming that the cause and effect variables are continuous and therefore are capable of being arbitrarily reduced in magnitude so that continuous differential changes are defined and meaningful. This assumption is useful in that it permits us to employ differential analysis to model and study the system behavior.

3.1.2 The System Kernel

All rational systems that satisfy the principle of causality exhibit a proportional relationship between the incremental change in input (cause) and incremental change in output (effect). The manner in which systems differ from one another is through the proportionality or change function defined by $g(e, c)$, which relates the full dependence between input and output for a given system. The proportionality relationship between Δe and Δc given by Eq. (3.1.1) can be written for discrete or incremental systems as a system equation of the form:

$$\Delta e = g(e, c)\Delta c \tag{3.1.3}$$

For continuous systems, we have for the proportionality relationship between de and dc, where $g(e, c)$ is defined as the system kernel, the following system equation:

$$de = g(e, c)\, dc \tag{3.1.4}$$

The system kernel is the most important relationship to be determined and studied in system analysis. It is the system kernel, $g(e, c)$, for "g or gain" that translates a given change of input or stimulus imposed on a particular system into the resulting change of output or response of that particular system. If a quantitative mathematical relationship is known for the system kernel, it is possible to assign a quantitative output for any quantitative input that is within the range of definition for the system kernel. In biology, a kernel is a seed from which the nucleus of the seed determines the growth or outcome of maturation of the kernel. Thus, the kernel input results in the output of the system. Mathematically, the basis for assigning the name "system kernel" to the quantitative relationship or proportionality between the incremental change in input Δc and output Δe can be seen by expressing the basic system equation as follows:

$$g(e, c) \equiv \frac{\Delta e}{\Delta c} = \frac{\text{incremental output change}}{\text{incremental input change}} \tag{3.1.5}$$

Thus, $g(e, c)$ defines the incremental change in system output that occurs per unit incremental change in input. Such functions are often defined as kernels in mathematics, and since our kernel is related to the output change of a system per unit input change, we will refer to $g(e, c)$ as the system kernel.

Observe that if the system kernel is a function of continuous input and output variables, then in the limit, as the incremental change in input Δc becomes arbitrarily small, the system kernel is equal to the derivative of the system output with respect to the system input. That is,

$$\lim_{\Delta c \to 0} \frac{\Delta e}{\Delta c} \equiv \frac{de}{dc} = \frac{\text{differential output change}}{\text{differential input change}} \tag{3.1.6}$$

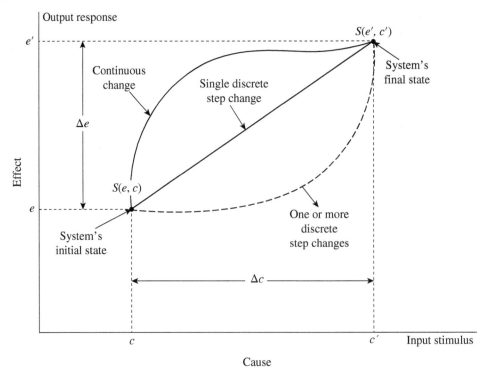

Fig. 3-1 Discrete and continuous change of a system's state.

Let us examine the characteristics of the system kernel in somewhat greater depth. Suppose we have a system whose output state at some instant in time, t, is given by the absolute state function $e = S(e, c)$, where $S(e, c)$ defines the system's complete instantaneous condition or state as a consequence of the system's entire previous history. In general, it is not possible to determine a system's absolute state function. As we will learn, only changes in this state can be observed from appropriate changes in the system's output response Δe due to appropriate changes in the system's input stimuli Δc. In Figure 3-1, if the system's absolute state function at some instant in time is defined as $S(e, c)$, and we subsequently perturb the system input by an amount $\Delta c = c' - c$, then the system will respond with an output response $\Delta e = e' - e$, given by

$$\Delta e = e' - e = S(e', c) - S(e, c) \tag{3.1.7}$$

where we have assumed that the system response is determined by the change in the present state of the system.

Observe that the path taken by the system as it changes its absolute state from $S(e, c)$ to $e = S(e', c')$ may be continuous or composed of a single or series of discrete or quantum step changes. In principle, the minimum change in the output response that could occur is that which results from a single, minimum incremental step or quantum change in the input, which would produce a minimum incremental step or quantum change in the output. Typically, the change in the output response for such a single quantum perturbation in the input would usually be very small, corresponding to a quantum change of state or response. Furthermore, in general, most changes that are imposed on systems will entail many minimum steps or quantum changes, and the output change in the system's response will usually appear to be continuously variable with the input changes.

Now since $c' = \Delta c + c$ and $e' = \Delta e + e$, Eq. (3.1.7) can be written as

$$\Delta e = S(e + \Delta e, c + \Delta c) - S(e, c) \tag{3.1.8}$$

By additional rearrangement and manipulation, Eq. (3.1.8) can also be written as follows, if Δc and $\Delta e \neq 0$:

$$\Delta e = [S(e + \Delta e, c + \Delta c) - S(e + \Delta e, c)]\frac{\Delta c}{\Delta c} + [S(e + \Delta e, c) - S(e, c)]\frac{\Delta e}{\Delta e} \tag{3.1.9}$$

where these modified arguments for the system's absolute state function represent the state function at these specific state conditions. Gathering terms containing Δe, Eq. (3.1.9) can then be expressed as follows:

$$\Delta e = \frac{(1/\Delta c)[S(e + \Delta e, c + \Delta c) - S(e + \Delta e, c)]}{1 - (1/\Delta e)[S(e + \Delta e, c) - S(e, c)]}\Delta c \tag{3.1.10}$$

Observe that Eq. (3.1.10) is not defined (i.e., the denominator is zero) if

$$\Delta e = S(e + \Delta e, c) - S(e, c) \tag{3.1.11}$$

However, this condition violates the basic premise of the principle of causality in that it implies that a change in output Δe occurs without a corresponding change in input Δc. The causality principle requires that for every output Δe from a system, there must exist one or more inputs Δc that produced that output Δe.

Now, if we define the system kernel $g(e, c)$ as the quantity in braces in Eq. (3.1.10), where

$$g(e, c) = \frac{(1/\Delta c)[S(e + \Delta e, c + \Delta c) - S(e + \Delta e, c)]}{1 - (1/\Delta e)[S(e + \Delta e, c) - S(e, c)]} \tag{3.1.12}$$

Then this equation for the change in system output Δe for a given change in input Δc is the basic system equation we have defined and described previously.

$$\Delta e = g(e, c)\Delta c$$

Observe that, as was stated previously, the system kernel as defined by Eq. (3.1.12) is found to depend on the differences of the absolute state function $S(e, c)$ between different input and output states. Also observe that if the state functions are continuous or can be assumed to be continuous so that Δe and de can be made arbitrarily small, then in the limit, as $\Delta e \to de$ and $\Delta c \to dc$, we have for the basic system equation that $de = g(e, c)\,dc$ and the system kernel then is given by

$$g(e, c) = \frac{\partial S/\partial c}{1 - \partial S/\partial e} \tag{3.1.13}$$

We observe that the definitions for the partial derivatives of the absolute system state function, $S(e, c)$, are the following mathematical limit definitions.

$$\frac{\partial S(e,c)}{\partial c} = Lim_{\Delta c \to 0} \left\{ \frac{1}{\Delta c} [S(e + \Delta e, c + \Delta c) - S(e + \Delta e, c)] \right\}$$

$$\frac{\partial S(e,c)}{\partial e} = Lim_{\Delta e \to 0} \left\{ \frac{1}{\Delta e} [S(e + \Delta e, c) - S(e, c)] \right\}$$

It is also interesting to observe that if the absolute state function $S(e, c)$ is not a function of the system output, e and Δe, so that $S(e, c) = S(c)$, then

$$S(e + \Delta e, c) - S(e, c) = 0$$

and the system kernel is given by

$$g(e, c) \Rightarrow g(c) = \frac{1}{\Delta c} [S(c + \Delta c) - S(c)]$$

This system kernel is independent of the output, e and Δe, and therefore exhibits no intrinsic or inherent feedback of the system's output. The change in the system's output is simply the change in the absolute system state due to changes in the system input. Thus

$$\Delta e = \frac{S(c + \Delta c) - S(c)}{\Delta c} \Delta c = S(c + \Delta c) - S(c) \tag{3.1.14}$$

As mentioned previously, however, the absolute state function for a system is generally not known, and only changes in the system's state are discernible. Therefore, whether $S(e, c)$ is a function of e or not is not usually known, and we will assume conservatively that, in general, the system kernel $g(e, c)$ is a function of both the output e and the input c.

3.2 Classification of System Equations

There are many ways of classifying systems, but perhaps the simplest and for many purposes, the most effective classification is based on the number of independent input variables and the number of dependent output variables associated with the system. For continuous systems, we will see that there is a close relationship between single-input systems and ordinary differential equations and between multiple-input systems and partial differential equations. Indeed, any ordinary or partial differential equation can be associated with a related continuous system equation. Furthermore, for discrete systems, there is a direct relationship between single-input systems and ordinary difference equations and between multiple-input systems and partial difference equations. These relationships become apparent when we consider the independent variables as system inputs and the dependent variables as the system outputs.

In this section, we will develop the various general forms of system equations with primary stress on continuous systems where changes in output and input variables are expressed in differential form. This is done only for convenience, however, and the reader should recognize that all system equations developed here apply equally well to discrete systems. The discrete system equation is obtained by simply replacing the continuous differential variables, such as de and dc, by the corresponding discrete difference variables Δe and Δc, respectively.

3.2.1 Single-Input, Single-Output Systems

The simplest system equation, and usually the beginning point for the study of any system, is that system equation has one input variable (designated as c for input cause) and one output variable (designated as e for output effect). Recalling the general system equation, the single-input, single-output system can be expressed mathematically as follows, respectively, for a continuous or discrete system.

$$\left.\begin{array}{l} de \\ \Delta e \end{array}\right\} = g(e, c) \left\{\begin{array}{l} dc \\ \Delta c \end{array}\right. \tag{3.2.1}$$

This equation has an equivalent block diagram as follows:

$$\left.\begin{array}{l} dc \\ \Delta c \end{array}\right\} \xrightarrow[\text{(causes)}]{\text{Input}} \boxed{\begin{array}{c} g(e, c) \\ \text{System kernel} \end{array}} \xrightarrow[\text{(effects)}]{\text{Output}} \left\{\begin{array}{l} de \\ \Delta e \end{array}\right.$$

The block diagram is a convenient graphical presentation of a system equation. The rectangular box (or "block" from which the diagrammatic representation draws its name) that encloses the system kernel is suggestive of the system boundary that isolates the system from its external environment. The input(s) to the system from the outside environment is indicated by an arrow(s) directed into the block, while the output(s) is indicated by an arrow(s) directed out of the block into the environment where the output can be observed. We will examine such block diagrams in detail in Chapter 6. The function $g(e, c)$ is called the system kernel and can be thought of as the response of the system, which expresses the system's output response de (continuous) or Δe (discrete) for a given input dc (continuous) or Δc (discrete), respectively.

It is important to recognize that Eq. (3.2.1) is also the general expression for the first-order ordinary differential or difference equation, as appropriate, in differential or difference form. In this form, either e or c may be considered as the dependent variable for a single-input, single-output system. The specific choice is usually made to simplify or promote solving the resulting differential or difference equation. We will consider generalized methods for solving first-order ordinary differential and difference equations in Chapter 4.

If external feedback of the output response occurs, then we have as a block diagram for the system the representation shown in Figure 3-2, where $f(e, c)$ is called the feedback kernel, and feedback of the output de or Δe occurs in such a manner that the effect on the system input is proportional to the feedback kernel $f(e, c)$ and proportional to the system output de or Δe. This system block diagram is often referred to as the generalized or canonical block diagram since all systems considered in this text can be represented by this generalized form.

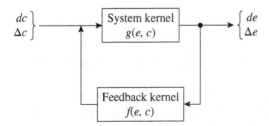

Fig. 3-2 Canonical block diagram for a system.

To determine the overall system relationship with feedback present, we observe for the continuous system, for example, that the total input to the system kernel now comes from two sources: the external input source dc and the system feedback source. These two sources add together to provide as a total input to the system the following:

$$dc + f(e, c)\, de$$

This input may be multiplied by the system kernel $g(e, c)$ to yield the complete system output de as follows:

$$de = g(e, c)[dc + f(e, c)de]$$

Grouping on common differential variables, we have that

$$[1 - g(e, c)f(e, c)]de = g(e, c)dc$$

and assuming that $[1 - gf]de$ does not vanish (otherwise, we also required then that $g(e, c)$ vanish) so that $g(e, c)f(e, c) \neq 1$, we find for the system equation with feedback that

$$\left.\begin{array}{c} de \\ \Delta e \end{array}\right\} = \frac{g(e, c)}{1 - g(e, c)\, f(e, c)} \left\{\begin{array}{c} dc \\ \Delta c \end{array}\right. \tag{3.2.2}$$

Equation (3.2.2) is a familiar form to those acquainted with feedback systems and is often referred to as the closed-loop system equation (i.e., systems with feedback). Observe that Eq. (3.2.2) may be redefined to provide an alternate system kernel $g_f(e, c)$ where

$$g_f(e, c) = \frac{g(e, c)}{1 - g(e, c)f(e, c)} \tag{3.2.3}$$

Then the system equation can be expressed as a simple open-loop system (i.e., without apparent feedback) as follows:

$$\left.\begin{array}{c} de \\ \Delta e \end{array}\right\} = g_f(e, c) \left\{\begin{array}{c} dc \\ \Delta c \end{array}\right. \tag{3.2.4}$$

Observe that if $0 < g(e, c)f(e, c) < 2$, then the denominator in Eq. (3.2.3) is less than 1, and the absolute values of $g_f(e, c) > g(e, c)$. The system is then said to exhibit positive feedback gain. Furthermore, if $g(e, c)f(e, c) \to 1$, then $g_f(\mathbf{e}, \mathbf{c}) \to \pm\infty$, which is obviously an unstable system response with infinite gain. We will discuss these general aspects of feedback in more detail in Section 8.3.

System equation (3.2.4) has for the simple open-loop system, the following block diagram representation without apparent feedback:

$$\left.\begin{array}{c} de \\ \Delta e \end{array}\right\} \longrightarrow \boxed{g_f(e,\ c)} \longrightarrow \left\{\begin{array}{c} dc \\ \Delta c \end{array}\right.$$

Finally, it is obvious that if the feedback kernel vanishes, $f(e, c) \to 0$, then the system kernel $g_f(e, c)$ reduces to $g(e, c)$.

It is important to recognize that the presence or absence of feedback in a given general system is not always obvious or even easily detectable in a real system, especially if the feedback occurs internally within the system. Thus, feedback of the system output may or may not occur inherently within a given system, and ascertaining the presence of such internal feedback may be difficult or even impossible.

3.2.2 Single-Input, Multiple-Output Systems

Perhaps the next simplest system equation to formulate and analyze is that system, which has one common input variable c, but several output variables $e_1, e_2, ..., e_n$. The single-input, multiple-output continuous system has a mathematical description as follows:

$$dc \rightarrow \begin{bmatrix} g_1(e_1, e_2, ..., e_n, c) \\ g_2(e_1, e_2, ..., e_n, c) \\ \vdots \\ g_n(e_1, e_2, ..., e_n, c) \end{bmatrix} \begin{matrix} \rightarrow de_1 \\ \rightarrow de_2 \\ \rightarrow \vdots \\ \rightarrow de_n \end{matrix}$$

The system equation assumes the form of a set of n equations in differential form as follows:

$$\begin{aligned} de_1 &= g_1(e_1, e_2, ..., e_n, c)\, dc \\ de_2 &= g_2(e_1, e_2, ..., e_n, c)dc \\ &\vdots \\ de_n &= g_n(e_1, e_2, ..., e_n, c)dc \end{aligned} \tag{3.2.5}$$

Mathematical representation of this system of equations can be greatly simplified by the use of matrix notation. If we define column matrices [i.e., matrices with n rows and 1 column or $(n \times 1)$ matrices] as follows:

$$de = \begin{bmatrix} de_1 \\ de_2 \\ \vdots \\ de_n \end{bmatrix}, \quad g(e, c) = \begin{bmatrix} g_1(e_1, e_2, ..., e_n, c) \\ g_2(e_1, e_2, ..., e_n, c) \\ \vdots \\ g_n(e_1, e_2, ..., e_n, c) \end{bmatrix}$$

then Eq. (3.2.5) can be written in the following equivalent matrix form:

$$de = g(e, c)dc \tag{3.2.6}$$

where de and $g(e, c)$ are column matrices ($n \times 1$ matrices) as previously defined.

Observe that a boldface symbol appearing throughout this text indicates that the symbol is to be considered as a multi-component scalar variable or vector quantity. Equation (3.2.6) is the matrix equation representation for a system with one input dc and n outputs $de_1, de_2, ..., de_n$ (or equivalently the column vector de).

In the presence of complete external feedback of the output de (i.e., a distinct feedback kernel for each output, outputs $de_1, de_2, ..., de_n$), the block diagram for the system then appears as follows:

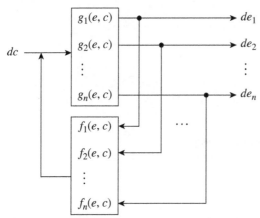

where $f_1, f_1, ..., f_n$ are the feedback kernels associated with their respective outputs $de_1, de_2, ..., de_n$. By grouping the feedback kernel components into a row matrix $(1 \times n)$, as follows:

$$f(e, c) = (f_1, f_1, ..., f_n)$$

It is possible to represent the system block diagram with feedback in the following simpler block diagram form (generally referred to as the "canonical block diagram form")[1] where de and $g(\mathbf{e}, c)$ are column matrices as previously defined. The similarity of this block diagram with the associated block diagram obtained for the single-input, single-output system is apparent.

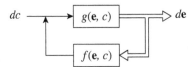

To determine the overall system relationship with feedback present, we observe that the total input to the system kernel $g(\mathbf{e}, c)$ can be expressed in matrix form as

$$dc + f(\mathbf{e}, c)de \tag{3.2.7}$$

where the matrix product $f(\mathbf{e}, c)de$ defines the following scalar quantity:

$$f(\mathbf{e}, c)dc = [f_1, f_2, ..., f_n] \begin{bmatrix} de_1 \\ de_2 \\ \vdots \\ de_n \end{bmatrix} = f_1 de_1 + f_2 de_2 + \cdots + f_n de_n$$

This equation is a result of using the standard definition of matrix multiplication. The total system input expression, Eq. (3.2.7), may be multiplied by the system kernel $g(\mathbf{e}, c)$ to provide as the system output de with feedback the following:

$$de = g(\mathbf{e}, c)[dc + f(\mathbf{e}, c)de]$$

This can then be rearranged into the following form grouping de terms:

$$[I - g(\mathbf{e}, c)f(\mathbf{e}, c)]de = g(\mathbf{e}, c)dc \tag{3.2.8}$$

where I is the $(n \times n)$ identity or unity matrix defined as follows:

$$I = \begin{bmatrix} 1 & 0 & ... & 0 \\ 0 & 1 & & 0 \\ \vdots & & & \\ 0 & 0 & ... & 1 \end{bmatrix}$$

---

[1] Double width line paths indicate multiple inputs or outputs.
In this figure, the double-line output represents multiple outputs $de_1, de_2, ..., de_n$.

The unity matrix serves the role of unity (i.e., the number 1) in matrix arithmetic. Since matrix division is not defined, to express Eq. (3.2.8) explicitly in terms of *de*, we must use the concept of an inverse matrix. The coefficient matrix in Eq. (3.2.8) is given by

$$I - g(\mathbf{e}, c)f(\mathbf{e}, c) = \begin{bmatrix} 1 - g_1 f_1 & -g_1 f_2 & \cdots & -g_1 f_n \\ -g_2 f_1 & 1 - g_2 f_2 & & -g_2 f_n \\ \vdots & & & \\ -g_n f_1 & -g_n f_2 & \cdots & 1 - g_n f_n \end{bmatrix} \tag{3.2.9}$$

This matrix is well defined if the matrix is square (i.e., has the same number of rows and columns), which Eq. (3.2.9) satisfies, and is not singular, and the determinant of $[I - gf]$ is defined and does not vanish, that is, if the

$$\det|I - gf| = (1 - g_1 f_1 - g_2 f_2 \cdots, -g_n f_n) \neq 0$$

If this determinant is not zero (i.e., $\det|I - gf| \neq 0$), then the inverse matrix of $[I - g(\mathbf{e}, c) f(\mathbf{e}, c)]$ is defined as $[I - gf]^{-1}$ and is given by

$$[I - g(\mathbf{e}, c) f(\mathbf{e}, c)]^{-1} = \frac{1}{\det|I - gf|} \text{adjoint}[I - g(\mathbf{e}, c) f(\mathbf{e}, c)] \tag{3.2.10}$$

The adjoint matrix of $[I - gf]$ is defined as the matrix of cofactors of $[I - gf]$ and is given in any standard textbook on matrices or linear analysis. Utilizing the definition of an inverse matrix, we have that, for any nonsingular square matrix A, the matrix product of A and its inverse A^{-1} in any order of multiplication gives the unity matrix I; that is

$$AA^{-1} = A^{-1}A = I$$

Furthermore, the product of any square matrix A and the appropriate unity matrix I is equivalent to the original matrix A; that is

$$AI = IA = A$$

Thus, if we pre multiply (i.e., multiply from the left) Eq. (3.2.8) by

$$[I - g(\mathbf{e}, c) f(\mathbf{e}, c)]^{-1}$$

we have that

$$[I - gf]^{-1}[I - gf]de = [I - gf]^{-1}gdc$$

and simplifying, we obtain the following explicit expression for the system output *de*

$$de = [I - g(\mathbf{e}, c) f(\mathbf{e}, c)]^{-1} g(\mathbf{e}, c)dc \tag{3.2.11}$$

It can be shown that system Eq. (3.2.11) can also be expressed in the simpler form

$$de = \frac{g(\mathbf{e}, c)}{1 - g(\mathbf{e}, c)f(\mathbf{e}, c)} = \frac{g(\mathbf{e}, c)}{1 - \sum\limits_{i=1}^{n} g_i f_i} \tag{3.2.12}$$

since the matrix equivalency is evident

$$[I - g(\mathbf{e}, c)f(\mathbf{e}, c)]^{-1}g(\mathbf{e}, c) = \frac{g(\mathbf{e}, c)}{1 - g_1 f_1 - g_2 f_2 \cdots - g_n f_n} = \frac{g(\mathbf{e}, c)}{1 - \sum\limits_{i=1}^{n} g_i f_i}$$

The discrete system equation for a single-input, multiple-output system can be expressed as follows from the obvious similarity with Eq. (3.2.12):

$$\Delta \mathbf{e} = \frac{g(\mathbf{e}, c)}{1 - \sum\limits_{i=1}^{n} g_i f_i} \Delta c$$

where the functions f and g are as previously defined for the continuous system.

Equation (3.2.11) or equivalently (3.2.12) is the general system equation with feedback for continuous systems with a single input and multiple outputs. Observe that if the matrix feedback kernel vanishes, then Eqs. (3.2.11) and (3.2.12) reduce to Eq. (3.2.6).

To demonstrate the application of the system equations developed for single-input, multiple-output systems, consider a system with one input and two outputs. Then we have for the matrix system equations that

$$g(\mathbf{e}, c) = \begin{bmatrix} g_1 \\ g_2 \end{bmatrix} \quad d\mathbf{e} = \begin{bmatrix} de_1 \\ de_2 \end{bmatrix} \quad f(\mathbf{e}, c) = [f_1, f_2]$$

The system equation without feedback is given by

$$\begin{bmatrix} de_1 \\ de_2 \end{bmatrix} = \begin{bmatrix} g_1 \\ g_2 \end{bmatrix} dc \tag{3.2.13}$$

and in the case of feedback we have, from Eq. (3.2.8), that

$$\left[\begin{bmatrix} 1 & 0 \\ 0 & 1 \end{bmatrix} - \begin{bmatrix} g_1 \\ g_2 \end{bmatrix} [f_1, f_2] \right] \begin{bmatrix} de_1 \\ de_2 \end{bmatrix} = \begin{bmatrix} g_1 \\ g_2 \end{bmatrix} dc$$

The coefficient matrix of the column vector $d\mathbf{e}$, which we will call A for convenience, can be expressed as

$$A = \begin{bmatrix} 1 - g_1 f_1 & -g_1 f_2 \\ -g_2 f_1 & 1 - g_2 f_2 \end{bmatrix}$$

To determine the inverse of A, we must find the determinant of A, which is given by

$$\det A = (1 - g_1 f_1)(1 - g_2 f_2) - g_1 g_2 f_1 f_2 = 1 - g_1 f_1 - g_2 f_2$$

Assuming that the determinant of A is not zero i.e., $\det A \neq 0$, then we obtain for the inverse matrix of A that

$$A^{-1} = \frac{1}{\det A} \begin{bmatrix} 1 - g_2 f_2 & -g_1 f_2 \\ -g_2 f_1 & 1 - g_1 f_1 \end{bmatrix}$$

The matrix system equation for $d\mathbf{e}$ then is given by

$$d\mathbf{e} = A^{-1} g(\mathbf{e}, c) dc$$

and performing the matrix operations, we have as an equivalent equation for the system that

$$
\begin{bmatrix} de_1 \\ de_2 \end{bmatrix} = \begin{bmatrix} \dfrac{g_1}{1 - g_1 f_1 - g_2 f_2} \\ \dfrac{g_2}{1 - g_1 f_1 - g_2 f_2} \end{bmatrix} dc \tag{3.2.14}
$$

Observe, as expected, if the feedback kernel vanishes (i.e., $f_1 = f_2 = 0$), then Eq. (3.2.14) reduces to Eq. (3.2.13).

By obvious extension, the corresponding system equation for a discrete system with input Δc and outputs Δe_1 and Δe_2 with feedback is

$$
\begin{bmatrix} \Delta e_1 \\ \Delta e_2 \end{bmatrix} = \begin{bmatrix} \dfrac{g_1}{1 - g_1 f_1 - g_2 f_2} \\ \dfrac{g_2}{1 - g_1 f_1 - g_2 f_2} \end{bmatrix} \Delta c \tag{3.2.15}
$$

3.2.3 Multiple-Input, Single-Output Systems

Often it is necessary to consider more than one input to a system to account for the system's behavior. The simplest multiple-input system is that system that exhibits a single output in response to the multiple inputs. The multiple-input, single-output continuous system has a block diagram representation as follows:

The system equation, in this case, assumes the following form for m inputs:

$$
de = g_1 dc_1 + g_2 dc_2 + \cdots + g_m dc_m \tag{3.2.16}
$$

Again, this system equation can be simplified by using matrix notation. If we define an equivalent row matrix for the system kernels (i.e., the g's) and a column matrix for the differential input dc's where

$$
g(e, \mathbf{c}) = [g_1 g_2 \cdots g_m], \quad d\mathbf{c} = \begin{bmatrix} dc_1 \\ dc_2 \\ \vdots \\ dc_m \end{bmatrix},
$$

then the system equation can be written in the equivalent matrix form as follows:

$$
de = g(e, \mathbf{c}) d\mathbf{c} \tag{3.2.17}
$$

where $g(e, \mathbf{c})$ is a $(1 \times m)$ row matrix and $d\mathbf{c}$ is an $(m \times 1)$ column matrix.

It will be shown in Chapter 5 that Eq. (3.2.17) is the differential form of a partial differential equation whose dependent variable is e and whose independent variables are the c's. We will find that this observation will assist us in our efforts to obtain specific solutions for Eq. (3.2.17).

In the presence of complete feedback of the output de, the block diagram for the system assumes the following form:

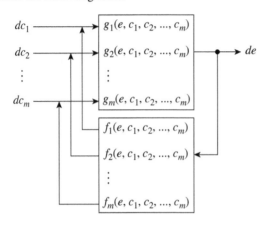

where $f_1, f_2, f_3, ..., f_m$ are the feedback kernels associated with the inputs $dc_1, dc_2, ..., dc_m$, respectively. By grouping the feedback kernel components in a column matrix as follows:

$$f(e, \mathbf{c}) = \begin{bmatrix} f_1 \\ f_2 \\ \vdots \\ f_m \end{bmatrix}$$

it is possible to portray the system block diagram with feedback in the simple canonical form

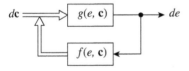

where $d\mathbf{c}$ and $f(e, \mathbf{c})$ are column matrices.

To determine the overall system relationship with feedback present in a multiple-input, single-output system, we find that the total input to the system kernel $g(e, \mathbf{c})$ is given in matrix form as

$$de = f(e, \mathbf{c})de]$$

This total input may then be multiplied by the system kernel $g(e, \mathbf{c})$ to provide the following matrix equation:

$$de = g(e, \mathbf{c})[d\mathbf{c} + f(e, \mathbf{c})de]$$

This result can be rearranged to give

$$[1 - g(e, \mathbf{c})f(e, \mathbf{c})]de = g(e, \mathbf{c})d\mathbf{c}$$

and if $g(e, \mathbf{c})f(e, \mathbf{c}) \neq 1$, we obtain as the system equation with feedback

$$de = \frac{g(e, \mathbf{c})}{1 - g(e, \mathbf{c})f(e, \mathbf{c})} = \frac{g(e, \mathbf{c})}{1 - \sum\limits_{i=1}^{m} g_i f_i} d\mathbf{c} \qquad (3.2.18)$$

and for the discrete system equation with feedback

$$\Delta e = \frac{g(e, \mathbf{c})}{1 - g(e, \mathbf{c})f(e, \mathbf{c})} \Delta \mathbf{c} = \frac{g(e, \mathbf{c})}{1 - \sum\limits_{i=1}^{m} g_i f_i} \Delta \mathbf{c} \qquad (3.2.19)$$

Equations (3.2.18) and (3.2.19) are the system equations for multiple-input, single-output systems with feedback. Of course, in the absence of feedback, the feedback kernel vanishes, and Eq. (3.2.18) reduces to Eq. (3.2.17).

To demonstrate a simple application for the multiple-input, single-output system, consider a system now with two inputs and one output. Then we have for the matrix system equations the following:

$$g(e, \mathbf{c}) = (g_1, g_2), \quad d\mathbf{c} = \begin{bmatrix} dc_1 \\ dc_2 \end{bmatrix}, \quad f(e, \mathbf{c}) = \begin{bmatrix} f_1 \\ f_2 \end{bmatrix}$$

The system equation without feedback is given by

$$de = [g_1 g_2] \begin{bmatrix} dc_1 \\ dc_2 \end{bmatrix} = g_1 dc_1 + g_2 dc_2 \qquad (3.2.20)$$

In the presence of feedback, we have that

$$\left[1 - [g_1 \; g_2] \begin{bmatrix} f_1 \\ f_2 \end{bmatrix} \right] de = [g_1 \;\; g_2] \begin{bmatrix} dc_1 \\ dc_2 \end{bmatrix}$$

and, if $(1 - g_1 f_1 - g_2 f_2) \neq 0$, then we have for *de* the following system equation:

$$de = \frac{g_1}{1 - g_1 f_1 - g_2 f_2} dc_1 + \frac{g_2}{1 - g_1 f_1 - g_2 f_2} dc_2 \qquad (3.2.21)$$

Again, if no feedback occurs, $f_1 = f_2 = 0$, and Eq. (3.2.21) reduces to Eq. (3.2.20).

3.2.4 Multiple-Input, Multiple-Output Systems

The most generalized system is that system that has both multiple inputs and multiple outputs. All rational systems, in principle, can be treated as multiple-input, multiple-output systems, which have a block diagram representation as follows for continuous systems.

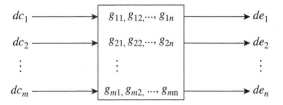

The system equation for this generalized system assumes the following scalar form:

$$de_1 = g_{11}dc_1 + g_{12}dc_2 + \cdots + g_{1m}dc_m$$
$$de_2 = g_{21}dc_1 + g_{22}dc_2 + \cdots + g_{2m}dc_m$$
$$\vdots$$
$$de_n = g_{n1}dc_1 + g_{n2}dc_2 + \cdots + g_{nm}dc_m$$

This system equation, because of its size and many components, is much simpler when represented in matrix form. Then the system equation can be represented as follows:

$$de = g(e, c)dc \qquad (3.2.22)$$

where

$$de = \begin{bmatrix} de_1 \\ de_2 \\ \vdots \\ de_n \end{bmatrix}, \quad g(e, c) = \begin{bmatrix} g_{11} & g_{12} & \cdots & g_{1m} \\ g_{21} & g_{22} & & g_{2m} \\ \vdots & & & \\ g_{n1} & g_{n2} & \cdots & g_{nm} \end{bmatrix}, \quad dc = \begin{bmatrix} dc_1 \\ dc_2 \\ \vdots \\ dc_m \end{bmatrix}$$

and de and dc are column matrices [i.e., $(n \times m)$ and $(m \times n)$ matrices, respectively], and the system kernel $g(e, c)$ is an $(n \times m)$ rectangular matrix. The system matrix equation (3.2.22) defines a set of n equations, which can be considered as a system of partial differential equations with n dependent variables (the e's) and m independent variables (the c's).

For the situation in which feedback of each of the m output variables occurs, the system block diagram appears as shown in the following canonical form:

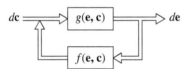

The generalized feedback kernel $f(e, c)$ has the form of an $(m \times n)$ matrix as follows:

$$f(e, c) = \begin{bmatrix} f_{11} & f_{12} & \cdots & f_{1n} \\ f_{21} & f_{22} & & f_{2n} \\ \vdots & & & \\ f_{m1} & f_{m2} & \cdots & f_{mn} \end{bmatrix}$$

With feedback present, the total input to the system kernel is given by

$$dc + f(e, c)de$$

The equation, for our general multiple-input, multiple-output system with feedback, then has the form

$$de = g(e, c)[dc + f(e, c)de]$$

Grouping on common terms, we find that

$$[I - g(e, c)f(e, c)]de = g(e, c)dc \qquad (3.2.23)$$

where I is the $(n \times n)$ unit matrix. If the square matrix $[I - gf]$ is nonsingular, that is, the determinant does not vanish, then the matrix $[I - gf]$ has an inverse given by

$$[I - gf]^{-1} = \frac{1}{\det|I - gf|} \operatorname{adjoint}[I - gf]$$

Equation (3.2.23) for our system may be expressed explicitly for de as follows:

$$de = [I - g(\mathbf{e}, \mathbf{c})f(\mathbf{e}, \mathbf{c})]^{-1} g(\mathbf{e}, \mathbf{c}) dc \qquad (3.2.24)$$

This is the general system equation with feedback for m inputs and n outputs. Again, if the matrix feedback kernel f vanishes identically, then Eq. (3.2.24) reduces to Eq. (3.2.22). The system equation corresponding to Eq. (3.2.24) for a discrete variable system is

$$\Delta \mathbf{e} = [I - g(\mathbf{e}, \mathbf{c})f(\mathbf{e}, \mathbf{c})]^{-1} g(\mathbf{e}, \mathbf{c}) \Delta \mathbf{c} \qquad (3.2.25)$$

where

$$\Delta \mathbf{e} = \begin{bmatrix} \Delta e_1 \\ \Delta e_2 \\ \vdots \\ \Delta e_n \end{bmatrix}, \quad \Delta \mathbf{c} = \begin{bmatrix} \Delta c_1 \\ \Delta c_2 \\ \vdots \\ \Delta c_m \end{bmatrix}$$

and other terms have been previously defined.

To demonstrate a simple example of the multiple-input, multiple-output continuous system, consider a system with two inputs, two outputs, and complete feedback. Then we have for the matrix system equations the following definitions:

$$g(\mathbf{e}, \mathbf{c}) = \begin{bmatrix} g_{11} & g_{12} \\ g_{21} & g_{22} \end{bmatrix}, \quad f(\mathbf{e}, \mathbf{c}) = \begin{bmatrix} f_{11} & f_{12} \\ f_{21} & f_{22} \end{bmatrix}$$

$$de = \begin{bmatrix} de_1 \\ de_2 \end{bmatrix}, \quad dc = \begin{bmatrix} dc_1 \\ dc_2 \end{bmatrix}$$

The system equation without feedback is given by

$$\begin{bmatrix} de_1 \\ de_2 \end{bmatrix} = \begin{bmatrix} g_{11} & g_{12} \\ g_{21} & g_{22} \end{bmatrix} \begin{bmatrix} dc_1 \\ dc_2 \end{bmatrix} \qquad (3.2.26)$$

In the presence of feedback, we have for the system equation that

$$\begin{bmatrix} de_1 \\ de_2 \end{bmatrix} = \frac{1}{\det A} \begin{bmatrix} g_{11} - f_{22}(g_{11}g_{22} - g_{12}g_{21}) & g_{12} - f_{12}(g_{11}g_{22} - g_{12}g_{21}) \\ g_{21} + f_{21}(g_{11}g_{22} - g_{12}g_{21}) & g_{22} + f_{11}(g_{11}g_{22} - g_{12}g_{21}) \end{bmatrix} \qquad (3.2.27)$$

where

$$\det A = 1 - (g_{11}f_{11} + g_{22}f_{22} + g_{12}f_{21} + g_{21}f_{12})$$
$$+ (g_{11}g_{22} - g_{12}g_{21})(f_{11}f_{22} - f_{12}f_{21})$$

Again, observe that Eq. (3.2.27) reduces to system equation (3.2.26) without feedback (i.e., when $f_{11} = f_{22} = f_{12} = f_{21} = 0$).

3.3 Summary

In this important, fundamental chapter, we have taken the causality principle and used it to formulate the formal mathematical representation for the general system equation. The system kernel, which couples the system's inputs and outputs, has been defined and generalized to include the occurrence of feedback of the system outputs.

Systems have been classified according to the number of inputs and number of outputs, and the general system equation for each system type (i.e., single-input, single-output; single-input, multiple-output; multiple-input, single-output; and multiple-input, multiple-output) has been formulated for systems with and without feedback. Table 3-1 provides a useful summary of these system equations.

Table 3-1 Table of System Equations (Bold **e** and **c** are multi-valued variables)

System classification	No feedback	With feedback
Single input Single output	$de = g(e, c)dc$ Eq. (3.2.1)	$de = g(e, c)dc/[1 - g(e, c)f(e, c)]$ Eq. (3.2.2)
Single input Multiple output	$de = g(\mathbf{e}, c)dc$ Eq. (3.2.6)	$de = g(\mathbf{e}, c)dc/[1 - f(\mathbf{e}, c)g(\mathbf{e}, c)]$ Eq. (3.2.12)
Multiple input Single output	$de = g(e, \mathbf{c})d\mathbf{c}$ Eq. (3.2.17)	$de = g(e, \mathbf{c})d\mathbf{c}/[1 - g(e, \mathbf{c})f(e, \mathbf{c})]$ Eq. (3.2.18)
Multiple input Multiple output	$d\mathbf{e} = g(\mathbf{e}, \mathbf{c})d\mathbf{c}$ Eq. (3.2.22)	$d\mathbf{e} = [I - g(\mathbf{e}, \mathbf{c})f(\mathbf{e}, \mathbf{c})]^{-1}g(\mathbf{e}, \mathbf{c})d\mathbf{c}$ Eq. (3.2.24)

Where

$$g(\mathbf{e}, c) = \begin{bmatrix} g_1 \\ g_2 \\ \vdots \\ g_n \end{bmatrix} = [g_1 \quad g_2 \quad \cdots \quad g_n]^T, \quad (n \times 1) \text{ matrix}$$

$$f(\mathbf{e}, c) = \begin{bmatrix} f_1 \\ f_2 \\ \vdots \\ f_n \end{bmatrix}^T = [f_1 \quad f_2 \quad \cdots \quad f_m], \quad (1 \times m) \text{ matrix}$$

$$g(e, \mathbf{c}) = [g_1 \quad g_1 \quad \cdots \quad g_m], \quad (1 \times m) \text{ matrix}$$
$$f(e, \mathbf{c}) = [f_1 \quad f_1 \quad \cdots \quad f_m], \quad (m \times 1) \text{ matrix}$$

$$g(\mathbf{e}, \mathbf{c}) = \begin{bmatrix} g_{11} & g_{12} & \cdots & g_{1m} \\ g_{21} & g_{22} & & g_{2m} \\ \vdots & & & \\ g_{n1} & g_{n2} & \cdots & g_{nm} \end{bmatrix}, \quad (n \times m) \text{ matrix}$$

$$f(\mathbf{e}, \mathbf{c}) = \begin{bmatrix} f11 & f12 & \cdots & f1n \\ f21 & f22 & & f2n \\ \vdots & & & \\ f_{m1} & f_{m2} & \cdots & fmn \end{bmatrix}, \quad (m \times n) \text{ matrix}$$

It is apparent for multiple-input, multiple-output systems with feedback that the resulting system Eqs. (3.2.24) and (3.2.25) are very complex and difficult to solve analytically. Furthermore, these system equations are nonlinear and can exhibit undefined and chaotic behavior if the determinant of the inverse matrix containing the feedback, $f(e, c)$, is present. A general conclusion that can be drawn for multiple-input, multiple-output systems with feedback is they generally defy analytical solutions. Only discrete variable systems using numerical methods coupled with computer processing offer possible solutions. If all terms in the system kernel, $g(e, c)$, and feedback kernel, $f(e, c)$, are constants, then analytical solutions can be found. Such multiple-input, multiple-output systems pose severe challenges to the system investigator. We will address the strategy for modeling and analyzing such multiple systems in Chapter 8.

The problems given at the end of this chapter will provide practice and reinforcement in recognizing and formulating system equations for any quantitative system.

Bibliography

ABRAMOWITZ, A. and I. STENGUN, *Handbook of Mathematical Functions with Formulas, Graphs, and Mathematical Functions*, National Bureau of Standards, Washington, DC, Applied Mathematics Series, **55**, 1964. Comprehensive collection of mathematical resources and tables by U.S. Government Printing Office.

FRAZER, R., W. DUNCAN, and A. COLLAR, *Elementary Matrices and Applications to Dynamics and Differential Equations*, Cambridge University Press, New York, 1938. Elementary matrices and applications to dynamics and differential equations.

KREYSZIG, E. *Advanced Engineering Mathematics*. 10th Edition. John Wiley & Sons, New York, 2010. Widely used standard textbook for advanced engineering mathematics applications.

NERING, E., *Linear Algebra and Matrix Theory*, John Wiley & Sons, New York, 1963. A useful resource for matrix applications in system science.

RUSSELL, B., *Principles of Mathematics*, W. W. Norton & Co., New York, 1961. Chapter 55 is devoted to a quantitative discussion of causality by this Nobel Laureate.

SAGE, A., and J. MELSA, *System Identification*, Academic Press, New York, 1967. An excellent resource text for establishing quantitative system models.

VON BERTALANFFY, L., *General Systems Theory*, George Braziller, New York, 1968. An informative treatment of foundations, developments, and applications in systems theory.

WEAST R. and S. SELBY, *Handbook of Tables for Mathematics*. 3rd Edition. The Chemical Rubber Co., Cleveland, OH, 1967. Comprehensive CRC Handbook for mathematics.

Problems

1 Write the system equation from the block diagram for the following systems. Where appropriate, express the system equation in matrix form and define each matrix. Ensure that all matrices are conformable (i.e., the number of rows and columns is appropriate to satisfy the basic matrix algebraic operations of addition, subtraction, and multiplication).

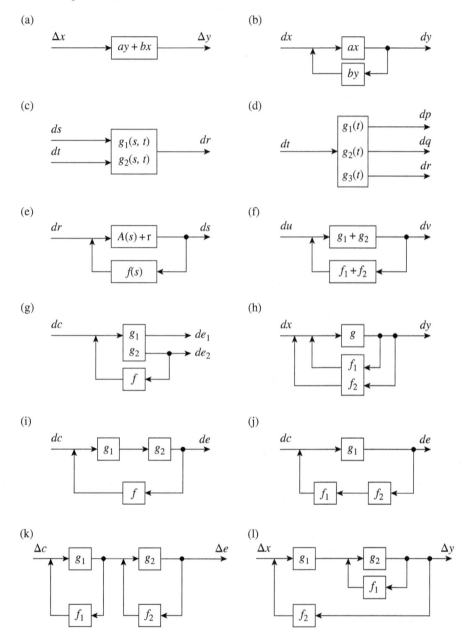

(*Hint:* Treat g_1 initially as a separate system, solve for its output, and then consider g_2 as a separate system, etc.)

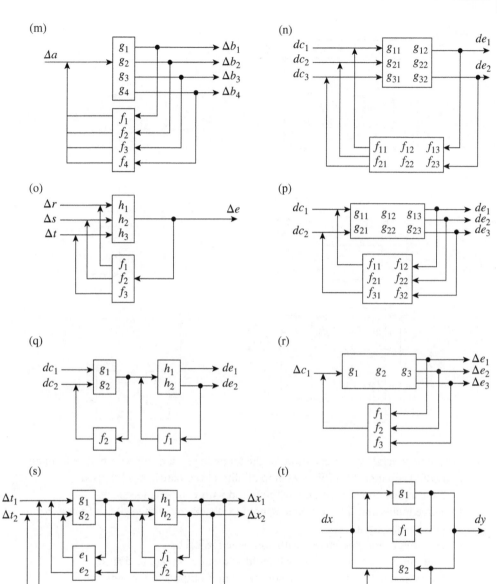

(Hint: Treat g_1 initially as a separate system, solve for its output, and then consider g_2 as a separate system, etc.)

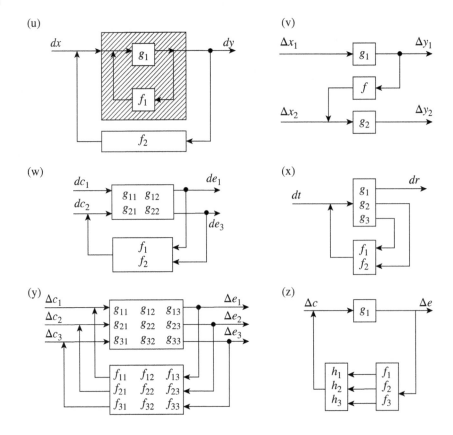

(u)

(v)

(w)

(x)

(y)

(z)

2 Write the general system equation for the following systems in both matrix form and individual equation form. Verify compatibility of the matrix algebra used.

a) Four inputs and two outputs without and with complete feedback.

b) Three inputs and three outputs without feedback and with feedback from two outputs to one input.

c) Two inputs and one output with complete feedback.

d) One input and three outputs with and without complete feedback.

e) N inputs and M outputs with feedback from K outputs to L inputs, where N, M, K, and L are nonzero positive integers.

f) Three inputs and four outputs with feedback from two outputs to two inputs.

3 Consider a system with n inputs and n outputs without feedback. What are the consequences if the determinant of the system kernel is zero?

4 Discuss the consequences for a single-input, single-output system equation with feedback as the system kernel $g(e, c)$ approaches 0 or $\pm\infty$.

5 Show that the multiple-input system kernel component with feedback g_{fij} is defined as K(constant).

$$g_{fij} = \frac{g_{ij}}{1 - \sum_1^m g_i f_i} = \begin{cases} \dfrac{\partial e_i}{\partial c_j} \\[2mm] \dfrac{\Delta e_i}{\Delta c_j} \end{cases}$$

In terms of the system's absolute state function,

$$S(e, c)[i.e., e = S(e, c) + K(\text{constant})]$$

Do we have that g_{fij} is independent of K?

$$g_{fij} = \frac{\dfrac{\partial S_i}{\partial c_j}}{1 - \dfrac{\partial S_i}{\partial e_j}} \quad \text{for continous systems}$$

What is the relationship, if any, between g_{fij} and $\dfrac{\partial S_i}{\partial c_j}$ and also between $\sum g_i f_i$ and $\dfrac{\partial S_i}{\partial e_i}$ in these two expressions? Under what conditions for S_i does feedback occur? Why?

6 For a continuous (or discrete) single-input, single-output system equation with feedback, where

$$de = \frac{g(e, c)}{1 - g(e, c) f(e, c)} dc$$

suppose that for some set of values of $e \to e^*$, $c \to c^*$, that

$$f(e^*, c^*) \to \frac{1}{g(e^*, c^*)}$$

Then we have that

$$\text{Lim}_{\substack{e \to e^* \\ c \to c^*}} [1 - g(e, c) f(e, c)] = 0 = \text{Lim}_{\substack{e \to e^* \\ c \to c^*}} g(e, c) \frac{dc}{de}$$

and we have that either

$$g(e^*, c^*) \to 0 \quad \text{or} \quad \frac{dc}{de}\bigg|_{\substack{e \to e^* \\ c \to c^*}} \to 0$$

This is called a singularity, a pole, or a critical point for the system, and unless $g(e^*, c^*) \to 0$ then the output change de becomes unbounded for any finite input, which is generally an undesirable operating condition!

7 Observe that the single input, multiple outputs and multiple inputs, single output with feedback have similar denominators (i.e., $[1 - gf]$). Establish that this is true. What happens for those values of e and c, if anything, when $[1 - gf] = 1$?

8 What happens to the multiple-input or multiple-output system equations if feedback is not complete and some of the feedback terms vanish?

9 Verify the results for a double-input, double-output system with feedback shown in Section 3.2.4. What are the consequences of $g_{11}g_{22} = g_{12}g_{21}$? What does that mean? Note:

$$\begin{bmatrix} de_1 \\ de_2 \end{bmatrix} = \frac{1}{\det A} \begin{bmatrix} g_{11} & g_{12} \\ g_{21} & g_{22} \end{bmatrix}$$

$$\det [A] = 1 - (g_{11}f_{11} + g_{22}f_{22} + g_{12}f_{21} + g_{21}f_{12})$$

10 Consider the multiple-input, multiple-output system equation with feedback as given by Eq. (3.2.24). What is the significance for the system equation if

$$\det[I - g(\mathbf{e}, \mathbf{c})f(\mathbf{e}, \mathbf{c})] = 0$$

Suppose $f(\mathbf{e}, \mathbf{c}) = \text{constant} = 1/\lambda$. Then we have that

$$\det[g(\mathbf{e}, \mathbf{c}) - \lambda I] = 0$$

where λ defines the eigenvalues of the system kernel $g(\mathbf{e}, \mathbf{c})$. Now, what is the significance of feedback when $f = 1/\lambda$?

11 Consider the following mixed-system block diagram, which exhibits both discrete and continuous inputs and outputs. Write the system equations. Interpret your result.

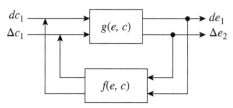

12 For the single-input, single-output system with repeated feedback shown in the figure, where

$$g_{n+1}(e, c) = \frac{g_n(e, c)}{1 - g_n(e, c)f_n(e, c)}$$

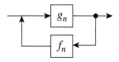

where $n = 1, 2, ..., N$, then for any finite number of feedback kernels N, show that we have

$$g_{N+1}(e, c) = \frac{g_1(e, c)}{1 - g_1(e, c) \displaystyle\sum_{i=1}^{N} f_i(e, c)}$$

Extend this result to include multiple-input and multiple-output systems.

13 Consider a three input (dc_1, dc_2, dc_3), three output (de_1, de_2, de_3) system with full feedback of the system outputs (de_1, de_2, de_3). Using the MATLAB Symbolic Math Toolbox shows that the final system equations, including the system kernel, are given by the following set of system equations.

$$de_1 = 1/D\ [\{g_{11}\ (f_{22}\ g_{22} + f_{33}\ g_{33} - f_{22}\ f_{33}\ g_{22}\ g_{33} + f_{23}\ f_{32}\ g_{23}\ g_{32} - 1)\}\ dc_1$$
$$- \{g_{12}\ (f_{12}\ g_{12} - f_{12}\ f_{33}\ g_{12}\ g_{33} + f_{13}\ f_{32}\ g_{13}\ g_{32})\}\ dc_2$$
$$- \{g_{13}\ (f_{13}\ g_{13} + f_{12}\ f_{23}\ g_{12}\ g_{23} - f_{13}\ f_{22}\ g_{13}\ g_{22})\}]\ dc_3$$

$$de_2 = 1/D\ [\{- g_{21}\ (f_{21}\ g_{21} - f_{21}\ f_{33}\ g_{21}\ g_{33} + f_{23}\ f_{31}\ g_{23}\ g_{31})\}\ dc_1$$
$$+ \{g_{22}\ (f_{11}\ g_{11} + f_{33}\ g_{33} - f_{11}\ f_{33}\ g_{11}\ g_{33} + f_{13}\ f_{31}\ g_{13}\ g_{31} - 1)\}\ dc_2$$
$$\{g_{23}\ (f_{23}\ g_{23} - f_{11}\ f_{23}\ g_{11}\ g_{23} + f_{13}\ f_{21}\ g_{13}\ g_{21})\}]\ dc_3$$

$$de_3 = 1/D\ [\{- g_{31}\ (f_{31}\ g_{31} + f_{21}\ f_{32}\ g_{21}\ g_{32} - f_{22}\ f_{31}\ g_{22}\ g_{31})\}\ dc_1$$
$$- \{g_{32}\ (f_{32}\ g_{32} - f_{11}\ f_{32}\ g_{11}\ g_{32} + f_{12}\ f_{31}\ g_{12}\ g_{31})\}\ dc_2$$
$$+ \{g_{33}\ (f_{11}\ g_{11} + f_{22}\ g_{22} - f_{11}\ f_{22}\ g_{11}\ g_{22} + f_{12}\ f_{21}\ g_{12}\ g_{21} - 1)\}]\ dc_3$$

Where

$$D = [\ f_{11}\ g_{11} + f_{22}\ g_{22} + f_{33}\ g_{33} - f_{11}\ f_{22}\ g_{11}\ g_{22} + f_{12}\ f_{21}\ g_{12}\ g_{21}$$
$$- f_{11}\ f_{33}\ g_{11}\ g_{33} + f_{13}\ f_{31}\ g_{13}\ g_{31} - f_{22}\ f_{33}\ g_{22}\ g_{33} + f_{23}\ f_{32}\ g_{23}\ g_{32}$$
$$+ f_{11}\ f_{22}\ f_{33}\ g_{11}\ g_{22}\ g_{33} - f_{11}\ f_{23}\ f_{32}\ g_{11}\ g_{23}\ g_{32} - f_{12}\ f_{21}\ f_{33}\ g_{12}\ g_{21}\ g_{33}$$
$$+ f_{12}\ f_{23}\ f_{31}\ g_{12}\ g_{23}\ g_{31} + f_{13}\ f_{21}\ f_{32}\ g_{13}\ g_{21}\ g_{32} - f_{13}\ f_{22}\ f_{31}\ g_{13}\ g_{22}\ g_{31} - 1]$$

Comment on the response of this system if D = 0 and the impact upon successful analysis of this system in the presence of full system feedback.

$$de_1 = 1/D\{[g_{11}(f_{22}g_{22} - 1)]dc_1$$
$$- [g_{12}(f_{12}g_{12})]dc_2$$
$$de_2 = 1/D\{[-g_{21}(f_{21}g_{21})]dc_1$$
$$+ [g_{22}(f_{11}g_{11} - 1)]dc_2$$

where

$$D = [f_{11}g_{11} + f_{22}g_{22} + f_{12}f_{21}g_{12}g_{21} - 1]$$

$$\begin{bmatrix} de_1 \\ de_2 \end{bmatrix} = \begin{bmatrix} g_{11} - f_{22}(g_{11}g_{22} - g_{12}g_{21}) & g_{12} - f_{12}(g_{11}g_{22} - g_{12}g_{21}) \\ g_{21} - f_{21}(g_{11}g_{22} - g_{12}g_{21}) & g_{22} - f_{11}(g_{11}g_{22} - g_{12}g_{21}) \end{bmatrix} \begin{bmatrix} dc_1 \\ dc_2 \end{bmatrix}$$

where

$$det\ A = 1 - (g_{11}f_{11} + g_{22}f_{22} + g_{12}f_{21} + g_{21}f_{12})$$
$$+ (g_{11}g_{22} - g_{12}g_{21})(f_{11}f_{22} - f_{12}f_{21})$$

4

Single-Input Systems

Because of their relative simplicity and universal importance in the study of systems, the major analytical solution methods associated with single-input systems will be examined in detail in this chapter. With access to adequate computing facilities and appropriate software (e.g., MATLAB and MAPLE) support, the system investigator can use computer-based software packages for analysis that may reduce the need for the analytical-based methods presented in this chapter for single-input systems and Chapter 5 for multiple-input systems. However, definite advantages do accrue from understanding and using the analytical methods described in these two chapters. Furthermore, the symbolic computational packages also provided by MATLAB and MAPLE permit quasi-analytical means of assessing and evaluating system equations, particularly those with single inputs.

4.1 Definition and Significance of a Single-Input System

The single-input system model is applicable to any system that can be reasonably assumed to have only one significant or dominant input or independent variable and one or more outputs or dependent variables. For a system with more than one input, it is necessary to reduce all inputs to the system to a single dominant or representative input in order to utilize the single-input system model. This reduction requirement may appear to be an unrealistic and incorrect assumption, but, as we will see in Chapter 8, it is usually possible to reduce multiple-inputs to a single input system by combining and/or neglecting the individual inputs to achieve a single, effective input and one or more outputs. Another means of solving multiple-input and multiple-output system equations is to hold sequentially all variables fixed except for a single input and single output set and systematically explore the behavior of the system equation under investigation.

Mathematically, all single-input systems can be directly related to ordinary (i.e., single independent variable) differential equations for continuous systems and to ordinary (again, single independent variable) difference equations for discrete systems. Since ordinary differential and difference equations are generally easier to formulate and solve than are partial differential and difference equations, which are related to multiple-input (or equivalently, multiple, independent variable) systems, single-input systems are the practical beginning point in the study of most systems. Furthermore, the mathematical

Introduction to System Science with MATLAB, Second Edition.
Gary Marlin Sandquist and Zakary Robert Wilde.
© 2023 John Wiley & Sons Ltd. Published 2023 by John Wiley & Sons Ltd.

understanding and development of ordinary differential and difference equations (and, thus, single-input system analysis) is extensive and well developed, with many powerful methods and techniques available for analyzing such equations.

The single-input, single-output system is, of course, the simplest nontrivial system to study; and the resulting system equation is equivalent to a first-order ordinary differential equation for continuous single-input, single-output systems and a first-order difference equation for discrete single-input, single-output systems. Therefore, the single-input, single-output systems are the obvious place to begin our study.

4.2 Single-Input, Single-Output Systems

Single-input, single-output systems comprise all those systems that can be acceptably treated as having only one dominant or significant input imposed on the system and similarly only one dominant or significant output issuing from the system. Obviously, then, the simplest nontrivial type of system to conceive and thus the easiest to formulate and study is a single input, single-output system. This initial approach to the analysis of most systems is usually advisable even for systems that may subsequently require analysis as multiple-input and/or output systems.

The general equation for the single-input, single-output system, Eq. (4.2.1), is equivalent to the general first-order differential equation in differential form for the continuous system kernel and to the general first-order difference equation for the discrete system kernel. In the single-input, single-output system equation

$$\left.\begin{array}{c} de \\ \Delta e \end{array}\right\} = g(e,c) \left\{\begin{array}{c} dc \\ \Delta c \end{array}\right. \tag{4.2.1}$$

where de and dc are infinitesimal (i.e., differential and continuously variable) changes and Δe and Δc are the finite (i.e., incremental or discrete) changes in the system output and input, respectively. The function $g(e, c)$ is defined as the system kernel, which defines the system output per unit input, and is generally a function of both the input variable c and the output variable e. If a single-input, single-output system exhibits feedback of the output, which is defined by the feedback kernel, $f(e, c)$ then the system equation with feedback becomes

$$\left.\begin{array}{c} de \\ \Delta e \end{array}\right\} = \frac{g(e,c)}{1 - g(e,c)f(e,c)} \left\{\begin{array}{c} dc \\ \Delta c \end{array}\right. \tag{4.2.2}$$

as we have seen previously in Section 3.2.1. However, it is always possible to express the system kernel with feedback as given by Eq. (4.2.2) without loss of generality as follows:

$$\frac{g(e,c)}{1 - g(e,c)f(e,c)} = g_f(e,c) \tag{4.2.3}$$

where the function of $g_f(e, c)$, which is the system kernel with feedback, can be treated as the associated open-loop system kernel (i.e., a system kernel without apparent feedback) for Eq. (4.2.2). We will consider general system kernel classifications and solution methods for both continuous and discrete systems where the applicable system kernel $g(e, c)$ is understood to have the general form given by either system Eq. (4.2.1) or (4.2.2) without loss of generality for the classification or solution method. Of course, the occurrence of feedback in a given system will generally make the resulting system kernel more difficult to analyze.

It is apparent from the form of Eq. (4.2.2) that the occurrence of a nonzero feedback function (i.e., $f(e, c) \neq 0$) generally makes the resulting system equation nonlinear. The only exceptions to this observation are the condition that both functions, $g(e, c)$ and $f(e, c)$ are functions of the independent variable c only, and the resulting system equation is then linear. The other event is the isolated, very unlikely condition that

$$f(e, c) = \frac{1}{g(e, c)} - \frac{1}{A(c) + B(c)e}$$

where the functions $A(c)$ and $B(c)$ are either constants or functions of the independent variable c only. Substitution of this specific feedback function readily confirms that the resulting system equation is now linear. Aside from these two exceptions, the presence of feedback makes the resulting system equation nonlinear and generally more difficult to solve. Finally, most real physical systems exhibit feedback features and should be considered for accurate modeling.

An important observation made earlier in Chapter 3 is the singular condition that results for Eq. (4.2.2) when the feedback function, $f(e, c)$, assumes a value such that the denominator goes to zero.

$$g(e, c) f(e, c) = 1$$

For this condition for feedback, the system equation becomes unbounded and undefined. Of course, a real physical system cannot sustain such an operating condition for that value of feedback, and the physical will collapse. Obviously, such an operating condition for the feedback imposed upon a real system is to be avoided or at least recognized for potential system failure. There are possible physical conditions in which this class of feedback value may occur, and this singularity results in an extreme response by the system. A nuclear explosion, the collision of the earth with a massive asteroid, or a "black hole" event in cosmology are extreme examples of such unbounded output resulting in the destruction of the system.

4.2.1 Discrete Systems

The general equation for the single-input, single-output discrete system is equivalent to the general first-order ordinary difference equation and can be expressed as follows for the discrete system with feedback or without feedback:

$$\Delta e = g(e, c) \, \Delta c \tag{4.2.4}$$

where Δe is the incremental change in the system output arising from the incremental change Δc in the system input. The function $g(e, c)$ is defined as the discrete system kernel and is, in general, a function of both the discrete input and output variables c and e.

Equation (4.2.1) is also the general form for the first-order ordinary difference equation, and it is apparent that the solutions for this equation are determined by the system kernel $g(e, c)$ and by the associated initial conditions, $c(k)$ and $e(k)$, for the initial step for $k = 0$ or 1 for the system.

For the general elementary difference equation with non-uniform incremental stepping or spacing of the independent discrete variable [i.e., $c(1)$, $c(2)$, $c(3)$, ..., $c(n)$], where the incremental changes in c are not necessarily equal or uniform), Eq. (4.2.4) may be treated as a non-uniformly incremented difference equation. The general solution is obtained using the sequential solution algorithm as follows:

Since the incremental change in Δc and Δe is equivalent to the difference in the input and output states, respectively, then for changes from the initial states $c(1)$ and $e(1)$ to $c(2)$ and $e(2)$, respectively, we have that

$$\Delta e(1) = e(2) - e(1) = g[e(1), c(1)]\Delta c = g[e(1), c(1)] [c(2) - c(1)]$$

so

$$e(2) = e(1) + g[e(1), c(1)][c(2) - c(1)] \tag{4.2.5}$$

For the incremental change from states $e(2)$ and $c(2)$ to $e(3)$ and $c(3)$, respectively, we have that

$$\Delta e(2) = e(3) - e(2) = g[e(2), c(2)]\Delta c(2)$$

or

$$e(3) = e(2) + g[e(2), c(2)] [c(3) - c(2)] \tag{4.2.6}$$

and for the nth step change, we find that

$$e(n + 1) = e(n) + g[e(n), c(n)] [c(n + 1) - c(n)] \tag{4.2.7}$$

Equations (4.2.5), (4.2.6), and (4.2.7) provide a sequential set of equations for determining the output $e(n + 1)$ for a discrete system after any sequence of n sequential steps in the system input. By combining Eqs. (4.2.5), (4.2.6), and (4.2.7), the output $e(n+1)$ can also be expressed as follows for any incremental step n, such that $n \geq 1$ and $i \leq n$:

$$e(n + 1) = e(n) + g[e(n), c(n)] [c(n + 1) - c(n)] \tag{4.2.8}$$

The argument of the discrete system kernel, $g[e(i), c(i)]$, can be evaluated by knowing both the system output $e(i)$ and input $c(i)$ for all previous states for $i \leq n$. But these variables are readily obtained by sequentially specifying the changes in c, that is, $e(1)$, $e(2)$, $e(3)$, ..., $e(m)$, using Eq. (4.2.8).

Thus, for any well-defined discrete system kernel, $g(e, c)$, and a consistent sequence of changes in input, $c(1)$, $c(2)$, $c(3)$, ..., $c(m)$, it is always possible to analytically determine the resulting changes in system output, $e(1)$, $e(2)$, $e(3)$, ..., $e(m)$. The general solution for the single input, single-output discrete system equation is then known. Unfortunately, it is not always possible to find an analytic solution for the corresponding continuous system equation.

The demonstration of the fact that any single-input, single-output discrete system equation (and any discrete system equation whatsoever, including multiple-input, multiple-output discrete systems) can be solved sequentially as given by Eq. (4.2.8) can be shown as follows. Assume that the system kernel $g[e(i), c(i)]$ given in Eq. (4.2.8) is continuous and bounded over the range of definition and that $g[e(i), c(i)]$ remains defined as each $c(i + 1) - c(i) = \Delta c(i)$ approaches zero, and the number of increments becomes infinite such that

$$\text{Lim}_{n \to \infty} e(n + 1) = e(1) + \text{Lim}_{\substack{n \to \infty \\ \Delta c(i) \to 0}} \sum_{i=1}^{n} g[e(i), c(i)] \, \Delta c(i) \tag{4.2.9}$$

and Eq. (4.2.9) is defined, and the appropriate limits exist. Then, we have in the limit that

$$e = e(1) + \int_{c(1)}^{c} g(e, c) dc \tag{4.2.10}$$

This is the quadrature form for the general solution of the continuous single input, single output system equation

$$de = g(e, c) \, dc$$

with the initial condition $e[c(1)] = e(1)$.

Example Problem for Evaluating a Discrete System Kernel.
Consider the following set of tabulated data for a given system model.

N	1	2	3	4	5	6	7	8
$c(N)$	0	1.50	2.30	3.71	4.15	6.21	7.30	7.50
$\Delta(N)$	1.50	0.80	1.41	0.44	2.06	1.09	0.20	–
$g(c, e)$	0	6.00	9.20	14.84	16.60	24.84	29.20	–
$\Delta e(N)$	0	4.80	12.97	6.53	34.20	27.08	5.84	–
$e(N)$	3.00	3.00	7.80	20.77	27.30	61.50	88.58	94.42

where entries for $\Delta c(N)$, $\Delta e(N)$, and $e(N)$ have been determined as follows:

$$\Delta c(N) = c(N + 1) - c(N)$$
$$\Delta e(N) = g[c(N), e(N)] \Delta c(N)$$
$$e(I + 1) = e(I) + g[c(I), e(I)] \Delta c(I)$$

and we have assumed as an initial condition that

$$e(1) = 3.00 \quad \text{when } c(1) = 0.00$$

In the remainder of the material in this section on single-input, single-output systems, we will stress methods for classifying and solving continuous system equations since we have obtained the general solution for the single-input, single-output discrete system equation given by Eq. (4.2.8). The principal challenge for analyzing discrete systems is determining the size of incremental steps for the independent variable, $\Delta c(n)$, to achieve efficient and accurate solutions for the overall system equations.

4.2.2 Continuous Systems

The general equation for the single-input, single-output continuous system is identical with the general first-order differential equation in differential form and can be expressed as follows:

$$de = g(e, c) dc \tag{4.2.11}$$

where *de* is the infinitesimal change in the system output arising from the imposition of the infinitesimal input *dc* on the system. The function $g(e, c)$ is defined as the continuous system kernel, which in general, is a function of both the input and output variables c and e, respectively.

Recalling elementary differential equations, Eq. (4.2.11) has the form of the general first-order differential equation in differential form. Thus, we see that our expression for the first-order single-input continuous system is identical with the mathematical notation for differential equations. This single-input, single-output may be treated as would any first-order differential equation. Since Eq. (4.2.11) is the general form of the first-order differential equation, it is apparent that the solutions of the first-order system, or any system for that matter, are determined by the system kernel $g(e, c)$ and by the associated initial condition, c_0 and $e(c_0) = e_0$, selected for the system.

If the complete solution for Eq. (4.2.11) is known, then the response of the system, within the range of accuracy and applicability of the kernel chosen, can be predicted for any value of the cause or input variable for which the solution and the system kernel are valid.

The general solution of Eq. (4.2.11) usually can be expressed as follows:

$$\emptyset(e, c) = k$$

where $\emptyset(e, c)$ is a continuous function of its arguments e and c, and k is an arbitrary constant determined by the initial conditions. Actually, the most general continuous solution for Eq. (4.2.11) has the functional form $F(e, c, k) = 0$, where it is assumed that the arbitrary solution constant k may appear in the function F in a transcendental way so that F cannot be explicitly solved for k and expressed as $k = \emptyset(e, c)$. However, such functions entail considerations in the advanced theory of functions, which we will not pursue here.

Assuming that the solution $\emptyset(e, c)$ is a continuous function, where the partial derivatives a $\partial\emptyset/\partial e$ and $\partial\emptyset/\partial c$ exist and are defined, then the total differential of $d\emptyset$ is given by

$$d\emptyset = \frac{\partial\emptyset}{\partial e} de + \frac{\partial\emptyset}{\partial c} dc = 0$$

which can be expressed equivalently as follows, if $\partial\emptyset/\partial c \neq 0$:

$$\frac{de}{dc} = -\frac{\partial\emptyset/\partial c}{\partial\emptyset/\partial e} = g(e, c)$$

If $\partial\emptyset/\partial e = 0$, then $\emptyset$ is not a function of the output e. However, this is a meaningless and unacceptable result for a rational system since it implies that the output e of a system does not depend on the input c, which produces the output. Furthermore, such a result is a direct violation of our basic system equation, which requires that any change in the output Δe or de of a rational system be a function of the input Δc or dc.

It would be satisfying to assert that an analytic solution exists and is easily determined for every first-order continuous system kernel that satisfies a minimum set of mathematical requirements for boundedness and continuity. However, unlike discrete system equations, this is not the case, and although it is possible to solve analytically many first-order continuous system equations, it is not possible to solve all of them analytically. Nevertheless, we shall systematically examine many of the kernel types for continuous systems that can be solved analytically.

4.2.3 Constant System Kernels

If the system kernel $g(e, c)$ is or can be considered approximately constant so that $g(e, c) = g_0$ over the range of variation of interest for both the input variable c and corresponding output variable e, then the system equation has the following form for a continuous system with a constant kernel.

$$de = g_0 \, dc \tag{4.2.12}$$

This equation may be directly integrated over the range of variation for e and c if the system kernel is constant or approximately constant. Integration of Eq. (4.2.12) over the limits $e_1 < y < e$ and $c_1 < x < c$, where x and y are dummy variables of integration, gives

$$\int_{e_1}^{e} dy = \int_{c_1}^{c} g_0 \, dx$$

and

$$e = e_1 + g_0[c - c(1)] \tag{4.2.13}$$

which is the solution for the continuous system equation when the system kernel is constant, and $e(c_1) = e_1$.

If the discrete system kernel is a constant, say g_0, then from Eq. (4.2.8), the output of the system at $e(n + 1)$ is given by

$$e(n + 1) = e(1) + g_0 \sum_{i=1}^{n} [c(i + 1) - c(i)] = e(1) + g_0[c(n + 1) - c(1)] \tag{4.2.14}$$

It is apparent that this solution is the same as that for the continuous system with a constant system kernel at corresponding values of c and e. Actually, for any continuous or discrete system equation with single or multiple-inputs and single or multiple-outputs, if the system kernels are constant, then the solutions for the system's response are identical.

Figure 4-1 is a plot of both system solution Eqs. [(4.2.13) and (4.2.14)] for the same constant system kernel g_0. Note that the discrete output variable $e(i)$ is defined only for the discrete values, $e(1)$, $e(2)$, ..., $e(i)$, ..., $e(n)$, $e(n + 1)$, and the corresponding discrete input variable values, $c(1)$, $c(2)$, ..., $c(i)$, ..., $c(n)$, $c(n + 1)$. The continuous output variable e and input variable c are defined for all values, such that $c(1) \leq c \leq c(n + 1)c$ and $e(1) \leq e \leq e(n + 1)$.

4.2.4 Linear System Kernels

If has either of the following two linear forms for the system variables, the resulting differential equation for the system variables is linear and may be easily solved.

$$g(e, c) = \begin{cases} A(c) + B(c)e \\ \dfrac{1}{D(e) + E(e)c} \end{cases}$$

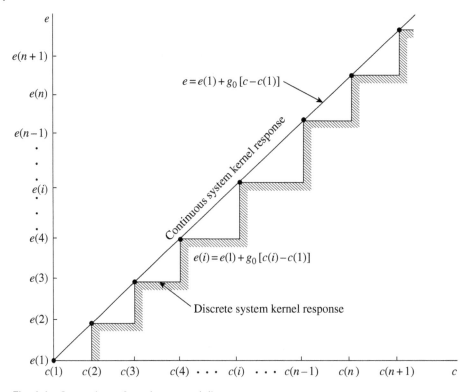

Fig. 4-1 Comparison of continuous and discrete system responses.

That the resulting differential equations are linear may be easily verified by rearranging the resulting system equations for the linear system kernel as follows:

$$\frac{de}{dc} = A(c) + B(c)e \text{ (linear equation in } e) \tag{4.2.15}$$

$$\frac{dc}{de} = D(e) + E(e)c \text{ (linear equation in } c) \tag{4.2.16}$$

Observe that in the first equation, we are considering the variable c to be the independent variable while e is the dependent variable. However, in the second equation, this role is reversed, and we treat e as the independent variable and c as the dependent variable. The linearity of a differential equation is based solely on the dependent variable and the degree to which it appears in each of the terms in the differential equation. Only if the dependent variable and its derivatives occur to either the first degree or the zero degree for each term of the equation, which is separated by a plus or minus sign, is the equation linear. Furthermore, the manner in which the independent variable appears within the equation has no bearing on linearity. Thus, it is often useful in solving system equations to be flexible in assigning the roles of dependent and independent variables and seek the form for which the system equations are linear.

The general solution method for these two linear differential equations is the same, and we will derive this solution by considering the following general linear differential equation, where x is the independent variable and y is the dependent variable:

$$\frac{dy}{dx} + f(x)y = h(x) \tag{4.2.17}$$

We observe that if we multiply both sides of Eq. (4.2.17) by the nonzero function $\exp\left(\int f dx\right)$, which is the integration factor for the equation, as we will learn in Section 4.2.5, then Eq. (4.2.17) becomes

$$\left[\frac{dy}{dx} + f(x)\right] \exp \int f\,dx = \frac{d}{dx}\left[y \exp \int f\,dx\right] = h(x) \exp \int f\,dx$$

which is now an exact differential equation (see Section 4.2.5) and can be directly integrated to give

$$y(x) = \exp\left(-\int f dx\right) \left[\int^{t=x} h(t)\exp\left(\int f dt\right) dt + k\right] \tag{4.2.18}$$

Note that k is the arbitrary integration constant and is determined by specifying a single value of y for a particular value of x, such as $y(x_0)$ where $x = x_0$.

Applying this general solution for the linear differential equation given by Eq. (4.2.18) to Eq. (4.2.15), we find, as the general solution for the system equation,

$$e(c) = \exp\left(\int B\,dc\right) \left[\int^{t=c} A(t)\left[\exp\left(-\int B\,dt\right)\right] dt + k_1\right]$$

For Eq. (4.2.16), the general solution for that system equation is

$$c(e) = \exp\left(\int E\,de\right) \left[\int^{t=e} D(t) \exp\left(-\int E\,dt\right) dt + k_2\right]$$

where k_1 and k_2 are the arbitrary integrating constants.

Example Problem for a Linear Continuous System Kernel.
Consider the system equation

$$dy = \frac{a^2 + y^2}{y(x+b)}\,dx \tag{4.2.19}$$

which is linear in x when expressed as follows:

$$\frac{dx}{dy} - \frac{y}{a^2 + y^2}\,x = \frac{by}{a^2 + y^2} \tag{4.2.20}$$

The integrating factor $I(x, y)$ is given by

$$I(x,y) = \exp\left[-\int \frac{y}{a^2 + y^2}\,dy\right] = \exp\left[-\frac{1}{2}\,\mathrm{Ln}(a^2 + y^2)\right] = \frac{1}{\sqrt{a^2 + y^2}} \tag{4.2.21}$$

Multiplying Eq. (4.2.20) by the integrating factor $I(x, y)$ given by Eq. (4.2.21), we have that

$$\frac{1}{\sqrt{a^2 + y^2}} \left[\frac{dx}{dy} - \frac{yx}{a^2 + y^2} \right] = \frac{d}{dy} \left[\frac{x}{\sqrt{a^2 + y^2}} \right] = \frac{by}{(a^2 + y^2)^{3/2}} \qquad (4.2.22)$$

Equation (4.2.22) is exact and can be integrated to give

$$\int \frac{d}{dy} \left\{ \frac{x}{\sqrt{a^2 + y^2}} \right\} dy = \frac{x}{(a^2 + y^2)^{1/2}} = \int \frac{by}{(a^2 + y^2)^{3/2}} dy + k \qquad (4.2.23)$$

We can complete the integration for the right side of Eq. (4.2.23) by letting $z = a^2 + y^2$ with $dz = 2y\,dy$, so

$$b \int \frac{y}{(a^2 + y^2)^{3/2}} dy = \frac{b}{2} \int \frac{dz}{z^{3/2}} = \frac{b}{2} \frac{(-2)}{\sqrt{z}} = \frac{-b}{(a^2 + y^2)^{1/2}} \qquad (4.2.24)$$

Thus, the solution for Eq. (4.2.19) from Eqs. (4.2.23) and (4.2.24) is given by the following function for x:

$$x = k\sqrt{a^2 + y^2} - b \qquad (4.2.25)$$

and, if $y = y_1$ when $x = x_1$, then the arbitrary integrating constant is given by

$$k = \frac{x_1 + b}{\sqrt{a^2 + y_1^2}}$$

and Eq. (4.2.25) becomes

$$\frac{y^2 + a^2}{y_1^2 + a^2} = \left(\frac{x + b}{x_1 + b} \right)^2$$

This is the solution for the system equation defined by Eq. (4.2.19).

4.2.5 Exact System Kernels

As we have seen, the general single-input, single-output system equation can be written as follows:

$$de = g(e, c)\,dc \qquad (4.2.26)$$

As was shown in Section 4.2.2, the most general solution for Eq. (4.2.26) has the form

$$\varphi(e, c, k) = 0 \qquad (4.2.27)$$

where k is the arbitrary constant of integration. Suppose that we can solve explicitly for k from Eq. (4.2.27) to give (this is the form of the general solution we considered in Section 4.2.2)

$$k = \varphi(e, c)$$

The total differential of $\varphi(e, c)$, that is, is then given by

$$d\varphi = \frac{\partial \varphi}{\partial e} de + \frac{\partial \varphi}{\partial c} dc = 0 \qquad (4.2.28)$$

Now, if we compare Eqs. (4.2.26) and (4.2.28), we observe that it may be possible to express Eq. (4.2.26) in the form given by Eq. (4.2.28) in terms of $\varphi\,(e, c)$, which is the general solution of Eq. (4.2.26). Suppose we multiply Eq. (4.2.26) by an arbitrary nonzero function, say $H(e, c)$, to give

$$H(e, c)\, de = H(e, c)\, g(e, c)\, dc \qquad (4.2.29)$$

Then for Eq. (4.2.29) to be equivalent to Eq. (4.2.28), we require that the function $H(e, c)$ satisfy the following equations:

$$H = \frac{\partial \varphi}{\partial e} \quad \text{and} \quad Hg = -\frac{\partial \varphi}{\partial c} \qquad (4.2.30)$$

Now, if $\varphi\,(e, c)$ is a continuous function (practically always satisfied) of e and c, such that

$$\frac{\partial^2 \varphi}{\partial e\, \partial c} = \frac{\partial^2 \varphi}{\partial c\, \partial e}$$

we can differentiate Eq. (4.2.30) to obtain the following requirement for our function $H(e, c)$

$$\frac{\partial^2 \varphi}{\partial c\, \partial e} = \frac{\partial H}{\partial c} = -\frac{\partial Hg}{\partial c} = \frac{\partial^2 \varphi}{\partial e\, \partial c} \qquad (4.2.31)$$

If we can find a function $H(e, c)$ such that Eq. (4.2.31) for H and g is satisfied, then $H(e, c)$ is called an integrating factor for Eq. (4.2.26); and Eq. (4.2.29) is said to be an exact differential equation, which can be immediately integrated to give

$$\int H\, de - \int H\, g dc = \int \frac{\partial \varphi}{\partial e}\, de + \int \frac{\partial \phi}{\partial c}\, dc = \varphi = k$$

where we have used Eq. (4.2.30) to express H and $H\,g$ in terms of φ.

As we will learn, it is always necessary to find an appropriate integrating factor and produce an exact differential equation before we can analytically solve any differential equation by integration. For example, for the linear differential equation given by Eq. (4.2.17), where the system kernel is given by $g = h\,(x) - f(x)\,y$, we have

$$dy + [\,f(x)\,y - h(x)\,]dx \qquad (4.2.32)$$

and the integrating factor was given by

$$H = \exp \int f(x)dx$$

which gave for Eq. (4.2.32) the result

$$\exp \left[\int f dx\right] dy + \exp \int f dx [\,f\,y - h\,]\, dx = 0 \qquad (4.2.33)$$

This result is an exact differential equation since we have that

$$\frac{\partial}{\partial x}\left[\exp\left(\int f\, dx\right)\right] = f \exp \int f\, dx$$

Table 4-1 Integrating Factors for System Equations

Differential expression	Integration factor	Resulting differential
$x\,dy + y\,dx$	$(xy)^n$	$\dfrac{1}{n+1}d\left[(xy)^{n+1}\right]$
$x\,dy - y\,dx$	$(y/x)^n$	$\dfrac{1}{n+1}d\left[(y/x)^{n+1}\right]$
$x\,dy - y\,dx$	$1/(x^2 + y^2)$	$d\left[\tan^{-1}(y/x)\right]$
$a\,dx + b\,dy$	$(ax + by)^n$	$\dfrac{1}{n+1}d\left[(ax+by)^{n+1}\right]$
$ax^p dx + by^m dy$	$\left[\dfrac{a}{p+1}x^{p+1} + \dfrac{b}{m+1}y^{m+1}\right]^n$	$\dfrac{1}{n+1}d\left[\dfrac{a}{p+1}x^{p+1} + \dfrac{b}{m+1}y^{m+1}\right]^{n+1}$
$px^{p-1}dx + amy^{m-1}dy$	$(x^p + a\,y^m)^n$	$\dfrac{1}{n+1}d\,(x^p + ay^m)^{n+1}$

General Test for Exactness and Form of Integrating Factor Given

$$p(x, y)\,dx + q(x, y)\,dy$$

Differential expression	Integrating factor
$\dfrac{\partial p}{\partial y} = \dfrac{\partial q}{\partial x}$	Equation is exact
$\dfrac{1}{q}\left(\dfrac{\partial p}{\partial y} = \dfrac{\partial q}{\partial x}\right) = F(x)$	Integrating factor function of "x" only
$\left(\dfrac{\partial p}{\partial y} = \dfrac{\partial q}{\partial x}\right)\Big/(px - qy) = F(x)$	Integrating factor function of "$x\,y$" only
$\left(\dfrac{\partial p}{\partial y} = \dfrac{\partial q}{\partial x}\right)\Big/p = F(x)$	Integrating factor function of "y" only

and

$$\frac{\partial}{\partial y}\left[\exp\int f dx(fy - h)\right] = f\,\exp\int f\,dx$$

are identical, as required by Eq. (4.2.31) for Eq. (4.2.33), to be exact.

Useful differential equivalencies that can be used in determining integrating factors for differential equations are given in Table 4-1.

Example Problem for an Exact System Kernel.
Given the system equation

$$[f(e) + Ac]de + [g(c) + Ae]dc = 0 \tag{4.2.34}$$

we find that the system kernel is exact since

$$\frac{\partial}{\partial c}[f(e) + Ac] = A = \frac{\partial}{\partial e}[g(c) + Ae]$$

If we express Eq. (4.2.34) as

$$f(e)\,de + A(c\,de + e\,dc) + g(c)\,dc = 0$$

we can integrate the result to obtain as the solution for the system equation given by Eq. (4.2.34)

$$\int f(e)\,de + A\,c\,e + \int g(c)\,dc = k$$

where k is the arbitrary integration constant.

4.2.6 Separable System Kernels

Suppose the system kernel $g(e, c)$ is separable and can be expressed by the following system equation:

$$de = g_e(e)\,g_c(c)\,dc \tag{4.2.35}$$

where $g_e(e)$ is a function of e only and $g_c(c)$ is a function of c only. Then if $g_e(e) \neq 0$, multiplication of Eq. (4.2.35) by the function $1/g_e(e)$ (which is the integrating factor for this separable system equation) results in an exact differential equation of the form

$$\frac{de}{g_e(e)} = g_c(c)\,dc \tag{4.2.36}$$

Integration of Eq. (4.2.36) provides the general solution of Eq. (4.2.35) in the form

$$\int \frac{1}{g_e(e)}\,de = \int g_c(c)\,dc + k$$

where k is the arbitrary integration constant.

Example Problem for a Separable Continuous System Kernel.
Given the system equation

$$de = \frac{(a + e)(b + c)}{(A + e)(B + c)}\,dc \tag{4.2.37}$$

which is separable and can be expressed as an exact differential equation as follows:

$$\frac{(A + e)}{(a + e)}\,de = \frac{(b + c)}{(B + c)}\,dc \tag{4.2.38}$$

Integrating Eq. (4.2.38), we have for the left side of the equation that

$$\int \frac{(A + e)}{(a + e)}\,de = \int \frac{a + e + (A - a)}{(a + e)}\,de = \int de + (A - a)\int \frac{de}{(a + e)} \tag{4.2.39}$$
$$= e + (A - a)\,\mathrm{Ln}(a + e)$$

And integrating the right side of Eq. (4.2.38), we find that

$$\int \frac{(b + c)}{(B + c)}\,dc = \int \frac{B + c + (b - B)}{(B + c)}\,de = \int dc + (b - B)\int \frac{dc}{(B + c)} \tag{4.2.40}$$
$$= c + (b - B)\,\mathrm{Ln}(B + c)$$

So the general solution for Eq. (4.2.37) using Eqs. (4.2.38), (4.2.39), and (4.2.40) is

$$e + (A - a)\,\mathrm{Ln}(a + e) = c + (b - B)\,\mathrm{Ln}(B + c) + k \tag{4.2.41}$$

If $e = e_1$ when $c = c_1$, then

$$k = e_1 + (A - a)\,\mathrm{Ln}(a + e_1) - c_1 + (b - B)\,\mathrm{Ln}(B + c_1)$$

and thus Eq. (4.2.41) can be expressed as the transcendental solution of system Eq. (4.2.37), with the condition that $e(c_1) = e_1$, as follows:

$$e - e_1 + (A - a)\,\mathrm{Ln}\!\left(\frac{a + e}{a + e_1}\right) = c - c_1 + (b - B)\,\mathrm{Ln}\!\left(\frac{B + c}{B + c_1}\right)$$

Observe that this resulting solution is transcendental in both e and c and therefore cannot be expressed explicitly for either variable. Unfortunately, this situation will arise frequently in system analysis. But the computer can provide rapid and accurate evaluation of this transcendental solution.

4.2.7 Homogeneous System Kernels

If the single-input, single-output system kernel $g(e, c)$ can be expressed as

$$de = g(e/c)dc \tag{4.2.42}$$

then the system kernel is said to be homogeneous. The system equation with a homogeneous kernel

$$de = g(e/c)dc \tag{4.2.43}$$

can be solved by defining the variable combination e/c as a new single variable such as $u = e/c$, so $e = u\,c$ and $de = u\,dc + c\,du$. Then for the homogeneous differential equation, Eq. (4.2.42), we have in terms of u and c that

$$u\,dc + c\,du = g(u)dc$$

This equation, with simple rearrangement of terms, becomes separable as follows:

$$c\,du = [g(u) - u]dc$$

Upon variable separation, the resulting exact equation can be integrated to give as the solution for Eq. (4.2.43) the following:

$$\int^{u\,=\,e/c} \frac{du}{g(u) - u} = \int \frac{dc}{c} + k = \mathrm{Ln}\,c + k \tag{4.2.44}$$

where k is the arbitrary integrating constant. Occasionally, the resulting integral for the solution of Eq. (4.2.43) is simplified if the substitution $v = c/e$ is used instead of $u = e/c$ in Eq. (4.2.43). Then the resulting solution for Eq. (4.2.43) is given by

$$\int^{v\,=\,c/e} \frac{g\,(1/v)\,dv}{1 - v\,g(1/v)} = \int \frac{de}{e} + k = \mathrm{Ln}\,e + k$$

which is equivalent to the solution given by Eq. (4.2.44).

Example Problem for the Homogeneous System Kernel.
Given the system equation

$$dy = \frac{ax + by}{Ax + By} dx \tag{4.2.45}$$

it is possible to express the system kernel as follows (if):

$$g(x,y) = \frac{ax + by}{Ax + By} = \frac{a + b(y/x)}{A + B(y/x)} = g(y/x)$$

Thus, Eq. (4.2.45) is homogeneous and can be made separable by the substitution

$$y/x = u, \quad \text{so } y = ux \quad \text{and} \quad dy = u \, dx + x \, du \tag{4.2.46}$$

Equation (4.2.45) then becomes the following, using Eq. (4.2.46) for $y = u \, x$ and $dy = u \, dx + x \, du$:

$$u \, dx + x \, du = \frac{(a + bu)x}{(A + Bu)x} dx = \frac{(a + bu)}{(A + Bu)} dx \tag{4.2.47}$$

And rearranging Eq. (4.2.47) into separable form, we have

$$x \, du = \left[\frac{(a + bu)}{(A + Bu)} - u \right] dx$$

Separating variables and integrating, we find

$$\int^{u = y/x} \frac{A + Bu}{a + bu - u(A + Bu)} \, du = \int \frac{dx}{x} + k = \text{Ln} \, x + k \tag{4.2.48}$$

The left side of Eq. (4.2.48) can be integrated using standard integral tables to provide for the first term in the integrand:

$$A \int^{u = y/x} \frac{du}{a + (b - A)u - Bu^2} = \frac{A}{q} \text{Ln} \left[\frac{b - A - 2B(y/x) - q}{b - A - 2B(y/x) + q} \right]$$

where

$$q = \sqrt{(b - A)^2 + 4aB}$$

And for the second term in the integrand we have

$$B \int^{u = y/x} \frac{u \, du}{a + (b - A)u - Bu^2} = \frac{b - A}{2q} \text{Ln} \left[\frac{b - A - 2B(y/x) - q}{b - A - 2B(y/x) + q} \right]$$
$$- \frac{1}{2} \text{Ln} \left[a + (b - A)(y/x) - B(y/x)^2 \right]$$

We then find as the general solution for the system equation, Eq. (4.2.45), where k is the usual integration constant,

$$\frac{1}{2q}(A+b)\,\mathrm{Ln}\left[\frac{b-A-2B(y/x)-q}{b-A-2B(y/x)+q}\right]$$

$$-\frac{1}{2}\,\mathrm{Ln}\left[a+(b-A)(y/x)-B(y/x)^2\right]=\mathrm{Ln}\,x+k$$

Again we observe that the solution for the system equation is transcendental. Neither x nor y can be solved explicitly.

4.2.8 Bernoulli-Type System Kernels

If the single-input, single-output system kernel $g(e, c)$ can be expressed as

$$g(e,c) = \begin{cases} A(c)\,e + B(c)\,e^n & (4.2.49) \\ 1/[D(e)\,c + E(e)\,c^m] & (4.2.50) \end{cases}$$

then the system kernel has the form of Bernoulli's equation and can be made into a linear equation with the appropriate substitution.

To demonstrate the substitution technique, consider the Bernoulli differential equation

$$\frac{dy}{dx} = P(x)\,y + Q(x)\,y^n \tag{4.2.51}$$

Now we define a new dependent variable z as follows:

$$y = z^{n/(1-n)} \quad \text{and} \quad dy = \frac{1}{1-n}\,z^{n/(1-n)}\,dz$$

So Eq. (4.2.51) becomes, in terms of z,

$$\frac{1}{1-n}z^{n/(1-n)}\frac{dz}{dx} = P(x)\,z^{n/(1-n)} + Q(x)\,z^{n/(1-n)}$$

Canceling the term $z^{n/(1-n)}$ for $z \neq 0$, we have

$$\frac{1}{1-n}\frac{dz}{dx} = P(x)\,z + Q(x)$$

which is a linear differential equation and has as a solution (see Section 4.2.4) the following equation, with k as the integration constant:

$$z = \exp\left[(1-n)\int P\,dx\right]\left\{(1-n)\int^{v=x} Q\exp\left[(n-1)\int P\,dv\right]dv + k\right\}$$

and in terms of y and x,

$$y(x) = \exp\left[(1-n)\int P\,dx\right]\left\{(1-n)\int^{v=x} Q\exp\left[(n-1)\int P\,dv\right]dv + k\right\}^{1/(1-n)}$$

For Eq. (4.2.49) in terms of the Bernoulli equation system kernel, we have

$$e(c) = \exp\left[(1-n)\int A\,dx\right]\left\{(1-n)\int^{v=c} B\exp\left[(n-1)\int A\,dv\right]dv + k\right\}^{1/(1-n)}$$

and for Eq. (4.2.50), we have

$$c(e) = \exp\left[(1-m)\int D\,dx\right] \left\{(1-m)\int^{v=e} E\exp\left[(m-1)\int D\,dv\right]dv + k\right\}^{1/(1-m)}$$

Example Problem for the Bernoulli Equation System Kernel.
Given the system equation

$$dy = \left(y^2 - y/x\right)dx \tag{4.2.52}$$

Since we have a Bernoulli-type system kernel with $n = 2$, then

$$y = z^{1/(1-2)} = 1/z \quad \text{and} \quad dy = -1/z^2\,dz$$

Substituting our variable change into Eq. (4.2.52), we have

$$-1/z^2\,dz = \left[1/z^2 - 1/(x\,z)\right]\,dx$$

and multiplying by z^2, we have after some rearrangement that

$$\frac{dz}{dx} = \frac{z}{x} - 1 \tag{4.2.53}$$

This is a linear equation with an integrating factor given by

$$\exp\left(-\int\frac{dx}{x}\right) = \exp\left(-\text{Ln}\,x\right) = 1/x$$

So Eq. (4.2.53) can be written as

$$\frac{dz}{dx} = \frac{z}{x} - 1$$

Integrating this equation directly, we have that

$$z = x\,\text{Ln}(k/x)$$

or, in terms of y, we have, as the solution for the system defined by Eq. (4.2.52)

$$y = \frac{1}{\text{Ln}(k/x)}$$

where k is the integrating constant. This solution for $y(x)$ as a function of x is traceable to our original Eqs. (4.2.29) and (4.2.50)

4.2.9   Ricatti-Type System Kernels

If the single-input, single-output system kernel $g(e, c)$ can be expressed as

$$g(e,c) = \begin{cases} A(c) + B(c)\,e + D(c)\,e^2 & (4.2.54) \\[2mm] \dfrac{1}{E(e) + F(e)\,c + G(e)\,c^2} & (4.2.55) \end{cases}$$

then the system kernel has the form of the Ricatti equation and can be solved for certain cases, depending on the coefficient functions A, B, and D or E, F, and G as appropriate. It is apparent, for example, that if $A = 0$ or $E = 0$, then the resulting Ricatti equation reduces to a special case of the Bernoulli equation and can be solved as shown in Section 4.2.8.

It is possible to express the system equation that has a kernel of the Ricatti equation form as an equivalent second-order, linear, variable coefficient, and ordinary differential equation. Furthermore, methods exist, such as the "Method of Frobenius," for obtaining analytical solutions for such differential equations. To establish this conclusion, consider the system equation that has the following Ricatti equation kernel:

$$\frac{de}{dc} = A(c) + B(c)e + D(c)e^2 \tag{4.2.56}$$

Now let us define a new dependent variable a such that

$$e = -\frac{1}{Du}\frac{du}{dc}$$

so

$$\frac{de}{dc} = \frac{1}{D^2u}\frac{dD}{dc}\frac{du}{dc} + \frac{1}{Du^2}\left(\frac{du}{dc}\right)^2 - \frac{1}{Du}\frac{d^2u}{dc^2}$$

Expressing our system equation as a function of e and c, we have that

$$\frac{1}{D^2u}\frac{dD}{dc}\frac{du}{dc} + \frac{1}{Du^2}\left(\frac{du}{dc}\right)^2 - \frac{1}{Du}\frac{d^2u}{dc^2} = A - \frac{B}{Du}\frac{du}{dc} + \frac{1}{Du^2}\left(\frac{du}{dc}\right)^2$$

We observe that we can cancel the terms containing $(du/dc)^2$ and, upon rearranging terms, we can express our system equation in terms of u and c as the following linear second-order differential equation:

$$\frac{d^2u}{dc^2} - \left(B + \frac{1}{D}\frac{dD}{dc}\right)\frac{du}{dc} + ADu = 0 \tag{4.2.57}$$

If the explicit functions are known for A, B, and D, and these functions satisfy certain requirements relative to their singularities, then certain infinite series can be employed to obtain the complete analytical solution for Eq. (4.2.57). Furthermore, as we will learn in Section 4.3.11, it is possible to express Eq. (4.2.57) as a single-input, double-output system equation with a linear, variable coefficient kernel. For example, in the case of Eq. (4.2.57), if we let

$$e_1 = u = \exp\left(-\int e\,Ddc\right)$$

$$e_2 = \frac{du}{dc} = -De_1$$

then we have for the single-input, double-output system equation, which is equivalent to Eq. (4.2.56), the following:

$$de_1 = e_2\,dc \tag{4.2.58.1}$$

$$de_2 = \left[\left(B + \frac{1}{D}\frac{dD}{dc} \right) e_2 - A\,De_1 \right] dc \tag{4.2.58.2}$$

We will learn in Section 4.3 how to treat such single-input, multiple-output system equations.

4.2.10 Other Special System Kernel Types

For the single-input, single-output system

$$de = g(e,c)dc \tag{4.2.59}$$

It is occasionally found that the kernel $g(e,\,c)$ can be expressed as a function of certain combinations of the system variables e and c, which can be readily solved with appropriate substitutions. We will consider several cases of combinations of variables for which solutions can be found.

Case I

If the system kernel can be expressed as a function of the single argument as

$$g(e,c) = g(Ae + Bc + C) \tag{4.2.60}$$

where A, B, and C are constants, then the variable transformation

$$u = Ae + Bc + C \quad \text{and} \quad du = A\,de + B\,dc$$

provides the following system equation:

$$du = [A\,g(u) + B]dc$$

which is separable and can be integrated to provide as a solution

$$c = \int^{u = Ae + Bc + C} \frac{du}{A\,g(u) + B} + k$$

Case II

If the system kernel can be expressed as a function of the argument ratio as

$$g(e,c) = g\left\{ \frac{Ae + Bc + C}{De + Ec + F} \right\} \tag{4.2.61}$$

where A, B, C, D, E, and F are constants and $AE \neq BD$, then the transformation

$$u = Ae + Bc + C \quad \text{and} \quad v = De + Ec + F$$

where

$$\frac{de}{dc} = \frac{E\,du - B\,dv}{A\,dv - D\,du} = g(u/v)$$

provides the following transformed system equation, which is homogeneous and can be treated by the method given in Section 4.2.7 for solving homogeneous system equations:

$$\frac{du}{dv} = \frac{B + Ag(u/v)}{E + Dg(u/v)} = G(u/v)$$

If $AE = BD$, then the system kernel in Eq. (4.2.61) can be written as

$$g(e, c) = g\left\{\frac{Ae + Bc + C}{E(Ae + Bc) + FB}\right\}$$

and the transformation

$$u = Ae + Bc \quad \text{and} \quad de = (1/A)\,(du - B\,dc)$$

provides the following system equation:

$$du = \left[Ag\left(\frac{u + C}{Eu + FB}\right) + B\right]dc$$

which is separable and can be integrated to give

$$c = \int^{u\,=\,Ae\,+\,Bc} \frac{du}{Ag[(u + c)/(Eu + FB)] + B} + k$$

Case III

If the system equation can be expressed in the form

$$g(e, c) = g(Ae + Bc + C, De + Ec + F) \tag{4.2.62}$$

where A, B, C, D, E, and F are constants, then the transformation

$$u = Ae + Bc + C$$
$$v = De + Ec + F$$

results in the following transformed equation:

$$[E + Dg(u, v)]du = [B + Ag(u, v)]dv \tag{4.2.63}$$

and it is possible that Eq. (4.2.63) may be solved by one of the methods previously developed in this chapter. For example, perhaps an integrating factor $I(u, v)$ can be determined that makes Eq. (4.2.63) exact:

$$E\frac{\partial}{\partial v}I(u, v) + \frac{\partial}{\partial v}[Dg(u, v)I(u, v)] = -\frac{\partial}{\partial u}[Ag(u, v)I(u, v)] - B\frac{\partial}{\partial u}I(u, v)$$

in which case Eq. (4.2.63) can be directly integrated after multiplication by the appropriate integrating factor.

Case IV

If the system kernel can be expressed as any one of the following special single argument forms

$$g(e, c) = \begin{cases} e^2 g(e, c) & \text{(4.2.64)} \\ 1/c^2\, g(e, c) & \text{(4.2.65)} \\ e/c\, g(e, c) & \text{(4.2.66)} \\ 1/c[g(e\,c) - e] & \text{(4.2.67)} \\ c^{m-1}\, g(e/c^m) & \text{(4.2.68)} \end{cases}$$

then the variable transformation

$$u = e\,c \quad \text{and} \quad du = e\,dc + c\,de$$

provides the following system equations for Eqs. (4.2.64) through (4.2.67).
For Eq. (4.2.64), we have

$$du = \frac{1}{e} \left[\frac{1}{g(u)} + u \right] de$$

which is separable and can be integrated to give

$$\text{Ln}\,e = \int^{u\,=\,ec} \frac{g(u)\,du}{1 + u\,g(u)} + k$$

For Eq. (4.2.65), we have

$$du = \frac{1}{c}[g(u) + u]dc$$

which is also separable and can be integrated to provide

$$\text{Ln}\,c = \int^{u\,=\,e\,c} \frac{du}{u + g(u)} + k$$

For Eq. (4.2.66), we obtain

$$du = u[1 + g(u)]\frac{dc}{c}$$

which is also separable and gives upon integration

$$\text{Ln}\,c = \int^{u\,=\,e\,c} \frac{du}{u\,[1 + g(u)]} + k$$

Furthermore, for Eq. (4.2.67), we have that

$$du = g(u)dc$$

and this is also separable, so we have upon integration that

$$c = \int^{u\,=\,ec} \frac{du}{g(u)} + k$$

For Eq. (4.2.68), we employ the following variable transformation:

$$e = uc^m \quad \text{and} \quad de = c^m du + muc^{m-1}dc$$

Then system Eq. (4.2.68) becomes

$$du = [g(u) - mu](1/c)dc$$

which is separable and, upon integration, gives

$$\text{Lnc} = \int^{u\,=\,ec^{-m}} \frac{du}{g(u) - mu} + k$$

Case V

Suppose that the general single-input, single-output system equation is expressed as

$$g(e, c) = \frac{de}{dc} = p \tag{4.2.69}$$

Equation (4.2.69) can be expressed in the related form

$$e = h(c, e, p) \tag{4.2.70}$$

Differentiating this function with respect to c, we have that

$$\frac{de}{dc} = p = \frac{\partial h}{\partial c} + \frac{\partial h}{\partial e} p + \frac{\partial h}{\partial p} \frac{dp}{dc} \tag{4.2.71}$$

If $\partial h / \partial e = 0$ in Eq. (4.2.71), which implies that $h(c, e, p)$ defined by Eq. (4.2.70) is not an explicit function of e, then Eq. (4.2.71) reduces to the following first-order differential equation in the variables p and c:

$$p = \frac{\partial h}{\partial c} + \frac{\partial h}{\partial p} \frac{dp}{dc} \tag{4.2.72}$$

Conversely, if $\partial h / \partial c = 0$ in Eq. (4.2.71), which implies that $h(c, e, p)$ is not an explicit function of c, then by observing that

$$\frac{dp}{dc} = \frac{dp}{de} \frac{de}{dc} = p \frac{dp}{de}$$

we find that Eq. (4.2.71) assumes the form of a first-order differential equation in the variables p and e, as follows

$$p = p \frac{\partial h}{\partial e} + p \frac{\partial h}{\partial p} \frac{dp}{de}$$

And if $p = de/dc \neq 0$, then canceling p from this equation, we have that

$$1 = \frac{\partial h}{\partial e} + \frac{\partial h}{\partial p} \frac{dp}{de} \tag{4.2.73}$$

We now observe that if system Eq. (4.2.69) can be expressed in the equivalent forms given by either Eq. (4.2.72) or (4.2.73), it may be possible to solve one of these first-order differential equations and thereby gain a solution for the system equation.

Example Problem with Special Kernel Form.
Given system equation where A, B, C, D, E, and F are constants and $AE \neq BD$

$$de = \frac{Ae + Bc + C}{De + Ec + F} dc$$

If we let

$$u = Ae + Bc + C, \quad du = A\,de + B\,dc$$
$$v = De + Ec + F, \quad dv = D\,de + E\,dc$$

then we have for de and dc that

$$\frac{de}{dc} = \frac{E\,du - B\,dv}{A\,dv - D\,du} = \frac{u}{v}$$

and rearranging, we have the following system equation with a homogeneous kernel:

$$du = \frac{A - Bv/u}{D - Ev/u}\,dv$$

If we now let $w = u/v$, $du = v\,dw + w\,dv$, then this system equation becomes

$$dw = \left[\frac{A\,w + B}{D\,w + E} - w\right]\frac{dv}{v}$$

which is separable and gives upon integration, recalling the original definitions of w, u, and v

$$\mathrm{Ln}(De + Ec + F) = \int \frac{Dw + E}{B + (A - E)w - Dw^2}\,dw + k \qquad (4.2.74)$$

where the dummy variable w is given by

$$w = \frac{Ae + Bc + C}{De + Ec + F}$$

Eq (4.2.74) can be integrated using Bibliography at the end of this chapter (see Abramowitz and Stengun).

Although many of the specialized system kernel types studied in this chapter may appear to have limited application, in fact, it is often possible to approximate a given system kernel by one of the specialized kernel types for which an analytical solution is known. This is an attractive procedure when attempting to model and analyze a nonlinear system kernel for which analytical solutions are often very limited.

4.3 Single-Input, Multiple-Output Systems

Single-input, multiple-output systems comprise all those systems that can be treated as having only one dominant or significant input imposed on the system and multiple-outputs issuing from the system because of this single input. Single-input, multiple-output systems are considered to be, after the single-input, single-output systems, the simplest systems to model and generally the easiest to analyze and evaluate. Examples of such single-input, multiple-output systems are numerous and particularly evident in the physical sciences. As we will see, these systems include all ordinary differential equations of second-order or greater and all ordinary differential equations whose dependent variables are vector quantities. Furthermore, although it is generally advisable to begin the analysis of a general system by assuming a single-input, single-output system model, the extension of the actual system to a single-input, multiple-output system is often desirable and required for accurate modeling of particular systems.

The general equation for the single-input, multiple-output system is

$$\left.\begin{array}{c} d\mathbf{e} \\ \Delta\mathbf{e} \end{array}\right\} = g(\mathbf{e}, c) \left\{\begin{array}{c} dc \\ \Delta c \end{array}\right. \tag{4.3.1}$$

where, for convenience, we have defined the following $(n \times 1)$ column matrices as follows:

$$d\mathbf{e} = \begin{bmatrix} de_1 \\ \vdots \\ de_n \end{bmatrix}, \quad \Delta\mathbf{e} = \begin{bmatrix} \Delta e_1 \\ \vdots \\ \Delta e_n \end{bmatrix}, \quad g(\mathbf{e}, c) = \begin{bmatrix} g_1(e_1, ..., e_n, c) \\ \vdots \\ g_1(e_1, ..., e_n, c) \end{bmatrix}$$

The variables $d\mathbf{e}$ and $\Delta\mathbf{e}$ are the matrix infinitesimal changes and incremental changes in the system output, respectively. The functions $g(e_1, ..., e_n, c)$ compose the column matrix system kernel $g(\mathbf{e}, c)$, which is generally a function of the input variable c and all the output variables $e_1, e_2, ..., e_n$.

If the single-input, multiple-output system exhibits feedback of the output as given by the system matrix feedback kernel $f(e, c)$, then the matrix system equation with feedback, which was developed in Section 3.2.2, is given as

$$\left.\begin{array}{c} d\mathbf{e} \\ \Delta\mathbf{e} \end{array}\right\} = \frac{g(\mathbf{e}, c)}{1 - f(\mathbf{e}, c)\, g(\mathbf{e}, c)} \left\{\begin{array}{c} dc \\ \Delta c \end{array}\right. \tag{4.3.2}$$

where the feedback kernel $f(e, c)$ is now a $(1 \times n)$ row matrix as follows:

$$f(\mathbf{e}, c) = f_1, ..., f_n.$$

Observe that the matrix product, $f(\mathbf{e}, c)\, g(\mathbf{e}, c)$, is a scalar quantity defined as

$$1 - f(\mathbf{e}, c)\, g(\mathbf{e}, c) = 1 - \sum_{i=1}^{n} f_i(\mathbf{e}, c)\, g_i(\mathbf{e}, c)$$

Now it is always possible to express the matrix system kernel with feedback, as given by Eq. (4.3.2), without loss of generality, as follows:

$$g_f(\mathbf{e}, c) = \frac{g(\mathbf{e}, c)}{1 - f(\mathbf{e}, c)\, g(\mathbf{e}, c)} = \frac{g(\mathbf{e}, c)}{1 - \sum\limits_{i=1}^{n} f_i(\mathbf{e}, c)\, g_i(\mathbf{e}, c)} \tag{4.3.3}$$

where the matrix function $g_f(\mathbf{e}, c)$, i.e., the system kernel with feedback, can be treated as the associated open-loop system kernel (i.e., without apparent feedback). For Eq. (4.3.2), we will consider general system kernel classifications and solution methods for both continuous and discrete single input, multiple-output systems. Furthermore, it is to be understood that the applicable system kernel, $g(\mathbf{e}, c)$, is the general form for either the open system Eq. (4.3.1) without feedback or the closed system Eq. (4.3.2) with feedback, without loss of generality of the classification or solution method.

4.3.1 Discrete System Kernels

The general equation for the single-input, multiple-output discrete system is equivalent to a set of first-order ordinary difference equations (or a reduced set of higher-order difference equations, as will be shown in Section 4.3.11), and can be expressed in the matrix form as follows for the discrete system with or without feedback:

$$\Delta\mathbf{e} = g(\mathbf{e}, c)\, \Delta c \tag{4.3.4}$$

where $\Delta\mathbf{e}$ is the incremental change in the system's multiple-output arising from the incremental change Δc in the system input. The discrete system kernel $g(\mathbf{e}, c)$ is, in general, a function of both the single discrete input c and the multiple-output $\mathbf{e}$, where

$$\mathbf{e} = \begin{bmatrix} e_1 \\ \vdots \\ e_n \end{bmatrix}, \quad g(\mathbf{e}, c) = \begin{bmatrix} g_1(e_1, ..., e_n, c) \\ \vdots \\ g_1(e_1, ..., e_n, c) \end{bmatrix}$$

and each component of the column matrix $g(\mathbf{e}, c)$ is defined in general as

$$g_i(\mathbf{e}, c) = g_i(e_1, e_2, ..., e_n, c) \quad \text{for} \quad i = 1, ..., n.$$

It is apparent that solutions for the discrete system equation are determined by the system kernel, $g_i(\mathbf{e}, c)$, and the associated initial conditions defined for multiple-outputs for the prescribed input, that is, $\mathbf{e}[c(1)]$ for $c = c(1)$. Now Eq. (4.3.4) may be considered to constitute a set of ordinary difference equations with non-uniform spacing (in general) of the discrete variable

$$\Delta c(1) = c(2) - c(1), \Delta c(2) = c(3) - c(2), ..., \Delta c(n) = c(n+1) - c(n)$$

where the incremental changes in c are not necessarily equal or uniform. The general solution for Eq. (4.3.4) is obtained from the sequential solutions obtained for each incremental step. For the incremental changes, $\Delta c(1)$ and $\Delta \mathbf{e}(1)$, we have from Eq. (4.3.4) that

$$\Delta\mathbf{e}(1) = \mathbf{e}(2) - \mathbf{e}(1) = g[\mathbf{e}(1), c(1)]\Delta c(1) = g[\mathbf{e}(1), c(1)] [c(2) - c(1)]$$

and solving for $\mathbf{e}(2)$, we have that

$$\mathbf{e}(2) = \mathbf{e}(1) + g[\mathbf{e}(1), c(1)][c(2) - c(1)] \tag{4.3.5}$$

For the incremental change $\Delta c(2)$ and $\Delta \mathbf{e}(2)$, we then have from Eq. (4.3.4) that

$$\Delta\mathbf{e}(2) = \mathbf{e}(3) - \mathbf{e}(2) = g[\mathbf{e}(2), c(2)] \Delta c(2)$$

and in terms of $\mathbf{e}(3)$, we have that

$$\mathbf{e}(3) = \mathbf{e}(2) + g[\mathbf{e}(2), c(2)] [c(3) - c(2)] \tag{4.3.6}$$

Continuing this procedure, we find for the nth incremental change $\Delta c(n)$ and $\Delta \mathbf{e}(n)$ that

$$\Delta\mathbf{e}(n) = \mathbf{e}(n+1) - \mathbf{e}(n) = g[\mathbf{e}(n), c(n)] \Delta c(n)$$

an for $\mathbf{e}(n+1)$, then

$$\mathbf{e}(n+1) = \mathbf{e}(n) + g[\mathbf{e}(n), c(n)] [c(n+1) - c(n)] \tag{4.3.7}$$

Now Eqs. (4.3.5), (4.3.6), and (4.3.7) provide a sequential set of equations for determining the multiple-output $\mathbf{e}(n+1)$ for a discrete system after any sequence of n sequential steps in the system input. The multiple-outputs, $\mathbf{e}(n+1)$, can thus be expressed by repetitive substitution for any set of incremental steps n, such that $n \geq 1$, as follows:

$$\mathbf{e}(n+1) = \mathbf{e}(1) + \sum_{i=1}^{n} g[\mathbf{e}(i), c(i+1) - c(i)] \tag{4.3.8}$$

where $\mathbf{e}(1)$ is the initial condition for the output as prescribed previously and must be specified to determine all $\mathbf{e}(n)$ for $n \geq 1$.

Thus, again we find, as we did for the discrete single-input, single-output system, that for any well-defined discrete system kernel, $g(e, c)$, and a consistent sequence of input changes, it is always possible to analytically determine the system output. That Eq. (4.3.8) is the general solution for system Eq. (4.3.4) can be demonstrated by assuming that the kernel is continuous and bounded over the range of interest so that we may allow each incremental input change to approach zero and the number of increments to become infinite so that we have the following limit established:

$$\mathbf{e}(n+1) = \mathbf{e}(1) + \operatorname{Lim}_{\substack{n \to \infty \\ \Delta c(i) \to 0}} \sum_{i=1}^{n} g[\mathbf{e}(i), c(i)] \, \Delta c$$

Now, if the equation is defined and the appropriate limits exist, then we can establish the following result:

$$\mathbf{e}(c) = \mathbf{e}(1) + \int_{c(1)}^{c} g(\mathbf{e}, c) \, dc$$

which is the quadrature form for the general solution of the continuous single-input, multiple-output system equation

$$d\mathbf{e} = g(\mathbf{e}, \, c) \, dc$$

with the initial condition that

$$\mathbf{e}[c(1)] = \mathbf{e}(1)$$

4.3.2 Continuous System Kernels

The general set of equations for the single-input, multiple-output continuous system is equivalent to a general set of first-order differential equations in differential form and can be expressed in matrix form as

$$d\mathbf{e} = g(\mathbf{e}, c) \, dc \tag{4.3.9}$$

where $d\mathbf{e}$ is the differential change in the system's multiple-output arising from the differential input dc being imposed on the system. The function $g(\mathbf{e}, c)$ is the continuous system kernel, which, in general, is a function of both the input variable c and output variables $\mathbf{e}$, where $g(\mathbf{e}, c)$ and $\mathbf{e}$ represent the column matrices:

$$\mathbf{e} = \begin{bmatrix} e_1 \\ \vdots \\ e_n \end{bmatrix}, \quad g(\mathbf{e}, c) = \begin{bmatrix} g_1(\mathbf{e}, c) \\ \vdots \\ g_1(\mathbf{e}, c) \end{bmatrix}$$

Since Eq. (4.3.9) represents a set of first-order differential equations, these system equations may be manipulated and solved as we would any set of first-order differential equations. Furthermore, it is apparent that these solutions are determined by the system kernels $g(\mathbf{e}, c)$, and the associated set of initial conditions are selected for the system such that $c = c_0$ and $\mathbf{e} = \mathbf{e}(c_0)$, where

$$\mathbf{e}(c_0) = \begin{bmatrix} e_1(c_0) \\ \vdots \\ e_n(c_0) \end{bmatrix}$$

The general solution of Eq. (4.3.9) can usually be expressed by the column matrix equation

$$\Phi(\mathbf{e}, c) = \mathbf{k} \tag{4.3.10}$$

where

$$\Phi(\mathbf{e}, c) = \begin{bmatrix} \Phi_1(e_1, ..., e_n, c) \\ \vdots \\ \Phi_1(e_1, ..., e_n, c) \end{bmatrix} = \begin{bmatrix} k_1 \\ \vdots \\ k_n \end{bmatrix}$$

and k is a column matrix of independent arbitrary constants determined by the initial conditions for the system. Assume that the solutions matrix $\Phi(\mathbf{e}, c)$ is composed of continuous functions where the partial derivatives

$$\frac{\partial \Phi_1}{\partial c} \quad \text{and} \quad \frac{\partial \Phi_1}{\partial e_i}$$

exist and are defined for all the outputs ($i = 1, 2, ..., n$) and the single input over the range of interest sought for the solutions. Then the total differential of Eq. (4.3.10) forms the following matrix equation:

$$d\Phi(\mathbf{e}, c) = \frac{\partial \Phi}{\partial \mathbf{e}}(\mathbf{e}, c)\, d\mathbf{e} + \frac{\partial \Phi}{\partial c}(\mathbf{e}, c)\, dc = 0 \tag{4.3.11}$$

where

$$\frac{\partial \Phi}{\partial \mathbf{e}} = \begin{bmatrix} \dfrac{\partial \Phi_1}{\partial e_1} & \cdots & \dfrac{\partial \Phi_1}{\partial e_n} \\ \vdots & & \\ \dfrac{\partial \Phi_n}{\partial e_1} & \cdots & \dfrac{\partial \Phi_n}{\partial e_n} \end{bmatrix}, \quad d\mathbf{e} = \begin{bmatrix} de_1 \\ \vdots \\ de_n \end{bmatrix}$$

and

$$\frac{\partial \Phi}{\partial c} = \begin{bmatrix} \dfrac{\partial \Phi_1}{\partial c} \\ \vdots \\ \dfrac{\partial \Phi_n}{\partial c} \end{bmatrix}, \quad d\Phi(\mathbf{e}, c) = \begin{bmatrix} d\Phi_1 \\ \vdots \\ d\Phi_n \end{bmatrix}$$

Matrix Eq. (4.3.11) may be solved for the differential change in output $d\mathbf{e}$ if the square matrix, $[\partial \Phi / \partial \mathbf{e}]$, is not singular and, thus, possesses an inverse denoted by $[\partial \Phi / \partial \mathbf{e}]^{-1}$. Assuming that this matrix is non-singular, we then find from Eq. (4.3.11) that

$$de = -\left[\frac{\partial \Phi}{\partial \mathbf{e}}\right]^{-1} \frac{\partial \Phi}{\partial c} \, dc$$

So we observe that the system kernel $g(\mathbf{e}, c)$ for the single-input, multiple-output system is defined by the system solutions as the following matrix equation:

$$g(\mathbf{e}, c) = -\left[\frac{\partial \Phi}{\partial \mathbf{e}}(\mathbf{e}, c)\right]^{-1} \frac{\partial \Phi}{\partial c}(\mathbf{e}, c) \tag{4.3.12}$$

To exemplify this result, consider a single-input, double-output system that has the general system equation

$$\begin{bmatrix} de_1 \\ de_2 \end{bmatrix} = \begin{bmatrix} g_1 \\ g_2 \end{bmatrix} dc \tag{4.3.13}$$

Let us assume that the general solution for Eq. (4.3.13) can thus be expressed in matrix form as

$$\Phi(\mathbf{e}, c) = \begin{bmatrix} \Phi_1 \\ \Phi_2 \end{bmatrix} = \begin{bmatrix} k_1 \\ k_2 \end{bmatrix} \tag{4.3.14}$$

where k_1 and k_2 are arbitrary constants. Forming the total differential of Eq. (4.3.14) yields

$$d\Phi = \begin{bmatrix} \dfrac{\partial \Phi_1}{\partial e_1} & \dfrac{\partial \Phi_1}{\partial e_2} \\ \dfrac{\partial \Phi_2}{\partial e_1} & \dfrac{\partial \Phi_2}{\partial e_2} \end{bmatrix} \begin{bmatrix} de_1 \\ de_2 \end{bmatrix} + \begin{bmatrix} \dfrac{\partial \Phi_1}{\partial c} \\ \dfrac{\partial \Phi_2}{\partial c} \end{bmatrix} dc$$

We find that the inverse matrix, assuming non-singularity, $[\partial \Phi / \partial \mathbf{e}]^{-1}$, is given by

$$\left[\frac{\partial \Phi}{\partial \mathbf{e}}\right]^{-1} = \frac{1}{D} \begin{bmatrix} \dfrac{\partial \Phi_2}{\partial e_2} & -\dfrac{\partial \Phi_1}{\partial e_2} \\ -\dfrac{\partial \Phi_2}{\partial e_1} & \dfrac{\partial \Phi_1}{\partial e_1} \end{bmatrix}$$

where D is the determinant of $\partial \Phi / \partial e$ given by

$$D = \frac{\partial \Phi_1}{\partial e_1} \frac{\partial \Phi_2}{\partial e_2} - \frac{\partial \Phi_1}{\partial e_2} \frac{\partial \Phi_2}{\partial e_1}$$

Thus, we can express the system equation for a single-input, double-output system in terms of its general solution matrix as follows:

$$\begin{bmatrix} de_1 \\ de_2 \end{bmatrix} = \frac{1}{D} \begin{bmatrix} \dfrac{\partial \Phi_2}{\partial e_2} & -\dfrac{\partial \Phi_1}{\partial e_2} \\ -\dfrac{\partial \Phi_2}{\partial e_1} & \dfrac{\partial \Phi_1}{\partial e_1} \end{bmatrix} \begin{bmatrix} \dfrac{\partial \Phi_1}{\partial c} \\ \dfrac{\partial \Phi_2}{\partial c} \end{bmatrix} dc$$

And the system kernel is defined as

$$g(\mathbf{e}, c) = \begin{bmatrix} g_1 \\ g_2 \end{bmatrix} = \frac{1}{D} \begin{bmatrix} \dfrac{\partial \Phi_2}{\partial e_2} & -\dfrac{\partial \Phi_1}{\partial e_2} \\ -\dfrac{\partial \Phi_2}{\partial e_1} & \dfrac{\partial \Phi_1}{\partial e_1} \end{bmatrix} \begin{bmatrix} \dfrac{\partial \Phi_1}{\partial c} \\ \dfrac{\partial \Phi_2}{\partial c} \end{bmatrix}$$

where the elements g_1 and g_2 of the system equation are found to be

$$g_1 = \dfrac{\dfrac{\partial \Phi_2}{\partial e_2} \dfrac{\partial \Phi_1}{\partial c} - \dfrac{\partial \Phi_1}{\partial e_2} \dfrac{\partial \Phi_2}{\partial c}}{\dfrac{\partial \Phi_1}{\partial e_1} \dfrac{\partial \Phi_2}{\partial e_2} - \dfrac{\partial \Phi_1}{\partial e_2} \dfrac{\partial \Phi_2}{\partial e_1}}$$

$$g_2 = \dfrac{\dfrac{\partial \Phi_1}{\partial e_1} \dfrac{\partial \Phi_2}{\partial c} - \dfrac{\partial \Phi_2}{\partial e_1} \dfrac{\partial \Phi_1}{\partial c}}{\dfrac{\partial \Phi_1}{\partial e_1} \dfrac{\partial \Phi_2}{\partial e_2} - \dfrac{\partial \Phi_1}{\partial e_2} \dfrac{\partial \Phi_2}{\partial e_1}}$$

Thus, if the set of complete solutions for Eq. (4.3.13) is known, then the response of the single-input, multiple-output system can be determined for any value of the input variable for which the system kernel is defined, and the solution is valid. Unfortunately, Eq. (4.3.12), which relates the single-input, multiple-output system kernel in terms of the system's general solutions, is not particularly useful in determining the specific form of the solution matrix defined by Eq. (4.3.10). The principal value of Eq. (4.3.12) is in establishing the relationship between the system kernel and general solutions and providing an analytical means for verifying that the system kernel obtained by whatever methods employed actually is a solution for the system.

As with the single-input, single-output continuous system equation, it is not possible to determine analytical solutions for every single-input, multiple-output continuous system kernel even if the kernel satisfies the necessary requirements for boundedness and continuity. However, there are many single-input, multiple-output system kernels for which analytical solutions can be formed. We will systematically examine some of these kernels.

4.3.3 Constant System Kernels

If the system kernel for a single-input, multiple-output $g(\mathbf{e}, c)$ is or can be considered approximately constant, so that

$$g(\mathbf{e}, c) = \begin{bmatrix} g_{01} \\ \vdots \\ g_{0n} \end{bmatrix} = g(\mathbf{e}_0) \tag{4.3.15}$$

where each g_{0i} is constant over the range of variation of interest for both the single input variable c and the corresponding range of output variable $\mathbf{e}$, then the system equation has the following matrix form for a continuous system with a constant kernel:

$$d\mathbf{e} = g(\mathbf{e}, c)\, dc \tag{4.3.16}$$

Equation (4.3.16) is an exact differential equation and may be directly integrated over the range of variation for c and e for which the system kernel is constant. Integration over the range $\mathbf{e}_0$ to $\mathbf{e}$ and c_0 to c gives in matrix form the system solution

$$\mathbf{e} = \mathbf{e}_0 + \int_{c_0}^{c} g(\mathbf{e}_0)\, dc = \mathbf{e}_0 + g(\mathbf{e}_0)\, (c - c_0)$$

or, equivalently, in expanded form for each component of **e**,

$$e_i = g_{0i}(c - c_0) + e_{0i}, \quad \text{for} \quad i = 1, \ldots, n$$

Because so many single-input, multiple-output systems can, for at least sufficiently small changes in input, be treated as having constant system kernels, system Eqs. (4.3.16) serve as preliminary estimates of the system output, are very useful, and generally are recommended as beginning points for the analysis of many multiple-input and output systems.

4.3.4 Linear System Kernels with Constant Coefficients

Consider a single-input, multiple-output continuous system equation that has the general form

$$d\mathbf{e} = g(\mathbf{e}, c)dc \tag{4.3.17}$$

where the system kernel is given by the following linear expression in *e*:

$$g(\mathbf{e}, c) = A(c) + B\mathbf{e} \tag{4.3.18}$$

and where

$$A(c) = \begin{bmatrix} A_1(c) \\ \vdots \\ A_n(c) \end{bmatrix} \quad \text{and} \quad B = \begin{bmatrix} B_{11} & \cdots & B_{1n} \\ \vdots & & \\ B_{n1} & \cdots & B_{nn} \end{bmatrix}$$

The matrix elements, A_i's, may be functions of c, but each B_{ij} term in the B matrix is a constant that is independent of both c and **e**. Then the resulting system kernel, Eq. (4.3.18), is linear with constant coefficients, and the general solution can be readily obtained.

Since the system kernel is linear, it is possible to determine the solution for the system in two steps. First, we will solve Eq. (4.3.17) for the associated homogeneous linear constant-coefficient kernel [i.e., Eq. (4.3.18) with $A(c) = 0$] and, second, we will determine a particular solution for the system equation with the nonhomogeneous term [i.e., $A(c)$] present. The complete solution for the system Eq. (4.3.17) will then be given as the sum of the homogeneous solution and a particular solution.

Let us consider the homogeneous component (for which we let $\mathbf{e} = \mathbf{e}_h$) of system Eq. (4.3.17) so that we have

$$d\mathbf{e}_h = B\mathbf{e}_h \, dc \tag{4.3.19}$$

To solve this equation, let us assume that the solution has the general form

$$\mathbf{e}_h = D \exp(c) \quad \text{and} \quad d\mathbf{e}_h = D \exp(c)dc$$

where the constant matrix D is given by

$$D = \begin{bmatrix} D_1 \\ \vdots \\ D_n \end{bmatrix}$$

Then the homogeneous system equation becomes

$$\lambda D \exp(\lambda c) dc = BD \exp(c) dc$$

which can be expressed in the equivalent form:

$$[B - \lambda I] D \exp(\lambda c) dc = 0 \tag{4.3.20}$$

The matrix I is the unity matrix defined previously in Chapter 3. Since we want nontrivial solutions for Eq. (4.3.17) for all values of c in some interval of interest, we ignore the possibility that $D = 0$. Instead, we require that matrix Eq. (4.3.20) be satisfied by the following condition to provide nontrivial solutions:

$$[B - \lambda I] = 0$$

This matrix equation will vanish if the matrix $[B - \lambda I]$ is singular, and therefore its determinant is zero. Thus, we have that

$$\det [B - \lambda I] = \begin{bmatrix} B_{11} - \lambda & B_{12} & \cdots & B_{1n} \\ B_{21} & B_{22} - \lambda & & B_{2n} \\ \vdots & & & \\ B_{n1} & B_{n2} & \cdots & B_{nn} - \lambda \end{bmatrix} \tag{4.3.21}$$

Equation (4.3.21) is the eigenvalue or characteristic equation associated with the homogeneous matrix system Eq. (4.3.19). The values of λ that satisfy this equation are called the eigenvalues or characteristic roots for matrix B. The expansion of the determinant in Eq. (4.3.21) defines a polynomial expression
$P(\lambda)$ of nth degree in λ of the form

$$P(\lambda) = \det [B - \lambda I] = \sum_{i=0}^{n} Ci \; \lambda^i = 0$$

where the b's are algebraic functions of the elements B_{ij} that compose the matrix B. In general, any polynomial of nth degree with real coefficients possesses exactly n roots, which, in general, are complex-valued. However, since the coefficients B_{ij} are all assumed to be real, then all complex roots of $P(\lambda)$ must occur in complex conjugate pairs, and repeated roots, if they occur, must be real-valued. These various possibilities for the characteristic roots can be accounted for by the following matrix equation form for the general homogeneous solution $\mathbf{e}_h$

$$\mathbf{e}_h = D \exp[\lambda c]$$

or equivalently

$$\mathbf{e}_h = \begin{bmatrix} E_{11} & E_{12} & \cdots & E_{1n} \\ E_{21} & E_{22} & \cdots & E_{2n} \\ \vdots & & & \\ E_{n1} & E_{n2} & \cdots & E_{nn} \end{bmatrix} \begin{bmatrix} f_1 \exp(a_1 c)(\cos b_1 c + i \sin b_1 c) \\ f_2 \exp(a_2 c)(\cos b_2 c + i \sin b_2 c) \\ \vdots \\ f_n \exp(a_n c)(\cos b_n c + i \sin b_n c) \end{bmatrix}$$

where for each root n in the set $\lambda_1, \lambda_2, ..., \lambda_n$ we have the complex value that

$$\lambda_j = a_j + i b_j (bj = 0 \text{ if } \lambda_j \text{ is real})$$

$$f_j = \exp\left[\text{Ln } c \sum_{k=1}^{j-1} \delta(\lambda_j - \lambda_k)\right]$$

and

$$\delta(\lambda_j - \lambda_k) = \begin{cases} 0 \text{ if } \lambda_j \neq \lambda_k \\ 1 \text{ if } \lambda_j = \lambda_k \end{cases}$$

With the homogeneous solution known for Eq. (4.3.19), we have that the complete solution for system equation (4.3.17) is given as follows:

$$\mathbf{e} = \mathbf{e}_h + \mathbf{e}_p$$

where $\mathbf{e}_p$ is any particular solution of Eq. (4.3.17). In later sections, we will develop systematic methods for determining $\mathbf{e}_p$. For the special case where the nonhomogeneous matrix A in Eq. (4.3.18) is constant, say A_0, then the particular solution is given by

$$\mathbf{e}_h = -B^{-1} A_0$$

and the complete solution for the system is then given by

$$\mathbf{e} = \mathbf{e}_h - B^{-1} A_0$$

Example Problem for the Linear Kernel with Constant Coefficients.
Suppose we have the following linear system equation for a single-input, double-output system:

$$\begin{bmatrix} de_1 \\ de_2 \end{bmatrix} = \left[\begin{Bmatrix} a_1 \\ a_2 \end{Bmatrix} + \begin{Bmatrix} b_{11} \, eh_1 + b_{12} \, eh_2 \\ b_{21} \, eh_1 + b_{22} \, eh_2 \end{Bmatrix}\right] dc \tag{4.3.22}$$

where all a's and b's are real constants. The homogeneous equation associated with Eq. (4.3.22) is given by

$$\begin{bmatrix} de_{h1} \\ de_{h2} \end{bmatrix} = \begin{bmatrix} b_{11} \, e_{h1} + b_{12} \, e_{h2} \\ b_{21} \, e_{h1} + b_{22} \, e_{h2} \end{bmatrix} dc \tag{4.3.23}$$

and if we let $e_h = D \exp(c)$, then we have for Eq. (4.3.23) that

$$\begin{bmatrix} (b_{11} - \lambda) + b_{12} \\ b_{21} + (b_{22} - \lambda) \end{bmatrix} \begin{bmatrix} D_1 \\ D_2 \end{bmatrix} \exp(\lambda c) \, dc = 0$$

Since we seek nontrivial solutions for Eq. (4.3.23), we require that

$$\det\begin{bmatrix} (b_{11} - \lambda) + b_{12} \\ b_{21} + (b_{22} - \lambda) \end{bmatrix} = (b_{11} - \lambda)(b_{22} - \lambda) - b_{12} \, b_{21} = 0$$

So

$$P(\lambda) = 0 = \lambda^2 - (b_{11} + b_{22}) + b_{11}b_{22} - b_{12}b_{21} \tag{4.3.24}$$

Solving the second-degree or quadratic equation in λ as defined by Eq. (4.3.24), we have, as our two roots,

$$\left.\begin{array}{c} \lambda_1 \\ \lambda_2 \end{array}\right\} = \frac{b_{11} + b_{22}}{2} \pm \frac{1}{2} \sqrt{(b_{11} + b_{22})^2 - 4(b_{11}\,b_{22} - b_{12}b_{21})}$$

Recalling that both λ's are real if

$$(b_{11} + b_{22})^2 > 4\,(b_{11}\,b_{22} - b_{12}\,b_{21})$$

then we have two real distinct roots, say α_1 and α_2, with zero imaginary components (i.e., $\beta_1 = \beta_2 = 0$). If we have that

$$(b_{11} + b_{22})^2 = 4\,(b_{11}\,b_{22} - b_{12}\,b_{21})$$

then both roots are real and equal, say $\alpha_1 = \alpha_2$, with zero imaginary components (again $\beta_1 = \beta_2 = 0$). Finally, if

$$(b_{11} + b_{22})^2 < 4\,(b_{11}\,b_{22} - b_{12}\,b_{21})$$

then we have two distinct complex roots, which form a complex conjugate pair, as follows:

$$\left.\begin{array}{c} \lambda_1 \\ \lambda_2 \end{array}\right\} = \alpha \pm i\,\beta$$

From Eq. (4.3.23), the homogeneous solution for our example has the generalized form

$$\mathbf{e}_h = \begin{bmatrix} e_{1h} \\ e_{2h} \end{bmatrix} = \begin{bmatrix} E_{11} \exp(\lambda_1\,c) & E_{12} \exp(\lambda_2\,c) \\ E_{21} \exp(\lambda_1\,c) & E_{22} \exp(\lambda_2\,c) \end{bmatrix}$$

Now since we know that the general solution has the form

$$\mathbf{e} = \mathbf{e}_h + \mathbf{e}_p$$

we will assume that $\mathbf{e}_p$ is given by the following constant matrix:

$$\mathbf{e}_p = \begin{bmatrix} F_1 \\ F_2 \end{bmatrix}$$

Substituting this form for the particular solution, we find that the general solution of Eq. (4.3.22) is given by

$$e_1 = E_{11} \exp(\lambda_1 c) + \frac{b_{12}}{\lambda_2 - b_{11}} E_{22} \exp(\lambda_2 c) - \frac{a_1\,b_{22} - a_2\,b_{12}}{b_{11}\,b_{22} - b_{12}\,b_{21}}$$

$$e_2 = E_{22} \exp(\lambda_2 c) + \frac{b_{21}}{\lambda_1 - b_{22}} E_{11} \exp(\lambda_1 c) - \frac{a_2\,b_{11} - a_1\,b_{21}}{b_{11}\,b_{22} - b_{12}\,b_{21}}$$

where E_{11} and E_{22} are arbitrary constants determined by the initial conditions for the system.

4.3.5 Linear System Kernels with Variable Coefficients

The single-input, multiple-output system equation for a linear system kernel with variable coefficients has the general form

$$d\mathbf{e} = [A(c) + B(c)\mathbf{e}]dc \tag{4.3.25}$$

This equation is similar to Eq. (4.3.17) in Section 4.3.4 except that the coefficient matrix $B(c)$ is now a general function of the input variable c.

For rather mild conditions on $B(c)$, that is, the functions comprising $B(c)$ are analytical functions (analytical functions are functions that can be expressed as simple power series expansions), solutions for Eq. (4.3.25) can also be expressed as an appropriate power series.

To demonstrate the application of such series methods, consider the following single-input, double-output linear system equation:

$$de_1 = [A_1(c) + B_1(c)e_1 + D_1(c)\, e_2]dc \tag{4.3.26a}$$

$$de_2 = [A_2(c) + B_2(c)e_1 + D_2(c)\, e_2]dc \tag{4.3.26b}$$

Now for $i = 1, 2$, we will assume that the A_i, B_i, and D_i functions can be described as power series, as follows:

$$A_i(c) = \sum_{n=0}^{\infty} a(i)_n c^n \tag{4.3.27a}$$

$$B_i(c) = \sum_{n=0}^{\infty} b(i)_n c^n \tag{4.3.27b}$$

$$D_i(c) = \sum_{n=0}^{\infty} d(i)_n c^n \tag{4.3.27c}$$

We also assume that the general solution for system equations (4.3.26) can be expressed as

$$e_i(c) = e_{i0} \sum_{m=0}^{\infty} a(i)_m c^m \tag{4.3.28}$$

where e_{10} and e_{20} are the initial condition constants, $a(1)_0 = a(2)_0 = 1$, and the other $a(i)_0$ terms will be determined shortly. If we substitute Eqs. (4.3.27) and (4.3.28) into system equation (4.3.26), we have

$$e_{10} \sum_{m=0}^{\infty} a(1)_m m\, c^{m-1} = \sum_{n=0}^{\infty} [a(1)_n c^n$$

$$+ \sum_{m=0}^{\infty} \{e_{10}\, b(1)_n a(1)_m + e_{20}\, d(1)_n a(2)_m\}\, c^{m+n}]$$

$$\tag{4.3.29.1}$$

$$e_{20} \sum_{m=0}^{\infty} a(2)_m mc^{m-1} = \sum_{n=0}^{\infty} [a(2)_n c^n$$

$$+ \sum_{m=0}^{\infty} \{e_{10}\, b(2)_n\, a(1)_m + e_{20}\, d(2)_n a(2)_m\}c^{m+n}]$$

$$\tag{4.3.29.2}$$

Now the summing variables, m and n, are arbitrary dummy variables and are eliminated upon evaluation of summation just as the integration dummy variables are eliminated upon evaluation of a definite integral. So if we let $m = k+1$ in the left summation terms in Eqs. (4.3.29.1) and (4.3.29.2), and $n = k$ in the single summation terms and $m = k-1$ in the double summation terms on the right side of Eqs. (4.3.29.1) and (4.3.29.2), then we obtain the following expressions after appropriate changes are made in the summation limits:

$$e_{10} \sum_{k=0}^{\infty} (k+1) \, a(1)_{k+1} c^k = \sum_{k=0}^{\infty} \left[a(1)_k c^k \right.$$

$$\left. + \sum_{n=k}^{\infty} \left\{ e_{10} \, b(1)_n a(1)_{k-n} + e_{20} \, d(1)_n a(2)_{k-n} \right\} c^k \right]$$

$$(4.3.30.1)$$

$$e_{20} \sum_{k=0}^{\infty} (k+1) a(2)_{k+1} c^k = \sum_{n=0}^{\infty} \left[a(2)_k c^k \right.$$

$$\left. + \sum_{n=k}^{\infty} \left\{ e_{10} \, b(2)_n a(1)_{k-n} + e_{20} \, d(2)_n a(2)_{k-n} \right\} c^k \right]$$

$$(4.3.30.2)$$

Now we observe that the first term in each left-hand series vanishes, since

$$(k+1) \, a(i)_{k+1} c^k = 0, \quad \text{for} \quad k = -1$$

and also, we observe for any arbitrary coefficient b_{kn}, we have for the following double summation result that

$$\sum_{k=0}^{\infty} \sum_{k-n=0}^{\infty} b_{kn} = \sum_{k=0}^{\infty} \sum_{n=0}^{k} b_{kn} \qquad (4.3.31)$$

since negatively subscripted values for b_{kn} are not permitted and are assumed to vanish. Then employing Eq. (4.3.31), we may reduce and rearrange Eqs. (4.3.30) to provide

$$\sum_{k=0}^{\infty} \left[(k+1) e_{10} \, a(1)_{k+1} - a(1)_k \right.$$

$$(4.3.32.1)$$

$$\left. - \sum_{n=0}^{k} \left\{ e_{10} \, b(1)_k a(1)_{k-n} + e_{20} \, d(1)_n a(2)_{k-n} \right\} \right] c^k = 0$$

$$\sum_{k=0}^{\infty} \left[(k+1) \, e_{20} \, a(2)_{k+1} - a(2)_k \right.$$

$$(4.3.32.2)$$

$$\left. - \sum_{n=0}^{k} \left\{ e_{10} \, b(2)_k a(1)_{k-n} + e_{20} \, d(2)_n a(2)_{k-n} \right\} \right] c^k = 0$$

Since c is a continuous variable over some specified range of values, and Eqs. (4.3.32) must be satisfied for all values of c within this range, we require that the coefficient expression of each power of c, (i.e., $h(\mathbf{e}, c) \, d\mathbf{e} = h(\mathbf{e}, c) \, g(\mathbf{e}, c) \, dc$) must vanish for $k = 0, 1, ...,$. Therefore, we obtain the following equations, which define the coefficients $a(i)_{k+1}$:

$$a(1)_{k+1} = \frac{1}{k+1} \left[\frac{a(1)_k}{e_{10}} + \sum_{n=0}^{k} \left\{ b(1)_n \, a(1)_{k-n} + \frac{e_{20}}{e_{10}} d(1)_n a(2)_{k-n} \right\} \right] \qquad (4.3.33.1)$$

$$a(2)_{k+1} = \frac{1}{k+1}\left[\frac{a(2)_k}{e_{10}} + \sum_{n=0}^{k}\left\{b(2)_n a(1)_{k-n} + \frac{e_{20}}{e_{10}}d(2)_n a(2)_{k-n}\right\}\right] \qquad (4.3.33.2)$$

for $k \geq 0$, where $a(1)_0 = a(2)_0 = 1$, and e_{10} and e_{20} are the initial values for $e_1(c)$ and $e_2(c)$, respectively, when $c = 0$.

Observe that Eqs. (4.3.33) define recursive relationships for each $a(1)_{k+1}$ and $a(2)_{k+1}$ in terms of the previously determined values $a(1)_1$, $a(1)_2$, ... and $a(2)_1$, $a(2)_2$, ... and the $a(i)$, $b(i)$, and $d(i)$ coefficients, which determine the system functions $A_i(c)$, $B_i(c)$, and $C_i(c)$ are defined by Eqs. (4.3.27). Thus, with all the $a(1)_i$ and $a(2)_i$ coefficients defined, the general system solutions given by Eq. (4.3.28) are now completely determined.

Although algebraic expansions for each a_k term defined by Eqs. (4.3.33) can be found for a limited number of sequential terms (i.e., $0 \leq k \leq K$) by repetitive substitution, unless the resulting series are rapidly convergent, it is generally advisable to evaluate the a_k coefficients numerically via the computer.

The power series solution approach we have used for the single-input, double-output system may be extended to a single-input system with any finite number of outputs. Consider the general linear system equation component i for N outputs given by

$$de_i = \left[A_i(c) + \sum_{j=1}^{N} B_{ij}(c)\, e_j\right] dc \qquad (4.3.34)$$

where the coefficient functions are assumed expressible as power series, as follows:

$$A_i(c) = \sum_{n=1}^{\infty} a(i)_n\, c^n \qquad (4.3.35.1)$$

and

$$B_{ij}(c) = \sum_{n=1}^{\infty} b(i,j)_n\, c^n \qquad (4.3.35.2)$$

Now we will assume that the general solution for system Eq. (4.3.34) is given by $e_i(c)$, where

$$e_i(c) = e_{i0} \sum_{m=0}^{\infty} a(i)_m\, c^m \qquad (4.3.36.1)$$

where e_{i0} is the initial condition for $e_i(c)$ and $a(i)_0 = 1$. We find in a similar fashion, as we did for the double-output system, that the $a(i)_m$ coefficients are determined by the following recursive relations for $i = 1, 2, ..., I$ and $k > 0$:

$$a(i)_{k+1} = \frac{1}{k+1}\left[a(i)_k/e_{i0} + \sum_{n=0}^{k}\sum_{j=1}^{N}\frac{e_{j0}}{e_{i0}}b(i,j)_n\, a(j)_{k-n}\right] \qquad (4.3.36.2)$$

Again, with all coefficients of the respective power series now defined, Eqs. (4.3.36) provide the general solution for the linear single-input, multiple-output system equation.

Occasionally, the system investigator may encounter a linear system model in which the coefficient function [i.e., the $A_i(c)$ and $B_{ij}(c)$ in Eq. (4.3.35)] cannot be expressed as a simple power series because of the presence of singular points. Singular points are finite values of c for which A_i or $B_{ij}(c)$ become undefined or infinite. If these singularities are limited to regular singular points within the range of interest sought, then the "Method of Frobenius" can provide solutions based on the following solution series expansions:

$$e_j(c) = e_{j0} \sum_{i=0}^{\infty} a(j)_i c^{i+s}$$

The parameters provide an additional parameter, s, generally not an integer, which is essential in establishing the solution for equations with regular singular points. However, such advanced solution techniques will not be considered here, and the investigator is referred to the pertinent literature.

4.3.6 Exact System Kernels

Consider the single-input, multiple-output system equation

$$d\mathbf{e} = g(\mathbf{e}, c) \, dc \tag{4.3.37}$$

where $d\mathbf{e}$ and $g(\mathbf{e}, c)$ are $(n \times 1)$ column matrices. Suppose we multiply this equation by a nonzero $(n \times n)$ matrix function $h(\mathbf{e}, c)$, given as

$$h(\mathbf{e}, c) = \begin{bmatrix} h_1(\mathbf{e}, c) & 0 & \cdots & 0 \\ 0 & & & \\ \vdots & & & \\ 0 & 0 & \cdots & h_n(\mathbf{e}, c) \end{bmatrix}$$

where $h(\mathbf{e}, c)$ are general functions of $\mathbf{e}$ and c and will be defined shortly. Then system equation (4.3.37) can be expressed as

$$h(\mathbf{e}, c) \, d\mathbf{e} = h(\mathbf{e}, c) \, g(\mathbf{e}, c) \, dc \tag{4.3.38}$$

We will now show that Eq. (4.3.38) can assume the matrix form for the total differential of a matrix function. Consider an arbitrary function of the variables $\mathbf{e}$ and c given by

$$\Phi(\mathbf{e}, c) = \mathbf{K} \tag{4.3.39}$$

where $\mathbf{K}$ is a matrix constant, and both Φ and $\mathbf{K}$ are $(n \times 1)$ matrices. If we form the total matrix differential of Eq. (4.3.39), we have

$$d\Phi(\mathbf{e}, c) = \frac{\partial \Phi}{\partial \mathbf{e}}(\mathbf{e}, c) \, d\mathbf{e} + \frac{\partial \Phi}{\partial c}(\mathbf{e}, c) \, dc = 0 \tag{4.3.40}$$

Now we observe that Eqs. (4.3.40) and (4.3.38) will be equivalent if

$$\frac{\partial \Phi}{\partial \mathbf{e}}(e, c) = h(e, c) \quad \text{and} \quad \frac{\partial \Phi}{\partial c}(e, c) = -h(e, c) \, g(e, c) \tag{4.3.41}$$

If a function $h(e, c)$, called the integrating factor for the system equation (4.3.37), can be found that satisfies Eq. (4.3.41), then Eq. (4.3.38) is exact and can be immediately integrated to provide the general solution for the system equation (4.3.37). Since the function Φ given by Eq. (4.3.39), which is a solution of Eq. (4.3.40) or (4.3.38) if Eq. (4.3.41) is satisfied, may be generally regarded as a continuous function in the variables $\mathbf{e}$ and c with continuous derivatives, we have that

$$\frac{\partial^2 \Phi}{\partial \mathbf{e}\, \partial c} = \frac{\partial^2 \Phi}{\partial c\, \partial \mathbf{e}} \tag{4.3.42}$$

and the order in which differentiation is taken on Φ is immaterial if the function is continuous. Thus, assuming that Eq. (4.3.42) is satisfied, then Eq. (4.3.41) provides the following necessary and sufficient condition for Eq. (4.3.38) to be exact:

$$\frac{\partial^2 \Phi}{\partial c\, \partial \mathbf{e}} = \frac{\partial h}{\partial c} = -h\frac{\partial g}{\partial \mathbf{e}} - \frac{\partial h}{\partial \mathbf{e}} g \tag{4.3.43}$$

Equation (4.3.43), which is a diagonal matrix equation, has for the jth diagonal element

$$\frac{\partial h_j}{\partial c} = -h_j\frac{\partial g_j}{\partial e_j} - \frac{\partial h_j}{\partial e_j} g_j$$

or, equivalently,

$$\frac{\partial \operatorname{Ln} h_j}{\partial c} = -\frac{\partial g_j}{\partial e_j} - g_j\frac{\partial \operatorname{Ln} h_j}{\partial e_j} \tag{4.3.44}$$

for $j = 1, \ldots, n$. Thus, if a set of functions $i_1, i_2, \ldots, i_n$ can be found that respectively satisfy Eq. (4.3.44) for the appropriate kernel component, then the i's constitute a set of integrating factors for the system equation, which then has as a general solution

$$\int h(\mathbf{e}, c)\, d\mathbf{e} = \int h(\mathbf{e}, c)\, g(\mathbf{e}, c)\, dc + \mathbf{K}$$

where $\mathbf{K}$ is the integration constant matrix.

4.3.7 Separable System Kernels

Suppose that the single-input, multiple-output system equation

$$d\mathbf{e} = g(\mathbf{e}, c)dc$$

has a system kernel of the following form for each component for $i = 1, \ldots, n$:

$$g_i(\mathbf{e}, c) = G_i(\mathbf{e})\, F(c)$$

And, furthermore, for all i and j such that $1 \le (i \text{ or } j) \le n$ and $i \ne j$,

$$\frac{g_i(\mathbf{e}, c)}{g_j(\mathbf{e}, c)} = \frac{G_i(\mathbf{e})}{G_j(\mathbf{e})} = \frac{\partial H_j/\partial e_j}{\partial H_i/\partial e_i}$$

where $H_i(e_i)$ and $H_j(e_j)$ are explicit functions of e_i and e_j, respectively. Then the system kernel is separable and results in a system of exact differential equations of the form (obtained by evaluating de_i/de_j for all i and j such that $1 \le i$ or $j \le n$ and $i = j$)

$$\frac{\partial H_i(e_i)}{\partial e_i} de_i = \frac{\partial H_j(e_j)}{\partial e_j} de_j$$

This set of equations can be integrated to give the intermediate solution

$$H_i(e_i) = H_j(e_j) + K_{ij}$$

Furthermore, for any of this set of equations for which we can solve explicitly for all e_i for $1 \le i \le n$ and $i = j$ in terms of e_j,

$$e_i = H_j^{-1}\left[H_j(e_j) + K_{ij}\right]$$

then the reduced system equation for that component e_i is given by

$$\frac{de_j}{M(e_j)} = F(c)\, dc \tag{4.3.45}$$

where

$$M(e_j) = G_j\left[H_1^{-1}\{H_j(e_j) + K_{1j}\}, H_2^{-1}\{H_j(e_j) + K_{2j}\},\right.$$
$$\left....., H_n^{-1}\{H_j(e_j) + K_{nj}\}\right]$$

Of course, Eq. (4.3.45) can now be immediately integrated to give, as the final solution for the exact single-input, multiple-output system, the following:

$$\int \frac{de_j}{M(e_j)} = \int F(c)\, dc + K_j$$

where K_j is the constant of integration.

Example Problem for a System with a Separable Kernel.
Consider the system equations

$$de_1 = e_1(a_1 + b_1 e_2)F(c)\, dc$$
$$de_2 = e_2(a_2 + b_2 e_1)F(c)\, dc$$

where the a's and b's are constants. If we form the ratio de_1/de_2 we find the $F(c)$ is eliminated, and we have that

$$\frac{de_1}{de_2} = \frac{e_1(a_1 + b_1 e_2)}{e_2(a_2 + b_2 e_1)}$$

This differential equation is separable, and we have that

$$(a_2 + b_2 e_1)\frac{de_1}{e_1} = (a_1 + b_1 e_2)\frac{de_2}{e_2}$$

which yields upon integration

$$a_2 \text{Ln } e_1 + b_2 e_1 = a_1 \text{Ln } e_2 + b_1 e_2 + K_1$$

where K_1 is the usual integration constant. To obtain a solution for the system outputs e_1 and e_2 in terms of c, we must solve the preceding equation explicitly for either e_1 or e_2. We could find an approximate, explicit solution for e_1 or e_2 if we knew the magnitudes of the a and b terms. Suppose that a_2 Ln e_1 is very small in comparison to the other terms in the equation. Then (setting $a_2 = 0$) we have that

$$e_1 = \frac{1}{b_2}[a_1 \text{ Ln } e_2 + b_1 e_2 + K_1]$$

and substituting this result into our original system equation for de_2, we have

$$de_2 = e_2 [a_1 \text{ Ln } e_2 + b_1 e_2 + K_1] F(c) dc$$

This system equation is separable and can be integrated as shown in the following equation:

$$\int \frac{de_2}{e_2 [a_1 \text{ Ln } e_2 + b_1 e_2 + K_1]} = \int F(c) dc + K_2$$

Unfortunately, evaluating the first integral analytically is difficult. The solutions of differential equations often define integrals that can be difficult or even impossible to integrate analytically. If we let $x = \text{Ln } e_2$, we find that the first integral expression becomes

$$\int \frac{de_2}{e_2 [a_1 \text{ Ln } e_2 + b_1 e_2 + K_1]} = \int \frac{dx}{[a_1 x + b_1 \exp(x) + K_1]}$$

But this integral is also not easily integrated in terms of elementary functions. The reader should use the Symbolic file of MATLAB and attempt to find a result for this integral.

Another means of determining an analytical expression for this integral is to expand either integrand in a uniformly convergent infinite series and integrate each term of the series. We then retain sufficient terms to provide an approximate solution for the equation. Such techniques are described in the mathematical literature and will not be considered here.

4.3.8 Homogeneous System Kernels

If the single-input, multiple-output system kernel $g(e, c)$ can be expressed in the form

$$g(\mathbf{e}, c) = g(\mathbf{e}/c) = \begin{bmatrix} g_1(e_1/c, ..., e_n/c) \\ \vdots \\ g_n(e_1/c, ..., e_n/c) \end{bmatrix} \tag{4.3.46}$$

then the system kernel is of the homogeneous type and can be reduced to a set of system equations that are independent of the input variable c. When any set of system equations has a system kernel that is not explicitly a function of the input variable(s), the system equation is said to be autonomous. The general autonomous system equation for single-input, multiple-output systems will be considered in Section 4.3.9.

The system equation with a homogeneous kernel given by Eq. (4.3.46) has the form

$$d\mathbf{e} = g(\mathbf{e}/c)dc$$
$$c d\mathbf{u} + \mathbf{u} dc = g(\mathbf{u})dc \tag{4.3.47}$$

This equation can be simplified by defining a new set of dependent variables as follows. Let

$$\mathbf{u} = \mathbf{e}/c, \quad d\mathbf{e} = c \, d\mathbf{u} + \mathbf{u} \, dc$$

where

$$\mathbf{e} = \begin{bmatrix} e_1 \\ \vdots \\ e_n \end{bmatrix}, \quad \mathbf{u} = \begin{bmatrix} e_1/c \\ \vdots \\ e_n/c \end{bmatrix}$$

Then the system equation (4.3.47) becomes in terms of $\mathbf{u}$ and c the following:

$$c \, d\mathbf{u} + \mathbf{u} \, dc = g(\mathbf{u})dc$$

With a simple rearrangement of terms, we have for $c \neq 0$

$$d\mathbf{u} = [\, g(\mathbf{u}) - \mathbf{u}\,] \, \frac{dc}{c}$$

which is the reduced system equation in terms of a and c. For the ith component of this matrix equation, we have, for $i = 1, ..., n$, that

$$du_i = [\, g_i(u_1, u_2, ..., u_n)\,] \frac{dc}{c}$$

We observe that this resulting set of equations can be expressed in terms of c and a function of $\mathbf{u}$ as follows:

$$\frac{d\,u_i}{[g_i(u_1, u_2, ..., u_n) - u_i]} = \frac{dc}{c} = d \operatorname{Ln} c \qquad (4.3.48)$$

Since the independent variable c appears as an exact derivative in Eq. (4.3.48) and is independent of i, then the remaining transformed system equations in terms of the u_i's can be expressed in the repetitive form

$$\frac{du_1}{g_1 - u_1} = \frac{du_2}{g_2 - u_2} = \cdots = \frac{du_n}{g_n - u_n}$$

The functions $g_i(u)$ are autonomous (i.e., independent of c) and can often be solved by appropriate integrating factor selection or some of the other methods covered in this chapter.

Example Problem with Homogeneous System Kernel.
Consider the following single-input, double-output system:

$$de_1 = \frac{e_1 + a \, e_2}{c} \, dc$$

$$de_2 = \frac{b \, e_1 + e_2}{c} \, dc$$

We observe that each respective system kernel is homogeneous since

$$g_1 = g_1 \left(\frac{e_1}{c}, \frac{e_2}{c} \right) = \frac{e_1}{c} + a \, \frac{e_2}{c}$$

$$g_2 = g_2 \left(\frac{e_1}{c}, \frac{e_2}{c} \right) = b \, \frac{e_1}{c} + \frac{e_2}{c}$$

Our reduced system equations with the variable transformations

$$u_1 = \frac{e_1}{c} \quad \text{and} \quad u_2 = \frac{e_2}{c}$$

become

$$c \, du_1 = a \, u_2 dc$$
$$c \, du_2 = b \, u_1 dc \tag{4.3.49}$$

or

$$\frac{du_1}{au_2} = \frac{du_2}{au_1} = \frac{dc}{c} = d \operatorname{Ln} c$$

We observe that the reduced system relating u_1, and u_2 is separable (see Section 4.2.6) and can be integrated to give

$$\int u_1 \, du_1 = \frac{a}{b} \int u_2 \, du_2 + \frac{K_1}{2}$$

which provides as an intermediate solution

$$u_1{}^2 = \frac{a}{b} u_2{}^2 + K_1 \tag{4.3.50}$$

Solving this equation for u and substituting the result into Eq. (4.3.49) gives

$$cdu_2 = \pm b \sqrt{\frac{a}{b} u_2{}^2 + K_1} \, dc$$

This equation is also separable and, upon integration, gives

$$\int \frac{d \, u_2}{\sqrt{u_2{}^2 + (b/a) \, K_1}} = \operatorname{Ln} K_2 \pm \sqrt{ab} \operatorname{Ln} c$$

and completing the integration, we have (recalling that $e_2 = u_2 c$)

$$e_2 = K_2 \, c^{\pm \sqrt{ab}} \left(\frac{e_2{}^2}{c^2} + \frac{b}{a} K_1 \right)^{1/2}$$

and using Eq. (4.3.50), we find that

$$e_1 = \pm \left(\frac{a}{b} \left(e_2{}^2 + K_1 c^2 \right) \right)^{1/2}$$

as solutions for our example's outputs.

4.3.9 Autonomous System Kernels

Autonomous system kernels are those kernels that are not explicit functions of the input variables (i.e., the c's). In other words, the input or independent variables dc not explicitly appear in the mathematical expressions for the kernels. For a single-input, multiple-output system, an autonomous kernel results if

$$g(\mathbf{e}, c) = g(\mathbf{e})$$

and the autonomous system equations, in matrix form, are

$$d\mathbf{e} = g(\mathbf{e})dc$$

If we express this matrix equation in expanded form, we have

$$de_1 = g_1(\mathbf{e})dc = g_1(e_1, ..., e_n)dc$$

$$\vdots$$

$$de_n = g_n(\mathbf{e})\, dc = g_n(e_1, ..., e_n)dc$$

Since each system kernel element $g_n(e)$ is not explicitly dependent on c, it is possible to eliminate the differential input variable dc by forming ratios for any desired set of output variables we wish as follows, for $1 \le i$ or $j \le n$

$$\frac{de_i}{de_j} = \frac{g_i(\mathbf{e})\, dc}{g_j(\mathbf{e})\, dc} = \frac{g_i(\mathbf{e})}{g_j(\mathbf{e})}$$

We now have a set of $n-1$ equations for the n variables, $e_1, e_2, ..., e_n$ and it may be possible to integrate this reduced system equation to determine each output e_i. The example below will demonstrate the method.

Example with Autonomous System Kernel.
Consider the following single-input, double-output autonomous system equation:

$$de_1 = g_1(e_1, e_2)dc$$
$$de_2 = g_2(e_1, e_2)dc$$

Here, formation of our ratio of individual output component system equations is obvious since there is only one unique combination. So we have that

$$\frac{de_2}{de_1} = \frac{g_2(e_1, e_2)}{g_1(e_1, e_2)} \tag{4.3.51}$$

We have now reduced our system of two first-order equations to a single first-order equation. Observe that Eq. (4.3.51) can be treated as a single input, single-output system equation of the form

$$de_2 = \frac{g_2(e_1, e_2)}{g_1(e_1, e_2)} de_1 = g(e_1, e_2)de_1 \tag{4.3.52}$$

where either e_1 or e_2 can be considered the system's input and e_2 or e_1, respectively, as the system's outputs. Since Eq. (4.3.52) is equivalent to a first-order differential equation, all the methods developed in Section 4.2 are directly applicable.

To complete this example, assume further that

$$g_1(e_1, e_2) = A\, e_2 + B \tag{4.3.53.1}$$
$$g_2(e_1, e_2) = D\, e_1 + F \tag{4.3.53.2}$$

where A, B, D, and F are constants and $A \neq D$. Then for Eq. (4.3.52), we have that

$$de_2 = \frac{De_1 + F}{A\,e_2 + B}\,de_1$$

which is separable and can be integrated to give

$$A\,(e_2 + B/A)^2 = D\,(e_1 + F/D)^2 + K_0 \tag{4.3.54}$$

where K_0 is the integration constant. Solving Eq. (4.3.54) for e_2, we have that

$$e_2 = -\frac{B}{A} \pm \left\{ \frac{D}{A}\left[e_1 + \frac{F}{D}\right]^2 + K_1 \right\}^{1/2} \tag{4.3.55}$$

and the first integrating constant is now defined as K_1. Substitution of Eq. (4.3.55) into Eq. (4.3.53.1) for kernel $g_1(e_1, e_2)$, gives for the system equation that

$$de_1 = \pm A \left\{ \frac{D}{A}\,(e_1 + F/D)^2 + K_1 \right\}^{1/2} dc$$

This equation is also separable, and we have upon integration that

$$e_1 + \frac{F}{D} + \left\{(e_1 + F/D)^2 + AK_1/D\right\}^{1/2} = K_2 \exp\left[\pm\sqrt{AD}\,c\right] \tag{4.3.56}$$

where K_2 is the second integration constant. Finally, using Eqs. (4.3.56) and (4.3.55), we have for e_2 as a function of c, with the initial conditions that $e_1(c_0) = e_{10}$ and $e_2(c_0) = e_{20}$,

$$e_2 + \frac{B}{A} + \left\{ \left[e_2 + \frac{B}{A}\right]^2 - K_1 \right\}^{1/2} = K_2 \sqrt{\frac{D}{A}} \exp\left[\pm\sqrt{AD}\,c\right]$$

where

$$K_1 = \left[e_{20} + \frac{B}{A}\right]^2 - \frac{D}{A}\left[e_{10} + \frac{F}{D}\right]^2$$

$$K_2 = \left[e_{10} + \frac{F}{D} + \left\{\left[e_{10} + \frac{F}{D}\right]^2 + \frac{AK_1}{D}\right\}^{1/2}\right] \exp\left[\pm\sqrt{\frac{D}{A}}\,c_0\right]$$

4.3.10 System Kernels Associated with Classical Second-Order Ordinary Differential Equations (ODEs)

Some of the most important and frequently encountered differential equations in mathematics and the physical sciences are those equations that can be classified as second-order, linear, ordinary differential equations with variable coefficients and have the standard form

$$\frac{d^2y}{dx^2} + P(x)\,\frac{dy}{dx} + Q(x)\,y = R(x) \tag{4.3.57}$$

where P, Q, and R are analytical functions of the independent variable x, with the possible exception of isolated, regular singular points within their range of definition of the variable x. Since Eq. (4.3.57) is linear (observe that each term contains the dependent variable, $y(x)$, to the first or zero power only), it can be shown that the complete solution of Eq. (4.3.57) is given by the sum of the solution of the homogeneous form of Eq. (4.3.57) [i.e., with $R(x) = 0$] and any particular solution of Eq. (4.3.57) with $R(x)$ present. Generally, when the homogeneous solutions are known, a particular solution can be determined using the method called a variation of parameters, as we will show later in this section.

The most general method available for solving differential equations of the form given by Eq. (4.3.57) is the Method of Frobenius, which we encountered in Section 4.3.5. However, we are primarily interested in utilizing the known solutions for various forms of Eq. (4.3.57) as they apply to single-input, double-output systems.

It is easily shown that a single-input, double-output system equation can be associated with Eq. (4.3.57) as follows by letting

$$de_1 = g_1 dc = e_2 d\,c \tag{4.3.58.1}$$

$$de_2 = g_2\, dc = -[P(c)\, e_2 + Q(c)\, e_1 - R(c)]dc \tag{4.3.58.2}$$

where we have defined

$$c = x, \quad e_1(c) = y(x), \quad \text{and} \quad e_2(c) = \frac{dy(x)}{dx}$$

Actually, Eq. (4.3.57) is a special case of the single-input, double-output linear system equation, which has the general form

$$de_1 = [A_{11}(c)e_1 + A_{12}(c)e_2 + B_1(c)]dc \tag{4.3.59.1}$$

$$de_2 = [A_{21}(c)e_1 + A_{22}(c)e_2 + B_2(c)]dc \tag{4.3.59.2}$$

It is apparent that Eqs. (4.3.59.1) and (4.3.59.2) reduce to Eqs. (4.3.58.1) and (4.3.58.2), respectively, if $A_{11}(e) = B_1(e) = 0$ and $A_{12}(e) = 1$. Thus, we observe that for any given second-order linear differential equation, there exists an equivalent single-input, double-output linear system equation. Indeed, as we will demonstrate elsewhere in this text, for any original consistent set of independent equations, whether algebraic, transcendental, or ordinary or partial differential equations, there exists an equivalent set of system equations associated with the original set of equations.

Table 4-2 lists some of the classical second-order linear differential equations with the descriptive title usually associated with each equation. The table also gives the corresponding single-input, double-output system equation equivalent to the differential equation. The table is not meant to be a complete listing of all important linear second-order differential equations, but rather it is a demonstration of the equivalencies between such differential equations and appropriate system equations. However, there is always the possibility that a single-input, double-output system kernel we wish to study will be found in Table 4-2 or be associated with a second-order differential equation for which the solution is known.

With both independent homogeneous solutions of Eq. (4.3.57) known [see $y_1(x)$ and $y_2(x)$ in Table 4-2], the complete nonhomogeneous solution for Eq. (4.3.57) as found using the method of variation of parameters is as follows:

$$y(x) = K_1 y_1(x) + K_2 y_2(x) + y_P$$

Table 4-2 Equivalence Second-Order Ordinary Differential Equations (ODEs) and Associated System Equations

General Second-Order Ordinary Differential Equation (ODE)	Solution for General Second ODE
$$\frac{d^2y}{dx^2} + P(x)\frac{dy}{dx} + Q(x)\,y = R(x)$$	$$y(x) = K_1\,y_1(x) + K_2\,y_2(x) + y_p(x)$$ $y_1(x)$ and $y_2(x)$ linearly independent solutions $y_p(x)$ particular solution (see Eqs. 4.3.60 and 4.3.61) K_1 and K_2 arbitrary constants

System Equations for General Second-Order ODE	System Equation Solutions for General Second-Order ODE:
$$de_1 = g_1(e_1, e_2, c)\,dc$$ $$de_2 = g_2(e_1, e_2, c)\,dc$$ $$g_1 = (A/D)[Fe_2(c) + G]$$ $$g_2 = -(A/F)\,[\{Fe_2(c) + G\}P(Ac + B)$$ $$+ \{D\,e_1(c) + E\}Q(Ac + B) + R(Ac + B)]$$	$$e_1(c) = \frac{1}{D}\,[y(A\,c + B) - E]$$ $$e_2(c) = \frac{1}{F}\left[\frac{D}{A}\frac{de_1}{dc} - G\right]$$ System equation transformations $$x = Ac + B, \quad y = De_1(c) + E, \quad dy/dx = Fe_2(c) + G$$ where A, B, D, E, F, G arbitrary constants

Constant Coefficient – Second-Order Ordinary Differential Equation	Solution for Constant Coefficient Second-Order ODE
$$a\frac{d^2y}{dx^2} + b\frac{dy}{dx} + f\,y = 0$$ where a, b, f arbitrary constants	$$y(x) = K_1\,\exp(m_1 x) + K_2\,\exp(m_2 x)$$ $$\left.\begin{array}{c} m_1 \\ m_2 \end{array}\right\} = \frac{b}{2a} \pm \frac{1}{2a}\sqrt{b^2 - 4ac}$$

System Equations Constant Coefficient Second-Order ODE	System Equation Solutions Constant Coefficient Second-Order ODE
(For g_1 see General Second-Order Differential Equation) $$g_2 = -[\{Fe_2(c) + G\}b/a + \{De_1(c) + E\}c/a]$$	(For e_2 see General Second-Order Differential Equation) $$e_1(c) = K_3\exp$$ $$(Am_1c)\left(1 + K4\left[\frac{\exp[(m_2 - m_1)Ac] - 1}{m_2 - m_1}\right]\right) + E/D$$ K_3, K_4, D, E arbitrary constants

Bessel Second-Order Ordinary Differential Equation	Solution Bessel Second-Order ODE
$$x^2\frac{d^2y}{dx^2} + x\frac{dy}{dx} + \left(x^2 - n^2\right)y = 0$$	$$y(x) = A_1 J_n(x) + A_2 Y_n(x)$$ $J_n(x)$ = Bessel function first kind order n $Y_n(x)$ = Bessel function second kind order n A_1 and A_2 arbitrary constants If $n \neq$ integer then $Y_n(x) \rightarrow J_{-n}(x)$

Table 4-2 (Continued)

System Equations for Bessel Second-Order ODE

(For g_1 see General Second-Order Differential Equation)

$$g_2 = -A/F[\{Fe_2(c) + G\}/(Ac + B) + \{De_1(c) + E\}\{1 - n^2/(Ac + B)^2\}]$$

System Equation Solutions for Bessel Second-Order ODE

(For e_2 see General Second-Order Differential Equation)

$$e_1(c) = A_1 J_n(Ac + B) + A_2 Y_n(Ac + B) - E/D$$

$A_1, A_2, D,$ and E are arbitrary constants

If $n \neq$ integer then $Y_n(x) \rightarrow J_{-n}(x)$

Modified Bessel Second-Order Ordinary Differential Equation

$$x^2 \frac{d^2y}{dx^2} + x \frac{dy}{dx} - (x^2 + n^2)y = 0$$

Solution for Modified Bessel Second-Order ODE

$$y(x) = A_1 I_n(x) + A_2 K_n(x)$$

$I_n(x)$ = Modified Bessel function of the first kind of order n

$K_n(x)$ = Modified Bessel function of the second kind or order n

A_1 and A_2 arbitrary constants

If $n \neq$ integer then $K_n(x) \rightarrow I_{-n}(x)$

System Equations for Modified Bessel ODE

(For g_1 see General Second-Order Differential Equation)

$$g_2 = -A/F[\{Fe_2(c) + G\}/(Ac + B) - \{De_1(c) + E\}\{1 + n^2/(Ac + B)^2\}]$$

System Equation Solutions for Modified Bessel ODE

(For e_2 see General Second-Order Differential Equation)

$$e_1(c) = A_1 I_n(Ac + B) + A_2 K_n(Ac + B) - E/D$$

$A_1, A_2, D,$ and E are arbitrary constants

If $n \neq$ integer, $K_{n \rightarrow} \rightarrow I_{-n})$

$$e_1(c) = A_1 G_{even}(Ac + B) + A_2 H_{odd}(Ac + B) - E/D$$

Legendre – Ordinary Differential Equation

$$(1 - x^2)\frac{d^2y}{dx^2} - 2x\frac{dy}{dx} + n(n + 1)y = 0$$

Solution for the Legendre ODE

$$y(x) = A_1 P_n(x) + A_2 Q_n(x)$$

System Equations for Legendre ODE

(For g_1 see General Second-Order Differential Equation)

$$g_2 = [2(Ac + B)(Fe_2(c) + G) - n(n + 1)(De_1(c) + E)]/\{1 - (Ac + B)^2\}$$

System Equation solutions for Legendre ODE

(For e_2 see General Second-Order Differential Equation)

$$e_1(c) = A_1 P_n(Ac + B) + A_2 Q_n(Ac + B) - E/D$$

$A_1, A_2, D,$ and E are arbitrary constants

Note: If $n \neq$ integer, $Q_{n \rightarrow} P_{-n}$

(Continued)

Table 4-2 (Continued)

Associated Legendre – Ordinary Differential Equation	Solution for the Associated Legendre ODE
$$(1-x^2)\frac{d^2y}{dx^2} - 2x\frac{dy}{dx} + \left[n(n+1) - \frac{m^2}{1-x^2}\right]y = 0$$	$$y(x) = A_1 P^m{}_n(x) + A_2 Q^m{}_n(x)$$

System Equations for Associated Legendre ODE	System Equation solutions for Associated Legendre ODE
(For g_1 see General Second-Order Differential Equation)	(For e_2 see General Second-Order Differential Equation)
$$g_2 = -\frac{A}{F}[2(Ac+B)(Fe_2(c)+G)/ \\ \{1+(Ac+B)^2\} - (De_1(c)+E)\{n(n+1) \\ -m^2\}/\{1(Ac+B)^2\}]$$	$$e_1(c) = A_1 P^m{}_n(Ac+B) + A_2 Q^m{}_n(Ac+B) - E/D$$ A_1, A_2, D, and E arbitrary constants

Euler Second-Order Ordinary Differential Equation	Solution for the Euler Second-Order ODE
$$ax^2\frac{d^2y}{dx^2} + bx\frac{dy}{dx} + fy = 0$$	$$y(x) = K_1 x^{m1}\left[1 + K_2\left\{x^{(m2-m1)} - 1\right\}/(m_2-m_1)\right]$$ $$\left.\begin{matrix}m_1\\m_2\end{matrix}\right\} = \frac{b-a}{2a} \pm \frac{1}{2a}\sqrt{(b-a)^2 - 4af}$$

System Equations for Euler Second-Order ODE	System Equation solutions for Euler Second-Order ODE
(For g_1 see General Second-Order Differential Equation)	(For e_2 see General Second-Order Differential Equation)
$$g_2 = -\frac{A}{F}[\{Fe_2(c)+G\}P(Ac+B) \\ + \{De_1(c)+E\}Q(Ac+B) + R(Ac+B)]$$	$$e_1(c) = K_1(Ac+B)^{m1}$$ $$\left[1 + K_2\left\{(Ac+B)^{(m2-m1)} - 1\right\}/(m_2-m_1)\right] - \frac{E}{D}$$

Hermite Second-Order Ordinary Differential Equation	Solution for Hermite Second-Order ODE
$$\frac{d^2y}{dx^2} - 2x\frac{dy}{dx} + 2ny = 0$$	$$y(x) = A_1 H_n(x) + A_2 H_{-n}(x)$$

System Equations for Hermite ODE	System Equation solutions for Hermite ODE solution
(For g_1 see General Second-Order Differential Equation)	(For e_2 see General Second-Order Differential Equation)
$$g_2 = -2A/F[\{Fe_2(c)+G\}(Ac+B) - n\{De_1(c)+E\}]$$	$$e_1(c) = A_1 H_n(Ac+B) + A_2 H_{-n}(Ac+B) - E/D$$

Table 4-2 (Continued)

Laguerre Second-Order Ordinary Differential Equation	Solutions for Laguerre Second-Order ODE

$$x\frac{d^2y}{dx^2} + (1-x)\frac{dy}{dx} + ny = 0$$

$$y(x) = A_1L_n(x) + A_2L_{-n}(x)$$

System Equations for Laguerre ODE	System Equation solutions for Laguerre ODE

(For g_1 see General Second-Order Differential Equation)

(For e_2 see General Second-Order Differential Equation)

$$g_2 = -A/F[\{Fe_2(c) + G\}P(Ac + B) + \{De_1(c) + E\}Q(Ac + B) + R(Ac + B)]$$

$$e_1(c) = A_1L_n(Ac + B) + A_2H_{-n}(Ac + B) - E/D$$

Mathieu Second-Order Ordinary Differential Equation	Solutions for Mathieu Second-Order ODE

$$(1-x^2)\frac{d^2y}{dx^2} - x\frac{dy}{dx} + \left[a - 2b(2x^2 - 1)\right]y = 0$$

(For g_1 see General Second-Order Differential Equation)

$$g_2 = -A/F\left[\{Fe_2(c) + G\}P(Ac + B) + \{De_1(c) + E\}Q(Ac + B) + R(Ac + B)\right]$$

System Equation solutions for Mathieu ODE

(For e_2 see General Second-Order Differential Equation)

$$e_1(c) = A_1G_{even}(Ac + B) + A_2H_{odd}(Ac + B) - E/D$$

Solutions in this form include conditions when $m_2 = m_1$.

Solution given by Lim $y(x)$ as $m_2 \rightarrow m_1$.

(1) System kernel $g_1 = (Fe_2 + G)A/D$ is same for any differential equation of form

$$y'' + Py' + Qy = 0 \text{ and for transformations given}$$
above for x, y, and dy/dx

(2) After $e_1(c)$ is determined, $e_2(c)$ is given by $e_2 = [(D/A) \, de/dc - G]/F$

---

where

$$y_p(x) = \int^{t=x} \frac{R(t)[y_2(x)\, y_1(t) - y_1(x)\, y_2(t)]}{y_1(t)\, y_2'(t) - y_2(t)\, y_1'(t)} \, dt \tag{4.3.60}$$

and K_1 and K_2 are arbitrary constants. We are guaranteed that if y_1 and y_2 are indeed linearly independent solutions of the homogeneous form of Eq. (4.3.57), then the denominator in

Eq. (4.3.60) [called the Wronskian for Eq. (4.3.57)] will not vanish. Furthermore, we now have for the complete solution for our single-input, double-output system equations [Eqs. (4.3.58a) and (4.3.58b)], with $R(c) \neq 0$, that

$$e_1(c) = K_1 y_1(x) + K_2 y_2(x) + y_P$$
$$e_2(c) = K_1 y_1'(x) + K_2 y_2'(x) + y_P'$$

where

$$y(x) = K_1 y_1(x) + K_2 y_2(x) + y_p(x) \tag{4.3.61}$$

In the actual application of Eqs. (4.3.60) and (4.3.61) to determine particular solutions, it is useful to observe that the Wronskian (or denominator of the integrand) can be expressed as

$$W(y_1, y_2) = y_2' y_1 - y_1' y_2 = \exp\left[-\int dx\right]$$

where $P(x)$ is defined as the coefficient in the original differential equation, Eq. (4.3.57), for which we sought a particular solution.

Example Problem with Linear Second-Order System Kernel.
Consider the following single-input, double-output system equation:

$$de_1 = e_2 dc$$
$$de_2 = [-e_1 - e_2/c + (2/\pi) bc]\, dc$$

We observe that both system kernels are linear in e_1 and e_2, the outputs; and upon review of Table 4-2, we see that if we set $b = 0$ [the homogeneous form for general equation (4.3.57)], we recognize that our system is equivalent to Bessel's differential equation and find that

$$e_1(c) = K_1 J_0(c) + K_2 Y_0(c) + y_p(c)$$
$$e_2(c) = -K_1 J_1(c) - K_2 Y_1(c) + y_p'(c)$$

Using Eq. (4.3.60) for y_p with $R(c) = b\, c$, we have that

$$y_p(x) = b\, Y_0(c) \int^{x=c} J_0(x)\, dx - b J_0(c) \int^{x=c} Y_0(x)\, dx$$

$$y_p'(x) = b J_1(c) \int^{x=c} Y_0(x)\, dx - b\, Y_1(c) \int^{x=c} J_0(x)\, dx$$

4.3.11 Equivalence of Single-Input System Equations with Ordinary Differential and Difference Equations of Any Order

It is possible to show that under very general conditions any finite nth order ordinary differential or difference equation may be transformed into a set of n first-order differential or difference equations, respectively. Furthermore, this set of transformed equations may then be associated with a single-input system with n outputs. To establish this observation for continuous systems, suppose that the general nth-order differential equation can be expressed explicitly in terms of the highest order derivative as follows:

$$\frac{d^n y}{dx^n} = H\left[x, y, \frac{dy}{dx}, \frac{d^2 y}{dx^2}, ..., \frac{d^{n-1} y}{dx^{n-1}}\right] \tag{4.3.62}$$

where x is the independent variable, y is the dependent variable, and H is an arbitrary function of the independent and dependent variables. Let us define a new set of dependent variables as follows:

$$e_1(x) = y(x)$$

$$\frac{de_1(x)}{dx} = \frac{dy(x)}{dx} = e_2(x)$$

$$\frac{de_2(x)}{dx} = \frac{d^2 y(x)}{dx^2} = e_3(x)$$

$$\frac{de_3(x)}{dx} = \frac{d^3 y(x)}{dx^3} = e_4(x)$$

Continuing, we have that

$$\frac{de_{n-1}(x)}{dx} = \frac{d^{n-1} y(x)}{dx^{n-1}} = e_n(x)$$

and finally that

$$\frac{de_n(x)}{dx} = \frac{d^n y(x)}{dx^n} = H\left[x, y, \frac{dy}{dx}, \frac{d^2 y}{dx^2}, ..., \frac{d^{n-1} y}{dx^{n-1}}\right]$$

With this set of defined dependent variables, we can now define a set of first-order differential equations equivalent to the nth-order differential equation:

$$c = x \quad \text{so} \quad dc = dx$$

$$e_1 = y$$

$$\frac{de_1}{dx} = e_2$$

$$\frac{de_2}{dx} = e_3$$

$$\frac{de_3}{dx} = e_4$$

and continuing we have

$$\frac{de_{n-1}}{dx} = e_n$$

$$\frac{de_n}{dx} = H[x, e_1, e_2, ..., e_n]$$

This set of first-order differential equations can now be expressed in a system matrix equation form as follows, where, for association convenience, we let $x = c$ and $dx = dc$

$$
\begin{bmatrix} de_1 \\ de_2 \\ \vdots \\ de_n \end{bmatrix} = \begin{bmatrix} e_2 \\ e_3 \\ \vdots \\ H(c, e_1, e_2, ..., e_n) \end{bmatrix} dc \tag{4.3.63}
$$

We see that this system matrix equation is simply a special case of the general single-input, multiple-output system equation where the system kernel now has the following special form:

$$
de = g(\mathbf{e}, c)\, dc, \quad \text{where} \quad g(\mathbf{e}, c) = \begin{bmatrix} e_2 \\ e_3 \\ \vdots \\ H(c, e_1, e_2, ..., e_n) \end{bmatrix}
$$

Thus, we conclude that any set of ordinary differential equations of any finite order that can be expressed within the general form of Eq. (4.3.62) also represents an associated single-input, multiple-output system where the combined order of the differential equations is equal to the number of system outputs. With a similar development, it is possible to show that under general conditions, any set of finite order ordinary difference equations is expressible as an associated set of single-input, multiple-output difference equations each of the first order.

Let us consider an example to demonstrate the equivalence of atypical higher-order ordinary differential equation and its associated single-input, multiple-output system equation. A second-order ordinary differential equation governs the displacement x of the mass m for the simple "mass-spring-friction" vibration system shown in Figure 4-2.

From Newton's second law, we have that the algebraic sum of the external forces imposed on the mass m is equal to the mass times the acceleration of the mass, so for the horizontal displacement of the mass, we have

$$
m\frac{d^2x}{dt^2} = F\left(x, \frac{dx}{dt}, t\right) - kx - c\frac{dx}{dt}
$$

where F is an external force applied to the mass, $k\,x$ is the resisting spring force, and $c\,(dx/dt)$ is the resisting frictional force. If we define new dependent variables as follows,

$$
e_1 = x \quad \text{(the mass displacement)}
$$
$$
e_2 = \frac{dx}{dt} = \frac{de_1}{dt} \quad \text{(the mass velocity)}
$$

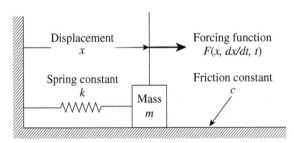

Displacement
x

Forcing function
$F(x, dx/dt, t)$

Spring constant
k

Friction constant
c

Mass
m

Fig. 4-2 Simple mass-spring-friction vibration system.

then with time as the input to the system, our system equations become

$$de_1 = e_2\, dt$$
$$de_2 = (1/m)[F(e_1, e_2, t) - c\, e_2 - k\, e_1]dt$$

In matrix form, we have that

$$\begin{bmatrix} de_1 \\ de_2 \end{bmatrix} = \begin{bmatrix} e_2 \\ \dfrac{1}{m}[F(e_1, e_2, t) - c\, e_2 - k\, e_1] \end{bmatrix} dt \tag{4.3.64}$$

where the single-input, double-output matrix system kernel is given by

$$g(\mathbf{e}, t) = \begin{bmatrix} g_1 \\ g_2 \end{bmatrix} = \begin{bmatrix} e_2 \\ \dfrac{1}{m}[F - c\, e_2 - k\, e_1] \end{bmatrix} \tag{4.3.65}$$

The block diagram representation of Eq. (4.3.64) for the system is as follows:

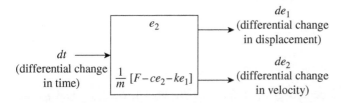

In this example, the system output variables e_1, and e_2 have physical identity and are the mass displacement and velocity, respectively. In other applications, the system output variables assigned to the dependent variables and higher-ordered derivatives may or may not be physically familiar or even have direct physical significance. However, this unfamiliarity with contrived variables should not impede our ability and efforts to model such systems.

It has been stated previously that it is difficult or even impossible to determine whether the observed system kernel for a real system inherently exhibits internal feedback. For example, for this simple mass-spring-friction vibration system, let us assume that the system components g_1 and g_2 defined by Eq. (4.3.65) actually dc exhibit feedback so that

$$g_1 = e_2 = g_{1f} \tag{4.3.66.1}$$
$$g_2 = (1/m)[F(e_1, e_2, t) - c\, e_2 - k\, e_1] = g_{2f} \tag{4.3.66.2}$$

where

$$g_{1f} = \frac{g_1*}{1 - g_1*f_1 - g_2*f_2} \tag{4.3.67.1}$$

$$g_{2f} = \frac{g_2*}{1 - f_1 g_1*f_1 - g_2*f_2} \tag{4.3.67.2}$$

and g_1* and g_2* are the system kernel components without feedback, and f_1 and f_2 are the feedback kernel components. The block diagram for our simple mass-spring-friction vibration system now appears as follows:

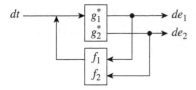

Now let us express our sets of equations [i.e., Eqs. (4.3.67.1) and (4.3.67.2)] for g_{1*} and g_{2*} as follows:

$$g_{1*}\left(1 + f_1 g_{1f}\right) + g_{2*} f_2 g_{1f} = g_{1f}$$

$$g_{1*} f_1 g_{2f} + g_{2*}\left(1 + f_2 g_{2f}\right) = g_{2f}$$

Solving this set of equations for g_{1*} and g_{2*}, we find that

$$g_{1*} = \frac{g_{1f}}{1 + f_1 g_{1f} + f_2 g_{2f}} \qquad (4.3.68.1)$$

$$g_{2*} = \frac{g_{2f}}{1 + f_1 g_{1f} + f_2 g_{2f}} \qquad (4.3.68.2)$$

Recalling the actual definitions for our vibration system kernels as given by Eqs. (4.3.66) and imposing these definitions on Eqs. (4.3.68), we finally find that

$$g_{1*} = \frac{m\, e_2}{m(1 + f_1\, e_2) + f_2(F - c\, e_2 - k\, e_1)}$$

$$g_{2*} = \frac{(F - c\, e_2 - k\, e_1)}{m(1 + f_1\, e_2) + f_2(F - c\, e_2 - k\, e_1)}$$

Now for any finite-valued functions, f_1 and f_2 for these feedback kernels, g_{1*} and g_{2*} are the appropriate components for our vibration system kernel and will produce the externally observed system equations [i.e., Eqs. (4.3.66)] when g_{1*} and g_{2*} are substituted into Eqs. (4.3.67) regardless of the specified feedback kernels f_1 and f_2. It is interesting to observe that the ratios of the respective system kernel components

$$\frac{g_{1f}}{g_{2f}} = \frac{g_{1*}}{g_{2*}} = \frac{m\, e_2}{(F - c\, e_2 - k\, e_1)}$$

are equal regardless of whether the feedback is present or not. We will more fully examine the influence of feedback on systems in subsequent chapters.

4.4 Treatment of Single-Input Systems Using MATLAB Symbolic Toolbox

We have examined the major manual analytical methods for solving system equations with a single input. However, the symbolic math toolbox in MATLAB can often provide equivalent analytical results equivalent to many of the analytical methods used in system science for single input system equations. The computation capability (algorithms) for these symbolic operations is provided primarily by MAPLE software developed by Waterloo Maple, Inc. A partial list of those symbolic operations that are of particular interest for modeling of systems and their analysis are shown in Table 4-3.

These symbolic tools provide the system investigator with significant capabilities in manipulating and solving the system equations encountered in system analysis. The results of these symbolic operation tools are of great value since the results are those that would be

Table 4-3 MATLAB Symbolic Operations for System Modeling

Subject	Operations
Calculus	Differentiation, integration, limits, summation, Taylor series
	Linear algebra
	Inverses, determinants, eigenvalues, singular value decomposition and canonical forms of symbolic matrices
Simplification	Simplification of algebraic expressions
Solutions	Symbolic solution of algebraic and differential equations
Special functions	Special functions in applied mathematics
Transforms	Laplace, Fourier, and z-transforms and their inverses

obtained by analytical processing by the system investigator. Indeed, much of the labor and skills normally required by the investigator are available using these symbolic tools. Of course, there are limitations to these symbolic tools, and they cannot analytically process or solve any system equation that may be attempted using these symbolic tools. Nevertheless, these tools greatly enhance the system investigators capabilities and generally provide independent verification of the investigator's analytical results obtained manually.

We restrict our use of these symbolic tools primarily to those operations of principal interest for system analysis, namely, the solution of systems of differential equations composing the system equations. Of course, the symbolic tools are very useful in manipulating and examining the system equations under investigation. For example, finding maximum and minimum values for the system equation, u is easily accomplished using the toolbox function "diff" to perform symbolic differentiation

$$v = \text{diff}(u)$$

where u is the function $u(x)$ to be differentiated with respect to x.

The resulting function v is set to zero to obtain the maximum, minimum, and points of infection for $u(x)$. This is easily accomplished using the toolbox function "solve" as follows

$$v = 0$$

and

$$w = \text{solve}(v)$$

Of course, the extremum found, i.e., the minimum, maximum, or point of inflection, can be assessed by finding the value of the derivative of v at the point identified for each extremum value.

MATLAB also provides functions for working with difference equations that are encountered in discrete model equations. The function "filter" exemplified below

$$y = \text{filter}(a, b, x)$$

processes data for the vector dependent variable $x(n)$ and discrete vector parameters $a(n)$ and $b(n)$. For example, consider the following difference equation

$$a(n)y(n) = a(1)y(1) + a(2)y(2) + \cdots + a(n-1)y(n-1)$$
$$+ b(1)x(1) + b(2)x(2) + \cdots + b(n-1)x(n-1)$$

The system investigator often encounters analytical expressions that required integration. If $u(x)$ is a symbolic expression, then

$$v = \text{int}(u,x)$$

attempts to find the related symbolic expression $v(x)$, which is the indefinite integral or anti-derivative of u for the variable of integration x provided that $v(x)$ exists in closed form and MATLAB can determine the indefinite integral. The constant of integration is implied must be and added by the investigator. If MATLAB is unable to integrate the expression $u(x)$, then the initial command is simply returned as the default function

$$v = \text{int}(u,x)$$

Undoubtedly, a very important symbolic operator is that for analytically solving differential equations. The symbolic function "dsolve" computes the symbolic solution for a system of differential equations. For example, the system of first-order differential equations

$$Df1 = g1(x), \quad Df2 = g2(x), \quad Df3 = g3(x), \tag{4.4.1}$$

where the symbol D denotes the differential operator $D = d/dx$ is solved using the symbolic operator for "dsolve" as follows

$$S = \text{dsolve}('Df1 = g1(x)', 'Df2 = g2(x)', 'Df3 = g3(x)', ...) \tag{4.4.2}$$

The computed solutions are returned in the symbolic structure S, and the values for the f 1, f 2, f 3,... are returned by entering each function

$$f1 = S.f1 \tag{4.4.3}$$

and MATLAB responds with the solution as

$$f1 = \text{solution for} f1 \tag{4.4.4}$$

and repeating for the remaining solutions for each subsequent fn. It is also possible to impose initial conditions on the solutions, as shown in the following example.

$$[f1, f2, ...] = \text{dsolve}('Df1 = g1(x)', 'Df2 = g2(x)', 'f1(0) = f10, f2(0) = f20, ...)$$

Symbolic matrix algebraic operations such as addition, subtraction, and multiplication are provided for matrices that satisfy the proper rules for conformance of rows and columns. Furthermore, transpose and inverse matrices (non-singular matrices) can also be evaluated symbolically. These matrix operations are particularly useful when processing multi-input and or multi-output system equations since the results preserve the symbolic expressions for the original system equations. Thus the symbolic parameters are preserved for easy identification.

Other symbolic toolbox operations that are very useful and will be demonstrated in examples later in this text are the transform operators for Laplace and Fourier transforms. For example, the Laplace transform, $L(s)$, of the function $F(t)$ with respect to the default variable t is found using the symbolic operator "laplace" as follows

$$L(s) = \text{laplace}(F)$$

and to invert the Laplace transform $L(s)$, the symbolic operator "ilaplace" returns the original function $F(t)$ as follows

$F(t) = \text{ilaplace}(L)$

Many other useful symbolic operators are available in both the MATLAB toolbox and the MAPLE symbolic operator menu. For the full complement of these operators the reader is referred to the appropriate literature for these symbolic packages. We shall demonstrate a few of these examples and problems presented throughout this text.

4.5 Summary

In this chapter, we have examined the major analytical methods for solving system equations with a single input. We observed that any discrete system equation could be solved by sequential steps if the system kernel was known. For continuous systems, system equations with kernels that were constant, linear, exact, separable, and homogeneous were examined. For the single-input, single-output system, special kernel forms that included the Bernoulli and Ricatti types were treated. Multiple-output, single-input systems were also examined with autonomous kernels (kernels not explicit functions of the single input).

The equivalence of sets of higher-order ordinary difference equations with multiple-output system equations was established. Finally, the utility of known solutions for various second-order ordinary differential equations was demonstrated for solving related single-input, double-output system equations.

Bibliography

ABRAMOWITZ, A. and I. STENGUN, *Handbook of Mathematical Functions with Formulas, Graphs, and Mathematical Tables*, National Bureau of Standards, Paperback, 1 June 1965. Preeminent handbook of mathematical materials.

ALEXANDER, J. E., and J. M. BAILEY, *Systems Engineering Mathematics*, Prentice-Hall, Englewood Cliffs, NJ, 1963. An introduction to mathematical methods for engineering systems.

BLAQUIERE, A., *Nonlinear System Analysis*, Academic Press, New York, 1966. A resource for the analysis of nonlinear system equations.

DAVIS, H. T., *Nonlinear Deferential and Integral Equations*, Dover Publications, New York, 1962. An excellent, indepth treatment of analytical methods for nonlinear differential and integral equations.

FREEMAN, H., *Discrete-Time Systems*, John Wiley & Sons, New York, 1965. The modeling and analysis of discrete system models are presented in this book.

HOCHSTADT, H., *Differential Equations*, Dover Publications, New York. 2012. An excellent advanced treatment of analytical methods in differential equations.

INCE, E. L., *Ordinary Differential Equations*, Dover Publications, New York, 1956. An essential reference for ordinary differential equations.

KALMAN, R., P. FALB, and M. ARBIB, *Topics in Mathematical System Theory*, McGraw-Hill Book Co., New York, 1969. An advanced treatment of mathematical methods in system science.

KAPLAN, W., *Ordinary Differential Equations*, Addison-Wesley Publishing Co., Reading, MA, 1958. A standard, well-respected treatment of analytical methods for solving ordinary differential equations.

KREYSZIG, E., *Advanced Engineering Mathematics*, 10th edition, John Wiley & Sons, New York, 2011. The 10th Edition of this comprehensive text is an excellent reference for analyzing mathematical problems in this text.

Symbolic Math Toolbox, *User's Guide*, The MathWorks, Inc., various versions, Provides the software and documentation for utilizing the Symbolic Math Toolbox.

SNEDDON, I. N., *Fourier Transforms*, 1st edition, McGraw-Hill Book Co., New York, 1951. A standard, respected introduction to Fourier transform methods.

THOMPSON, W. T., *Laplace Transformation, Theory and Engineering Applications*, Prentice-Hall, Englewood Cliffs, NJ, 1950. Many good applications of Laplace transforms to systems analysis.

WHITE, H. J., *Systems Analysis*, W. B. Saunders Co., Philadelphia, 1969. General reference for analysis in system science.

WIDDER, D. V., *The Laplace Transform*, Princeton University Press, Princeton, NJ, 1941. A classical standard primer to Laplace transform methods.

WYMORE, A. W., *A Mathematical Theory of Systems Engineering: The Elements*, John Wiley & Sons, New York, 1967. Useful reference for mathematical modeling and analysis of systems addressed in this text.

Problems

1 Attempt to solve following linear system equations either manually or use MATLAB Symbolic Toolbox referenced in Section 4.4. (These problems may defy analytic solution so identify them and suggest further action.)

a) $dy = (x + 1 + ay)\dfrac{dx}{x}$

b) $de = \dfrac{(a + be)}{(f + c)} dc$

c) $dy = \left(\dfrac{a + y}{x + x^2}\right) dx$

d) $de = [f(c) + g(c)e]dc$

e) $\dfrac{dr}{ds} + ar = f(s)$

f) $de = \dfrac{f(e)}{[g(e) + c]} dc$

g) $z\,dz + [ayz^2 + f(y)]dy = 0$

$\left(\text{Hint : let } z^2 = u\right)$

h) $de = e[A + B\operatorname{Ln}e]\dfrac{dc}{c}$

$(\text{Hint : let } \operatorname{Ln} e = u)$

i) $y + y\sin x = \cos x$

Recall $y' = \dfrac{dy}{dx}$

j) $de = \left[\dfrac{Ac}{(e + ac)^2} + Bc^2(e + ac)\right]dc$

$\left[\text{Hint : let}(e + ac)^2 = u^2\right]$

k) $de = (ac + be)\dfrac{dc}{c}$

l) $de = \dfrac{(a + e)}{(e^2 + bc)} dc$

m) $y' \exp(y) = [\exp(y) - x]\dfrac{1}{x^2}$

$[\text{Hint : let } \exp(y) = u]$

n) $\exp(y') = Ayf(x)$

(Try logs)

o) $de = \dfrac{g}{(1 - gf)} dc$

$[(\text{where } f = a + bc,$

what is g for linearity?])

p) $x\,dx = (ax^2 + b)\dfrac{dt}{t}$

q) $dy = y(a + bx\operatorname{Ln}y)dx$

(what if $ac + be = 0$?)

r) $(ac + be)de = (ac + be)^2 dc$

s) $dn = \left[m^2 + \dfrac{n}{(m)}\right]dm$

(Is this defined? Why?)

t) $\operatorname{Ln}[dy + (x + y)dx] = x$

u) $s\,d(\operatorname{Ln}s) + (a + bt)dt = 0$

(Is this equation defined?)

v) $Ln[dy + (x+y)dx] = Ln(dx)$

w) $d\left(\dfrac{1}{y}\right) = (ax + by^2)dx$

x) $d[f(y)] = [g(x) + h(x)f(y)]d[e(x)]$

y) $Ke\,ds + \left[K + b - \dfrac{ae}{c}\right]e\,dt$

$= [K(as + bt)]e\,dt$

(Try : let $as + bt = u, cs + bt = v$)

z) $dy = \left[E(x) + F(x)\int G(y)\,dy\right]dx$

[(See problem(x)above)]

2 Attempt to solve the following separable system equations either by manual calculations or using the MATLAB Symbolic Toolbox referenced in Section 4.4.

a) $dy = \dfrac{[(1+x)y]}{[(1+y)x]}dx$

b) $[(A+e)(B+c)]de = [(a+e)(b+c)]dc$

c) $xy\,dy = a(x\,dy + 2y\,dx)$

d) $de = ac^2\dfrac{(e+b)}{[e^2c+f)]}dc$

e) $ds = \dfrac{[(a+bs)t]}{[(c+t)s]}dt$

f) $dy = \{Ln\,f(x) + Ln\,g(y)\}dx$

g) $dy = (ax + by + c)^n dx$

(Hint : let $ax + by + c = u$)

h) $\exp\left(\dfrac{dy}{dx}\right) = F(x)G(y)$

i) $de = \dfrac{[A + Bc - Dc^2]}{[E + Fe + Ge^2]}dc$

j) $dx = (a + xy)^2 + \dfrac{a}{y^2}\,dy$

$\left(Hint : let\,u = x + \dfrac{a}{y}\right)$

k) $s\exp(s + t)ds = dt$

l) $de = \dfrac{A[c^a(B + e)]}{[e^b(D + c)^m]}dc$

m) $z' = 3u^2(1 + z)$

n) $(A + Bz^2)dy = Czy^2 dz$

o) $x^3 dx + (x)^{\frac{1}{2}}y\,dy = x^2 y\,dy$
    $+ \sin(c)\cos(e)\}]\,dc$

p) $de = [1 - \{\sin(e)\cos(c)$

3 Attempt to solve the following homogeneous system equations (where $y' = \dfrac{dy}{dx}$) either by manual calculations or using the MATLAB Symbolic Toolbox referenced in Section 4.4. (Note: The problems may defy solution.)

a) $(x^2 + y^2)dy = 4xy\,dx$

b) $de = \dfrac{Ae}{[Be + c]}dc$

c) $tds = \left[s + (s^2 + t^2)^{\frac{1}{2}}\right]dt$

d) $\sqrt{e}\,dc = \sqrt{e + c}\,de$

e) $y'[a\sqrt{xy} - x] + x + y = 0$

f) $(ax + by)dy = (cx + ey)dx$

g) $y' = Ln\left(\dfrac{y}{x}\right) + \dfrac{y}{x}$

h) $d(ax + by)^2 = d(cx + ey)^2$

i) $de = \left[\dfrac{e}{c} - 3\left(\dfrac{e}{c}\right)^2\right]dc$

j) $d\left(\dfrac{1}{y}\right) = \dfrac{dx}{[(x + y)(x - y)]}$

k) $(x + y)dy = (x - y)dx$

l) $de = \left[\dfrac{(Ae + Bc)}{(De + Ec)}\right]^2 dc$

m) $dy = \dfrac{(ax + by + c)}{(Ax + By + C)}dx$

n) $dp = \left[A + B\left(\dfrac{p}{q}\right)^a + C\left(\dfrac{p}{q}\right)^b\right]dq$

o) $xy' - y = 2\sqrt{x^2 - y^2}$

p) $(As^2 + Br)d\left(\dfrac{r}{s}\right) = (cs^2 + Dr)ds$

4 Attempt to solve system equations by determining an integrating factor if it exists. Identify integrating factor used.

a) $dy = x[y + x^2 + y^2]dx$

b) $dy = [xy + A(x-1)]dx$

c) $e\,dc - c\,de + Lnc\,dc = 0$

d) $(A + By)dy = (Cxz + Dxy + E)dx$

e) $(1 + uv)v\,du + (1 - uv)adv = 0$

f) $dy = (Ax^2y + B\,y^2x)dx$

g) $dy = \left[\dfrac{x^2y}{(x^3 + y^3)}\right]dx$

h) $de = \left[\dfrac{a + bc^2}{(e + ac)}\right]dc$

i) $de = \left[\dfrac{(ec - e^2)}{c^2}\right]dc$

j) $y^2dy = [A + Bx + Cx^2y]dx$

k) $(t + s^2)dt + 2st\,ds = 0$

l) $dy = \left[A + B\left\{1 + \dfrac{e}{(Ac)}\right\}^2\right]dc$

m) $(3x^2 + 6xy^2)dx$
$+ (6x^2y + 4y^3)dy = 0$

n) $de = \dfrac{\left(\dfrac{Bc}{e^n} + \dfrac{m}{c}\right)}{\left(\dfrac{Ae}{c^m} + \dfrac{n}{e}\right)}dc$

o) $0 = e(ex^2 + fx + ay^2)dx$
$+ (bex^2 + afy + aby^2)dy$

5 Attempt to solve these Bernoulli system equations manually or use MATLAB Symbolic Toolbox referenced in Section 4.4.

a) $\dfrac{dy}{dx} + \dfrac{y}{x} = y^2$

b) $dr = ar\left(1 + \dfrac{s}{r^2}\right)ds$

c) $dx = [x^4 \sec y + x \tan y]dy$

d) $de = \dfrac{f(e)}{\left(c + \dfrac{e}{c}\right)}dc$

e) $zy' + y = y^2 Ln\,z$

f) $du = \left[v^2u + \left(\dfrac{u}{v}\right)^{\frac{1}{2}}\right]dv$

g) $3y^2y' + y^3 = x - 1$

h) $du = v(u + a)(u + v)dv$

i) $dx = \dfrac{[2x - ax^2y^n]}{y}dy$

j) $\dfrac{du}{dv} = av(u + K) + u^2 - K^2$

k) $du + [u \cos v - u^n \sin 2v]dv = 0$

l) $de = (e + a)\,[e + f(c)]g(c)$

m) $de = \left[\cos c - \dfrac{e^2}{2}\right]dc$

n) $dx = 2xy\,(ax^2y^2 - 1)dy$

6 Attempt to solve system equations manually or use MATLAB Symbolic Toolbox referenced in Section 4.4.

a) $(y + 2xy^2 - x^2y^3)dx = -2x^2y\,dy$

b) $de = Ae^2(1 + ec)dc$

c) $ts' + 2(1 + t^2)s = 1$

d) $dm = \dfrac{(am + bn + c)}{(Am + Bn + C)}dn$

e) $de = \left(\dfrac{c}{e - e}\right)dc$

f) $dy = \left(A + \dfrac{By}{x} + \dfrac{Cx}{y}\right)dx$

g) $dy = \dfrac{(y - xy^2)}{(x^2y - x)}dx$

h) $dx = \left[1 + \dfrac{1}{(x^2y^2)}\right]\dfrac{dy}{y^2}$

i) $e^t(t^2 + s^2 + 2t)dt + 2s\,e^t ds = 0$

j) $dA = \exp\left[\dfrac{(A + aB)}{(A - aB)}\right]dB$

k) $(x^2 + y^2 + 2x)dy = 2y\,dx(l)ds =$
$a\dfrac{s}{t}(st^2 - b)dt$

m) $dW = \left[\left\{\dfrac{(2W + 3X - 4)}{(W - 2X + 3)}\right\}^2 + 4\right]dX$

n) $(2\ x\ y^4 + \sin\ y)dx + (4\ x^2y^3 + x$
$\cos\ y)dy$

o) $dx = \dfrac{x}{t}\left[1 + \dfrac{a}{(xt)^2}\right]dt$

7 Classify and attempt to solve the following single-input, multiple-output system equations either by manual calculations or using the MATLAB Symbolic Toolbox.

a) $de_1 = (Ae_1 + B)dc$

$de_2 = (De_1 + F)dc$

b) $dx = (Ay + B)dt$

$dy = (Cx + D)dt$

c) $du = (au + bv)dx$

$dv = (cu + ev)dx$

$dx = (Ax + By + Cz)dt$

d) $dy = (Dx + Ey)dt$

$dz = Gz\,dt$

e) $de_1 = \dfrac{(A + Bc)}{e_1^2}\,dc$

$de_2 = \dfrac{(D + Ec^2)}{e_2}\,dc$

f) $de_1 = (A + Be_1 + Ce_2)dc$

$de_2 = (D + Ce_2)dc$

$de_1 = (2 + 3e_1 + 2e_2)dc$

g) $de_2 = (3 + 4e_2 - 3e_3)dc$

$de_3 = (5 + 7e_1 + 4e_3)dc$

h) $de_1 = (A_1 + B_1e_1 + C_1e_2)f(c)dc$

$de_2 = (A_2 + B_2e_1 + C_2e_2)f(c)dc$

i) $\dfrac{dx}{dt} = (1 + ax)(1 + by)f(t)$

$\dfrac{dy}{dt} = (1 + cx)(1 + ey)f(t)$

j) $de_1 = f(e_1)f(e_3)F(c)dc$

$de_2 = (A + Be_1 + Ce_2)dc$

$de_3 = g(e_1)g(e_3)F(c)dc$

k) $de_1 = \left(\dfrac{e_1}{c} + 2\dfrac{e_2}{c}\right)dc$

$de_2 = \left(\dfrac{e_1}{c} - 3\dfrac{e_2}{c}\right)dc$

l) $du = \left(A_1 + B_1\dfrac{u}{x} + C_1\dfrac{v}{x}\right)dc$

$dv = \left(A_2 + B_2\dfrac{u}{x} + C_2\dfrac{v}{x}\right)dc$

m) $\dfrac{de_1}{dt} = 2e_1 + 3e_3$

$\dfrac{de_2}{dt} = e_1 + 2e_2 - 3e_3$

$\dfrac{de_3}{dt} = e_3 - 4e_1$

n) $de_1 = (2e_1 + 3e_2 - 4e_3)dc$

$de_2 = (5e_1 + 6e_2 - 3e_3)dc$

$de_3 = (4e_1 + 2e_2 - 4e_3)dc$

o) $du = v\,dx$

$dv = \left[\left(\dfrac{a}{x^2} - b\right)u - \dfrac{v}{x}\right]dx$

(Hint : Bessel's equation!)

p) $dx = y\,dt$

$dy = \left[\dfrac{(2ty - ax)}{(1 - t^2)}\right]dx$

(Hint : Mathieu's equation!)

q) $de_1 = e_2dc$

$de_2 = \left[\dfrac{Ae_2}{(c + D)} + \dfrac{Be_1}{(c + D)^2}\right]dx$

(Hint : Euler's equation!)

r) $du = A\left[\left(1 - \dfrac{1}{c}\right)u - 4\dfrac{v}{c}\right]dc$

$dv = Au\,dc$

(Hint : Laguerre's equation!)

s) $de_1 = (A_1 + e_1)f_1(c)dc$

$de_2 = (A_2 + e_2)f_2(c)dc$

$de_3 = (A_3 + e_3)f_3(c)dc$

t) $de_1 = (A_1 + B_1ce_1 + D_1c^2e_2)dc$

$de_2 = (A_2 + B_2c^2e_1 + D_2ce_2)dc$

u) $\dfrac{dw}{dt} = Aw$

$\dfrac{dx}{dt} = Cw + Dx$

$\dfrac{dy}{dt} = Ey + Fw + Gx$

$\dfrac{dz}{dt} = Hz + ly + Jw + Kx$

v) $de_1 = g(a_1e_1 + b_1e_2 + c_1)dt$

$de_2 = h(a_2e_1 + b_2e_2 + c_2)dt$

(g and h functions of arguments)

w) $de_1 = (A_1 + B_1ce_1 + D_1ce_2)dc$

$de_2 = (A_2 + B_2ce_1 + D_2ce_2)dc$

x) $\dfrac{dx}{dt} = f_1(x)g_1(y)h_1(z)$

$\dfrac{dy}{dt} = f_2(x)g_2(y)h_2(z)$

$\dfrac{dz}{dt} = f_3(x)g_3(y)h_3(z)$

y) $de_1 = (Ae_1^2 + Be_2^2)dc$

$de_2 = (Ce_2 + De_1)e_1dc$

z) $d\begin{bmatrix} e_1 \\ \vdots \\ e_5 \end{bmatrix} = \begin{bmatrix} A_1e_1 \\ \vdots \\ A_5e_5 \end{bmatrix}$

8 Verify that the following general solutions, where k_1 and k_2 are arbitrary constants.

$$k_1 = f_1(x) + g_2(y) - h_1(t)$$
$$k_2 = f_2(x) + g_2(y) - h_2(t)$$

address the following system equations for t as the single input:

$$d\begin{bmatrix} x \\ y \end{bmatrix} = \frac{1}{D} \begin{bmatrix} g_2' \, h_1' - g_1' \, h_2' \\ f_1' \, h_2' - f_2' \, h_1' \end{bmatrix} dt$$

where

$$D = f_1'(x)g_2'(y) - f_2'(x)g_1'(y)$$

and for any variable t [*note for any* $F(t)$, $F'(t) = \dfrac{dF}{dt}$]

9 Using the results of problem 8, solve the following systems:

a)

$$de_1 = \frac{[(A_1 + B_1 e_2)c - (D_1 + E_1 e_2)]}{F(e_1, e_2)} dc$$

$$de_2 = \frac{[(A_2 + B_2 e_1)c - (D_2 + E_2 e_1)]}{F(e_1, e_2)} dc$$

where $F = (A_2 + B_2 e_1)(D_1 + E_1 e_2) - (D_2 + E_2 e_1)(A_1 + B_1 e_2)$.

b)

$$d\begin{bmatrix} u \\ v \end{bmatrix} = \frac{1}{G(u,v)} \begin{bmatrix} u^3 x^2 - \dfrac{u}{v} \\ v^2 x^2 - \dfrac{1}{vx} \end{bmatrix}$$

where $G = v^2 u - \dfrac{u^3}{v}$.

10 Verify the following general solutions, where k_1 and k_2 are arbitrary constants

$$k_1 = f_1(x)g_1(y)h_1(t)$$
$$k_2 = f_2(x)g_2(y)h_2(t)$$

define the following system equations:

$$d\begin{bmatrix} x \\ y \end{bmatrix} = \frac{1}{G_1 F_2 - F_1 G_2} \begin{bmatrix} G_2 H_1 - G_1 H_2 \\ F_1 H_2 - F_2 H_1 \end{bmatrix} dt$$

where

$$G_1 = d\text{Ln}\frac{g_1(y)}{dy} \qquad G_2 = d\text{Ln}\frac{g_2(y)}{dy}, \qquad F_1 = d\text{Ln}\frac{f_1(y)}{dx}$$

$$F_2 = d\text{Ln}\frac{f_2(y)}{dx}, \qquad H_1 = d\text{Ln}\frac{h_1(y)}{dt}, \qquad H_{12} = d\text{Ln}\frac{h_2(y)}{dt}$$

Compare this result with that of problem 8.

11 Convert the following higher-order differential equations to appropriate sets of system equations in terms of outputs e_1, e_2,..., and input c. Attempt to solve these system equations either by manual calculations or using the MATLAB Symbolic Toolbox referenced in Section 4.4 in this chapter.

a) $\dfrac{d^2x}{dt^2} + a\dfrac{dx}{dt} + bx = 0$

b) $\dfrac{d^2e}{dc^2} + ce = A$

c) $\dfrac{d^3x}{dt^3} + x^2y = 0$

d) $\dfrac{d^2e}{dc^2} + c\dfrac{de}{dc} + e = 0$

e) $\dfrac{d^2u}{dy^2} + cu = f(y)$

f) $\dfrac{d^nw}{dt^n} = f(w)$
(where n is a positive integer)

g) $\dfrac{dx}{dt} = ax + by$

$\dfrac{d^2y}{dt^2} = cx + ey$

h) $\dfrac{d^2u}{dx^2} = Au + Bv$

$\dfrac{d^2v}{dx^2} = Cu + Dv$

i) $\dfrac{d^2s}{dt^2} + as\dfrac{ds}{dt} + bs^2 = 0$

j) $\left(\dfrac{1}{x^3}\right)\dfrac{d^3y}{dx^3} + \left(\dfrac{a}{x}\right)\dfrac{dy}{dx} + by = 0$

k) $\dfrac{d^3y}{dx^3} + P(x)\dfrac{d^2y}{dx^2}$

$Q(x)\dfrac{dy}{dx} + R(x)y = S(x)$

l) $\dfrac{dy}{\left[y + \dfrac{d_2y}{dx^2}\right]} = dx$

m) $\dfrac{du}{dt} + a\dfrac{dv}{dt} = bu + cv$

$\dfrac{du}{dt} + e\dfrac{dv}{dt} = fu + gv$

n) $0 = f(z) + g(t)\dfrac{dt}{dz}$

o) $Y + f(x) + dx = 0$

p) $\dfrac{d^2s}{dt^2} + a\dfrac{ds}{dt} + bs^3 = 0$

q) $\dfrac{dr}{dt}[a\exp(r)] + bt = f(t)$

r) $\dfrac{d^2m}{dn^2} = f(n)\left[\dfrac{dm}{dn}\right]^3$

s) $\dfrac{d^2x}{dt^2} + x^ny = 0$

t) $y'' - n(1 - y^2)y' + y = f(x)$

u) $dzy + Pn(dz\sim + xy = 0$

v) $\dfrac{d^2w}{dt^2} = Ax^2, \dfrac{d^2w}{dt^2} = By^2$

$\dfrac{d^2y}{dt^2} = Cz^2, \dfrac{d^2z}{dt^2} = Dwxyz$

w) $\dfrac{d^2y}{dx^2} + Ln\left(\dfrac{dy}{dx}\right) + xy = 0$

$\dfrac{d^2z}{dx^2} + \exp\left(\dfrac{dy}{dx}\right) + xz = 0$

x) $\dfrac{dX}{dt} = AX + BY + C$

where matrix $X = [x_1...x_n]^T$

$\dfrac{dY}{dt} = DX + EY + F$

where matrix $Y = [y_1...y_n]^T$

y) $F\left(x, y, \dfrac{dy}{dx}\right) = 0$

z) $G(x)\dfrac{d^2y}{dx^2} + \exp(y)H(x)\dfrac{dy}{dx} = 0$

$H(x)\dfrac{dy}{dx} + \exp(y)G(x) = 0$

12 Given that the general second-order differential equation

$$\frac{d^2y}{dx^2} + B\frac{dy}{dx} + Ay = 0$$

is equivalent to

$$\frac{de}{dc} = A - Be + e^2$$

where

$$e - \left(\frac{1}{y}\right)\frac{dy}{dx} \text{ and } c = x$$

Using the results of Section 4.3.10 show that the following first-order Ricatti differential equations are associated with the second-order differential equations specified. Obtain solutions for each system where possible.

a) $\dfrac{de}{dc} = a^2 - \dfrac{n^2}{c^2} - \dfrac{e}{c} + e^2$ Bessel

b) $\dfrac{de}{dc} = -a^2 - \dfrac{n^2}{c^2} - \dfrac{e}{c} + e^2$ modified Bessel

c) $\dfrac{de}{dc} = \left[\dfrac{n(n+1)}{(1-c)^2}\right] + \dfrac{2ce}{(1-c^2)} + e^2$ Legendre

d) $\dfrac{de}{dc} = \left[\dfrac{n(n+1)}{(1-c^2)}\right] = \dfrac{m^2}{(1-c^2)^2} + \dfrac{2ce}{(1-c^2)} + e^2$ associated Legendre

e) $\dfrac{de}{dc} = \dfrac{D}{[A(ac+b)^2]} - \dfrac{Be}{(ac+b)} + e^2$ Euler

f) $\dfrac{de}{dc} = 2n + 2ce + e^2$ Hermite

g) $\dfrac{de}{dc} = \dfrac{n}{c} - (1-c)\dfrac{e}{c} + e^2$ Laguerre

13 Using Table 4-2, determine solutions for the following single-input, double-output system equations:

a) $de_1 = (Ae_2 + 1)dc$

$$de_2 = -A\left[(e_2 + 1)\frac{b}{a} + (e_1 + 1)\frac{f}{a}\right]dc$$

b) $de_1 = ae_2dc$

$$de_2 = a\left[\frac{e_2}{(ac+b)}\right]dc - e_1\left\{1 + \frac{n^2}{(ac+b)^2}\right\}dc$$

c) $de_1 = a(e_1 + g)dc$

$$de_2 = \left[\frac{b(e_2 + g)}{(ac+b)} + \frac{d}{\{a(ac+b)^2\}}\right]dc$$

d) $de_1 = ae_2dc$

$$de_2 = -\left(\frac{a}{f}\right)\left[fe_2\left\{\frac{1}{(ac+b)} - 1\right\} + nf\frac{e_1}{(ac+b)}\right]dc$$

e) $de_1 = a(e_2 + 1)dc$

$$de_2 = \left[\frac{\{2a^2(c+1)(e_2+1) - n(n+1)a(e_1+1)\}}{\{1-a^2(c+1)^2\}}\right]dc$$

f) $de_1 = a[fe_2 + g]dc$

$$de_2 = \left(\frac{2a}{f}\right)[(fe_2 + g)(ac+b) - n(e_1+1)]dc$$

g) $de_1 = e_2 dc$

$$de_2 = \left[n e_2(1 - c^2) - e_1 - 1\right]dc$$

14 In Section 4.3.10, the general second-order ordinary differential equation

$$y'' + P(x)y' + Q(x)y = R(x)$$

is shown equivalent to the following single-input, double-output system equation:

$$dy = u\,dx$$
$$du = (R - Pu - Qy)dx$$

By defining new variables as follows:

$$x = F(c)$$
$$y = G_1(c)e_1 + G_2(c)e_2 + G_3(c)$$
$$u = H_1(c)e_1 + H_2(c)e_2 + H_3(c)$$

show that equation (a) is also equivalent to the following single-input, double-output system equation

$$d\begin{bmatrix} e_1 \\ e_2 \end{bmatrix} = \frac{1}{G_1 H_2 - G_2 H_1} \begin{bmatrix} H_2 - G_2 \\ -H_1 & G_1 \end{bmatrix} \left[\begin{pmatrix} A_{11} & A_{12} \\ A_{21} & A_{22} \end{pmatrix} \begin{pmatrix} e_1 \\ e_2 \end{pmatrix} + \begin{pmatrix} B_1 \\ B_2 \end{pmatrix} \right]$$

where for any function $f(c)$, $f'(c) = \dfrac{df}{dc}$.

$$A_{11} = H_1 F' - G'_1, \qquad A_{12} = H_2 F' - G'_2,$$
$$A_{21} = -[P\{F(c)\}H_1 + Q\{F(c)\}G_1]F' - H'_1$$
$$A_{22} = -[P\{F(c)\}H_2 + Q\{F(c)\}G_2)]F' - H'_2$$
$$B_1 = H_3 F' - G'_3$$
$$B_2 = [R\{F(c)\} - P\{F(c)\}H_3 - Q\{F(c)\}G_3]f' - H'_3$$

5

Multiple-Input Systems

Most real, rational systems are influenced and affected by many inputs or causes and, although it is generally possible to reduce or combine these inputs into a single dominant input, there are situations where multiple system inputs must be considered. In this chapter, we will examine in detail the analysis of multiple-input systems based on analytical methods available for a given system kernel. Although we will find that we are able to obtain analytical solutions for only a limited class of the possible continuous multiple-input system equations, through the general requirement for "exactness of a given total differential equation," we will be able to determine whether a given system equation is integrable with a solution of the form

$$\phi\left(e, \mathbf{c}\right) = \mathbf{K} \quad \text{where } \mathbf{K} \text{ is a constant}$$

for the multiple-input $\mathbf{c}$ and single-output e. This exactness test will eliminate unproductive searches and attempts at finding closed, analytical solutions for systems that do not have that solution form.

5.1 Definition and Mathematical Significance

Multiple-input system models are applicable to any system where it is necessary to assume that more than one significant input is required to adequately describe the system's output response. Although most systems are influenced by many inputs, it is usually desirable to minimize the number of these inputs to simplify the resulting system equations and reduce the analysis effort. If it is possible to reduce all the system inputs to a single dominant input, then the methods developed in Chapter 4 for single-input systems may be used to analyze the system equation. However, when it is practical or necessary to consider multiple inputs, then the methods of analysis developed in this chapter are applicable to these system equations.

We will find that multiple-input system equations can be associated with first-order partial differential equations for continuous systems and with first- order partial difference equations for discrete systems. When the multiple-input systems also exhibit multiple outputs, then the system equations may be associated with sets of partial differential and partial difference equations for continuous and discrete systems, respectively. The relationship

Introduction to System Science with MATLAB, Second Edition.
Gary Marlin Sandquist and Zakary Robert Wilde.
© 2023 John Wiley & Sons Ltd. Published 2023 by John Wiley & Sons Ltd.

between multiple-input systems and partial differential and difference equations will be shown in this chapter, and this equivalence will prove useful in formulating and analyzing multiple-input systems. Furthermore, an important consequence of this relationship between partial differential and difference equations and multiple-input system equations is that any set of partial differential or difference equations can be associated with a corresponding multiple-input system and related to a set of appropriate system equations.

The multiple-input, single-output is the simplest system to study in the multiple-input category of system equations and is the system model we will examine first.

5.2   Multiple-Input, Single-Output Systems

Multiple-input, single-output systems comprise all those systems that can be satisfactorily modeled as having multiple inputs influencing the system but only one dominant or significant output issuing from the system due to these multiple inputs. Many important systems are encountered in system science that can be considered to have only one output but require multiple inputs to account satisfactorily for this single output. Often the system investigator seeks to determine the causes for a single essential output from a system. For example, the investigator may seek to study the important inputs to a system that produce a change in the system's total energy output, temperature, pressure, or other physical or chemical states. Alternatively, the investigator may be seeking the significant causes for a single social, economic, or political output such as inflation, unemployment, famine, or even global war.

The general system equation for the multiple-input, single-output system has the matrix form

$$\left.\begin{array}{c} de \\ \Delta e \end{array}\right\} = g(e, \mathbf{c}) \left\{\begin{array}{c} d\mathbf{c} \\ \Delta\mathbf{c} \end{array}\right. \tag{5.2.1}$$

or equivalently,

$$\left.\begin{array}{c} de \\ \Delta e \end{array}\right\} = g_1 \left\{\begin{array}{c} dc_1 \\ \Delta c_1 \end{array}\right. + g_2 \left\{\begin{array}{c} dc_2 \\ \Delta c_2 \end{array}\right. + \cdots + g_n \left\{\begin{array}{c} dc_m \\ \Delta c_m \end{array}\right.$$

where for m inputs, we have defined the following matrix quantities and m is a positive integer such that $m \geq 2$:

$$g(e, \mathbf{c}) = [g_1\ g_2\ \cdots\ g_m], \quad d\mathbf{c} = \begin{bmatrix} dc_1 \\ dc_2 \\ \vdots \\ dc_m \end{bmatrix}, \quad \Delta\mathbf{c} = \begin{bmatrix} \Delta c_1 \\ \Delta c_2 \\ \vdots \\ \Delta c_m \end{bmatrix}$$

The variables $d\mathbf{c}$ and $\Delta\mathbf{c}$ are the differential and discrete matrix changes in the system's input respectively and result in the system's single output response de and Δe, respectively.

The component functions $g_i(e, c_1, c_2, ..., c_m)$ comprise the system matrix kernel $g(e, \mathbf{c})$, and are generally functions of both the output variable e and all input variables $c_1, c_2, ..., c_m$.

If the multiple-input, single-output system exhibits complete feedback of the input as given by the $(m \times 1)$ matrix feedback kernel $f(e, \mathbf{c})$, where

$$f(e, \mathbf{c}) = \begin{bmatrix} f_1(e, \mathbf{c}) \\ f_2(e, \mathbf{c}) \\ \vdots \\ f_m(e, \mathbf{c}) \end{bmatrix}$$

then the system equation with feedback (as developed in Section 3.2.3) is given by

$$\left. \begin{array}{c} de \\ \Delta e \end{array} \right\} = \frac{g(e, \mathbf{c})}{1 - g(e, \mathbf{c}) f(e, \mathbf{c})} = \begin{cases} d\mathbf{c} \\ \Delta \mathbf{c} \end{cases} \tag{5.2.2}$$

this is equivalent to a single scalar equation of the following form for a continuous system

$$de = \frac{g_1 \, d c_1 + g_2 \, d c_2 + \cdots + g_m \, d c_m}{1 - g_1 \, f_1 - g_2 \, f_2 - \cdots - g_m \, f_m}$$

And for the discrete system

$$\Delta e = \frac{g_1 \, \Delta c_1 + g_2 \, \Delta c_2 + \cdots + g_m \, \Delta c_m}{1 - g_1 \, f_1 - g_2 \, f_2 - \cdots - g_m \, f_m}$$

As we have done for single-input systems, it is always possible to express the system kernel with feedback as given by Eq. (5.2.2) without loss of generality as the following:

$$g_f(e, \mathbf{c}) = \frac{g(e, \mathbf{c})}{1 - g(e, \mathbf{c}) f(e, \mathbf{c})}$$

where the function $g_f(e, \mathbf{c})$ is the kernel for multiple-input, single-output systems with or without feedback. Thus, in the classification and solution methods examined in this chapter, it is understood that the system kernel under consideration can exhibit external feedback as given by Eq. (5.2.2) or be without feedback as expressed by Eq. (5.2.1).

5.2.1 Discrete Systems

The general equation for the multiple-input (assume m inputs), single-output discrete system equations is equivalent to the general first-order partial difference equation with m independent variables and can be expressed as follows for a discrete system:

$$\Delta e = g(e, \mathbf{c}) \, \Delta \mathbf{c} \tag{5.2.3}$$

where $g(e, \mathbf{c})$ is a $(1 \times m)$ row matrix and $\Delta \mathbf{c}$ is a $(m \times 1)$ column matrix:

$$g(e, \mathbf{c}) = [g_1, g_2, ..., g_m], \quad \Delta \mathbf{c} = \begin{bmatrix} \Delta c_1 \\ \Delta c_2 \\ \vdots \\ \Delta c_n \end{bmatrix} \tag{5.2.4}$$

The function $g(e, \mathbf{c})$, which depends in general on both $\mathbf{c}$ and e, is the discrete system kernel, and the solutions for Eq. (5.2.3) are determined by the form of the system kernel and by the associated initial conditions for e and $\mathbf{c}$; that is,

$$
e\left[c_1(1), c_2(1), ..., c_m(1)\right], \quad \mathbf{c}(1) = \begin{bmatrix} c_1(1) \\ c_2(1) \\ \vdots \\ c_n(1) \end{bmatrix}
\tag{5.2.5}
$$

Equation (5.2.3) may be treated as a single partial difference equation with nonuniform spacing, in general, of the discrete independent variables, $\Delta\mathbf{c}$. For example, let

$$
\Delta\mathbf{c}(j) = \begin{bmatrix} \Delta c_1(j) \\ \Delta c_2(j) \\ \vdots \\ \Delta c_m(j) \end{bmatrix} = \begin{bmatrix} c_1(j+1) - c_1(j) \\ c_2(j+1) - c_2(j) \\ \vdots \\ c_m(j+1) - c_m(j) \end{bmatrix}
$$

for $j = 1, ..., N$. The incremental changes for different independent variables, c_i or for different steps, $c_i(j)$ or $c_i(j+k)$, are not necessarily equal or uniform in size. The general solution for the partial difference equation, Eq. (5.2.3), is obtained by determining the sequential solutions at each incremental step for $j = 1, ..., N$. For the first incremental step, $\Delta e(1)$ and $\Delta\mathbf{c}(1)$, we have

$$
\Delta e(1) = e(2) - e(1) = g[e(1), \mathbf{c}(1)]\, \Delta\mathbf{c}(1) - g[e(1), \mathbf{c}(1)]\, [\mathbf{c}(2) - \mathbf{c}(1)]
$$

Solving this expression for $e(2)$ and performing our matrix multiplication, we find that

$$
e(2) = e(1) + \sum_{i=1}^{m} g_i[e(1), \mathbf{c}(1)]\, [c_i(2) - c_i(1)]
\tag{5.2.6}
$$

For the next incremental step, $\Delta e(2)$ and $\Delta\mathbf{c}(2)$, we have from Eq. (5.2.3) that

$$
\Delta e(2) = e(3) - e(2) = g\,[e(2), \mathbf{c}(2)]\Delta\mathbf{c}(2)
$$

and in terms of $e(3)$, then

$$
e(3) = e(2) + \sum_{i=1}^{m} g_i[e(2), \mathbf{c}(2)]\, [c_i(3) - c_i(2)]
\tag{5.2.7}
$$

Observe that Eq. (5.2.6) provides the expression for $e(2)$, which can be used in Eq. (5.2.7) to express $e(3)$ without explicit dependence on $e(1)$. Continuing this sequential process, for the Nth incremental change, $\Delta e(N)$ and $\Delta\mathbf{c}(N)$, we find that

$$
\Delta e(N) = e(N+1) - e(N) = g[e(N), \mathbf{c}(N)]\, \Delta\mathbf{c}(N)
$$

and in terms of $e(N+1)$, we have that

$$
e(N+1) = e(N) + \sum_{i=1}^{m} g_i[e(N), \mathbf{c}(N)]\, [c_i(N+1) - c_i(N)]
\tag{5.2.8}
$$

Using the previous sequential determination for $e(2)$, $e(3)$, ..., $e(N)$, then Eq. (5.2.8) can finally be written as

$$e(N + 1) = e(1) + \sum_{i=1}^{m} \sum_{j=1}^{N} g_i[e(j), c_1(j)...c_m(j)] \, [c_i(j + 1) - c_i(j)] \tag{5.2.9}$$

where $e(1)$ is the initial condition imposed on the output as given by Eq. (5.2.5). Equation (5.2.9) gives the general solution of the multiple-input, single-output discrete system for any incremental step N such that $N \geq 2$. The complexity of the expressions resulting from the double summation term will, of course, depend on the form of the discrete system kernel, $g(e, c)$. We observe again, as we did for single-input systems, that unlike continuous systems, we can always develop the general discrete step, analytical solution, as given by Eq. (5.2.9), for any multiple-input, single-output discrete system. Although Eq. (5.2.9) is formidable to evaluate manually for many steps and input variables, evaluation by the computer is possible, and generally, the preferred solution means.

Thus, for any well-defined discrete system kernel, $g(e, c)$ and a known, consistent sequence of changes in the discrete matrix input changes, $c(1)$, $c(2)$,..., $c(N)$,..., it is always possible to analytically determine the resulting changes in the systems single output over the related range, $e(1)$, $e(2)$, ..., $e(N)$. Therefore, we conclude that the general solution for the multiple-input, single-output discrete system can always be found and is therefore known. To demonstrate that Eq. (5.2.9) is the general solution of any discrete single-output, multiple-input system equation, assume that the matrix system kernel given by Eq. (5.2.4) is continuous and bounded over the range of solutions defined by Eq. (5.2.9). Then as each component of the discrete matrix input $[c(I + 1) - c(1)]$ approaches zero, the number of increments becomes infinite, so that

$$\lim_{N \to \infty} e(N + 1) = e(1) + \lim_{N \to \infty} \sum_{j=1}^{N} g[e(j), c(j)] \, \Delta e(j) \tag{5.2.10}$$

Then if Eq. (5.2.10) is defined and all the appropriate limits exist, we have that the discrete solution limit approaches the continuous solution.

$$e = e(1) + \sum_{i=1}^{m} \int_{c(1)}^{c} g_i(e, c) dc_i = e(1) + \int_{c(1)}^{c} g(e, c) dc$$

This is the quadrature form for the general solution of the continuous multiple-input, single-output system equation

$$de = g(e, c) \, dc$$

with the initial condition that

$$e[c(1)] = e(1)$$

In the remainder of this section, we will examine methods for classifying and solving continuous system equations with multiple inputs since we have obtained the complete solution for the discrete system.

5.2.2 Continuous Systems

The general equation for the multiple-input, single-output continuous system has the form of an ordinary total differential equation in m independent or input variables, where

$$de = g(e, \mathbf{c}) \, d\mathbf{c} = g_1(e, \mathbf{c}) \, dc_1 + \cdots + g_m(e, \mathbf{c}) \, dc_m \tag{5.2.11}$$

and de is the differential change in output resulting from the differential changes in the input $d\mathbf{c}$ imposed on the system. The function $g(e, \mathbf{c})$ is the continuous system kernel for multiple-input, single-output systems and is generally a function of both the input variables $\mathbf{c}$ and the output variable e.

The system kernel $g(e, \mathbf{c})$ is a $(1 \times m)$ column matrix, and the differential input $d\mathbf{c}$ is a $(m \times 1)$ row matrix for m inputs, where

$$g(e, \mathbf{c}) = \begin{bmatrix} g_1(e, \mathbf{c}) \\ g_2(e, \mathbf{c}) \\ \vdots \\ g_m(e, \mathbf{c}) \end{bmatrix}, \quad \mathbf{c} = [c_1, c_1, ..., c_m]$$

In Section 5.2.9, we will show that Eq. (5.2.11) is also related to a single first-order partial differential equation with e serving as the dependent variable and $\mathbf{c}$ (i.e., $c_1, c_2, ..., c_m$) serving as the m independent variables. Since Eq. (5.2.11) represents a first-order partial differential equation, the methods available for solving such partial differential equations are available to us. Furthermore, as we will learn later in this chapter, most of these methods entail manipulating and solving coupled sets of ordinary differential equations, which we considered in Chapter 4.

The general solution for Eq. (5.2.11) can usually be expressed by the single equation

$$\phi(e, c_1, c_2, ..., c_m) = K \tag{5.2.12}$$

where K is an arbitrary constant determined by the initial conditions for the system. If we assume that $\phi(e, c)$ is a continuous function where the partial derivatives

$$\frac{\partial \phi}{\partial e} \quad \text{and} \quad \frac{\partial \phi}{\partial c_i} \quad \text{for } i = 1, ..., m$$

exist and are defined over the range of interest sought for the solutions, then the total differential of Eq. (5.2.12) is given by

$$d\phi(e, \mathbf{c}) = \frac{\partial \phi}{\partial e} de + \frac{\partial \phi}{\partial c_1} dc_1 + \frac{\partial \phi}{\partial c_2} dc_2 + \cdots + \frac{\partial \phi}{\partial c_m} dc_m = 0 \tag{5.2.13}$$

If we assume that $\partial \phi / \partial e \neq 0$, which cannot vanish, otherwise our solution ϕ is independent of the output e, then we may express Eq. (5.2.13) as follows:

$$de = -\frac{1}{\partial \phi / \partial e} \left[\frac{\partial \phi}{\partial e} de + \frac{\partial \phi}{\partial c_1} dc_1 + \frac{\partial \phi}{\partial c_2} dc_2 + \cdots + \frac{\partial \phi}{\partial c_m} dc_m \right] = 0 \tag{5.2.14}$$

Comparing Eq. (5.2.14) with our single-input, multiple-output system Eq. (5.2.11), we observe for equivalence of the two equations we require that

$$g_i(e, c) = -\frac{\partial\phi/\partial c_i}{\partial\phi/\partial e} = \frac{\partial e}{\partial c_i}$$

for $i = 1, ..., m$. These results are not useful in general for determining the general solution, $\phi(e, \mathbf{c}) = k$, for a single-input, multiple-output system equation, but they do establish the relationship between such system equations and first-order partial differential equations.

5.2.3   Constant System Kernels

If the multiple-input, single-output system kernel $g(e, \mathbf{c})$ is or can be considered to be approximately constant over a range of interest for both the input variables $c_1, c_2, ..., c_m$ and the single-output variable e, then the system equation has the following form for the continuous system:

$$de = g_{10}\, dc_1 + g_{20}\, dc_2 + \cdots + g_{m0}\, dc_m \tag{5.2.15}$$

where each g_{i0} for $i = 1, 2, 3, ..., m$ is a constant. Equation (5.2.11) may be directly integrated over the range of variation for e and $\mathbf{c}$ for which the system kernel is constant. Integration of Eq. (5.2.11) gives as the solution the following:

$$e = e_o + g_{10}(c_1 - c_{10}) + g_{20}(c_2 - c_{20}) + \cdots + g_{m0}(c_m - c_{m0})$$

where $e = e_1$ when $c_1 = c_{10}, c_2 = c_{20},..., c_m = c_{m0}$.

Because of the difficulty generally encountered in attempting to solve multiple-input systems, a good beginning point in the analysis of multiple-input systems is to assume that the multiple-input kernel is constant. We will examine the various methodologies for formulating such constant system kernels in Chapter 7.

Example Problem for Constant System Kernel.
Consider the continuous system equation with two inputs and one output and with complete feedback, as follows:

$$de = \frac{1}{1 - g_{10}f_1 - g_{20}f_2}\, (g_{10}\, d\,c_1 + g_{20}\, d\,c_2) \tag{5.2.16}$$

where g_{10} and g_{20} are constant and $f_1 = a_1 + b_1 e$ and $f_2 = a_2 + b_2 e$. This system equation is separable and with separation of variables gives upon integration the following solution for this system equation,

$$\int_{e_0}^{e} [1 - g_{10}(a_1 + b_1 e) - g_{20}(a_2 + b_2 e)]\, de = \int_{c_{10}}^{c_1} g_{10}\, dc_1 + \int_{c_{20}}^{c_2} g_{20}\, dc_2$$

where $e = e_0$ when $c_1 = c_{10}$ and $c_2 = c_{20}$. Completing the integration of the equation, we have

$$(e - e_0)\left[1 - a_1 g_{10} - a_2\, g_{20} - \frac{1}{2}\, (e - e_0)\, (b_1\, g_{10} + b_2 g_{20})\right]$$
$$= g_{10}\, (c_1 - c_{10}) + g_{20}\, (c_2 - c_{20})$$

This equation is quadratic in the system output and the general solution for the system equation given by Eq. (5.2.16).

5.2.4 Exact System Kernels

The general matrix equation for the multiple-input, single-output system

$$de = g(e, \mathbf{c})\, d\mathbf{c}$$

can also be expressed in expanded form for m inputs as

$$de = g_1 dc_1 + g_2 dc_2 + \cdots + g_m\, dc_m \tag{5.2.17}$$

Now for Eq. (5.2.17) to be exact and therefore directly integrable, we require that

$$g_1 = \frac{\partial e}{\partial c_1}, \quad g_2 = \frac{\partial e}{\partial c_2}, \dots, \quad g_m = \frac{\partial e}{\partial c_m} \tag{5.2.18}$$

If the kernel components of Eq. (5.2.17) satisfy the conditions imposed by Eq. (5.2.18), then Eq. (5.2.17) assumes the exact form

$$de = \frac{\partial e}{\partial c_1} dc_1 + \frac{\partial e}{\partial c_2} dc_2 + \cdots + \frac{\partial e}{\partial c_m} dc_m$$

and is immediately integrable.

However, the conditions prescribed by Eq. (5.2.18) require knowledge of the analytical solution, $e(\mathbf{c})$, which is presumably unknown until Eq. (5.2.17) has been solved. However, we can eliminate this explicit dependence on $e(\mathbf{c})$ by observing that the following condition is satisfied

$$\frac{\partial^2 e(\mathbf{c})}{\partial c_i\, \partial c_j} = \frac{\partial^2 e(\mathbf{c})}{\partial c_j\, \partial c_i} \quad (\text{for } i, \ j = 1, 2, \dots, m) \tag{5.2.19}$$

if $e(\mathbf{c})$ is continuous and has continuous first and second derivatives as necessary. Imposing Eq. (5.2.19) on Eq. (5.2.18) provides the following:

$$\frac{\partial g_i}{\partial c_j} + \frac{\partial g_i}{\partial e} \frac{\partial e}{\partial c_j} = \frac{\partial g_j}{\partial c_i} + \frac{\partial g_j}{\partial e} \frac{\partial e}{\partial c_i} \tag{5.2.20}$$

But recalling from Eq. (5.2.18) that for all values of k

$$\frac{\partial e}{\partial c_k} = g_k$$

we can eliminate the occurrence of these partial derivatives from Eq. (5.2.20) and obtain the following:

$$\frac{\partial g_i}{\partial c_j} + g_j \frac{\partial g_i}{\partial e} = \frac{\partial g_j}{\partial c_i} + g_i \frac{\partial g_j}{\partial e} \tag{5.2.21}$$

for $i, j = 1, 2, \dots, m$ and $i \neq j$. Equation (5.2.21) constitute the requirements for the exactness of system Eq. (5.2.17). We will show later in this section that necessary and sufficient conditions for exactness can be determined even if the integrating factor for the system is unknown. Furthermore, we will show that Eq. (5.2.21) can be used to provide a test for integrability of the multiple-input, single-output system equation, which has a general solution of the form.

$$\phi\,(e, \mathbf{c}) = K$$

If the system Eq. (5.2.17) is not exact, it is possible that multiplication of Eq. (5.2.17) by the integrating factor $I\,(e, c_1, c_2, ..., c_m)$ may make the system equation exact. To assess the influence of such an integrating factor, consider a double-input, single-output system equation that has been multiplied through by the arbitrary integrating factor, I. Then

$$I\,de - I\,g_1 d\,c_1 - I\,g_2 d\,c_2 = 0 \tag{5.2.22}$$

For Eq. (5.2.22) to be exact, we require that

$$\frac{\partial I}{\partial c_1} = -I\,\frac{\partial g_1}{\partial e} - g_1\,\frac{\partial I}{\partial e} \tag{5.2.23.1}$$

and that

$$\frac{\partial I}{\partial c_2} = -I\,\frac{\partial g_2}{\partial e} - g_2\,\frac{\partial I}{\partial e} \tag{5.2.23.2}$$

$$I\,\frac{\partial g_1}{\partial c_2} + g_1\,\frac{\partial I}{\partial c_2} = I\,\frac{\partial g_2}{\partial c_1} + g_2\,\frac{\partial I}{\partial c_1} \tag{5.2.23.3}$$

Now we would like to determine whether Eqs. (5.2.23) have a solution for any arbitrary integrating factor. If we multiply Eq. (5.2.23.1) by g_2, multiply Eq. (5.2.23.2) by g_1, and add the three resulting equations, we have the following result after canceling I, where $I \neq 0$, from each term:

$$\frac{\partial g_1}{\partial c_2} + g_2\,\frac{\partial g_1}{\partial e} = \frac{\partial g_2}{\partial c_1} + g_1\,\frac{\partial g_2}{\partial e} \tag{5.2.24}$$

This equation, which is independent of the integrating factor, is identical with Eq. (5.2.21) for two inputs, which was the condition for exactness without consideration of the existence of an integrating factor. Thus, either Eq. (5.2.21) or (5.2.24) may be used to establish whether any integrating factor exists for the system Eq. (5.2.17) and whether the system equations possess a solution of the form $\phi\,(e, c_1, c_2) = K$. Therefore, Eq. (5.2.24) is called the condition of integrability for Eq. (5.2.22) and serves as a useful test for the existence of a solution for this type. It is important to recognize that even if Eq. (5.2.24) is satisfied by a given multiple system kernel, the problem of determining an integrating factor for the equation still remains.

This test for integrability of the double-input system equation may be extended to a multiple-input system equation of any order. For convenience in notation, we express the system equation for the single-input, multiple-output in the following form:

$$g_0 dc_0 + g_1 dc_1 + \cdots + g_m\,dc_m = 0 \tag{5.2.25}$$

where

$$dc_0 = de \quad \text{and} \quad g_0 = -1$$

If we multiply Eq. (5.2.25) by an arbitrary nonzero integrating factor $I(e, \mathbf{c})$ and then impose the previous requirement for exactness, we have $(m - 1)$ equations of the following form, after dividing each equation by I:

$$\frac{\partial g_i}{\partial c_j} + g_i\,\frac{\partial \mathrm{Ln}\,I}{\partial c_j} = \frac{\partial g_j}{\partial c_i} + g_j\,\frac{\partial \mathrm{Ln}\,I}{\partial c_i} \tag{5.2.26}$$

for $i = 0, 1, ..., m$. Multiplying Eq. (5.2.26) by g_k and then progressively eliminating all derivatives of the integrating factor, we obtain the following set of equations for $i, j, k = 0, 1, ..., m$ and $i \neq j \neq k$.

$$g_i \left(\frac{\partial g_j}{\partial c_k} - \frac{\partial g_k}{\partial c_j} \right) + g_j \left(\frac{\partial g_k}{\partial c_i} - \frac{\partial g_i}{\partial c_k} \right) + g_k \left(\frac{\partial g_i}{\partial c_j} - \frac{\partial g_j}{\partial c_i} \right) = 0 \qquad (5.2.27)$$

which must be satisfied identically and simultaneously. The total number of these equations is $m(m-1)(m+1)/6$, of which any set of $m(m-1)/2$ of the equations is independent and sufficient to establish integrability.

Equation (5.2.27) represent the condition of integrability for the single-input, multiple-output system equation to have an analytical solution of the form

$$\phi(e, c_1, ..., c_m) = K$$

where K is an arbitrary constant. Recognize that Eq. (5.2.27) is not a sufficient condition if the analytical solution has the form $\phi(e, c_1, ..., c_m, K) = 0$. It may not be possible to express this equation in the form of Eq. (5.2.12).

When the conditions of integrability as prescribed by Eq. (5.2.27) are not satisfied, then the total differential equation or system equation is not derivable from a single primitive equation of the form $\phi(e, c) = K$. If Eq. (5.2.27) is not satisfied, then a total differential equation or system equation in $2N$ or $2N - 1$ variables is generally equivalent to a set of not more than N algebraic equations known as integral equivalents. The task of determining the integral equivalents of a given total differential equation is known as "Pfaff's Problem" in the literature and is beyond the scope of this text. See "Schouten and Kulk" and "Thomas" in the Bibliography section for this chapter.

Now before attempting to analytically solve any multiple-input, single-output system equation, it is advisable to determine if an analytical solution with the form $\phi(e, c) = K$ exists. Equation (5.2.27) may be used to ensure that solutions of this form exist, and then efforts may be expended in determining a specific integrating factor. If Eq. (5.2.27) is not satisfied by the given system equation, the system is probably a candidate for the solution as a difference equation.

Example of Exact System Kernel.
Consider the double-input, single-output system equation

$$de = \frac{(A c_2 + B) e}{e - D} dc_1 + \frac{(A c_1 + B) e}{e - D} dc_2 \qquad (5.2.28)$$

Using Eq. (5.2.24) to determine the integrability of Eq. (5.2.28), we find that

$$\frac{\partial g_1}{\partial c_2} = \frac{A e}{e - D}, \qquad \frac{\partial g_1}{\partial e} = \frac{A c_2 + B}{e - D} - \frac{(A c_2 + B) e}{(e - D)^2}$$

and

$$\frac{\partial g_2}{\partial c_1} = \frac{A e}{e - D}, \qquad \frac{\partial g_2}{\partial e} = \frac{A c_1 + B}{e - D} - \frac{(A c_1 + B) e}{(e - D)^2}$$

Employing these relationships in Eq. (5.2.24), we do find that Eq. (5.2.28) is exact and therefore integrable. Indeed, we can separate Eq. (5.2.28) to give

$$(e - D) \frac{de}{e} = B(dc_1 + dc_2) + A(c_2 \, dc_1 + c_1 \, dc_2)$$

which can be directly integrated to give

$$e - D \operatorname{Ln} e = B(c_1 + c_2) + A c_1 c_2 + K$$

This is the general solution of our example system equation. Observe that it is not possible to express this transcendental solution in terms of e alone. The solution for the output e is transcendental, which is a frequent occurrence in system science.

5.2.5 Linear System Kernels

If the multiple-input, single-output system equation

$$de = g(e, \mathbf{c}) \, d\mathbf{c} \tag{5.2.29}$$

has a system kernel of the following form for each component of the system kernel,

$$g_i(e, \mathbf{c}) = G_i(\mathbf{c}) + H_i(\mathbf{c}) \, e \tag{5.2.30}$$

Then the system kernel is said to be linear. The form of the general solution for the multiple-input, single-output, linear system equation is

$$e = [K + g(\mathbf{c})] \exp f(\mathbf{c}) \tag{5.2.31}$$

Forming the total differential of Eq. (5.2.31) and eliminating the arbitrary constant K from the result, we have

$$de = \left[\frac{\partial g}{\partial c_1} \exp(f) + e \frac{\partial f}{\partial c_1} \right] dc_1 + \cdots + \left[\frac{\partial g}{\partial c_m} \exp(f) + e \frac{\partial f}{\partial c_m} \right] dc_m \tag{5.2.32}$$

Comparing Eqs. (5.2.32) and (5.2.30), we see that for our linear system equation to have a solution given by Eq. (5.2.31), we require for $i = 1, ..., m$

$$H_i(\mathbf{c}) = \frac{\partial f(\mathbf{c})}{\partial c_i} \tag{5.2.33}$$

and

$$G_i(\mathbf{c}) = \frac{\partial g(\mathbf{c})}{\partial c_i} \exp[f(\mathbf{c})] \tag{5.2.34}$$

Now from Eq. (5.2.33) for H_i we have, differentiating with respect to c_i,

$$\frac{\partial H_i}{\partial c_j} = \frac{\partial^2 f}{\partial c_j \, \partial c_i} \tag{5.2.35}$$

and again from Eq. (5.2.33) for H_j we have, differentiating with respect to c_i.

$$\frac{\partial H_j}{\partial c_i} = \frac{\partial^2 f}{\partial c_i \, \partial c_j} \tag{5.2.36}$$

But if f is a continuous function with continuous derivatives as required, then from Eqs. (5.2.35) and (5.2.36), we have

$$\frac{\partial^2 f}{\partial c_i \partial c_j} = \frac{\partial^2 f}{\partial c_j \partial c_i} = \frac{\partial H_j}{\partial c_i} = \frac{\partial H_i}{\partial c_j} \tag{5.2.37}$$

Equation (5.2.37) is the condition for the kernel components, H_i, to constitute an exact differential expression, so that for some function H, we have

$$dH = \frac{\partial H}{\partial c_1} dc_1 + \frac{\partial H}{\partial c_2} dc_2 + \cdots + \frac{\partial H}{\partial c_m} dc_m \tag{5.2.38}$$

then

$$\frac{\partial H}{\partial c_i} = \frac{\partial H_i}{\partial c_i}$$

Similarly, from Eq. (5.2.34) for $\partial g/\partial c_i$ we have, upon differentiation with respect to c_j,

$$\frac{\partial}{\partial c_j} [G_i \exp(-f)] = \frac{\partial^2 g}{\partial c_j \partial c_i} \tag{5.2.39}$$

and again from Eq. (5.2.34) for $\partial g/\partial c_j$ we have, differentiating with respect to c_i, that

$$\frac{\partial}{\partial c_i} [G_j \exp(-f)] = \frac{\partial^2 g}{\partial c_i \partial c_j} \tag{5.2.40}$$

Assuming that g is also a continuous function with continuous derivatives, then Eqs. (5.2.39) and (5.2.40) imply that

$$\frac{\partial^2 g}{\partial c_i \partial c_j} = \frac{\partial^2 g}{\partial c_j \partial c_i} = \frac{\partial}{\partial c_i} [G_j \exp(-f)] = \frac{\partial}{\partial c_j} [G_i \exp(-f)] \tag{5.2.41}$$

and Eq. (5.2.41) also is the condition for exactness of the differential expression for some function G that

$$d[G \exp(-f)] = \frac{\partial}{\partial c_1} [G \exp(-f)] dc_1 + \cdots + \frac{\partial}{\partial c_m} [G \exp(-f)] dc_m$$

where

$$\frac{\partial}{\partial c_i} [G \exp(-f)] = \frac{\partial}{\partial c_j} [G \exp(-f)]$$

Observe then that our linear system equation as defined by Eqs. (5.2.29) and (5.2.30) will have an analytical solution of the form defined by Eq. (5.2.31) only if Eqs. (5.2.37) and (5.2.41) are satisfied for all components of the linear system kernel.

Example of a Linear System Equation

Consider the following double-input, single-output system kernel.

$$de = [\lambda a + b (c_1 + c_2) e] dc_1 + \left[a + b \left(c_1 + \frac{c_2}{\lambda} \right) e \right] dc_2 \qquad (5.2.42)$$

Using Eqs. (5.2.37) and (5.2.41), we check for exactness of the system kernel components, and we find for H_1 and H_2 that

$$\frac{\partial H_1}{\partial c_2} = b \quad \text{and} \quad \frac{\partial H_2}{\partial c_1} = b$$

which is satisfied. Determining $f(\mathbf{c})$ as prescribed by Eq. (5.2.33), we find that

$$f(c_1,\ c_2) = b \left[\frac{\lambda c_1{}^2}{2} + c_1 c_2 + \frac{c_2{}^2}{2\lambda} \right] \qquad (5.2.43)$$

Now to determine the exactness of G_1 and G_2, we find from Eq. (5.2.41) that

$$\frac{\partial}{\partial c_2} [G_1 \exp(-f)] = -ab (\lambda c_1 + c_2) \exp(-f) = \frac{\partial}{\partial c_1} [G_2 \exp(-f)]$$

so G_1 and G_2 form an exact differential expression. Using Eq. (5.2.34), we then find g to be given by

$$g(c_1, c_2) = a \int \exp \left[-b \left(\frac{\lambda c_1{}^2}{2} + c_1 c_2 + \frac{c_2{}^2}{2\lambda} \right) \right] d(\lambda c_1 + c_2) \qquad (5.2.44)$$

It is apparent that Eq. (5.2.44) poses some difficulty to integrate, although it is possible to show that the integrals can be expressed in terms of error functions. Finally, with the functions $f(c_1, c_2)$ given by Eq. (5.2.43) and $g(c_1, c_2)$ given by Eq. (5.2.44), we know that the general solution for system Eq. (5.2.42) is given by

$$e(c_1, c_2) = \exp \{f(c_1, c_2)\} [g(c_1, c_2) + K]$$

5.2.6 Separable System Kernels

Consider the general multiple-input, single-output system equation

$$de = g_1 dc_1 + g_2 dc_2 + \cdots + g_m dc_m$$

where we suppose that each of the kernel functions g_i can be expressed as follows:

$$g_i(e, \mathbf{c}) = F(e) \frac{\partial}{\partial c_i} G(\mathbf{c}) \qquad (5.2.45)$$

Observe that $G(\mathbf{c})$ is a function of $\mathbf{c}$ only. Also, we assume that the kernel component $F(e)$ is either a function of e or constant but is independent of the input $\mathbf{c}$. Then Eq. (5.2.45) is of separable form and can be expressed as follows, if $F(e) \neq 0$:

$$\frac{1}{F(e)} de = \frac{\partial G}{\partial c_1} d c_1 + \frac{\partial G}{\partial c_2} d c_2 + \cdots + \frac{\partial G}{\partial c_m} d c_m$$

This equation can be integrated to provide as a general solution

$$\int \frac{de}{F(e)} = \int \frac{\partial G}{\partial c_1} dc_1 + \int \frac{\partial G}{\partial c_2} dc_2 + \cdots + \int \frac{\partial G}{\partial c_m} d c_m + K = G(\mathbf{C}) + K \qquad (5.2.46)$$

where K is the arbitrary integration constant. It is possible to show that Eq. (5.2.45), as we would expect, defines a system equation that is exact. Furthermore, it is also possible to show that if a multiple-input, single-output system equation has a separable kernel and is integrable, then the kernel must have the form given by Eq. (5.2.45).

Example of a Separable System Equation

Suppose for a double-input, single-output system we have that

$$de = (A + Be) c_2 dc_1 + (A + Be) c_1 dc_2 \qquad (5.2.47)$$

We find that Eq. (5.2.45) is satisfied, so the system equation is separable:

$$g_1(e, c) = F(e) \frac{\partial G}{\partial c_1}$$

$$g_2(e, c) = F(e) \frac{\partial G}{\partial c_2}$$

where

$$F(e) = A + Be$$
$$G = c_1 c_2$$

The solution for system Eq. (5.2.47), as given by Eq. (5.2.46), is

$$\int \frac{de}{A + Be} = \frac{1}{B} \text{Ln} (A + Be) = \int (c_2 dc_1 + c_1 dc_2) + K = c_1 c_2 + K$$

where K is the arbitrary integration constant.

5.2.7 Homogeneous System Kernels

If the multiple-input, single-output system equation

$$de = g(e, \mathbf{c}) \, d\mathbf{c} \qquad (5.2.48)$$

has a system equation of the following form for each component of the system kernel, for $i = 1, ..., m$,

$$g_i(e, \mathbf{c}) = g_i\left(\frac{c_1}{e}, \frac{c_2}{e}, ..., \frac{c_m}{e}\right) \qquad (5.2.49)$$

then the system kernel is said to be homogeneous. The following variable transformations,

$$u_i = \frac{c_i}{e} \quad \text{and} \quad u_i \, de + e \, du_i = dc$$

for all i, will transform Eqs. (5.2.48) and (5.2.49) into the following quasi-separable equation:

$$\frac{de}{e} = d \, \text{Ln} \, e = G(\mathbf{u}) \, d\mathbf{u} \qquad (5.2.50)$$

where for each du_i, we have that

$$G_i(\mathbf{u}) = \frac{g_i(\mathbf{u})}{1 - \sum\limits_{i=1}^{m} u_i\, g(\mathbf{u})}$$

If the conditions for integrability of the multiple-input, single-output system equation derived in Section 5.2.4 [Eq. (5.2.27)] are imposed on Eq. (5.2.50), we find then that the multiple-input, single-output homogeneous system equation is integrable if (for $i, j, k = 0, ..., m$ and $i \neq j \neq k$)

$$G_i\left(\frac{\partial G_j}{\partial u_k} - \frac{\partial G_k}{\partial u_j}\right) + G_j\left(\frac{\partial G_k}{\partial u_i} - \frac{\partial G_i}{\partial u_k}\right) + G_k\left(\frac{\partial G_i}{\partial u_j} - \frac{\partial G_j}{\partial u_i}\right) = 0$$

5.2.8 Inversion of the System Kernel

Consider the double-input, single-output system equation

$$de = g_1\, dc_1 + g_2\, dc_2 \tag{5.2.51}$$

Now let us assume that c_1 and c_2 can be expressed explicitly in terms of the kernel components g_1 and g_2, as follows:

$$c_1 = f_1(g_1, g_2) \quad \text{and} \quad c_2 = f_2(g_1, g_2)$$

so we have that

$$dc_1 = \frac{\partial f_1}{\partial g_1}\, dg_1 + \frac{\partial f_1}{\partial g_2}\, dg_2 \tag{5.2.52.1}$$

and

$$dc_2 = \frac{\partial f_2}{\partial g_1}\, dg_1 + \frac{\partial f_2}{\partial g_2}\, dg_2 \tag{5.2.52.2}$$

Using Eqs. (5.2.52.1) and (5.2.52.2) to eliminate the differential inputs, dc_1 and dc_2, from Eq. (5.2.51), we have

$$de = \left(g_1\frac{\partial f_1}{\partial g_1} + g_2\frac{\partial f_2}{\partial g_1}\right) dg_1 + \left(g_1\frac{\partial f_1}{\partial g_2} + g_2\frac{\partial f_2}{\partial g_2}\right) dg_2 \tag{5.2.53}$$

Now Eq. (5.2.53) is the transformed system equation that expresses the output e in terms of the transformed inputs g_1 and g_2. Although Eq. (5.2.53) appears rather formidable, it is useful to establish necessary conditions on the coefficients of the differential input variables, dc_1 and dc_2, so that the right side of Eq. (5.2.53) represents an exact, total differential. Thus, we will require that f_1 and f_2 satisfy the relationships

$$\frac{\partial e}{\partial g_1} = g_1\frac{\partial f_1}{\partial g_1} + g_2\frac{\partial f_2}{\partial g_1} \quad \text{and} \quad \frac{\partial e}{\partial g_2} = g_1\frac{\partial f_1}{\partial g_2} + g_2\frac{\partial f_2}{\partial g_2} \tag{5.2.54}$$

A necessary and sufficient condition that Eq. (5.2.54) be satisfied, assuming that a is a continuous function of g_1 and g_2, is that

$$\frac{\partial^2 e}{\partial g_2 \, \partial g_1} = \frac{\partial^2 e}{\partial g_1 \, \partial g_2}$$

So we have from Eq. (5.2.54) that

$$\frac{\partial}{\partial g_2}\left(g_1 \frac{\partial f_1}{\partial g_1} + g_2 \frac{\partial f_2}{\partial g_1}\right) = \frac{\partial}{\partial g_1}\left(g_1 \frac{\partial f_1}{\partial g_2} + g_2 \frac{\partial f_2}{\partial g_2}\right) \tag{5.2.55}$$

and, performing the indicated differentiation, we find

$$g_1 \frac{\partial^2 f_1}{\partial g_2 \, \partial g_1} + \frac{\partial f_2}{\partial g_1} + g_2 \frac{\partial^2 f_2}{\partial g_2 \, \partial g_1} = g_1 \frac{\partial^2 f_1}{\partial g_1 \, \partial g_2} + \frac{\partial f_1}{\partial g_2} + g_2 \frac{\partial^2 f_2}{\partial g_1 \, \partial g_2} \tag{5.2.56}$$

But if we assume that both f_1 and f_2 are continuous functions of g_1 and g_2 with continuous derivatives, then we observe that

$$\frac{\partial^2 f_1}{\partial g_1 \, \partial g_2} = \frac{\partial^2 f_1}{\partial g_2 \, \partial g_1} \quad \text{and} \quad \frac{\partial^2 f_2}{\partial g_1 \, \partial g_2} = \frac{\partial^2 f_2}{\partial g_2 \, \partial g_1} \tag{5.2.57}$$

and Eq. (5.2.56) reduces to the requirement that

$$\frac{\partial f_2}{\partial g_1} = \frac{\partial f_1}{\partial g_2} \tag{5.2.58}$$

This method of kernel inversion may be extended to multiple-input, multiple-output systems.

Example of System Kernel Inversion
Suppose we assume from Eq. (5.2.58) that

$$\frac{\partial f_2}{\partial g_1} = \frac{\partial f_1}{\partial g_2} = \lambda \tag{5.2.59}$$

where λ is a constant. Then we have from Eq. (5.2.59) upon integration that

$$f_1(g_1, g_2) = \lambda g_2 + h_1(g_1) \tag{5.2.60.1}$$

and

$$f_2(g_1, g_2) = \lambda g_1 + h_2(g_2) \tag{5.2.60.2}$$

where $h_1(g_1)$ and $h_2(g_2)$ are arbitrary functions of their indicated arguments.
Using Eqs. (5.2.60.1) and (5.2.60.2) in Eq. (5.2.53) gives for the system equation

$$d e = \left(g_1 \frac{\partial h_1}{\partial g_1} + \lambda g_2\right) d g_1 = \left(g_2 \frac{\partial h_2}{\partial g_2} + \lambda g_1\right) d g_2 \tag{5.2.61}$$

Equation (5.2.61) is exact and may be integrated to give the following solution for the system equation:

$$e = \lambda g_1 g_2 + g_1 h_1 + g_2 h_2 - \int h_1(g_1) \, dg_1 - \int h_2(g_2) \, dg_2 + K$$

where K is the integration constant.

5.2.9 Equivalence of Multiple-Input, Single-Output System Equations and First-Order Partial Differential Equations

We will now show that the first-order partial differential equation with two independent variables that have the general form

$$P(x,y,z)\frac{\partial z}{\partial x} + Q(x,y,z)\frac{\partial z}{\partial y} = R(x,y,z) \tag{5.2.62}$$

can be expressed in the equivalent system equation form

$$dz = g_1(x,y,z)\,dx + g_2(x,y,z)\,dy \tag{5.2.63}$$

where the system kernel components $g_1(x, y, z)$ and $g_2(x, y, z)$ are related to the coefficient functions $P(x, y, z)$, $Q(x, y, z)$, and $R(x, y, z)$ in Eq. (5.2.62). Equation (5.2.63) is the general form of the continuous system equation associated with two differential inputs, dx and dy, and one differential output, dz.

To show the equivalence between Eqs. (5.2.62) and (5.2.63), we will employ the "Method of Lagrange[1]" that is based on the observation that any solution of Eq. (5.2.62) that has the general form

$$u(x,y,z) = K \quad \text{where } K \text{ is an arbitrary constant}$$

is also a solution of the following set of differential equations:

$$\frac{dx}{P(x,y,z)} = \frac{dy}{Q(x,y,z)} = \frac{dz}{R(x,y,z)} \tag{5.2.64}$$

And it is also a solution of the related partial differential equation

$$P(x,y,z)\frac{\partial u}{\partial x} + Q(x,y,z)\frac{\partial u}{\partial y} + R(x,y,z)\frac{\partial u}{\partial z} = 0 \tag{5.2.65}$$

To establish this observation, given that $u(x, y, z) = K$ is a solution of Eq. (5.2.65), the total differential of u is given by the following equation:

$$du = \frac{\partial u}{\partial x}\,dx + \frac{\partial u}{\partial y}\,dy + \frac{\partial u}{\partial z}\,dz = 0 \tag{5.2.66}$$

And since u satisfies Eq. (5.2.65), the property of proportions gives for the ratio of Eqs. (5.2.65) and (5.2.66) the following:

$$\frac{\dfrac{\partial u}{\partial x}\,dx + \dfrac{\partial u}{\partial y}\,dy + \dfrac{\partial u}{\partial z}\,dz}{\dfrac{\partial u}{\partial x}P + \dfrac{\partial u}{\partial y}Q + \dfrac{\partial u}{\partial z}R} = \frac{dx}{P} = \frac{dy}{Q} = \frac{dz}{R}$$

---

[1] The reader who is unfamiliar with this method for solving first-order partial differential equations is referred to any standard textbook on partial differential equations.

Thus $u(x, y, z) = K$ satisfies Eqs. (5.2.64), (5.2.65), and (5.2.66). To show that u is also a solution of Eq. (5.2.62) if we differentiate $u(x, y, z) = K$ partially with respect to x and y, respectively, we have

$$\frac{\partial u}{\partial x} + \frac{\partial u}{\partial z}\frac{\partial z}{\partial x} = 0 \text{ and } \frac{\partial u}{\partial y} + \frac{\partial u}{\partial z}\frac{\partial z}{\partial y} = 0 \tag{5.2.67}$$

Using Eq. (5.2.67) to eliminate $\partial u/\partial x$ and $\partial u/\partial y$ from Eq. (5.2.65), we have that

$$-\frac{\partial u}{\partial z}\frac{\partial z}{\partial x}P - \frac{\partial u}{\partial z}\frac{\partial z}{\partial y}Q + \frac{\partial u}{\partial z}R = \frac{\partial u}{\partial z} \pm 0 \tag{5.2.68}$$

Canceling $\partial u/\partial z$ and rearranging Eq. (5.2.68), we have that

$$P\frac{\partial z}{\partial x} + Q\frac{\partial z}{\partial y} = R$$

which is identically Eq. (5.2.62). Thus we have shown that associated with any partial differential equation of the form given by Eq. (5.2.62) or (5.2.65), there exists a set of ordinary differential equations given by Eq. (5.2.64) whose solution satisfies Eqs. (5.2.62) and (5.2.65). We will now examine the means for solving Eq. (5.2.64).

Consider as an equation relating dz to dx the following expression obtained from Eq. (5.2.64):

$$dz = \frac{R}{P}dx \tag{5.2.69}$$

And for dz and dy, consider Eq. (5.2.64) again, so

$$dz = \frac{R}{Q}dy \tag{5.2.70}$$

If we multiply Eq. (5.2.69) by an arbitrary nonzero function, $a(x, y, z)$, and Eq. (5.2.70) by an arbitrary nonzero function, $b(x, y, z)$, and then add the resulting expressions, we have that

$$(a + b)\,dz = \frac{a\,R}{P}dx + \frac{b\,R}{Q}dy \tag{5.2.71}$$

Equation (5.2.71) can also be expressed as

$$dz = \frac{R}{P}f\,dx + \frac{R}{Q}(1 - f)\,dy \tag{5.2.72}$$

where

$$f = \frac{a}{a + b}$$

Now Eq. (5.2.72) is equivalent to the system Eq. (5.2.63) if we define the following relationships for the system kernels:

$$g_1 = \frac{R}{P}f \text{ and } g_2 = \frac{R}{P}(1 - f) \tag{5.2.73}$$

where f is an arbitrary function usually chosen as an integrating factor of Eqs. (5.2.69) and (5.2.70).

If we solve Eq. (5.2.73) for P and Q, respectively, then the general first-order partial differential equation with two independent variables for the system kernels g_1 and g_2, as given by Eq. (5.2.62), can be written as follows:

$$\frac{f}{g_1}\frac{\partial z}{\partial x} + (1-f)\frac{1}{g_2}\frac{\partial z}{\partial y} = 1$$

Observe that there exist many equivalent first-order partial differential equations of the form given by Eq. (5.2.62) depending on the form of the integrating factor f selected.

The equivalence between a first-order partial differential equation in standard form and an associated system equation can be established for any finite number of independent variables. The standard form for such a first-order partial differential equation is as follows:

$$\sum_{i=1}^{m} P_i(c_1, c_2, ..., c_m, e)\frac{\partial e}{\partial c_i} = R(c_1, c_2, ..., c_m, e)$$

This equation can also be expressed in the equivalent differential form as

$$\frac{dc_1}{P_1} = \frac{dc_2}{P_2} = \cdots = \frac{dc_m}{P_m} = \frac{de}{R}$$

where $P_1, P_2, ..., P_m, R \neq 0$.

For any of the independent differential variables, we have

$$de = \frac{R}{P_j}\,dc_j, \text{ for } j = 1, 2, ..., m \tag{5.2.74}$$

Multiplying these equations by an arbitrary function Q_j and adding the resulting system of equations for all the independent differential variables, we obtain the result that

$$de = \frac{R}{Q_1 + Q_2 + \cdots + Q_m}\left(\frac{Q_1}{P_1}\,dc_1 + \frac{Q_2}{P_2}\,dc_2 + \cdots + \frac{Q_m}{P_m}\,dc_m\right)$$

We see that this equation has the form of a single-input, multiple-output system equation

$$de = g_1 dc_1 + g_2 dc_2 + \cdots + g_m dc_m$$

When we make the following associations that

$$g_j = \frac{Q_j}{P_j}R\left\{\sum_{i=1}^{m} Q_i\right\}^{-1} \text{ for } j = 1, 2, ..., m$$

where the Q_i's are arbitrary functions again selected as integrating factors for Eq. (5.2.74).

Example of Equivalence between Partial Differential and System Equation.
Let us determine the solution and the related system equation for the following first-order partial differential equation:

$$xu\,\frac{\partial u}{\partial x} + yu\,\frac{\partial u}{\partial y} = 1 \tag{5.2.75}$$

The LaGrangian differential form of this equation is

$$\frac{dx}{xu} = \frac{dy}{yu} = du$$

For the equation between dx and du, we choose and solve as follows:

$$\frac{dx}{x} = u\,du \quad \text{so}\, u^2 = 2\,\text{Ln}\,x + K_1 \tag{5.2.76}$$

where we have selected for our arbitrary multiplicative function u, which is an integrating factor for the differential equation. Furthermore, for the equation between dy and du, we have

$$\frac{dy}{y} = u\,du \quad \text{so}\, u^2 = 2\,\text{Ln}\,y + K_2 \tag{5.2.77}$$

where the function u is also an integrating factor for this differential equation.
The related system equation for Eq. (5.2.75) has the form

$$du = \frac{f}{xu} + \frac{(1-f)}{yu}\,dy \tag{5.2.78}$$

where f is any function of x, y, and u. The solution for Eq. (5.2.75) is given by Eqs. (5.2.76) and (5.2.77) in the form

$$\phi\,(K_1, K_2) = 0$$

or, equivalently,

$$u\,(x,y) = \pm\left[2\,\text{Ln}\,x + F\left(\text{Ln}\,\frac{x}{y}\right)\right]^{1/2} \tag{5.2.79}$$

where F is an arbitrary function of its argument and is determined from the boundary conditions for the problem.
Observe that the partial differential equation, Eq. (5.2.75), has the equivalent system equation form given by Eq. (5.2.78); and the solution given by Eq. (5.2.79) is a solution for both the partial differential equation, Eq. (5.2.75), and the system equation, Eq. (5.2.78), where

$$f = 1 + \frac{1}{2}\,\frac{dF\,[\text{Ln}\,(x/y)]}{d\,\text{Ln}\,(x/y)}$$

5.3 Multiple-Input, Multiple-Output Systems

The general system equation that will describe any rational system is the multiple-input, multiple-output system equation. Although it is usually advisable to begin the study of a given system by restricting the inputs or outputs to a single dominant variable that controls the input or output, it is often necessary that the system eventually be considered to exhibit both multiple inputs and multiple outputs. We will see that the multiple-input, multiple-output discrete system equation can be related to sets of first-order or higher-order partial difference equations, while multiple-input, multiple-output continuous system equations can similarly be associated with sets of first- or higher-order partial differential equations. However, unlike multiple-input, single-output continuous equations, it is not always possible to express all higher-order partial differential equations as equivalent system equations or similarly as sets of ordinary differential equations. But, in general, this restriction will not be of great importance in our considerations here since our system equations are naturally formulated as sets of total differential equations.

The general equation for the multiple-input, multiple-output system is given by

$$\left.\begin{matrix} d\mathbf{e} \\ \Delta\mathbf{e} \end{matrix}\right\} = g(\mathbf{e}, \mathbf{c}) \left\{ \begin{matrix} d\mathbf{c} \\ \Delta\mathbf{c} \end{matrix} \right. \tag{5.3.1}$$

where we have defined the following matrix relationships:

$$d\mathbf{e} = \begin{bmatrix} de_1 \\ de_2 \\ \vdots \\ de_n \end{bmatrix}, \quad \Delta\mathbf{e} = \begin{bmatrix} \Delta e_1 \\ \Delta e_2 \\ \vdots \\ \Delta e_n \end{bmatrix}, \quad d\mathbf{c} = \begin{bmatrix} dc_1 \\ dc_2 \\ \vdots \\ dc_m \end{bmatrix}, \quad \Delta\mathbf{c} = \begin{bmatrix} \Delta c_1 \\ \Delta c_2 \\ \vdots \\ \Delta c_m \end{bmatrix}$$

and

$$g(\mathbf{e}, \mathbf{c}) = \begin{bmatrix} g_{11} & g_{12} & \cdots & g_{1m} \\ g_{21} & g_{22} & & g_{2m} \\ \vdots & & & \\ g_{n1} & g_{n2} & \cdots & g_{nm} \end{bmatrix}$$

The differential variables, $d\mathbf{e}$ and $d\mathbf{c}$, and the difference variables, $\Delta\mathbf{e}$ and $\Delta\mathbf{c}$, represent the infinitesimal and incremental changes in the system's outputs and inputs, respectively. The $(n \times m)$ matrix system kernel, $g(\mathbf{e}, \mathbf{c})$, is generally a function of both inputs, $\mathbf{c}$, and outputs, $\mathbf{e}$.

If the multiple-input, multiple-output system exhibits feedback of the output as defined by the system matrix feedback kernel, $f(\mathbf{e}, \mathbf{c})$, the resulting system equation becomes the following, as was shown in Section 3.2.4.

$$\left.\begin{matrix} d\mathbf{e} \\ \Delta\mathbf{e} \end{matrix}\right\} = [I - g(\mathbf{e}, \mathbf{c}) f(\mathbf{e}, \mathbf{c})]^{-1} g(\mathbf{e}, \mathbf{c}) \left\{ \begin{matrix} d\mathbf{c} \\ \Delta\mathbf{c} \end{matrix} \right. \tag{5.3.2}$$

where we have the following $(m \times n)$ matrix definition for the feedback kernel, $f(\mathbf{e}, \mathbf{c})$:

$$f(\mathbf{e}, \mathbf{c}) = \begin{bmatrix} f_{11} & f_{12} & \cdots & f_{1n} \\ f_{21} & f_{22} & & f_{2n} \\ \vdots & & & \\ f_{m1} & f_{m2} & \cdots & f_{mn} \end{bmatrix}$$

The matrix $[I - gf]^{-1}$ is the inverse matrix of $[I - gf]$ shown below, where j ranges from 1 to m.

$$[I - gf] = \begin{bmatrix} 1 - \sum_j g_{1j}f_{j1} & -\sum_j g_{1j}f_{j2} & \cdots & -\sum_j g_{1j}f_{jm} \\ -\sum_j g_{2j}f_{j1} & 1 - \sum_j g_{2j}f_{j2} & & -\sum_j g_{2j}f_{jm} \\ \vdots & & & \\ -\sum_j g_{nj}f_{j1} & -\sum_j g_{nj}f_{j2} & \cdots & 1 - \sum_j g_{nj}f_{jm} \end{bmatrix}$$

It is convenient to express Eq. (5.3.2) in terms of a generalized matrix system kernel as $g_f(\mathbf{e}, \mathbf{c})$, where

$$g_f(\mathbf{e}, \mathbf{c}) = [I - g(\mathbf{e}, \mathbf{c})f(\mathbf{e}, \mathbf{c})]^{-1}g(\mathbf{e}, \mathbf{c})$$

We will not usually distinguish in this section whether the system kernel is without feedback, that is, $g(\mathbf{e}, \mathbf{c})$ or with feedback, $g_f(\mathbf{e}, \mathbf{c})$. We will only express the system kernel as $g(\mathbf{e}, \mathbf{c})$ and understand that the system may exhibit feedback if appropriate. It is important to observe that if the matrix $[I - gf]$ is singular, i.e., the determinant vanishes, then the system equation, Eq. (5.3.2), is undefined and the response of the system becomes arbitrarily large (i.e., approaches $\pm$ infinity) which is an unachievable state for a real system. Generally, this is an unacceptable condition for the feedback kernel and is to be avoided.

In general, multiple-input, multiple-output system equations pose formidable challenges and difficulties in obtaining analytic solutions. A very limited set of solution methods will be discussed here, and a condition of integrability or test for exactness will be presented to allow the system investigator to at least determine if analytic solutions for these systems exist. The usual general approach for treating such multiple input and/or output systems will be to employ the computer to provide numerical solutions for the prescribed system equation. In very limited cases, the symbolic toolboxes for MATLAB my provide analytical solutions for such multiple-input, multiple-output systems.

5.3.1 Discrete Systems

The general equation for the multiple-input, multiple-output system is equivalent to a set of first-order, partial differential equations. This matrix equation for discrete systems has the form

$$\Delta\mathbf{e} = g(\mathbf{e}, \mathbf{c})\,\Delta\mathbf{c} \tag{5.3.3}$$

where $g(\mathbf{e}, \mathbf{c})$ is an $(n \times m)$ matrix, $\Delta\mathbf{e}$ is an $(n \times 1)$ column matrix, and $\Delta\mathbf{c}$ is an $(m \times 1)$ column matrix. The discrete system kernel, $g(\mathbf{e}, \mathbf{c})$, and the initial conditions for the system, $\mathbf{e}(1)$ and $\mathbf{c}(1)$, determine the form of the solutions of the system equation, Eq. (5.3.3). The solution process for the discrete system treats Eq. (5.3.3) as a set of partial difference equations with nonuniform spacing, in general, associated for each input or independent variable. For example, for the input variable change at step j, we have

$$\Delta\mathbf{c}\,(j) = \begin{bmatrix} \Delta c_1(j) \\ \Delta c_2(j) \\ \vdots \\ \Delta c_m(j) \end{bmatrix} = \begin{bmatrix} \Delta c_1(j+1) - \Delta c_1(j) \\ \Delta c_2(j+1) - \Delta c_2(j) \\ \vdots \\ \Delta c_m(j+1) - \Delta c_m(j) \end{bmatrix} = \mathbf{c}(j+1) - \mathbf{c}(j) \qquad (5.3.4)$$

where, in general, $\Delta c_j \neq \Delta c_i$ for any i or j except when $i = j$.

The general solution for partial difference equations is obtained by determining sequentially the output changes, $\Delta\mathbf{e}$, resulting from the incremental input changes, $\Delta\mathbf{c}$. So for the first output step, $\Delta\mathbf{e}(1)$ and $\Delta\mathbf{c}(1)$, we have from Eq. (5.3.3) that

$$\Delta\mathbf{e}(1) = \mathbf{e}(2) - \mathbf{e}(1) = g[\mathbf{e}(1), \mathbf{c}(1)]\,\Delta\mathbf{c}(1) \qquad (5.3.5)$$

Solving Eq. (5.3.5) for $\mathbf{e}(2)$ and recalling Eq. (5.3.4) for $\Delta\mathbf{c}(1)$, we have

$$\mathbf{e}(2) = \mathbf{e}(1) + g[\mathbf{e}(1), \mathbf{c}(1)]\,[\mathbf{c}(2) - \mathbf{c}(1)] \qquad (5.3.6)$$

For the next incremental steps, $\Delta\mathbf{c}(2)$ and $\Delta\mathbf{e}(2)$, then we similarly find

$$\mathbf{e}(3) = \mathbf{e}(2) + g[\mathbf{e}(2), \mathbf{c}(2)]\,[\mathbf{c}(3) - \mathbf{c}(2)] \qquad (5.3.7)$$

Observe that Eq. (5.3.6) provides an expression for $\mathbf{e}(2)$ that can be used in Eq. (5.3.7) to express $\mathbf{e}(3)$ in terms of the initial conditions on $\mathbf{e}$, that is, $\mathbf{e}(1)$. Thus, Eq. (5.3.7) becomes

$$\mathbf{e}(3) = \mathbf{e}(1) + \sum_{I=1}^{2} g[\mathbf{e}\,(I), \mathbf{c}(I)]\,[\mathbf{c}(I+1) - \mathbf{c}(I)]$$

Continuing this sequential process, we find for the Nth incremental change, $\Delta\mathbf{e}(N)$ and $\Delta\mathbf{c}(N)$, that

$$\mathbf{e}(N+1) = \mathbf{e}(1) + \sum_{I=1}^{N} g[\mathbf{e}(I),\ \mathbf{c}(I)]\,[\mathbf{c}(I+1) - \mathbf{c}(I)] \qquad (5.3.8)$$

Equation (5.3.8) is the general solution for the multiple-input, multiple-output discrete system equation with the initial condition that $\mathbf{e} = \mathbf{e}(1)$ when $\mathbf{c} = \mathbf{c}(1)$ for $N \geq 2$. Observe that for N steps Eq. (5.3.8) requires that $2N$ matrix multiplications be performed on $g[\mathbf{e}, \mathbf{c}]$, an $(n \times m)$ matrix, and $\mathbf{c}$, an $(m \times 1)$ matrix. So, for example, for 100 steps of a 10-input, 10-output discrete system about 20,000 scalar multiplications and 20,000 additions are required to evaluate Eq. (5.3.8). Obviously, such a task is performed by a computer for the accuracy and speed required for practical analysis.

So, recalling Sections 4.2.1, 4.3.1, and 5.2.1 for discrete systems, we see it is always possible to solve the discrete system equation for any finite number of inputs and outputs. As we have done in those earlier sections, it is also possible to show that Eq. (5.3.8) under the appropriate conditions of convergence assumes the quadrature form for the matrix solution of the continuous multiple-input, multiple-output system equation:

$$\mathbf{e}(\mathbf{c}) = \mathbf{e}\,[\mathbf{c}(1)] + \int_{\mathbf{c}(1)}^{\mathbf{c}} g(\mathbf{e}, \mathbf{c})\,d\mathbf{c}$$

5.3.2 Continuous System Kernels

The general set of equations for the multiple-input, multiple-output continuous system is equivalent to a general set of coupled first-order differential equations in the following form

$$de = g(\mathbf{e}, \mathbf{c}) \, d\mathbf{c} \tag{5.3.9}$$

where $d\mathbf{e}$ is an $(n \times 1)$ differential output column vector, $g(\mathbf{e}, \mathbf{c})$, is the continuous system $(n \times m)$ matrix kernel, and $d\mathbf{c}$ is an $(m \times 1)$ differential input column vector. We will show later in this chapter that system Eq. (5.3.9) is also related to a coupled set of n first-order partial differential equations in m independent variables.

The general solution for system Eq. (5.3.9) can under reasonable assumptions be expressed by the algebraic matrix or vector equation

$$\Phi(\mathbf{e}, \mathbf{c}) = \mathbf{k} \tag{5.3.10}$$

where the two-column matrices are given as

$$\Phi(\mathbf{e}, \mathbf{c}) = \begin{bmatrix} \varphi_1(e_1, ..., e_n, c_1, ..., c_m) \\ \varphi_2(e_1, ..., e_n, c_1, ..., c_m) \\ \vdots \\ \varphi_n(e_1, ..., e_n, c_1, ..., c_m) \end{bmatrix}, \quad \mathbf{k} = \begin{bmatrix} k_1 \\ k_2 \\ \vdots \\ k_n \end{bmatrix}$$

where $\mathbf{k}$ is a column vector of independent arbitrary constants determined by the initial conditions of the system. Let us assume that the matrix $\Phi(\mathbf{e}, \mathbf{c})$ is continuous in $\mathbf{e}$ and $\mathbf{c}$, and the first partial derivatives of each matrix element for $1 \le i, j \le 1$ exist and are defined over some range of interest. If we form the total differential of Eq. (5.3.10), we obtain the following matrix equation:

$$d\Phi = 0 = \frac{\partial}{\partial \mathbf{e}} [\Phi(\mathbf{e}, \mathbf{c})] \, d\mathbf{e} + \frac{\partial}{\partial \mathbf{c}} [\Phi(\mathbf{e}, \mathbf{c})] \, d\mathbf{c} \tag{5.3.11}$$

where

$$\frac{\partial \Phi}{\partial \mathbf{e}} = \begin{bmatrix} \dfrac{\partial \phi_1}{\partial e_1} & \dfrac{\partial \phi_1}{\partial e_2} & \cdots & \dfrac{\partial \phi_1}{\partial e_n} \\ \dfrac{\partial \phi_2}{\partial e_1} & \dfrac{\partial \phi_2}{\partial e_2} & & \dfrac{\partial \phi_2}{\partial e_n} \\ \vdots & & & \\ \dfrac{\partial \phi_n}{\partial e_1} & \dfrac{\partial \phi_n}{\partial e_2} & \cdots & \dfrac{\partial \phi_n}{\partial e_n} \end{bmatrix}, \quad d\mathbf{e} = \begin{bmatrix} de_1 \\ de_2 \\ \vdots \\ de_n \end{bmatrix}$$

and

$$\frac{\partial \Phi}{\partial \mathbf{c}} = \begin{bmatrix} \dfrac{\partial \phi_1}{\partial c_1} & \dfrac{\partial \phi_1}{\partial c_2} & \cdots & \dfrac{\partial \phi_1}{\partial c_m} \\ \dfrac{\partial \phi_2}{\partial c_1} & \dfrac{\partial \phi_2}{\partial c_2} & & \dfrac{\partial \phi_2}{\partial c_m} \\ \vdots & & & \\ \dfrac{\partial \phi_n}{\partial c_1} & \dfrac{\partial \phi_n}{\partial c_2} & \cdots & \dfrac{\partial \phi_n}{\partial c_m} \end{bmatrix}, \quad d\mathbf{c} = \begin{bmatrix} dc_1 \\ dc_2 \\ \vdots \\ dc_n \end{bmatrix}$$

Now matrix Eq. (5.3.11) may be solved for the differential matrix output $d\mathbf{e}$ if the square matrix $[\partial\Phi/\partial\mathbf{e}]$ is not singular so that the inverse matrix denoted by $[\partial\Phi/\partial\mathbf{e}]^{-1}$ exists. Then we have

$$d\mathbf{e} = -\left[\frac{\partial\Phi}{\partial\mathbf{e}}\right]^{-1}\left[\frac{\partial\Phi}{\partial\mathbf{c}}\right]d\mathbf{c}$$

And as expected, we observe that the continuous kernel for the multiple-input, multiple-output system is given by the following $(n \times m)$ matrix:

$$g(\mathbf{e},\mathbf{c}) = -\left[\frac{\partial\Phi}{\partial\mathbf{e}}\right]^{-1}\left[\frac{\partial\Phi}{\partial\mathbf{c}}\right]$$

To reinforce this result, let us consider a double-input, double-output system equation as follows:

$$\begin{bmatrix} de_1 \\ de_2 \end{bmatrix} = \begin{bmatrix} g_{11} & g_{12} \\ g_{21} & g_{22} \end{bmatrix}\begin{bmatrix} dc_1 \\ dc_2 \end{bmatrix} \tag{5.3.12}$$

Now the general solution for Eq. (5.3.12) can be assumed to have the form

$$\Phi = \begin{bmatrix} \varphi_1(e_1,e_2,c_1,c_2) \\ \varphi_2(e_1,e_2,c_1,c_2) \end{bmatrix} = \begin{bmatrix} k_1 \\ k_2 \end{bmatrix} \tag{5.3.13}$$

Taking the total differential of Eq. (5.3.13), we have that

$$d\Phi = 0 = \begin{bmatrix} \dfrac{\partial\varphi_1}{\partial e_1} & \dfrac{\partial\varphi_1}{\partial e_2} \\ \dfrac{\partial\varphi_2}{\partial e_1} & \dfrac{\partial\varphi_2}{\partial e_2} \end{bmatrix}\begin{bmatrix} de_1 \\ de_2 \end{bmatrix} + \begin{bmatrix} \dfrac{\partial\varphi_1}{\partial c_1} & \dfrac{\partial\varphi_1}{\partial c_2} \\ \dfrac{\partial\varphi_2}{\partial c_1} & \dfrac{\partial\varphi_2}{\partial c_2} \end{bmatrix}\begin{bmatrix} dc_1 \\ dc_2 \end{bmatrix}$$

Inverting the coefficient matrix of the matrix $d\mathbf{e}$ and solving for $d\mathbf{e}$, we have that

$$\begin{bmatrix} de_1 \\ de_2 \end{bmatrix} = -\frac{1}{D}\begin{bmatrix} \dfrac{\partial\varphi_2}{\partial e_2} & -\dfrac{\partial\varphi_1}{\partial e_2} \\ -\dfrac{\partial\varphi_2}{\partial e_1} & \dfrac{\partial\varphi_1}{\partial e_1} \end{bmatrix}\begin{bmatrix} \dfrac{\partial\varphi_1}{\partial c_1} & \dfrac{\partial\varphi_1}{\partial c_2} \\ \dfrac{\partial\varphi_2}{\partial c_1} & \dfrac{\partial\varphi_2}{\partial c_2} \end{bmatrix}\begin{bmatrix} dc_1 \\ dc_2 \end{bmatrix}$$

where the determinant D, assumed not to vanish, is given by

$$D = \frac{\partial\varphi_1}{\partial e_1}\frac{\partial\varphi_2}{\partial e_2} - \frac{\partial\varphi_1}{\partial e_2}\frac{\partial\varphi_2}{\partial e_1} \neq 0$$

Performing the matrix multiplication for $[\partial\phi/\partial\mathbf{e}]^{-1}[\partial\phi/\partial\mathbf{c}]$ we obtain the double-input, double-output system equation, Eq. (5.3.12), where

$$g(\mathbf{e},\mathbf{c}) = -\frac{1}{D}\begin{bmatrix} \dfrac{\partial\varphi_2}{\partial e_2}\dfrac{\partial\varphi_1}{\partial c_1} - \dfrac{\partial\varphi_1}{\partial e_2}\dfrac{\partial\varphi_2}{\partial c_1} & \dfrac{\partial\varphi_2}{\partial e_2}\dfrac{\partial\varphi_1}{\partial c_2} - \dfrac{\partial\varphi_1}{\partial e_2}\dfrac{\partial\varphi_2}{\partial c_2} \\ \dfrac{\partial\varphi_1}{\partial e_1}\dfrac{\partial\varphi_2}{\partial c_1} - \dfrac{\partial\varphi_2}{\partial e_1}\dfrac{\partial\varphi_1}{\partial c_1} & \dfrac{\partial\varphi_1}{\partial e_1}\dfrac{\partial\varphi_2}{\partial c_2} - \dfrac{\partial\varphi_2}{\partial e_1}\dfrac{\partial\varphi_1}{\partial c_2} \end{bmatrix}$$

and we observe how the double-input, double-output system kernel is related to its general solution matrix defined by Eq. (5.3.13).

5.3.3 Constant System Kernels

If the multiple-input, multiple-output system kernel is a constant and has the form

$$d\mathbf{e} = G_o \, d\mathbf{c} \tag{5.3.14}$$

where

$$G_o = \begin{bmatrix} g_{11} & \cdots & g_{1m} \\ \vdots & & \\ g_{11} & \cdots & g_{nm} \end{bmatrix} \tag{5.3.15}$$

and each matrix element g_{ij} is a constant, then system Eq. (5.3.14) is exact and can be directly integrated to provide the following system solution:

$$\mathbf{e} = G_o \, \mathbf{c} + \mathbf{k} \tag{5.3.16}$$

where $\mathbf{k}$ is column matrix of arbitrary constants, one constant for each output e_j.

The assumption that a given multiple-input, multiple-output system kernel is constant as given by Eq. (5.3.15) is usually a wise choice for initial modeling and analysis of such multiple-input, multiple-output systems. Such analysis and evaluation of the solution, Eq. (5.3.16), will often prove adequate to acceptably describe the system's behavior at least for restricted ranges of the inputs.

5.3.4 Exact System Kernels

If a multiple-input, multiple-output system equation is integrable, then the equation must be expressed in "exact" form, as was demonstrated in Section 5.2.4 for single-output systems. The requirements for the exactness of the system kernel may be established by considering the ith output system equation for a general multiple-input, multiple-output system. We have then that

$$de_i = g_{i1}(\mathbf{c}) \, dc_1 + g_{i2}(\mathbf{c}) \, dc_2 + \cdots + g_{im}(\mathbf{c}) \, dc_m \tag{5.3.17}$$

for $i = 1, ..., n$. Let us express Eq. (5.3.17) in the following form:

$$g_{i0} \, dc_0 + g_{i1} dc_1 + \cdots + g_{im} dc_m = 0 \tag{5.3.18}$$

where for convenience we have set

$$g_{i0} = -1, \quad dc_0 = de_i$$

Now, if we multiply Eq. (5.3.18) by an arbitrary integrating factor $I_i(\mathbf{c}, \mathbf{e})$ and then impose the standard requirement for exactness for each kernel output component, we have, after dividing by $I_i(\mathbf{c}, \mathbf{e})$, that $n(m-1)$ equations of the form

$$\frac{\partial g_{ij}}{\partial c_k} + g_{ij} \frac{\partial \mathrm{Ln} \, I_i}{\partial c_k} = \frac{\partial g_{ik}}{\partial c_j} + g_{ik} \frac{\partial \mathrm{Ln} \, I_i}{\partial c_j} \tag{5.3.19}$$

for $i = 1, ..., n, j = 0, ..., m$, and $k = 0, ..., m$. Multiplying Eq. (5.3.19) by the kernel g_{il}, where $i \neq j \neq k$, and then progressively eliminating all derivatives of the integrating factor I_i, we obtain the following equation set, where $i, j, k = 0, 1, ..., m$ and $i \neq j \neq k$,

$$g_{ij}\left(\frac{\partial g_{ik}}{\partial c_l} - \frac{\partial g_{il}}{\partial c_k}\right) + g_{ik}\left(\frac{\partial g_{il}}{\partial c_j} - \frac{\partial g_{ij}}{\partial c_l}\right) + g_{il}\left(\frac{\partial g_{ij}}{\partial c_k} - \frac{\partial g_{ik}}{\partial c_j}\right) \qquad (5.3.20)$$

Equation (5.3.20) for each output e; represents the condition for exactness and, thus, the condition for integrability of the multiple-input, multiple-output system equation. Equation (5.3.20) must be satisfied identically and simultaneously for the $nm(m - 1)(m + 1)/6$ equations which arise for $i = 1, ..., n$. However, any set of $m(m - 1)/2$ of these equations which are linearly independent are sufficient to establish the integrability of the system equation. Thus, before attempting to solve any multiple-input, multiple-output system equation, it is advisable to determine that a solution of the form $G(\mathbf{e}, \mathbf{c}) = \mathbf{K}$ exists by ensuring that the system is integrable. If Eq. (5.3.20) is satisfied, then the investigator is justified in seeking an integrating factor. For example, the requirement that system equation, Eq. (5.3.17) be exact and therefore directly integrable without the use of an integrating factor is

$$\frac{\partial g_{ij}}{\partial c_k} + g_{ik}\frac{\partial g_{ij}}{\partial e} = \frac{\partial g_{ik}}{\partial c_j} + g_{ij}\frac{\partial g_{ik}}{\partial e}$$

for $i = 1, ..., n$ and $j, k = 1, ..., m$ and $j \neq k$.

5.3.5 Linear System Kernels

If the multiple-input, multiple-output system equation

$$d\mathbf{e} = g(\mathbf{e}, \mathbf{c})\, d\mathbf{c} \qquad (5.3.21)$$

has a linear kernel, then each component of the system kernel matrix $g(\mathbf{e}, \mathbf{c})$ has the general form

$$g_{ij}(\mathbf{e}, \mathbf{c}) = G_{ij}(\mathbf{c}) + \sum_{k=1}^{n} H_{ijk}(\mathbf{c})e_k \qquad (5.3.22)$$

The multiple-input, multiple-output linear system equation can then be expressed by the following matrix equation:

$$d\mathbf{e} = [G(\mathbf{c}) + H(\mathbf{c})\,E(\mathbf{c})]\, d\mathbf{c} \qquad (5.3.23)$$

where the nonhomogeneous matrix $G(\mathbf{c})$ is an $(n \times m)$ matrix of the form

$$G(\mathbf{c}) = \begin{bmatrix} G_{11}(\mathbf{c}) & G_{12}(\mathbf{c}) & \cdots & G_{1m}(\mathbf{c}) \\ G_{21}(\mathbf{c}) & G_{22}(\mathbf{c}) & & G_{2m}(\mathbf{c}) \\ \vdots & & & \\ G_{n1}(\mathbf{c}) & G_{n2}(\mathbf{c}) & \cdots & G_{nm}(\mathbf{c}) \end{bmatrix}$$

The coefficient matrix $H(\mathbf{c})$ is an $(n \times nm)$ matrix, as follows:

$$H(\mathbf{c}) = \begin{bmatrix} H_{111}\,H_{112}\cdots H_{11n} & \cdots & H_{m11}\,H_{m12}\cdots H_{m1n} \\ \vdots & & \\ H_{1n1}\,H_{1n2}\cdots H_{1n1} & \cdots & H_{mn1}\,H_{mn2}\cdots H_{mnn} \end{bmatrix}$$

and finally the modified output matrix $E(\mathbf{e})$ is an $(nm \times m)$ matrix with the special linear form

$$
E(\mathbf{e}) =
\begin{bmatrix}
e_1 & 0 & \cdots & 0 \\
\vdots & \vdots & & \vdots \\
e_n & 0 & \cdots & 0 \\
0 & e_1 & \cdots & 0 \\
& \vdots & & \\
0 & e_n & \cdots & 0 \\
\vdots & \vdots & & \vdots \\
0 & 0 & \cdots & e_1 \\
\vdots & \vdots & & \vdots \\
0 & 0 & \cdots & e_n
\end{bmatrix}
$$

There are alternative forms to express the system Eq. (5.3.21) when the system kernel is linear. For example, an alternative system matrix equation is given by

$$
d\mathbf{e} = G(\mathbf{c})\,d\mathbf{c} + H_1(\mathbf{c})\mathbf{e}\,dc_1 + \cdots + H_m(\mathbf{c})\mathbf{e}\,dc_m \tag{5.3.24}
$$

where the nonhomogeneous matrix $G(\mathbf{c})$ is the same as previously defined and

$$
H_i(\mathbf{c}) =
\begin{bmatrix}
H_{1i1}(\mathbf{c}) & H_{1i2}(\mathbf{c}) & \cdots & H_{1im}(\mathbf{c}) \\
H_{2i1}(\mathbf{c}) & H_{2i2}(\mathbf{c}) & & H_{2im}(\mathbf{c}) \\
\vdots & & & \\
H_{ni1}(\mathbf{c}) & H_{ni2}(\mathbf{c}) & \cdots & H_{nim}(\mathbf{c})
\end{bmatrix},
\quad
\mathbf{e} =
\begin{bmatrix}
e_1 \\
e_2 \\
\vdots \\
e_n
\end{bmatrix}
$$

However, Eq. (5.3.23) exhibits a matrix format similar to the general solution form, as we will show in Section 7.4.

If the coefficient matrix $H(\mathbf{c})$ has components that are functions of the inputs c, then series solution methods as described in Section 4.3.5 are applicable. We will demonstrate the series solution methodology for a double-input, double-output system.

Consider the following double-input, double-output linear system equation:

$$
de_1 = (G_{11} + H_{111}e_1 + H_{112}e_2)dc_1 + (G_{12} + H_{121}e_1 + H_{122}e_2)dc_2 \tag{5.3.25.1}
$$

$$
de_2 = (G_{21} + H_{211}e_1 + H_{212}e_2)dc_1 + (G_{22} + H_{221}e_1 + H_{222}e_2)dc_2 \tag{5.3.25.2}
$$

Equations (5.3.25) can also be expressed in the equivalent matrix form

$$
\begin{bmatrix} de_1 \\ de_2 \end{bmatrix} =
\begin{bmatrix} g_{11} & g_{11} \\ g_{11} & g_{11} \end{bmatrix}
\begin{bmatrix} dc_1 \\ dc_2 \end{bmatrix} \tag{5.3.26}
$$

We assume that all coefficient functions can be described as single power series functions as follows, for $i, k, l, m = 1, 2$:

$$
G_{kl} = \sum_{i,j=0}^{\infty} a_{kl}(i,j)\, c_1{}^i c_2{}^j \tag{5.3.27}
$$

and

$$H_{klm} = \sum_{i,j=0}^{\infty} b_{klm}(i,j)\, c_1{}^i c_2{}^j \tag{5.3.28}$$

where all the a's and b's are known constants. Now let us assume that both e_1 and e_2 can be expressed as follows, for $i = 1,\ 2$:

$$e_{ki} = \sum_{k,l=0}^{\infty} d_i(k,l)\, c_1^k c_2^l \tag{5.3.29}$$

We have for the total derivative of e_i that

$$de_i = \frac{\partial e_i}{\partial c_1}\, dc_1 + \frac{\partial e_i}{\partial c_2}\, dc_2 \tag{5.3.30}$$

So we have for Eqs. (5.3.25) and (5.3.30) for $i = 1,\ 2$ that

$$\left(\frac{\partial e_i}{\partial c_1} - G_{i1} - H_{i11}e_1 - H_{i12}e_2\right)dc_1 + \left(\frac{\partial e_i}{\partial c_2} - G_{i2} - H_{i21}e_1 - H_{i22}e_2\right)dc_2 \tag{5.3.31}$$

Now dc_1 and dc_2 are independent differentials, so the coefficients of each must vanish. Thus, we have four independent equations of the following form, for $i = 1,\ 2$ and $j = 1,\ 2$:

$$\frac{\partial e_i}{\partial c_j} = G_{ij} - H_{ij1}e_1 - H_{ij2}e_2 \tag{5.3.32}$$

If Eq. (5.3.32) is expressed in terms of our series expansions given by Eqs. (5.3.27), (5.3.28), and (5.3.29), we obtain the following requirements for the arbitrary constants $d_i(m,\ n)$:

$$d_i(m+1,n) = \frac{1}{m+1}$$
$$\left\{ a_{i\,1}(m,n) + \sum_{k,j=0}^{m,n} [b_{i11}(k,j)d_1(m-k,n-j) + b_{i12}(k,j)d_2(m-k,n-j)] \right\} \tag{5.3.33.1}$$

$$d_i(m,n+1) = \frac{1}{n+1}$$
$$\left\{ a_{i2}(m,n) + \sum_{k,j=0}^{m,n} [b_{i21}(k,j)d_1(m-k,n-j) + b_{i22}(k,j)d_2(m-k,n-j)] \right\} \tag{5.3.33.2}$$

for $i = 1,\ 2$ and $m,\ n = 0,\ 1,\ \ldots$.

Equations (5.3.33) provide recursion formulas for all $d_i(k,l)$ coefficients in terms of the two arbitrary constants $d_1(0,0)$ and $d_2(0,0)$. Thus, the output solutions e_1 and e_2, as defined by Eq. (5.3.29), are known in terms of the two independent constants. However, the requirement that Eqs. (5.3.27) and (5.3.28) be exact imposes the additional requirement, as determined in Section 5.3.4, that the series coefficients $a_{kl}(i,j)$ and $b_{klm}(i,j)$ satisfy the following relations:

$$\frac{\partial g_{12}}{\partial c_1} + g_{11}\frac{\partial g_{12}}{\partial e_1} = \frac{\partial g_{11}}{\partial c_2} + g_{12}\frac{\partial g_{11}}{\partial e_1} \tag{5.3.34.1}$$

and

$$\frac{\partial g_{22}}{\partial c_1} + g_{21} \frac{\partial g_{22}}{\partial e_1} = \frac{\partial g_{21}}{\partial c_2} + g_{22} \frac{\partial g_{21}}{\partial e_2} \tag{5.3.34.2}$$

In terms of the functions G_{ij} and H_{ijk}, Eqs. (5.3.34.1) and (5.3.34.2) provide

$$\frac{\partial G_{12}}{\partial c_1} + e_1 \frac{\partial H_{121}}{\partial c_1} + e_2 \frac{\partial H_{122}}{\partial c_1} + [G_{11} + H_{111}e_1 + H_{112}e_2] H_{121}$$

$$= \frac{\partial G_{11}}{\partial c_2} + e_1 \frac{\partial H_{111}}{\partial c_2} + e_2 \frac{\partial H_{112}}{\partial c_2} + [G_{12} + H_{121} e_1 + H_{122} e_2] H_{111}$$

and

$$\frac{\partial G_{22}}{\partial c_1} + e_1 \frac{\partial H_{221}}{\partial c_1} + e_2 \frac{\partial H_{222}}{\partial c_1} + [G_{21} + H_{211} e_1 + H_{212} e_2] H_{222}$$

$$= \frac{\partial G_{21}}{\partial c_2} + e_1 \frac{\partial H_{211}}{\partial c_2} + e_2 \frac{\partial H_{212}}{\partial c_2} + [G_{22} + H_{221} e_1 + H_{222} e_2] H_{212}$$

The application of series solution methods to higher-order input, output linear system equations entails the same steps we have just completed for the double-input, double-output linear system equation. However, it should be apparent that the series solution method becomes increasingly complex as the system order increases.

Now let us consider the multiple-input, multiple-output linear system equation with constant coefficients. It is convenient to consider the system equation in the matrix form given by Eq. (5.3.24), where

$$d\mathbf{e} = G(\mathbf{c}) \, d\mathbf{c} + H_1 \mathbf{e} \, dc_1 + \cdots + H_m \mathbf{e} \, dc_m \tag{5.3.35}$$

and where the H_i coefficients are now constant matrices. We treat the homogeneous equation [i.e., with $G(\mathbf{c})$ set to zero], recognizing that for any linear system, a particular solution can be added to the homogeneous solution to provide the complete solution.

Now we seek a solution of the following form for Eq. (5.3.35) with $G(\mathbf{c}) = 0$:

$$\mathbf{e} = \mathbf{e}_0 \exp(\lambda \, \mathbf{c})$$

where

$$\mathbf{e}_0 = \begin{bmatrix} e_{10} \\ e_{20} \\ \vdots \\ e_{n0} \end{bmatrix}, \quad \lambda = [\lambda_1 \cdots \lambda_m], \quad \mathbf{c} = \begin{bmatrix} c_1 \\ c_2 \\ \vdots \\ c_m \end{bmatrix}$$

If we differentiate our trial solution, we have

$$d\mathbf{e} = \mathbf{e}_0 \exp(\lambda \, \mathbf{c}) \sum_1^n \lambda_i \, dc_i = \mathbf{e} \, \lambda \, d\mathbf{c} \tag{5.3.36}$$

Now, if we substitute Eq. (5.3.36) into the homogeneous form of Eq. (5.3.35), we find after some rearrangement that

$$0 = (H_1 - \lambda_1 \, I) \, \mathbf{e} \, d \, c_1 + \cdots + (H_m - \lambda_m I) \, \mathbf{e} \, d \, c_m$$

Since the c's are independent variables, each coefficient of a dc_i term for $i = 1, ..., m$ must vanish. Furthermore, we are not interested in the trivial solution when $e = 0$. So we obtain an eigenvalue matrix equation of the general form

$$(H_i - \lambda_i I) = 0 \tag{5.3.37}$$

for $i = 1, ..., m$. For each eigenvalue λ_i, Eq. (5.3.37) will yield an nth degree polynomial equation in λ_i. For each polynomial equation, there will be n complex roots in general of the form

$$\lambda_{1i}, \lambda_{2i}, ..., \lambda_{ni}$$

So the general homogeneous solution of the linear, multiple-input, multiple-output system equation then has the form

$$
\mathbf{e} =
\begin{bmatrix}
e_{110} & \cdots & e_{1n0} \\
\vdots & & \\
e_{n10} & \cdots & e_{nn0}
\end{bmatrix}
\exp \left\{
\begin{bmatrix}
\lambda_{11} & \cdots & \lambda_{1m} \\
\vdots & & \\
\lambda_{n1} & \cdots & \lambda_{nm}
\end{bmatrix}
\begin{bmatrix}
c_1 & \cdots & 0 \\
\vdots & & \\
0 & \cdots & c_m
\end{bmatrix}
\right\}
\tag{5.3.38}
$$

Only n of the constants e_{ij0} in Eq. (5.3.38) are independent and are determined by the initial conditions [i.e., $\mathbf{e}(c_0)$ for the system equation].

5.3.6 Separable System Kernels

If the multiple-input, multiple-output system equation

$$d\mathbf{e} = g(\mathbf{e}, \mathbf{c}) \, d\mathbf{c}$$

has a system kernel with components that can be made to satisfy the functional form

$$g_{ij}(\mathbf{e}, \mathbf{c}) = F_i(e_i) \frac{\partial G_i}{\partial c_j}(\mathbf{c})$$

for $i = 1, ..., n$ and $j = 1, ..., m$, then the system equation can be separated as follows for any output:

$$\frac{1}{F_i(e_i)} \, d e_i = \frac{\partial G_i}{\partial c_1}(\mathbf{c}) \, dc_1 + \cdots + \frac{\partial G_i}{\partial c_m}(\mathbf{c}) \, dc_m \tag{5.3.39}$$

Equation (5.3.39) is now exact and is directly integrable to provide the following solution for the system equation:

$$\int \frac{d e_i}{F_i(e_i)} = G_i(\mathbf{c}) + k_i \tag{5.3.40}$$

where k_i; is the arbitrary integration constant. In general, for a given multiple-input, multiple-output system equation, it requires considerable effort to determine and develop an integrating factor that will make the system equation separable. However, if the system investigator recognizes the requirement for separability and integrability specified by Eq. (5.3.39), then it may be possible, if appropriate and justified, to quantitatively model the system kernel so that it will, indeed, be separable and admit a solution as defined by Eq. (5.3.40).

5.3.7 Equivalence of Multiple-Input, Multiple-Output Systems and Partial Differential Equations

Consider the following multiple input, output system equation with two inputs and three outputs:

$$\begin{bmatrix} de_1 \\ de_2 \\ de_3 \end{bmatrix} = \begin{bmatrix} g_{11} & g_{12} \\ g_{21} & g_{22} \\ g_{31} & g_{32} \end{bmatrix} \begin{bmatrix} dc_1 \\ dc_2 \end{bmatrix} \tag{5.3.41}$$

Now let us assume that there exists a function $z = F(x, y)$ that has well defined partial derivatives for z, as follows:

$$p = \frac{\partial z}{\partial x}, \quad q = \frac{\partial z}{\partial y}, \quad r = \frac{\partial^2 z}{\partial x^2}, \quad s = \frac{\partial^2 z}{\partial x \partial y} = \frac{\partial^2 z}{\partial y \partial x}, \quad t = \frac{\partial^2 z}{\partial y^2} \tag{5.3.42}$$

We can then define the total differential

$$dz = \frac{\partial z}{\partial x} dx + \frac{\partial z}{\partial y} dy = p \, dx + q \, dy \tag{5.3.43}$$

and for the second-order total differentials, we have z

$$d\left(\frac{\partial z}{\partial x}\right) = \frac{\partial^2 z}{\partial x^2} dx + \frac{\partial^2 z}{\partial x \partial y} dy \tag{5.3.44.1}$$

$$d\left(\frac{\partial z}{\partial y}\right) = \frac{\partial^2 z}{\partial y \partial x} dx + \frac{\partial^2 z}{\partial y^2} dy \tag{5.3.44.2}$$

Equations (5.3.44) can be expressed in terms of variables defined in expressions (5.3.42) as follows:

$$dp = r \, dx + s \, dy \tag{5.3.45}$$

$$dq = s \, dx + t \, dy \tag{5.3.46}$$

If we now assemble Eqs. (5.3.43), (5.3.45), and (5.3.46) as a matrix set, we have that

$$\begin{bmatrix} dz \\ dp \\ dq \end{bmatrix} = \begin{bmatrix} p & q \\ r & s \\ s & t \end{bmatrix} \begin{bmatrix} dx \\ dy \end{bmatrix} \tag{5.3.47}$$

Equation (5.3.47) is identical with system Eq. (5.3.41) for two inputs and three outputs if we define variable equivalences as follows:

$$e_1 = z, \quad e_2 = p = \frac{\partial z}{\partial x} = g_{11}, \quad e_3 = q = \frac{\partial z}{\partial y} = g_{12}$$

$$r = \frac{\partial^2 z}{\partial x^2} = g_{21}, \quad s = \frac{\partial^2 z}{\partial x \partial y} = \frac{\partial^2 z}{\partial y \partial x} = g_{22} = g_{31}$$

$$t = \frac{\partial^2 z}{\partial y^2} = g_{32}, \quad dx = dc_1, \quad dy = dc_2$$

Now suppose we have a general second-order partial differential equation that can be expressed as

$$R \frac{\partial^2 z}{\partial x^2} + S \frac{\partial^2 z}{\partial x \, \partial y} + T \frac{\partial^2 z}{\partial y^2} = V \tag{5.3.48}$$

where R, S, T, and V may be functions of x, y, z, $\frac{\partial z}{\partial x}$, and $\frac{\partial z}{\partial y}$. If we solve for any second partial derivative term explicitly, such as $\frac{\partial^2 z}{\partial x^2} = r$, then from Eq. (5.3.48), we have that

$$R r + S s + T t = V \tag{5.3.49}$$

If we replace r in the system equation (5.3.47) by $(V - S s - T t)/R$ then the z solution component for system equation. (5.3.47) will provide a general solution for Eq. (5.3.48), and our system equation (5.3.41) is associated with a partial differential equation of the form defined by Eq. (5.3.48). It is interesting to observe that Eq. (5.3.47) can be expressed in several equivalent reduced forms. For example,

$$d z = \left\{ \int (r \, dx + s \, dy) \right\} dx + \left\{ \int (s \, dx + t \, dy) \right\} dy \tag{5.3.50}$$

which expresses the single-output dz in terms of inputs dx and dy, and system kernels that are integral expressions. Another reduced form for Eq. (5.3.47) is

$$d z = \left(\frac{p t - q s}{r t - s^2} \right) dp + \left(\frac{r q - s p}{r t - s^2} \right) dq \tag{5.3.51}$$

which expresses the output dz in terms of inputs dp and dq. Again, either system Eq. (5.3.50) or (5.3.51) can be associated with a second-order partial differential equation such as Eq. (5.3.48) by employing it in solving the respective system equation.

5.3.8 Reduction of Multiple-Input, Multiple-Output Systems Equations

For the general multiple-input, multiple-output system equation, it is the occurrence of more than one input to the system that particularly complicates both analytic and even computer-based analysis of the system equation. This observation is probably to be expected since multiple-input systems are associated with partial differential and difference equations, and their difficulty for analysis is well known. However, there is an option available that allows the system investigator to reduce the number of inputs to any lesser numbered desired. To demonstrate this reduction method, consider the general multiple-input, multiple-output system equation

$$d\mathbf{e} = \mathbf{g}(\mathbf{e}, \mathbf{c}) \, d\mathbf{c} \tag{5.3.52}$$

Now let us transform the input or independent matrix variable using the following linear transformation:

$$\mathbf{c} = \mathbf{c}_0 + \lambda \mathbf{t} \tag{5.3.53}$$

And we have for the variable differentials, since $\mathbf{c}_0$ is a constant,

$$d\mathbf{c} = \lambda \, d\mathbf{t}$$

where

$$
\mathbf{c_0} = \begin{bmatrix} c_{10} \\ c_{20} \\ \vdots \\ c_{m0} \end{bmatrix}, \quad
\lambda = \begin{bmatrix} \lambda_{11} & \lambda_{12} & \cdots & \lambda_{1m} \\ \lambda_{21} & \lambda_{22} & & \\ \vdots & & & \\ \lambda_{m1} & \lambda_{m2} & \cdots & \lambda_{mm} \end{bmatrix}, \quad
\mathbf{t} = \begin{bmatrix} t_1 \\ t_2 \\ \vdots \\ t_m \end{bmatrix}
$$

We observe that $\mathbf{c_0}$ is the initial value of $\mathbf{c}$ and results if $t = 0$. The $(m \times m)$ square matrix λ is an arbitrary constant matrix, and $\mathbf{t}$ is the transformed input variable set, which to some extent is also arbitrary.

Suppose, for example, we wish to reduce system equation (5.3.52) to a single-input system. Then our linear transform set given by Eq. (5.3.53) would reduce to

$$
\mathbf{c} = \mathbf{c_0} + \lambda t, \quad d\mathbf{c} = \lambda\, dt \tag{5.3.54}
$$

where now

$$
\lambda = \begin{bmatrix} \lambda_1 \\ \lambda_2 \\ \vdots \\ \lambda_m \end{bmatrix}
$$

and t is a scalar variable. If we wished to span the input vector space from $\mathbf{c_0}$ (minimum value of $\mathbf{c}$) to $\mathbf{c}(\max)$, then we would define λ such that

$$
\lambda = [\mathbf{c}\,(\max) - \mathbf{c_0}]
$$

and t would be defined over the range $0 \le t \le 1$.

Employing Eq. (5.3.54) to reduce the inputs to one input, the multiple-input, multiple-output system equation becomes

$$
d\mathbf{e} = g(\mathbf{e}, \mathbf{c_0} + \lambda\, t)\ dt
$$

On the other hand, if we sought the response of the system to a single specific input, say $\mathbf{c}_j$, then we would let λ become max

$$
\lambda = \begin{bmatrix} 0 \\ \vdots \\ \lambda_j \\ \vdots \\ 0 \end{bmatrix} \quad \text{where } \lambda_j = \mathbf{c}_j(\max) - \mathbf{c}_{j0}
$$

and our input variable transformation would then become

$$
\mathbf{c} = \mathbf{c}^* + \begin{bmatrix} 0 \\ \vdots \\ \lambda_j \\ \vdots \\ 0 \end{bmatrix} t, \quad \text{where } \mathbf{c}^* = \begin{bmatrix} c_1{}^* \\ \vdots \\ c_{j0} \\ \vdots \\ c_m{}^* \end{bmatrix}
$$

The constant $\mathbf{c}^*$ would be chosen as a constant value to represent all other values of the inputs by a selected constant $\mathbf{c}^*$ for each $i = 1, ..., m$ with $\mathbf{c}^* = \mathbf{c}_{0j}$.

Of course, two or more specific inputs could be selected to assess the system's behavior for a larger set of inputs. Thus, the system investigator by judicious selection of representative inputs can sample the behavior of the system by solving the system equation sequentially for these representative inputs. Often this reduction procedure will prove to be the only practical means by which systems with many inputs and outputs can be analyzed by either analytical or computer methods.

By a similar transformation of the output or dependent variable, it is also possible to reduce the number of outputs of a system. However, the benefits of such an action are not nearly as evident as is the reduction of system inputs we have examined in this section.

5.3.9 Integral Equation Form of the System Equation

All continuous system equations with a finite number of inputs and outputs can also be expressed in an equivalent integral equation form. To establish this assertion, consider the general multiple-input, multiple-output system equation with or without feedback.

$$d\mathbf{e} = g(\mathbf{e}, \mathbf{c}) \, d\mathbf{c}$$

Now, if we simply integrate both sides of this equation over the interval $\mathbf{c}_0$ to $\mathbf{c}$, then we have that

$$\int_{e(c_0)}^{e(c)} d\mathbf{e} = \int_{c_0}^{c} g(\mathbf{e}, \mathbf{c}) \, d\mathbf{c}$$

and we obtain for the matrix output a that

$$\mathbf{e} = \int_{c_0}^{c} g(\mathbf{e}, \mathbf{c}) \, d\mathbf{c} + \mathbf{e}_0 \tag{5.3.55}$$

where $\mathbf{e}_0$ is a column matrix representing the initial values of the outputs for the input values $\mathbf{c}_0$.

It is interesting to examine the meaning and significance of Eq. (5.3.55). This equation defines the system output state $\mathbf{e}$ as the continuous sum (or integral) of all differential inputs times their respective instantaneous system kernel components. Thus, the system output state $\mathbf{e}$ is a function of the total input history from input state $\mathbf{c}_0$ to $\mathbf{c}$.

Equation (5.3.55) is characterized as a nonlinear matrix integral equation if the system kernel $g(\mathbf{e}, \mathbf{c})$ is a nonlinear function of any of the output components e_i composing $\mathbf{e}$. Unfortunately, there is no closed-form method for solving Eq. (5.3.55) in general. There is, however, a method, called Picard's method that entails successive iterations of Eq. (5.3.55) and thereby provides a sequence of functions whose limit approaches the solution of Eq. (5.3.55). To briefly examine Picard's method, suppose in Eq. (5.3.55) we evaluate the kernel at the initial output value $\mathbf{e}_0$ (a constant) and assume we are able to perform the integration over $\mathbf{c}$ to find $\mathbf{e}_1$, where

$$\mathbf{e}_1(\mathbf{c}) = \int_{c_0}^{c} g(\mathbf{e}_0, \mathbf{c}) \, d\mathbf{c} + \mathbf{e}_0$$

Now, if we employ the function $e_1(c)$ (in general, a function c) to evaluate the kernel and are again able to integrate the equation over c, we obtain $e_2(c)$, where

$$e_2(c) = \int_{c_0}^{c} g(e_1, c) \, dc + e_0$$

Continuing this procedure, we find, for the nth integration,

$$e_n(c) = \int_{c_0}^{c} g(e_{n-1}, c) \, dc + e_0$$

Now under appropriate conditions we will find that, as the number of integration steps n becomes large, $e_n(c)$, will approach, in the limit, the previous output $e_{n-1}(c)$, so we have that $e_n(c) \to e_\infty(c)$ and

$$e_\infty(c) = \int_{c_0}^{c} g(e_\infty, c) \, dc + e_0$$

But it is apparent that e_∞ is also the solution of Eq. (5.3.55), and we conclude that $e_\infty(c) \to e$. If only a few integrations are required for $e_n(c) \to e_{n-1}$, then Picard's method will provide an analytic approximation for the multiple-input, multiple-output system equation. However, although Picard's method has important theoretical value, the method often proves impractical as a solution method because of the successively more difficult integrations required with each iteration step.

Some other rather powerful analytical and numerical methods for solving Eq. (5.3.55) exist if the kernel is linear in e. A linear system equation has the general form

$$g(e, c) = g_c(c) + g_e(c) \, e$$

The integral equation form of the system equation, Eq. (5.3.55), with a linear system kernel now assumes the general form of a set of Volterra integral equations, as follows:

$$e = \int_{c_0}^{c} g_e(c) \, e \, dc + f(c)$$

where

$$f(c) = e_0 + \int_{c_0}^{c} g_c(c) \, dc$$

However, we will not undertake these advanced methods here, but refer the reader to the literature on integral equation methods.

5.3.10 Elimination of Individual Output Solutions to Reduce the System Equation

Consider the general system equation for either a discrete or continuous system:

$$\left. \begin{array}{c} de \\ \Delta e \end{array} \right\} = g(e, c) \left\{ \begin{array}{c} dc \\ \Delta c \end{array} \right. \tag{5.3.56}$$

with m inputs, $\mathbf{c}$, and n outputs, $\mathbf{e}$. Suppose that a general solution for the ith output, e_i, is known that satisfies the ith system equation as appropriate:

$$\left.\begin{array}{c} de_i \\ \Delta e_i \end{array}\right\} = g_i(\mathbf{e}, \mathbf{c}) \left\{\begin{array}{c} d\mathbf{c} \\ \Delta\mathbf{c} \end{array}\right.$$

where

$$e_i = G_i(\mathbf{e}, k_i) \tag{5.3.57}$$

Equation (5.3.57) may be used in system equation (5.3.56) to eliminate the dependence of e_i. Thus, system equation (5.3.56) may be reduced to $(n-1)$ outputs. If additional solutions can be found for other system outputs, the system equation may be further reduced. This method for reducing the number of system outputs is frequently of practical value, particularly for systems with small number of outputs. However, if the system equation is nonlinear, it should be remembered that singular or isolated solutions for a given output may not be included in the general solution and, thus, the reduced system equation would not reflect these singular solutions.

5.4 Summary

In this chapter we have examined some major analytical methods and related requirements for solving system equations with multiple inputs. We found that all discrete multiple-input systems could be analyzed analytically by successive calculations of the resulting output response for each stage of input changes.

Continuous multiple-input systems admit a variety of analytical methods for analysis, which are dependent upon the form of the system kernel. The principal multiple-input system kernels that were recognized and for which appropriate analysis methods were considered included constant, linear, separable, homogeneous, and invertible kernels. The direct correspondence of multiple-input systems with partial differential equations was also established, and the application of solution methods for partial differential equations to related system equations is evident.

Because the general multiple-input system equation usually proves to be intractable for analytical methods for analysis, the use of numerical methods and some special methods for reducing system inputs are very useful and important for successful analysis of the system. Some of these methods for reducing the number of system inputs are addressed in Chapter 8. These reduction methods find application both for analytical and computer-based analysis methods and are often imposed to gain insight into the behavior of the system modeled. Furthermore, the capability of rapid and effective graphical display of these solution attempts offered by MATLAB is of particular value in assessing these solutions for system science. We will fully demonstrate this graphical capability for system study and evaluation in Chapter 10.

Bibliography

ALEXANDER, J. E., and J. M. BAILEY, *Systems Engineering Mathematics*, Prentice Hall, Englewood Cliffs, NJ, 1963. A useful reference for engineering analysis methods.

BATEMAN, H. J., *Partial Differential Equations of Mathematical Physics*, Cambridge University Press, 1958. A classic treatment of partial differential equations.

BROWN, B. M., *The Mathematical Theory of Linear Systems*, Chapman & Hall, London, 1965. A standard reference for linear system analysis.

COURANT, R., and D. HILBERT, *Methods of Mathematical Physics*, Vol. **1 & 1**, Revised 1953, Wiley on Line. The classic treatment of mathematical analysis in Science. Chpt. I, Vol. 1 treats first-order systems with multiple inputs.

DEUTSCH, R., *System Analysis Techniques, Prentice-Hall*, Englewood Cliffs, NJ, 1969. This reference provides some advanced analytical methods.

INCE, E. L., *Ordinary Differential Equations*, Dover Publications, NY, 1956. An essential, encyclopedic reference on ordinary differential equations.

MILNE-THOMSON, L. M., *The Calculus of Finite Differences*, Macmillan, NY, 1951. A standard reference for the mathematical methods of finite difference equations.

MORRIS, M., and O. E. BROWN, *Differential Equations*, Prentice-Hall, Englewood Cliffs, NJ, 1952. A very readable treatment of partial differential equations.

PERLIS, S., *Theory of Matrices*, Addison-Wesley Publishing Co., Reading, MA, 1952. A classical treatment of matrix theory.

RICHARDSON, C. H., *An Introduction to the Calculus of Finite Differences*, Van Nostrand, Reinhold, NY, 1954. A useful introduction to finite difference equations.

SCHOUTEN, J. A., and W. V. D. KULK, *Pfaff's Problem and Its Generalizations*, Oxford University Press, NY, 1949. A general reference on Pfaff's problem.

THOMAS, J. M., *Differential Systems*, American Mathematical Society, Providence, RI, 1937. Displays the relationship between Pfaffian sets and sets of partial differential equations.

VOLTERRA, V., *Theory of Functionals and of Integral and Integro-Differential Equations*, Dover Publications, NY, 1959. Introduces the use of integral equations for systems.

Problems

1 Test the following system equations for exactness. If the equation is not exact, find an integrating factor if possible and solve the system equation.

a) $dz = \dfrac{(2xy - 1)\, dx + (x^2 + \cos t)\, dy}{y \sin z + 2z}$

b) $de = \left[c_1 - \dfrac{c_2}{c_1{}^2 + c_2{}^2}\right] dc_1 + \left[c_2 + \dfrac{c_1}{c_1{}^2 + c_2{}^2}\right] dc_2$

c) $de = \dfrac{(2c_1 e + 1)\, e}{c_1 - e}\, dc_1 + \dfrac{2c_2 e^2}{c_1 - e}\, dc_2$

d) $de = (a + c_2)\, dc_1 + \dfrac{e}{a + c_2}\, dc_2$

2 Test the following system equations for integrability and attempt to solve those that are integrable either manually or with the MATLAB symbolic toolbox.

a) $(y^3 - xy^2) dx + x^2 y \, dy + 2z \, dz = 0$

b) $de = \dfrac{c_2}{e} dc_1 - \dfrac{c_1}{e} dc_2$

c) $dx = -\dfrac{dy + dz}{x + y + z + 1}$

d) $(x + z) dx + x^2 z \, dy + (xzy - x) dz = 0$

e) $de = (2c_1 c_2 + 2c_1^2 c_3 - 2c_1 e + c_3) dc_1 + dc_2 + c_1 dc_3$

f) $(t + y) dx + (x + z) dy + (y + t) dz + (z + x) dt = 0$

g) $de = -\dfrac{c_2}{c_1} dc_1 - \dfrac{c_3}{c_1} dc_2 - \dfrac{e}{c_1} dc_3$

h) $3x^2 y^2 dx + 2(z + x^3 y) dy + (2y + t) dz + z \, dw = 0$

3 Show that the necessary condition for integrability of the differential equation

$$W \, dw + X \, dx + Y \, dy + Z \, dz = 0$$

is the following:

$$W X \left(\frac{\partial Y}{\partial z} - \frac{\partial Z}{\partial y} \right) + X Y \left(\frac{\partial Z}{\partial w} - \frac{\partial W}{\partial z} \right)$$
$$+ Y Z \left(\frac{\partial W}{\partial x} - \frac{\partial X}{\partial w} \right) + Z W \left(\frac{\partial X}{\partial y} - \frac{\partial Y}{\partial x} \right) = 0$$

4 Solve the following system equations:

a) $de = (2bc_1 + ac_2) dc_1 + (ac_1 + 2bc_2) dc_2$

b) $\begin{aligned} de_1 &= a \, e_1/c_1 dc_1 + b \, e_1/c_2 \, dc_2 \\ de_2 &= f e_2/c_1 dc_1 + g \, e_2/c_2 \, dc_2 \end{aligned}$

c) see equations below

$$de_1 = \frac{1}{c_2^2 - c_1^2} [\{c_2(ac_1 - e_2) - c_1(be_2 - e_1)\} dc_1 + \{c_2(ac_2 - e_1) - c_1(be_1 - e_2)\} dc_2]$$

$$de_2 = \frac{1}{c_1^2 - c_2^2} [\{c_1(ac_1 - e_2) - c_2(be_2 - e_1)\} dc_1 + \{c_1(ac_2 - e_1) - c_2(be_1 - e_2)\} dc_2]$$

d) $\begin{aligned} de_1 &= (Ac_2 + Aac_1 c_2 - ae_1) dc_1 + (Ac_1 + Abc_1 c_2 - be_1) dc_2 \\ de_2 &= (Bc_2 + Bbc_1 c_2 - be_2) de_1 + (Bc_1 + Bac_1 c_2 - ae_1) dc_2 \end{aligned}$

5 Determine the conditions required for integrability to solve these equations and attempt to obtain solutions either manually or with the MATLAB symbolic toolbox.

a) $\begin{aligned} de_1 &= (1 + ae_1) de_1 + be_2 dc_2 \\ de_2 &= f e_2 de_1 + (1 + ge_1) dc_2 \end{aligned}$

b) $de = (Ac_1^2 + Bc_2^2) de_1 + (De_1 c_2) dc_2$

c) $de = \dfrac{Ac_1 + Bc_2}{c_1} dc_1 + \dfrac{De_1}{c_1 + E c_2} dc_2$

d) $\begin{aligned} de_1 &= f_1(c_1 + Ac_2) dc_1 + f_1(c_1 - Bc_2) e_1 dc_2 \\ de_2 &= g_1(c_1 - Bc_2) e_1 dc_1 + g_1(c_1 + Ac_2) dc_2 \end{aligned}$

6 Determine the system equation kernels associated with the following system solutions. Regard the e's as output variables, c's as input variables, and k's as the arbitrary integration constants.

a) $e = 1 + Ac_1 + Bc_2 + Dc_1c_2 + Ec_1^2 + Fc_2^2 + k$

b) $e = A + Bc_1 + Dc_2 + E\,\mathrm{Ln}c_1 + F\,\mathrm{Ln}c_2 + k$

c) $e = A + Bc_1 + Dc_2 + E\exp(ac_1) + F\exp(bc_2) + k$

d) $e_1 = A(c_1 - c_2) + B\,\mathrm{Ln}\,(c_1/c_2) + k_1$

   $e_2 = D(c_1 + c_2) + E\exp(c_1 - c_2) + k_2$

e) $e_1 + Ac_1 + Bc_2 + D + k_1 = 0$

   $e_1 + Ee_2 + Fc_1 + Gc_2 + k_2 = 0$

f) $e_1 = f_1(c_1) + e_2f_2(c_1, c_2) + k_1f_3(c_2)$

   $e_2 = g_1(c_2) + e_1g_2(c_1, c_2) + k_2g_3(c_1)$

   where the f's and g's are continuous functions of their augments

g) $e = f(c_1, c_2) + \dfrac{g(c_1, c_2)}{h(c_1, c_2) + k}$

   where f, g, and h are continuous functions of their augments

7 Attempt to solve the following partial differential equations and determine the associated system equations:

a) $(x + y)\dfrac{\partial z}{\partial x} + (x - y)\dfrac{\partial z}{\partial y} = 1 + z$

b) $Ac_2\dfrac{\partial e}{\partial c_1} + Bc_1\dfrac{\partial e}{\partial c_2} = e$

c) $xyz\dfrac{\partial z}{\partial x} + \dfrac{\partial z}{\partial y} = \text{constant}$

d) $\dfrac{a\,c_1}{c_2}\dfrac{\partial e}{\partial c_1} + \dfrac{b\,c_2}{c_1}\dfrac{\partial e}{\partial c_2} = 0$

e) $y^a\dfrac{\partial z}{\partial x} + x^b\dfrac{\partial z}{\partial y} = z^c$

f) $(A + e)\dfrac{\partial e}{\partial c_1} + (B + e)\dfrac{\partial e}{\partial c_2} = Dc_1$

g) $\dfrac{\partial u}{\partial x} - \dfrac{\partial u}{\partial y} = x + y$

h) $A_1\dfrac{\partial e_1}{\partial c_1} + B_1\dfrac{\partial e_1}{\partial c_2} = D_1$

   $A_2\dfrac{\partial e_2}{\partial c_1} + B_2\dfrac{\partial e_2}{\partial c_2} = D_2$

i) $c_1\dfrac{\partial e}{\partial c_1} + c_2\dfrac{\partial e}{\partial c_2} + c_3\dfrac{\partial e}{\partial c_3} = e$

j) $Ac_1\dfrac{\partial e_1}{\partial c_1} + Bc_2\dfrac{\partial e_1}{\partial c_2} = e_2$

   $Dc_2\dfrac{\partial e_2}{\partial c_1} + Ec_1\dfrac{\partial e_2}{\partial c_1} = e_1$

k) $y\dfrac{\partial z}{\partial x} + x\dfrac{\partial z}{\partial y} = x - y$

l) $A\dfrac{\partial e_1}{\partial c_1} + B\dfrac{\partial e_2}{\partial c_2} = 0$

   $D\dfrac{\partial e_1}{\partial c_1} + E\dfrac{\partial e_2}{\partial c_2} = 0$

m) $2x\dfrac{\partial z}{\partial x} + (y - x)\dfrac{\partial z}{\partial y} = (x - y)$

n) $c_1\dfrac{\partial e_1}{\partial c_1} + c_2\dfrac{\partial e_2}{\partial c_2} = A\,e_1$

   $c_2\dfrac{\partial e_1}{\partial c_1} + c_1\dfrac{\partial e_2}{\partial c_2} = B\,e_2$

o) $\dfrac{\partial u}{\partial x} - \dfrac{\partial u}{\partial y} = x + y$

p) $Ac_1\dfrac{\partial e_1}{\partial c_1} + Bc_2\dfrac{\partial e_1}{\partial c_2} = e_2$

   $Dc_2\dfrac{\partial e_2}{\partial c_1} + Ec_1\dfrac{\partial e_2}{\partial c_2} = e_1$

q) $\dfrac{\partial u}{\partial x} - \dfrac{\partial u}{\partial y} = x + y$

r) $Ac_1\dfrac{\partial e_1}{\partial c_1} + Bc_2\dfrac{\partial e_1}{\partial c_2} = e_2$

   $Dc_2\dfrac{\partial e_2}{\partial c_1} + Ec_1\dfrac{\partial e_2}{\partial c_2} = e_1$

s) $xy\dfrac{\partial z}{\partial x} - x^2\dfrac{\partial z}{\partial y} = yz$

t)
$$Ac_1 \frac{\partial e_1}{\partial c_1} + Bc_2 \frac{\partial e_1}{\partial c_2} = e_2$$

$$Dc_2 \frac{\partial e_2}{\partial c_1} + Ec_1 \frac{\partial e_2}{\partial c_2} = e_2$$

u) $x^2 \frac{\partial z}{\partial x} - xy \frac{\partial z}{\partial y} = yz$

v) $x^2 \frac{\partial z}{\partial x} - y^2 \frac{\partial z}{\partial y} = xy$

w) $x^2 z \frac{\partial z}{\partial x} + y^2 z \frac{\partial z}{\partial y} = x + y$

x) $(z-y)^2 \frac{\partial z}{\partial x} + x z \frac{\partial z}{\partial y} = xy$

y) $y \frac{\partial z}{\partial x} + z \frac{\partial z}{\partial y} = \frac{y}{x}$

z) $(y + z) \frac{\partial u}{\partial x} + (x + z) \frac{\partial u}{\partial y} +$

$(x + y) \frac{\partial u}{\partial z} = u$

8 Solve the following multiple-input equations for one input variable at a time (use procedure provided in Section 5.3.8).

a) $de = (Ac_1 e + Bc_1^2 c_1)dc_1 + F(e)[Dc_1^2 + E(c_2 + 2c_1)]dc_2$

b) $dw = \frac{Ay + Bz}{wxyz} dx + \frac{Cxyzw}{Dx + Ey + z} dy + \left(\frac{G}{x} + \frac{H}{y} + \frac{1}{z}\right)dz$

c) $du = (x + yz)dx + (y + xz)dy + xydz$

d) $de = e\left(\frac{A}{c_1} dc_1 + \frac{B}{c_2} dc_2 + \frac{D}{c_3} dc_3\right)$

e) $de = (e/a)(-(a/c_1) \operatorname{Ln} edc_1 + (c_2 a + Ac_2)dc_2)$

f)
$$du_1 = Axyz \, dx + (Bx + u_1)dy + (Cy + xz)dz$$
$$du_2 = Dxz \, dx + (Ey + u_2)dy + (Fu_2 + xyz)dz$$

9 Determine if the multiple-input, multiple-output system equations are exact. If they are, then find their general solutions.

a) $de_1 = D\left(\frac{x-y}{e_1 - A}\right)(dx - dy)$

$de_2 = E\left(\frac{x+y}{e_2 - B}\right)(dx + dy)$

b) $de_1 = \frac{1}{e_2 - e_1}[(A - De_1 c_2)de_1 + (B - De_1 c_1)dc_2]$

$de_2 = \frac{1}{e_2 - e_1}[(Dc_2 e_2 - A)de_1 + (Dc_2 e_2 - B)dc_2]$

c) $du = (Ax + By)dx + (Bx + Cy)dy$

$dv = (Dxy^2 + Ey + F)dx + (Dx^2 y + Ex + G)dy$

d) $de_1 = \frac{1}{e_1}\left[f(x) + \frac{A}{y}\right]dx + e_1\left[h(y) - A\frac{x}{y^2}\right]dy$

$de_2 = e_1[g(x) + By]dx + e_1[h(y) + Bx]dy$

$(A + Be_1)de_1 = (De_1 + c_1^3 + c_1 c_2^2)dc_1 + (E/c_2 + c_1^2 c_2 + c_2^3)dc_2$

e) $e_2 de_2 = (Gc_1^2 + F/c_2)dc_1 + (Ac_2 - Fc_1/c_1^2)dc_2$

$(A + Ee_3)de_3 = (De_1 + c_2)dc_1 + (Ec_2^2 + c_1)dc_2$

f) $de_1 = (A \exp c_1 + Bc_2c_3)dc_1 + (Ec_2{}^2 + Bc_1c_3)dc_2 + (Be_1c_2)dc_3$

$$de_2 = \left(Dc_2 - F\frac{c_3}{c_1{}^2}\right)dc_1 + \left(De_1 + \frac{1}{c_3}\right)dc_2 + \left(\frac{F}{c_1} - \frac{c_2}{c_3{}^2}\right)dc_3$$

g) Find the values of the constants (A, B, C, D) for these equations to be exact.

$$(A_{11} + A_{12}e_2)de_1 = \left[B_{11}f(c_1) + B_{12}c_1c_2{}^2\right]dc_1 + \left[D_{11}g(c_2) + D_{12}c_1{}^2c_2\right]dc_2$$

$$(A_{21} + A_{22}e_2)de_2 = \left[B_{21}g(c_2) + \frac{B_{22}}{c_2}\right]dc_1 + \left[D_{21}f(c_2) + D_{22}\frac{c_1}{c_2{}^2}\right]dc_2$$

6

System Modeling

In this chapter, we undertake the important task of identifying and quantitatively modeling systems. Frequently, a very valuable aid in the initial identification of the inputs and outputs and various subsystems of a given system is to model graphically the overall system, including explicit designation of all subsystems and internal inputs and outputs. The powerful influence of visual and spatial conception and recognition of the model in a graphical format can be very revealing and productive for both model synthesis and analysis. We will examine block diagrams, signal flow graphs, and organizational diagrams as graphical modeling tools in this chapter.

Several major specific system identification and modeling techniques are examined. The basic single-input, single-output model is examined and justified as an excellent beginning model for many systems. Then the modeling techniques for physical and nonphysical science-based systems are examined, and the modeling differences are identified. The use of experimental methods for system modeling and verification is given, including the valuable aids provided by dimensional analysis and least-squares methods. A method called weighted input; output modeling provides a rational means for ranking and weighting inputs and outputs based on their contribution to the system's behavior. Stochastic system modeling is briefly addressed, and some of the essential features of this important field are assessed. Finally, some heuristic methods are examined.

6.1 Graphical Representation of Systems

A useful and sometimes essential tool in the initial modeling and formulation stage of a system study is the graphical representation of the system, including its inputs and outputs and possible feedback loops. The act of graphically and schematically portraying the system is conducive to accurate identification and improved understanding of what inputs interact with the system components and how these interactions produce the outputs anticipated. It is in the graphical representation stage of modeling that the system investigator should be as thorough and critical of all the known or anticipated system factors as possible. The investigator should attempt to detail the system and individually "componentized" the system elements as much as possible. A major strength of graphical representations lies in the pictorial, panoramic overview of the total system offered to the investigator.

Introduction to System Science with MATLAB, Second Edition.
Gary Marlin Sandquist and Zakary Robert Wilde.
© 2023 John Wiley & Sons Ltd. Published 2023 by John Wiley & Sons Ltd.

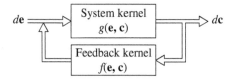

Fig. 6-1 Basic canonical system block diagram.

Once the investigator is satisfied with the graphical representation of the model, the methodologies offered by block diagramming and signal-flow graphs can be imposed. These methodologies provide a rational, systematic mechanism for reducing the system to the basic generalized canonical system representation shown in Figure 6-1.

The general techniques offered by graphical methods permit any compound system composed of subsystems, intermediate inputs, outputs, and feedback conditions to be transformed into a single canonical system diagram described by the basic system equation

$$d\mathbf{e} = g_f(\mathbf{e}, \mathbf{c})d\mathbf{c}$$

where $g_f(\mathbf{e}, \mathbf{c}) = [I - gf]^{-1}g$ if feedback is present. The graphical methods will also assist in identifying the significant inputs $\mathbf{c}$, outputs $\mathbf{e}$, and the composition of the system kernel $g_f(\mathbf{e}, \mathbf{c})$.

Graphical representation methods are useful for both qualitative and quantitative system modeling and analysis. The graphical representation methods we will consider are (1) block diagramming, (2) signal-flow graphs, and (3) organizational charting. Each method is useful in developing accurate and complete system models.

6.1.1 Block Diagramming

A block diagram is a graphical representation of the ordered, sequential pattern by which the system's causes and effects appear and interact with the components of the system. An accurately constructed block diagram for a system provides a simple but powerful means for examining, characterizing, manipulating, and simplifying the functional relationships that exist between the components of the system. Block diagramming is similar to the flowcharting used in computer coding and is usually an early and essential step in scoping and modeling any system.

A block diagram representation for a generalized system model consists of four types of graphical elements as follows:

1) Blocks or rectangular enclosures, which serve as representative system or subsystem boundaries for the system kernel specified within the block.
2) Summing points and takeoff points at which inputs and outputs are combined (added algebraically) or split (separated without any change in signal value).
3) Signal paths along which inputs and outputs are transmitted unidirectional without loss or change.
4) Arrows which assign the direction of transmission of signals along signal paths.

Figure 6-2 is a representative sample of a typical block diagram for a compound system. Observe that signals Δc_1 (input) and Δe_2 (feedback output) are transmitted along their respective signal paths and added at the summing point to form input Δc_2 (i.e., $\Delta c_2 = \Delta c_1 + \Delta e_2$), which is the combined input transmitted to system kernel G_1.

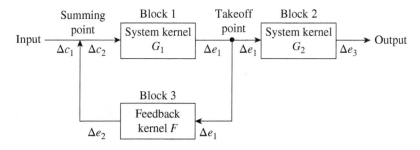

Fig. 6-2 Typical block diagram for a compound system.

Output Δe_1, which is determined by the system equation $\Delta e_2 = G_1 \Delta e_2$ is transmitted through the takeoff point as an input to the feedback kernel F and to the system kernel G_2 without loss or change.

The output Δe_2 from the feedback kernel is determined by the system equation $\Delta e_2 = F \Delta e_1$, and the output Δe_3 is given by $\Delta e_3 = G_2 \Delta e_1$. The overall system equation, when combined and reduced using the algebra of block diagrams, is found to be

$$\Delta e_3 = \frac{G_2 G_1}{1 - G_1 F} \Delta c_1$$

We will demonstrate the procedures for obtaining this canonical system equation form next.

It is always possible to convert any system block diagram into the canonical system form with feedback, as shown in Figure 6-1. It is important to recognize the universality of the canonical form since the development of the basic system equations given in Chapter 3 is based on the canonical system form, and our universal view and treatment of general system models will be based on the canonical system format. Reduction of a given block diagram representation to canonical form may be accomplished with algebraic theorems or equivalences for block diagrams. Some of the more important equivalences are given in Table 6-1. With the skillful application of these block diagram operations, reduction of any system to canonical form is generally possible.

Block diagrams of complicated systems may be simplified using the algebra of block diagrams. The letter G is used to represent any system kernel, and Δc and Δe denote any input or output signal. Observe that if Δc represents a multiple-input $\Delta \mathbf{c}$ or Δe represents a multiple output $\Delta \mathbf{e}$, then the associated system kernels $G(\mathbf{e}, \mathbf{c})$ are defined as conformable matrix kernels, and I is the unit matrix of appropriate size. Furthermore, if the inverse matrix G^{-1} appears in a block diagram, it is assumed that G is a nonsingular square matrix with an equal number of inputs and outputs.

The following generalized sequential procedure for the reduction to canonical form of compound systems composed of various subsystems and diverse signal paths is useful for most applications.

1) Combine system blocks in series.
2) Combine system blocks in parallel.
3) Eliminate all simple feedback loops.
4) Shift summing and takeoff points as necessary.
5) Recycle this procedure as necessary to achieve canonical form desired.

Table 6-1 Manipulation of Block Diagrams

Operation	System equation	System block diagram	Equivalent system block diagrams
1. Combine blocks in series	$\Delta e = (G_2 G_1)\Delta c$		
2. Combine blocks in parallel	$\Delta e = (G_1 + G_2)\Delta c$		
3. Remove block from forward path	$\Delta e = (G_1 + G_2)\Delta c$		
4. Eliminate feedback loop	$\Delta e = G_1(\Delta c + G_2\Delta e)$		
5. Remove block from feedback loop	$\Delta e = G_1(\Delta c + G_2\Delta e)$		
6. Shift sum points	$\Delta e = \Delta c_1 + \Delta c_2 + \Delta c_3$		

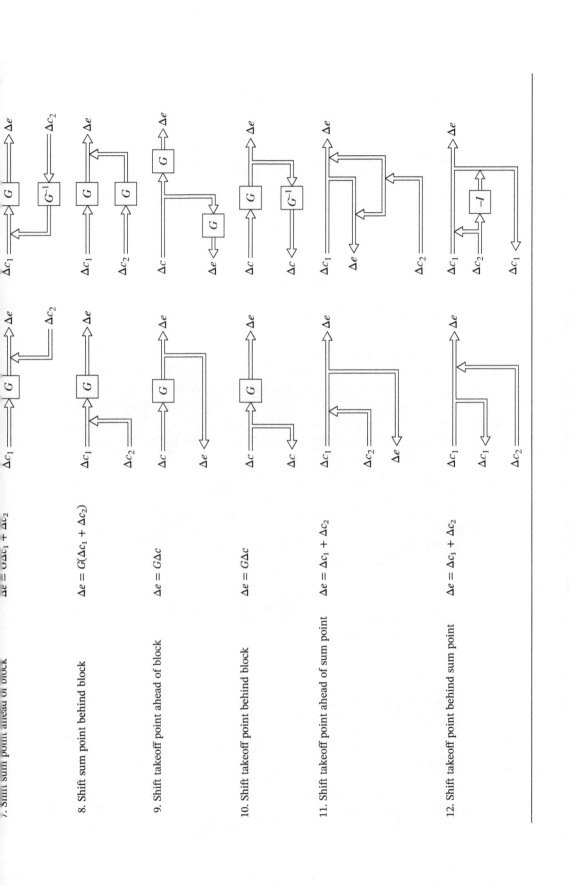

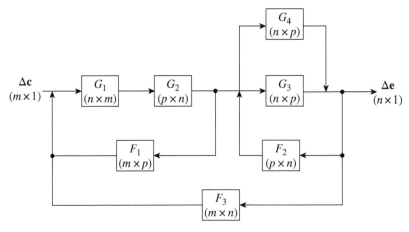

Fig. 6-3 Sample compound block diagram.

Example of Graphical Reduction Process.
As a demonstration of the reduction procedure, consider Figure 6-3 which illustrates a typical, compound, multiple-input ($\Delta\mathbf{c}$), multiple-output ($\Delta\mathbf{e}$) block diagram that we wish to reduce to canonical form, where the matrix sizes (i.e., number of rows and columns) are shown in parentheses.

Step 1. We combine series systems blocks G_1 and G_2 as follows:

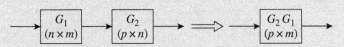

Observe the ordering for matrix multiplication (i.e., $G_2 G_1$) required for conformability.

Step 2. We combine parallel system blocks G_3 and G_4:

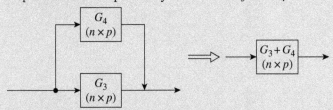

Step 3. We eliminate the simple feedback blocks F_1 and F_2, respectively:

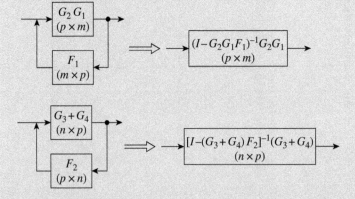

Step 4. We combine the series blocks formed in step 3:

Thus, we have reduced Figure 6-3 to the following canonical block system form, where

$$G = [I - (G_3 + G_4)F_2]^{-1}[G_3 + G_4][I - G_2G_1F_1]^{-1}G_2G_1$$

Recall that it is always possible to combine the feedback block with the general combined system kernel block and realize the following open-loop system form:

where

$$G_f = [I - GF_3]^{-1}G$$

Using block algebra, it is also generally possible to isolate any block, input, or output as desired for individual analysis or evaluation.

6.1.2 Signal-Flow Graphs

A signal-flow graph is similar to a system block diagram but provides an abbreviated graphical representation of the spatial or sequential relationship and ordering of the kernels, inputs, and outputs for a system or subsystem. Originally developed by S. J. Mason, signal-flow graphs are particularly simple to construct and are well suited for modeling, combining, and reducing large, complex systems with coupled subsystems. Mason also developed an algebraic gain formula that can be employed to determine the overall system kernel for the canonical system. The signal-flow graph is a "graphical" method for representing sets of algebraic equations and is equivalent to Cramer's rule for the solution of simultaneous algebraic equations that are linear in both input changes (i.e., dc or Δc) and output changes (i.e., de or Δe).

To effectively employ the signal-flow graph technique, certain graphical symbols and terms must be defined. Consider the following simple feedback block diagram and its equivalent signal-flow graph network.

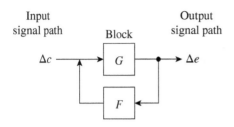

Input signal path — Block — Output signal path

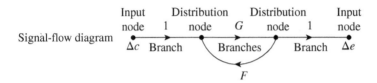

Now we define the following signal-flow graph terms:

Branch: a directed line segment that connects two nodes and acts as a one-way signal multiplier with a gain given by the system kernel 1, G, and F.

Node: an input (source), distribution, or output (sink) point for input, distribution, or output signals existing at a branch.

Input or source node: a node possessing only outgoing branches, which serves as a source for an input to the system.

Output or sink node: a node possessing only incoming branches, which serves as a sink for an output from the system.

Distribution or mixed node: a node possessing both incoming and outgoing branches for the system.

In comparing the block diagram and signal-flow diagram, we observe that each signal path of the block diagram represents a node in the signal flow graph, and each system block represents a branch. Observe that signal flow along a branch can occur only in the direction indicated by the branch arrow.

Three important algebraic rules are associated with signal-flow graphs:

1) The addition rule states that the value of a variable represented by any node is equal to the sum of all signals transmitted by branches entering that node. Therefore, we have

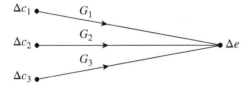

where

$$\Delta e = G_1 \Delta c_1 + G_2 \Delta c_2 + G_3 \Delta c_3$$

2) The transmission rule states that the value of a variable represented by any node is transmitted along every branch leaving that node. So we have

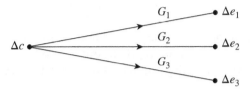

where

$$\Delta e_1 = G_1 \Delta c, \quad \Delta e_2 = G_2 \Delta c, \quad \Delta e_3 = G_3 \Delta c$$

3) The multiplication rule states that a series of connected branches is equivalent to a single branch whose kernel is the product of all kernels along the path formed by the branches. So we have for

$$\overset{G_1}{\underset{\Delta c_1}{\bullet \longrightarrow}} \quad \overset{G_2}{\underset{\Delta c_2}{\bullet \longrightarrow}} \quad \overset{G_3}{\underset{\Delta c_3}{\bullet \longrightarrow}} \quad \underset{\Delta e}{\bullet}$$

That

$$G_1 \Delta c_1 = \Delta c_2, \quad G_2 \Delta c_2 = \Delta c_3, \quad G_3 \Delta c_3 = \Delta e$$

and thus

$$\Delta e = G_3 G_2 G_1 \Delta c_1$$

Additional terminology used with signal-flow graphs and reinforcement of the earlier terminology are illustrated by the signal-flow graph shown in Figure 6-4.

- **Path:** a continuous succession of branches along which signal flow is permitted and no node is passed through more than once (e.g., in Figure 6-4, Δc_1 via G_1 to Δc_2 via G_2 to Δc_3 via G_3 to Δe, and Δc_2 via G_2 to Δc_3 via F_2 to Δc_2)
- **Node:** an input and/or output signal for branches, where an input node or source node has only outgoing branches (e.g., Δc_1), an output node or sink node has only incoming branches (e.g., Δe), and a mixed node or distribution node has both incoming and outgoing branches (e.g., Δc_2 and Δc_3).
- **Forward path:** any path from input to output node (e.g., Δc_1, via G_1 to Δc_2 via G_4 to Δe, and Δc_1, via G_1 to Δc_2 via G_2 to Δc_3 via G_3 to Δe.

Fig. 6-4   Typical signal-flow graph.

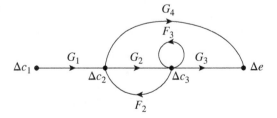

- **Feedback path or feedback loop:** a path that begins and ends on the same node (e.g., Δc_2 via G_2 to Δc_3 via F_2 to Δc_2).
- **Self-loop:** a feedback path composed of a single branch (e.g., Δc_3 via F_3 to Δc_3).
- **Branch gain:** the system kernel associated with a branch (e.g., G_1, G_2, G_3, G_4, F_2, and F_3).
- **Path gain:** the product of the branch gains or kernels encountered in traversing a path (e.g., the path gain for path Δc_1, via G_1 to Δc_2 via G_2 to Δc_3 via G_3 to Δe is $G_3 G_2 G_1$).
- **Loop gain:** the product of the branch gains of a loop (e.g., the loop gain for path Δc_2 via G_2 to Δc_3 via F_2 to Δc_2 is $G_2 F_2$).

The standard block diagram for the general multiple (m) input, multiple (n) output canonical system with feedback is shown by the following block diagram:

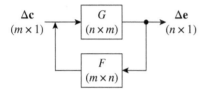

and the matrix system equation for this block diagram is

$$\Delta e = [I - GF]^{-1} G \Delta c$$

The signal-flow graph that is equivalent to this canonical system block diagram is given by

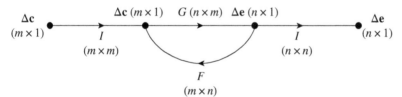

in addition, of course, the signal-flow graph has the same system equation as the block diagram.

Note: To preserve the definition of Δc as an input node and Δe as an output node and for convenience, extension branches with a unity matrix gain can be added to the signal-flow graph wherever selected.

The system kernel for the canonical system form can also be obtained from the Mason formula expressed in matrix form, as follows:

$$g(e, c) = \Delta^{-1} \sum_i P_i \Delta_i$$

where P_i is the ith forward path gain matrix and

$$\Delta = I - (-1)^{k+1} \sum_k \sum_j P_{ij}$$

$$\Delta = I - \sum_j P_{j1} + \sum_j P_{j2} - \sum_j P_{j3} \cdots$$

$\Delta = I - (\text{sum of all loop gains}) + (\text{sum of all gain products of two non-touching loops}) -$
$(\text{sum of all gain products of three non-touching loops}) + \cdots$

Δ_i is evaluated with all loops touching loop P_i eliminated

Observe that P_{jk} is the jth possible product of k non-touching loop gains and Δ is called the signal-flow graph matrix or characteristic function. Furthermore, loops and/or paths are non-touching if they have no common nodes. Care must be exercised in forming the gain products to ensure that the resulting matrix products are conformable and well defined.

Example Application of the Mason Formula for a Canonical System Kernel.
To demonstrate the use of the Mason formula, consider the multiple-input, multiple-output system shown in Figure 6-5, which is represented by both a signal-flow graph and an equivalent block diagram. This system has four loops with gains given by (observe the multiplication order)

$$\sum_{j=1} P_{j1} = G_2F_1 + G_5F_2 + G_5G_3G_2G_1F_3 + G_5G_4G_2G_1F_3$$

The sum of the gain products of two non-touching loops is

$$\sum_{j=1} P_{j2} = (G_5F_2)(G_2F_1)$$

There are no sets of three or more loops that are non-touching, so we have for the signal-flow graph matrix $(n \times n)$

$$\Delta = I - G_2F_1 - G_5F_2 - G_5G_3G_2G_1F_3 - G_5G_4G_2G_1F_3 + G_5F_2G_2F_1$$

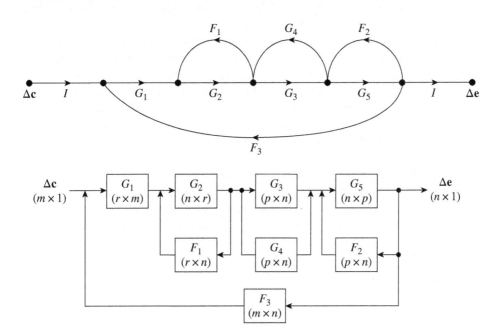

Fig. 6-5 Equivalent block diagram and signal-flow graph.

There are two forward paths between the source and sink nodes with gains given by $G_5G_3G_2G_1$ and $G_5G_4G_2G_1$. If either path is removed from the signal-flow graph, no loops remain, so both Δ_1 and Δ_2 are unity matrices; and we thus have for the matrix equation for the canonical system kernel

$$g(e, c) = \Delta^{-1}(G_5G_4G_2G_1 + G_5G_3G_2G_1)$$

and, of course, the system equation is given by

$$\mathbf{e} = g(\mathbf{e}, \mathbf{c})\mathbf{c}$$

6.1.3 Organization Diagrams

The organization diagram is an ordered sequence of subsystems displayed, usually in vertically descending and expanding block format. The subsystems may represent people (by function, title, and/or name), sub-organizations, command responsibilities, functional levels, hierarchy, and so on. Organizational diagrams are frequently encountered in business organizations (e.g., from chairperson of the board to part-time custodians), governmental agencies (e.g., president of the United States to U.S. taxpayers), military commands (e.g., commanding officer to enlisted recruit), and even academic disciplines (e.g., college president to college freshman). Usually, these organization diagrams list the principal parts or head of the organizational chain at the highest level or top of the diagram. The remaining components of the organization are sequenced in some vertically descending order of functional or organizational level or importance in an expanding treelike branching fashion occurring at each change of level.

Figure 6-6 is a simple organization diagram for a hypothetical business firm that designs and manufactures some product that is sold on the retail market. Typical inputs to the business firm are employees (engineers, technicians, production workers, administration, and staff), equipment and supplies, and income from sales. Since it is assumed that the business firm is owned by stockholders, an important output is the dividends paid on each share of stock. Observe that each block in the organizational diagram can be effectively isolated and considered as an independent subsystem, with inputs from lower organizational levels and outputs to higher levels. Thus, the organization diagram is simply a schematic block diagram of the subsystems of a single composite system and can be manipulated and combined as we would any other system block diagram composed of two or more subsystems. To provide a systematic way of identifying each subsystem's inputs and outputs at each level, we will designate the inputs and outputs at each level by the notation

Input : Δc_{ij}

Output : Δe_{ij}

Subsystem kernel : g_{ij}

where

i = level number

j = sequence count from left margin to right margin of the diagram

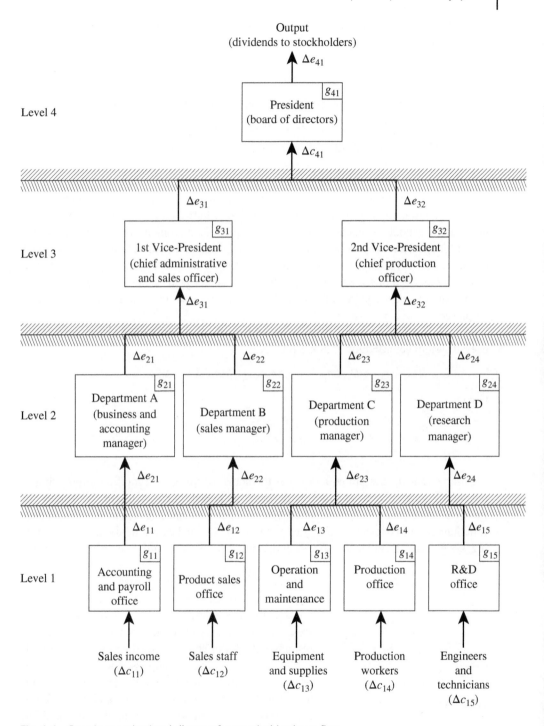

Fig. 6-6 Sample organizational diagram for a typical business firm.

We observe that the output at a given level can be combined with other outputs to serve as inputs to the next higher level. Thus, in Figure 6-6 at level 3, for example, we have as system equations

$$\Delta e_{31} = g_{31}\Delta c_{31} \quad \text{and} \quad \Delta e_{32} = g_{32}\Delta c_{32}$$

We observe that at the transition line between levels 2 and 3, we have

$$\Delta c_{31} = \Delta e_{21} + \Delta e_{22} \quad \text{and} \quad \Delta c_{32} = \Delta e_{23} + \Delta e_{24}$$

By describing our organization diagram in terms of system equations, we can now model, quantify, analyze, and evaluate organizations as we would any other system. Observe that in our example organizational system model displayed in Figure 6-6, we have no occurrence of feedback within the system. Obviously, intrinsic feedback can and does exist between sub-components in many organizations, and we will consider such examples later in this section. Also, we should recognize that extrinsic feedback for an organizational system can and usually does exist within the system's external environment. In our example system defined by Figure 6-6, a change in the dividends paid to the stockholders could very reasonably be externally recycled or fed back through the stockholders to influence some of the systems inputs, such as a change in equipment and supply purchases or a change in the company employees.

It is a simple demonstration of block diagram algebra (see Section 6.1.1) to show that the overall system equation for Figure 6-6 is

$$\Delta e_{41} = G_{11}\Delta c_{11} + G_{12}\Delta c_{12} + G_{13}\Delta c_{13} + G_{14}\Delta c_{14} + G_{15}\Delta c_{15}$$

where

$$G_{11} = g_{41}g_{31}g_{21}g_{11}, \quad G_{12} = g_{41}g_{31}g_{22}g_{12}$$
$$G_{13} = g_{41}g_{32}g_{23}g_{13}, \quad G_{14} = g_{41}g_{32}g_{23}g_{14}$$
$$G_{15} = g_{41}g_{32}g_{24}g_{15}$$

Observe that the system equation has the standard form for a multiple-input, single-output discrete system equation and has a reduced block diagram, as shown in Figure 6-7. We have assumed a discrete system for convenience. If values or mathematical functions can be consistently assigned to the system kernel components, then the resulting system equation can be solved and evaluated like any other quantitative system equation.

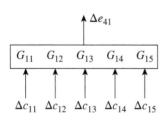

Fig. 6-7 Reduced block diagram for organization diagram in Figure 6-6.

However, we wish to demonstrate another useful aspect of system science as it can be applied to organization diagrams. Suppose in Figure 6-6 that we wish to assess the fractional contribution made by various subsystems within the company to the company's overall performance. For example, suppose the president (subsystem g_{41}) wishes to assess quantitatively the contribution that the two vice-presidents (g_{31} and g_{32}) make to the president's output, Δe_{41}. Then the system equation for the level 3 and 4 subsystem is

$$\Delta e_{41} = g_{41}(g_{31}\Delta c_{31} + g_{32}\Delta c_{32})$$

A useful measuring scheme that can be imposed here is that within each level of the organization, the sum of the entering inputs, the sum of the kernels, and the sum of the outputs is unity, respectively. Thus, we have that

Inputs (level 3):	$\Delta c_{31} + \Delta c_{32} = 1$
Kernels (level 3):	$g_{31} + g_{32} = 1$
Outputs (level 3):	$\Delta e_{31} + \Delta e_{32} = 1$
Inputs (level 4):	$\Delta c_{41} = 1 = \Delta e_{31} + \Delta e_{32}$
Kernel(s) (level 4):	$g_{41} = 1$
Output (level 4):	$\Delta e_{41} = 1$

Observe that, since level 4 has only one input, one kernel, and one output, each of these is unity. (That explains why presidents and board of directors can safely perform this analysis scheme without having to evaluate or be intimidated by their own performance!) Now the company president must assign in a rational way the percent contribution to the president's output Δe_{41} arising from the inputs Δc_{31} and Δc_{32}. Then, by comparative ranking and evaluation, the president must assign a percent contribution for the effectiveness of the two vice-presidents, g_{31} and g_{32}. The normalized contribution made by these corporate officers and their reporting subsystems to the president's output is given by

$$\Delta e_{41} = 100\% = \%(g_{31}) + \%(g_{32})$$

where

$$\%(g_{31}) = \% \left[\frac{g_{31}\Delta c_{31}}{g_{31}\Delta c_{31} + g_{32}\Delta c_{32}} \right]$$
$$\%(g_{32}) = \% \left[\frac{g_{32}\Delta c_{32}}{g_{31}\Delta c_{31} + g_{32}\Delta c_{32}} \right]$$

For example, if the inputs are valued at $\Delta c_{31} = 65\%$ and $\Delta c_{32} = 35\%$, and the system kernels (vice-presidents' performances) are $g_{31} = 60\%$ and $g_{32} = 40\%$, then

$$\%(g_{31}) = \frac{(0.60)(0.65)}{(0.60)(0.65) + (0.40)(0.35)} = 74\%$$
$$\%(g_{31}) = \frac{(0.40)(0.35)}{(0.60)(0.65) + (0.40)(0.35)} = 26\%$$

This result implies that the first vice-president accounts for about three quarters of the president's output, while the second vice-president accounts for the remaining one-quarter. Some may object to the use of such "cold" numerical assessments of human performance, but in retrospect the salaries paid to business executives usually reflect a numerical evaluation of their worth to the business.

As an example of the occurrence and use of intrinsic feedback in an organizational structure, let us for diversion consider the following "hypothetical management and domestic crisis," which has arisen for the president of the company portrayed in Figure 6-6.

While at dinner with his father and mother-in-law, the president's lovely young wife unexpectedly suggests that her husband should appoint her older brother, Charles II, as

Research Manager of the company to fill this recent vacancy. This vacancy in the company was created when the former Research Manager decided to "seek her fortune" teaching at a renowned college President. Furthermore, to compound this sudden, unexpected, delicate situation for the company, the president's father-in-law, Charles I (who is the former owner of the company and is the company's major stockholder), thinks his daughter's idea is excellent and necessary since Charles II is unmarried, and over 40 years old and he should assume his proper role in establishing and maintaining the family's prominent name. However, our president realizes that appointment of Charles II as Research Manager will be a disaster for the department and will effectively end any useful output from the company's Research Department. Conversely, the president also realizes that failure to appoint Charles II to this position may mark the end of his tenure as both a devoted husband and the company's president. Fortunately, our hero (who didn't become president only because he married the owner's beautiful daughter) has successfully completed this course in system science, and he understands the utility of "intrinsic system feedback." To gain time to formulate an equitable response to this unexpected dilemma, our president shrewdly shows his in-laws the most recent wallet pictures of their only grandson (wisely named Charles III) while he quickly scribbles some system calculations on the back of his napkin. Then, with a confident demeanor, the president proposes the following action:

- Appoint Charles II as Research Manager for the company, and Charles II will assume his proper role with both the family and the company.
- Establish an additional office under Charles II that will more fully utilize the output of the R&D office to increase productivity of the company. This action will result in increased production at lower unit cost and higher stock dividends, and happier stockholders.

Warm acceptance of this proposal greets our president who is ardently kissed by his beautiful wife and receives a warm, confirming handshake from his father-in-law. The father-in-law suggests that the president's stock option might be increased at the next stockholders meeting. Safely placed in the president's pocket is the crumpled napkin with the following partial system diagram and analysis (see Figure 6-6 for notation).

Example Partial System Diagram and Analysis for Example.

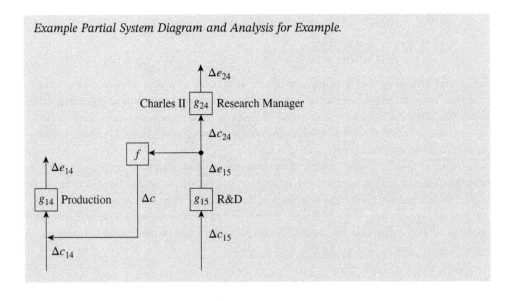

If Charles II is appointed as research manager, then $g_{24} \to 0$ and output $\Delta e_{24} = g_{24}g_{15}$ $\Delta c_{15} \to 0$. Now with feedback f from Δe_{15} (R&D output) to Δc_{14}, we have from the production department

$$g_{14}(\Delta c_{14} + f\Delta e_{15}) = \Delta e_{14}$$

and since

$$\Delta e_{15} = g_{15}\Delta c_{15}$$

then production can provide the combined output

$$\Delta e_{14} = g_{14}(\Delta c_{14} + fg_{15}\Delta c_{15})$$

and we can replace $g_{24}g_{15}$ by fg_{15}.

Ergo: Appoint Charles II as Research Manager but carefully select and appoint f to preserve the output from $g_{15}\Delta c_{15}$.

Thus, knowledge of system science saves the president's marriage, keeps the company solvent, and the president retains his six-figure salary and position. Fortunately, the president's father-in-law, Charles I, who is "a self-made, college dropout," has never heard of system science, let alone "intrinsic system feedback." We apologize for this tale of crisis management, but it is one example from this text that the reader may remember.

6.2 Modeling System Inputs, Outputs, and Kernels

A reasonable plan for the initial investigation of a difficult or unfamiliar system is to begin simply with a single-input, single-output system without feedback that has been carefully defined and isolated from its environment. Such a generalized but simple "plan of attack," as we outline in this section, is particularly advisable when investigating systems where background information and previous system studies are limited, and the system is one that is difficult to quantify and assign measurable inputs and outputs. Obviously, such a generalized plan is meant to be suggestive only, and the system investigator will undoubtedly develop individualized procedures with experience.

The beginning of any system study is the recognition of the rational system that is responsible for the inputs or outputs of interest. Often, identifying the essential components of the system that collectively undergo the cause and effect action associated with the system is obvious, such as the illness (the output) that results when a person (the system) consumes toxic food or water (the input). However, the identification and isolation of other systems, such as a study of the causes of inflation where the general system is the world economic system, is undoubtedly complex, diverse, and presents a serious modeling challenge. Here the investigator should carefully evaluate and limit the system to those system elements that are the primary causes of the effect sought. For example, the cost of crude oil might be considered as the major input to inflation. Obviously, the best advice here is for the system investigator to learn enough about the factors (inputs, outputs, system, and elements) that might contribute to the system so that the system investigation is accurate and worthwhile. Many system investigations arise over an initial interest in a particular cause or effect observed or believed to may occur. Then the task before the investigator is to identify and isolate the system responsible for that cause or effect and to determine the important contributing causes or resulting effects related to the initial interest. Some useful questions that may assist in identifying and later quantifying the system are as follows:

- What are the system elements that contribute significantly to the cause(s) and effect(s) sought?
- What are the interactions of these system elements with each other?
- Can the systems be grouped into distinct subsystem collections in which the behavior is more easily or better described?
- Can the elements assigned to the system be isolated from the system's environment? How? And if they can be isolated, what are the significant inputs (causes) and outputs (effects) for these elements that must pass through the system boundary?
- What does existing information and knowledge provide about the system, its elements and inputs and outputs, or cause and effect relationships?

After the essential system elements have been identified and grouped and the system isolated, the modeling task now turns to identifying the major inputs and outputs and, if appropriate, ordering and reducing these signals to the major inputs and outputs. Although extensive experience with the system and accurate experimental data is best for distinguishing these dominant signals, other alternatives to ordering the inputs and outputs identified exist to determine the dominant inputs and outputs. For example, an appeal may be made to individual or collective opinions, expert or otherwise, on the ranking or ordering of inputs and outputs.

When the essential task of identifying and defining those input and output variables that will be employed for a given system model is completed, the system investigator then must quantitatively formulate each kernel component for a given input and output. A useful tool that may be used for kernel formulation is a tabulated listing of the system inputs, outputs, and kernel components as typified by Table 6-2, which provides for up to

Table 6-2 Tabulation Form for System Inputs and Outputs and Kernel Components: $de = g(e, c)dc$

INPUTS OUTPUTS	Variable 1 (label) $dc_1 = (\ )$	Variable 2 (label) $dc_2 = (\ )$	Variable 3 (label) $dc_3 = (\ )$	Variable 4 (label) $dc_4 = (\ )$
Variable 1 (label) $de_1 = (\ )$	$g_{11} = \dfrac{\partial e_1}{\partial c_1}$	$g_{12} = \dfrac{\partial e_1}{\partial c_2}$	$g_{13} = \dfrac{\partial e_1}{\partial c_3}$	$g_{14} = \dfrac{\partial e_1}{\partial c_4}$
Variable 2 (label) $de_2 = (\ )$	$g_{21} = \dfrac{\partial e_2}{\partial c_1}$	$g_{22} = \dfrac{\partial e_2}{\partial c_2}$	$g_{23} = \dfrac{\partial e_2}{\partial c_3}$	$g_{24} = \dfrac{\partial e_2}{\partial c_4}$
Variable 3 (label) $de_3 = (\ )$	$g_{32} = \dfrac{\partial e_3}{\partial c_2}$	$g_{32} = \dfrac{\partial e_3}{\partial c_2}$	$g_{33} = \dfrac{\partial e_3}{\partial c_3}$	$g_{34} = \dfrac{\partial e_3}{\partial c_4}$
Variable 4 (label) $de_4 = (\ )$	$g_{43} = \dfrac{\partial e_4}{\partial c_3}$	$g_{42} = \dfrac{\partial e_4}{\partial c_2}$	$g_{43} = \dfrac{\partial e_4}{\partial c_3}$	$g_{44} = \dfrac{\partial e_4}{\partial c_4}$

four inputs and four outputs. Observe that each significant independent input variable is identified and defined, and a symbol assigned; and each significant output that depends on those inputs are also identified and defined and a symbol assigned. The system kernel component designator, g_{ij}, is indicated in the upper-left corner of each workspace, which is reserved for the system investigator to quantify and formulate the mathematical expression developed for that kernel component. Such a tabulation form for inputs and outputs for a given system, as shown in Table 6-2, is useful in ranking and weighting the inputs and outputs by the qualitative and quantitative methods presented in this text.

When the essential input and output variables have been identified for the given model, the next step is the quantification and formulation of the system kernel. To understand the meaning and significance of each kernel component, consider the general multiple-input, multiple-output system equation:

$$d\mathbf{e} = g(\mathbf{e}, \mathbf{c})d\mathbf{c}$$

where the system kernel is an $(n \times m)$ matrix function of $\mathbf{e}$ and $\mathbf{c}$ defined as follows:

$$g(\mathbf{e}, \mathbf{c}) = \begin{bmatrix} g_{11} & g_{12} & \cdots & g_{1m} \\ g_{21} & g_{23} & \cdots & g_{2m} \\ \vdots & & & \\ g_{n1} & g_{n2} & \cdots & g_{nm} \end{bmatrix}$$

As we have shown in Section 5.3.2, if the system equation is exact and directly integrable, then each component of the system kernel $g(\mathbf{e}, \mathbf{c})$ will satisfy the equation

$$g_{ij}(\mathbf{e}, \mathbf{c}) = \frac{\partial e_i}{\partial c_j} \tag{6.2.1.1}$$

If this equation is satisfied, then the system equation will have the following form for $i = 1, \ldots, n$,

$$de_i = \sum_{j=1}^{m} \frac{\partial e_i}{\partial c_j} dc_j \tag{6.2.1.2}$$

and can be directly integrated to give as the system solution $\mathbf{e}$'s

$$e_i(\mathbf{c}) = k_i$$

for $i = 1, \ldots, n$, where the k_i's are arbitrary integration constants.

Although few general system kernels are exact as defined by Eq. (6.2.1.1), this equation is the basis of a useful scheme for providing a rational method for quantifying and formulating system kernels. Let us carefully examine the mathematical significance of Eq. (6.2.1.2). A formal mathematically based word description of the partial derivative $\dfrac{\partial e}{\partial c_i}$ is as follows:

$$\frac{\partial e_i}{\partial c_j} = \begin{cases} \text{Net differential change of } e_i \text{ per unit differential} \\ \text{change of } c_j \text{ with all other } e_i\text{'s and } c_i\text{'s held fixed} \end{cases}$$

Now, if we recall the significance of the variables a and c, we observe that the preceding expression is also equivalent to the following statement for continuous systems:

$$\frac{\partial e_i}{\partial c_j} = \begin{cases} \text{Net differential increase in } i\text{th system output associated} \\ \text{with unit differential increase in the } j\text{th input or cause} \\ \text{with all other inputs and outputs held fixed.} \end{cases}$$

Thus, $\dfrac{\partial e_i}{\partial c_j}$ can be considered to represent a "balance" on the system variables and parameters that contribute to an increase of the ith output (effect) per unit change in the jth input (cause), minus the system factors that result in a decrease of the ith output per unit change in the jth output. In other words, $\dfrac{\partial e_i}{\partial c_j}$ is the net sum of the sources minus the sinks within the system that produce a change in the ith output per unit change in the jth input. The other system inputs and outputs are held fixed to determine this balance statement.

Therefore, we have as another word statement for $\dfrac{\partial e_i}{\partial c_j}$ that

$$\frac{\partial e_i}{\partial c_j} = \begin{cases} \text{Sum of all sources that increase } e_i \text{ minus sum of} \\ \text{all sinks that decrease } e_i \text{ per unit change in } c_j \end{cases}$$

Thus, the system investigator, in formulating the system kernel, must seek to quantify each kernel component by determining those system variables and parameters that contribute to that system kernel component. Obviously, successful kernel formulation requires sound knowledge of the system and broad experience in system synthesis and analysis.

An important concept in quantitatively modeling system kernels is the use of generalized elasticity functions for each of the kernel components. System elasticities are frequently used in economic models. To describe these elasticity functions, consider the general multiple-input, multiple-output system equation:

$$d\mathbf{e} = g(\mathbf{e}, \mathbf{c})d\mathbf{c} \tag{6.2.2.1}$$

If the ith output and all inputs of a system kernel are expressed as

$$g_{ij}(\mathbf{e}, \mathbf{c}) = \frac{e_i}{c_j} a_{ij}(\mathbf{e}, \mathbf{c}) \tag{6.2.2.2}$$

for $j = 1, ..., m$, then the ith component of the system equation can be expressed in the following differential fractional form:

$$\frac{de_i}{e_i} = \sum_{j=1}^{m} a_{ij}(\mathbf{e}, \mathbf{c}) \frac{dc_j}{c_j} \tag{6.2.3.1}$$

or equivalently,

$$d\mathrm{Ln}e_i = \sum_{j=1}^{m} a_{ij}(\mathbf{e}, \mathbf{c})d\mathrm{Ln}c_j \tag{6.2.3.2}$$

Assume that we can evaluate the coefficient functions a_{ij} at arbitrary system points $\mathbf{e}_*$, $\mathbf{c}_*$ so that $a_{ij}(\mathbf{e}^*, \mathbf{c}^*) = a_{ij}^*$ (a constant). Then system equations (6.2.2) or (6.2.3) are exact and can be integrated to give as the approximate system solution for the output e_i

$$e_i = K_i \prod_{j=1}^{m} c_j^{a_{ij}}$$

where K_i is the arbitrary integration constant.

The exponents a_{ij}^* used to approximate the system equation at $\mathbf{e}^*$ and $\mathbf{c}^*$ are referred to as the partial elasticity functions or elasticities for the system. From common use in economics, a partial elasticity a_{ij}^* may be categorized as follows:

$$a_{ij}^* = \begin{cases} > 1 \text{ elastic response} \\ = 1 \text{ unitary elastic respons} \\ < 1 \text{ inelastic response} \\ = 0 \text{ null response} \end{cases}$$

Thus, we observe from the categories and Eq. (6.2.3.2) that an elastic response implies that the fractional change in output is greater than the associated fractional change in input (i.e., $dLne_i > dLnc_j$). A unitary elastic response implies that the fractional output and input changes are equal, while an inelastic response implies that the fractional output change is less than the fractional input change. If the elasticity is zero, then no change in the output results from a change in that input.

To provide a simple example to illustrate these concepts and methods for quantifying the system kernel, consider the familiar perfect gas law, which describes a system composed of an ideal gas within an isolating boundary:

$$PV = NRT$$

The system variables are the absolute pressure P of the gas, the volume V, the number of moles of gas, N, and the absolute temperature, T. The parameter R is the universal gas constant. If we treat P as the system output and V; N, and T as system inputs, then we have for the system equation

$$dP = \frac{\partial P}{\partial V} dV + \frac{\partial P}{\partial N} dN + \frac{\partial P}{\partial T} dT$$

or, equivalently, in terms of system kernels

$$dP = g_V dV + g_N dN + g_T dT$$

where

$$g_V = \frac{\partial P}{\partial V} = -\frac{NRT}{V^2} = -\frac{P}{V} \text{ (Net pressure change per unit volume)}$$

$$g_N = \frac{\partial P}{\partial N} = \frac{RT}{V} = \frac{P}{N} \text{ (Net pressure change per unit mass)}$$

$$g_T = \frac{\partial P}{\partial T} = \frac{NR}{V} = \frac{P}{T} \text{ (Net pressure change per unit temperature)}$$

We observe that $g_V = \partial P/\partial V$ represents the expression for the net differential change in gas volume. We observe that this term is directly proportional to the pressure of the gas and inversely proportional to the volume of the gas. Thus g_v has the form of an elasticity expression given by Eq. (6.2.2.2). The elasticity value associated with g_v is -1.0, so the kernel component is said to produce an inelastic response. Thus, the net change in gas pressure arising from g_v is negative, which implies that, as the volume of the gas increases, the gas pressure reduces if N and T are held constant. Similar descriptions of system dependence and behavior are associated with the other kernel components, g_N and g_T. It is easily shown that the elasticities for both g_N and g_T are 1.0, so these kernels are said to exhibit unitary elastic responses.

Obviously, a major goal in quantifying and formulating system kernels is to achieve an accurate description of the system that, if possible, also results in an "exact" system equation that can be solved analytically. The effort to realize an exact system equation should not, however, result in an inadequate or incorrect system model. Since means for analysis of the system are available using the computer to process the system equations, the investigator should stress adequacy and accuracy of the system equation integrability.

6.2.1 Single-Input, Single-Output System

The use of a single-input, single-output model for the initial study of most systems is very desirable. The advantages include the following:

- The investigator must identify the major input and major output to the system. This action focuses attention on the primary single cause and single effect relationship exhibited by the system.
- A very broad class of quantitative kernels for single-input, single-output systems can be modeled, and the resulting system equation can often be solved analytically.
- Comparison and assessment of the single-input, single-output model's behavior with the actual behavior of the real system is generally straightforward, and evaluation of the system model is simplified.
- System model parameters are generally easily assessed and controlled, and their impact on the model's performance is usually readily determined.
- Analysis and evaluation of a single-input, single-output system is relatively easy and rapid to perform and can, therefore, provide an expeditious assessment and a comparison of the system model with the actual system.

In view of these considerations, the system investigator is well advised to begin the analysis of a system, if possible, with a single-input, single-output model. Of course, the first step here is to recognize and define the system so that the essential elements of the system can be identified and isolated from the environment. Then a system boundary is established to isolate the system and control the movement of causes and effects across this boundary. The next step is to determine and rank all significant causes and effects that are acted on or produced by the system. If a study of a single cause or effect is the basis for the system investigation, then this task reduces to the identification of the major resulting effect and inciting cause for the given system, respectively.

Now the final step in our simplified modeling approach before beginning analysis of the system is the quantification of the system kernel. Since we have reduced the system to a single-input and a single-output, we have for the system equation that

$$\left.\begin{array}{c} \Delta e \\ de \end{array}\right\} = g(e,c) \left\{ \begin{array}{c} \Delta c \\ dc \end{array}\right.$$

where we assume that the investigator has established whether to model the system as a discrete or continuous system model.

Several alternative methods are available to the system investigator for quantifying and formulating the system kernel. From the previous section, we found that the system kernel $g(e, c)$ represents the following quantities:

> $g(e,c)$ = Sum of all terms (functions of e and c) that increase the output e per unit increase in the input c, minus sum of all terms (as functions of e and c) that decrease the output e per unit increase in the input c

So for the single-input, single-output system, we simply need to quantify the sources and sinks that affect the system output. The ideal beginning point from the viewpoint of simplicity is to assume that the system kernel is linear. For the single-input, single-output system, we may assume that the kernel is linear either in terms of the output or the input; that is,

$$g(e,c) = \begin{cases} A(c) + B(c)e & \text{(linear in } e) \\ \dfrac{1}{D(e) + E(e)c} & \text{(linear in } c) \end{cases}$$

depending on our assessment of the actual system and its observed behavior. Now we need to quantify the functions $A(c)$ and $B(c)$ for the system's rate of change of a per unit change in c or the functions $D(e)$ and $E(e)$ for the system's rate of change of c per unit change in e.

The simplest linear model to employ is that in which both coefficient functions (e.g., A and B) are constants. In this case, the solution for the single-input, single-output system equation for the output e becomes

$$e = A + K \exp(Bc)$$

where K is our integration constant.

For the single-input, single-output system, we can consider more complex nonlinear system kernels and still have some confidence in obtaining an analytical solution for the system equation. The analytical solutions obtained for the single-input; single-output system model can then be effectively used by the investigator to sample the predicted behavior of the system. Armed with this increased understanding of the system, the investigator can use these earlier models to improve the study by expanding the number of inputs or outputs or by improving the description and accuracy of the single-input, single-output system kernel. Some of the methods we might also consider for quantitatively establishing the kernel are contained in chapter sections in this text as follows:

- Linear models (see Chapter Sections 4.2.4; 4.3.4 and 4.3.5; 5.2.5; 5.3.5; 7.3, 7.4)
- Theoretically derived models (see Chapters 7 and 8)

- Curve fitting to experimental data (see Chapter 8)
- Experimental simulation (see Chapter Section 6.2.4)
- Dimensional analysis (see Chapter 6)
- Expert opinion (see Chapter Section 6.4)
- Heuristic models (see Chapter Section 6.2.6)
- Models with feedback (see Chapter Section 8.3)

We will briefly consider each of these methods in this chapter and examine their applicability.

6.2.2 Physical Systems

The investigation of physical systems is greatly aided by the extensive base of data and knowledge that exists because of previous research and quantitative studies performed on physical systems in general. Important inputs commonly encountered such as time, space, matter, and energy are easily identified and measured for most physical systems, and the basic cause and effect relationships (e.g., conservation of mass and energy) governing these inputs are generally well established and quantitatively known. With an adequate background and understanding of the physical processes and reactions that a given physical system model can exhibit, it is generally possible with careful analysis to establish accurate system models descriptive of most physical systems. Usually, these system models are based on a mathematical statement of the conservation of an appropriate physical entity, such as the conservation of energy, mass, momentum, electrical charge, or number of nucleons. Of course, many other physically based cause and effect relationships are possible, and the essential ones should be identified and mathematically modeled for the given system as necessary.

An important consideration in the development of any quantitative system kernel is the *absolutely essential requirement* that all terms of the system equation be consistent and compatible in dimensions or units. For example, consider the general multiple-input, multiple-output system equation with feedback

$$d\mathbf{e} = [I - g(\mathbf{e}, \mathbf{c})f(\mathbf{e}, \mathbf{c})]^{-1}g(\mathbf{e}, \mathbf{c})d\mathbf{c}$$

For dimensional consistency, the units, or dimensions of each component of the matrix system kernel must be the same as those units of the corresponding differentials or differences (a discrete system), as follows:

$$\text{Units of } g_{ij}(\mathbf{e}, \mathbf{c}) = \text{units of } \frac{de_i}{dc_j} \quad \text{or} \quad \frac{\Delta e_i}{\Delta c_j}$$

Furthermore, for the system equation with feedback, the product of the feedback kernel and system kernel must be dimensionless; that is,

$$\text{Units of } g(\mathbf{e}, \mathbf{c})f(\mathbf{e}, \mathbf{c}) = \text{dimensionless or unitless}$$

where in general $f(\mathbf{e}, \mathbf{c})$ is an $(m \times n)$ matrix and $g(\mathbf{e}, \mathbf{c})$ is an $(n \times m)$ matrix. Although this requirement for dimensional consistency of any rational system equation may not seem particularly useful, we will show in Section 6.2.4, when dimensional analysis is presented, that the use of dimensional concepts can prove very useful in the quantitative modeling of many system kernels.

The International System of Units, usually abbreviated as SI units, is a prescribed set of units deemed adequate for expressing the values of all physical quantities. These SI units have been proposed by the General Conference of Weights and Measures and accepted by most nations throughout the world. In the SI system, seven basic units are defined for the quantities: length, mass, time, electric current, temperature, luminous intensity, and the atomic amount of a substance. Also, two auxiliary units for the measurement of a plane and solid angle in geometry are specified. From these basic units and auxiliary units, the units of all physical quantities can be derived and specified as rational functions of the basic and auxiliary units. Table 6-3 provides a listing of the basic and auxiliary units in the SI system and a brief listing of some derived units.

The present number of distinct physical variables and parameters that have been recognized and acknowledged, have a unique combination of basic units within the SI system and have been assigned a name and a related physical concept is small. Only about a hundred units have been identified. However, the number of distinct physical quantities that can be defined within the SI system unit base is large and impressive. It is interesting and informative to speculate on the potential number of distinct physical quantities that can be defined using the basic unit set specified by the SI system. For example, let us restrict our estimate to the units of length (m for meter), mass (kg for kilogram), time (s for second), electric current (A for ampere), temperature (K for Kelvin), and luminous intensity (cd for candela) and ignore the other unit of mole (since this unit permits mass to be considered as an equivalent unit), Kelvin (this unit may be associated with energy) and radian and steradian (since these units are actually dimensional ratios of length and area, respectively). Now, if we permit an integral power range of 1, 0, −1 for the basic units of mass, ampere, Kelvin, and candela and an integral power range of 4, 3, 2, 1, 0, −1, −2, −3, −4 for the basic units of length and 3, 2, 1, 0, −1, −2, −3 time, we find there are 945 unique unit combinations, discounting reciprocals of the same unit combination (which are thus not distinct physical entities from their reciprocal). The combinations range from the dimensional quantity [kg A cd m^{-3} s^{3}] to the dimensionless parameters and numbers and include all quantities presently recognized as physically meaningful. The physical existence and usefulness of many of the other quantities in these 945 unique combinations may be questionable, but what is provoking to speculate on is how many potentially significant, undiscovered, physical system equations involving these 945 unique quantities remain to be modeled, analyzed, and evaluated. Apparently, even in the physical sciences, system science has an opportunity for major growth and progress in the further study of our physical universe (dark energy, dark matter, parallel universes, etc.?). A particularly intriguing dimensional concept is the correct description and units for "matter" in dark matter.

A useful handbook compiled by C. F. Hix and R. P. Alley and published by General Electric in 1955 under the Creative Engineering Program provides a description of 109 physical laws and effects observed and recognized in the physical sciences. An understanding and appreciation of these important system equations (i.e., cause and effect relationships) provide a valuable background resource for the investigation of any physical science system. Thus, if a physical system under investigation exhibits one of these physical laws or effects, it would be

Table 6-3 International System of Units (SI)

Type	Quantity	Unit (word)	Symbol	Dimension
Basic units	Length	Meter	m	m
	Mass	Kilogram	kg	kg
	Time	Second	s	s
	Electric current	Ampere	A	A
	Temperature	Kelvin	K	K
	Luminous intensity	Candela	cd	cd
	Amount of substance	Mole	mol	mol
Auxiliary units	Plane angle	Radian	rad	rad
	Solid angle	Steradian	sr	sr
Derived units	Acceleration	Meter/square second	m/s^2	$m\,s^{-2}$
	Activity (radioactive source)	Becquerel (events/second)	Bq	s^{-1}
	Angular acceleration	Radian/square second	rad/s^2	$rad\,s^{-2}$
	Angular velocity	Radian/second	rad/s	$rad\,s^{-1}$
	Area	Square meter	m^2	m^2
	Density	Kilogram/cubic meter	kg/m^3	$m^{-3}kg$
	Dynamic viscosity	Newton - second/square meter	$N\,s/m^2$	$m^{-1}\,kg\,s^{-1}$
	Electric capacitance	Farad	F	$A^2\,m^{-2}\,kg^{-1}\,s^4$
	Electric charge	Coulomb	C	$A\,s$
	Electric field strength	Volt/meter	V/m	$m\,kg\,s^{-3}\,A^{-1}$
	Electric resistance	Ohm	Ω	$m^2\,kg\,s^{-3}\,A^{-2}$
	Entropy	Joule/Kelvin	J/K	$m^2\,kg\,s^{-2}\,K^{-1}$
	Force	Newton	N	$m\,kg\,s^{-2}$
	Frequency	Hertz (cycles/second)	Hz	s^{-1}
	Illumination	Lux	lx	$m^{-2}\,cd\,sr$
	Inductance	Henry	H	$m^2\,kg\,s^{-2}\,A^{-2}$
	Kinematic viscosity	Square meter/second	m^2/s	$m^2\,s^{-1}$
	Luminance	Candela/square meter	cd/m^2	$cd\,m^{-2}$
	Luminous flux	Lumen	lm	$cd\,sr^{-1}$
	Magnetomotive force	Ampere turn	A	A
	Magnetic field strength	Ampere/meter	A/m	$m^{-1}\,A$
	Magnetic flux	Weber	Wb	$m^2\,kg\,s^{-2}\,A^{-1}$
	Magnetic flux density	Tesla	T	$kg\,s^{-2}\,A^{-1}$
	Power	Watt	W	$m^2\,kg\,s^{-3}$
	Pressure	Pascal (Newton/square meter)	Pa (N/m^2)	$m^{-1}\,kg\,s^{-2}$

Table 6-3 (Continued)

Type	Quantity	Unit (word)	Symbol	Dimension
	Radiant intensity	Watt per steradian	W/sr	$m^2\,kg\,s^{-3}\,sr^{-1}$
	Specific heat	Joule/Kilogram-Kelvin	J/kg-K	$m^2\,s^{-1}\,K^{-1}$
	Thermal conductivity	Watt/meter-Kelvin	W/m-K	$m\,kg\,s^{-3}\,K^{-1}$
	Velocity	Meter/second	m/s	$m\,s^{-1}$
	Voltage	Volt	V	$m^2\,kg\,s^{-3}\,A^{-1}$
	Volume	Cubic meter	m^3	m^3
	Wave number	Cycles/meter	m^{-1}	m^{-1}
	Work, energy, heat	Joule	J	$m^2\,kg\,s^{-2}$

important to recognize the existence of the effect and assess its importance in the modeling effort.

Table 6-4 provides a cross-reference by scientific field for these physical laws and effects. The reader is referred to the literature in the respective field for a full description and the system equations pertinent to each listing.

6.2.3 Nonphysical Systems

Extensive training in the physical sciences and mathematics is generally required to develop the background, knowledge, and skills to model, analyze, and evaluate the precise, quantitative systems identified in the physical sciences and technology such as physics, chemistry, and engineering. Thus, it may appear to be impractical and even naive for a beginning system investigator to attempt to model and study any complex or unfamiliar system in the physical or nonphysical science fields, particularly if the system lies outside the investigator's primary area of experience. However, we should recognize that in the so-called "exact physical sciences" the basic system equations in well-developed, complex areas, such as atomic and nuclear physics, mechanics, thermodynamics, and physical chemistry are the result of long-term, intensive investigation, and experimentation. In these areas, even very small, but consistent errors that arise in experimental investigations are grounds for reevaluation of the appropriate system model and modification of the theoretical basis. In the study of system science envisioned here, with our desire for a generalized, broad field approach to system analysis and evaluation, we will generally be satisfied with much less accurate and precise system models. Our major goal is to realize, initially, a general understanding of a given system's behavior and the practical predictive capability of the system's behavior for known input conditions. In general, our modeling and analysis efforts are not critically dependent on achieving highly precise models with elaborate descriptive capability for the system. What we do seek are useful, but probably limited and approximate, quantitative system models in a complex universe. Generally, we will find that even elementary system models based on simple linear system kernels for the major input and output variables can provide reasonably accurate and useful descriptions of the behavior of many systems of interest over a practical range of responses.

Table 6-4 Physical Laws and Effects by Scientific Field

ACOUSTICS
Cavitation
Diffraction
Doppler (Fizeau) effect
Interference
Magnetic dispersion of sound
Organ pipe resonance
Weber-Fechner law

ATOMIC PHYSICS
Absorption
Boltzmann distribution law
Breit-Wigner theory
Bremsstrahlung radiation
Compton effect
Constant heat capacity law
Cosmic rays
Diffraction
Fission
Geiger-Nutall rule
Group diffusion law
Photoelectric effect
Quantum theory
Radiation pressure
Radioactivity
Richardson's (Edison) effect
Schrödinger's wave equation
Stark effect
Stefan-Boltzmann (Kirchhoff) law
Stoke's law for fluorescence
Wiedemann-Franz's law
Wien's displacement law
Wood & Ellett effect (Resonance radiation)
Zeeman effect

CHEMISTRY
Absorption
Adsorption
Arrhenius theory of electrolytic dissociation
Brownian movement
Debye frequency effect
Electrocapillarity
Electrokinetic phenomena
Faraday's law of electrolysis
Fick's law
Gibb's theory of equilibria
Graham's law
Henry's law
Ideal gas laws
Joule-Thompson (Joule–Kelvin effect)
Kohlrausch's law
Le Chatelier's law
Periodic table

Raoult's law
Snell's law of refraction
State & change of state
Surface tension
Volta effect
Wien effect

ELECTRICITY AND MAGNETISM
Ampere's law
Ampere (Biot-Savart) law
Barkhausen effects
Capacity of dielectric
Corbino effect
Coulomb's law
Curie-Weiss law
Debye frequency effect
Electrocapillarity
Electrokinetic phenomena
Faraday effect (Magneto-optical rotation)
Faraday's law of electrolysis
Faraday's law of induction
Ferranti effect
Galvanomagnetic & thermomagnetic effects
Gauss's law
Impedance/frequency
Johnson-Rahbek (Winslow effect)
Joule's law
Kerr electrostatic effect
Kirchoff's law
Kohlrausch's law
Magnetic behavior of materials
Magnetic dispersion of sound
Magnetic effects
Magnetic susceptance (Gauss effect)
Magnetostriction & allied effects
Ohm's law
Paschen's law
Photoelectric effect
Photomagnetic effect
Piezoelectric effect
Pinch effect
Proximity effect
Residual voltages
Resistance/dimension
Resistance/temperature
Richardson' (Edison) effect
Schrödinger wave equation
Shot effect or Skin effect
Thermoelastic effect
Thermoelectric effect
Thompson & Benedick effects
Triboelectricity
Volta effect

Table 6-4 (Continued)

Wiedemann-Franz's law
Wien effect
Wood & Ellett effect (Resonance radiation)
Zeeman effect

FLUID MECHANICS
Archimedes's principal
Bernoulli's theorem
Cavitation
Electrocapillarity
Electrokinetic phenomena
Graham' law
Henry's law
Ideal gas laws
Joule-Thompson (Joule-Kelvin effect)
Kelvin's principle for fluid drops
Stoke's law
Surface tension

GENERAL PHYSICS
Absorption
Archimedes' principal
Baushinger effect
Bernoulli theorem
Brownian movement
Constant heat capacity law
Electrocapillarity
Eotvos effect
Expansion/temperature
Hooke's law
Ideal gas laws
Joule-Thompson (Joule-Kelvin effect)
Kelvin's principle for fluid drops
Kepler's laws
Lambert's laws (Absorption coefficient)
Le Chatelier's law
Periodic table laws
Snell's (law of refraction) law
State & change of state
Stoke's law

MECHANICS
Bauschinger's effect
Coulomb law
Elastic limit
Hooke's law
Johnson-Rahbek (Winslow) effect
Magnetic dispersion of sound
Newton's laws
Piezoelectric effect
Poisson's ratio
Resistance/dimension
Thermoelastic effect
Triboelectricity

OPTICS
Absorption
Brewester's law
Christiansen effect
Diffraction
Doppler (Fizeau) effect
Faraday effect (Magneto-optical rotation)
Gladstone & Dald's law
Interference
Kerr effect (electrostatic)
Kerr magneto-optic effect
Lambert's law (absorption coefficient)
Optical rotary power
Purkinje effect
Radiation pressure
Snell's law of refraction
Stoke's law on fluorescence
Total reflection
Weber-Fechner law
Wood & Ellet effect (Resonance radiation)
Zeeman effect

PSYCHOLOGY
Purkinje effect
Weber-Fechner law

Thus, an excessive concern for high accuracy in system modeling of nonphysical and even physical systems should not become an impediment in formulating quantitative system models. Some of the major barriers that arise in formulating quantitative system models are temerity of the investigator upon confronting a new, major system, restricted imagination and innovation, unfamiliarity with the basic causal relationships of the model, and the identification of the essential elements of the system and separation of the system from its environment.

Often the inability to reduce complex systems to manageable system models and the investigator's desire to preserve great detail results in an overly detailed, difficult, or intractable model. Indeed, if the model is irresolvable with the resources and abilities available to the investigator, then that model is impractical and should be abandoned or modified to provide a simpler model that can be analyzed. The system investigator is generally well advised to begin with simple, limited, conservative models and then to build on a successful base of increasingly accurate quantitative models of the system under investigation.

6.2.4 Experimental Modeling

For systems that exist in the physical world and whose inputs and outputs can be quantitatively measured, experimental modeling and verification are obviously desirable. From experimental tests with the actual system or a similar experimental model, it may be possible to sample the change in output response from known input changes to produce a tabular data set or a graphical curve for each system kernel. If careful planning and data collection have occurred, generally, it is possible to obtain an accurate representation of the system's kernel for the measured inputs and outputs. To formulate an accurate quantitative mathematical representation from these experimental data, a process called curve fitting is often used. Although many general techniques are available for the fitting of graphical or tabular data, we will consider only three techniques here: least-squares fitting, dimensional analysis, and analytical estimation or curve fitting.

For most real system models, perhaps the best test of the acceptability and accuracy of the system equations is to compare and evaluate the behavior predicted by the model with the actual, experimentally observed behavior. Occasionally, experimental verification or testing is not possible, or practical, or entails unacceptable costs or consequences. But when experimental data, even limited data, can be obtained, they generally should be used. Of course, scrutiny of the experimental methods employed to monitor, measure, and report the data is important. Occasionally, good models have been abandoned and poor models adopted because of the acceptance of faulty or misinterpreted experimental data. However, if a set of experimental data is valid, it should serve as the "acid test" of a system model since we generally want models of real systems to predict real-world behavior.

Least Squares

Least squares is an effective way for establishing the "best values" of the parameters (in the least squares error sense) for a set of analytical expressions used to fit tabular or graphical data. The method is particularly convenient if the independent parameters used to fit the experimental data appear in a linear fashion in the set of expressions chosen to quantitatively describe the system. Suppose we wish to fit a given system data set with n independent, analytical functions of the form $h_1(x)$, $h_2(x)$, ..., $h_n(x)$, and n independent parameters p_1, p_2, ..., p_n. Then we have the task of determining the best values of p_i such that the function $g(x)$, where

$$g(x) = \sum_{i=1}^{n} h_i(x_i)p_i \tag{6.2.4}$$

will provide an acceptable continuous functional representation for the set of data. If we want to fit $g(x)$ to m different, distinct values of the independent variable x, that is,

$x_1, x_2, ..., x_m$, for which the dependent values $y_1, y_2, ..., y_m y_1$ are known, then we have the following set of equations:

$$\sum_{i=1}^{n} h_i(x_i) p_i = y_i$$

for $j = 1, ..., m$. This equation can be expressed in matrix form as follows:

$$HP = Y \tag{6.2.5}$$

where

$$H = \begin{bmatrix} h_1(x_1) & h_2(x_1) & \cdots & h_n(x_1) \\ h_1(x_2) & h_2(x_2) & \cdots & h_n(x_2) \\ \vdots & & & \\ h_1(x_m) & h_2(x_m) & \cdots & h_n(x_m) \end{bmatrix} \tag{6.2.6.1}$$

$$P = \begin{bmatrix} p_1 \\ p_2 \\ \vdots \\ p_n \end{bmatrix} \quad Y = \begin{bmatrix} y(x_1) \\ y(x_2) \\ \vdots \\ y(x_m) \end{bmatrix} \tag{6.2.6.2}$$

Observe that we have n unknowns ($p_1, p_2, ..., p_n$) in terms of m equations $y(x_1), y(x_2), ..., y(x_m)$. To form a square matrix equation (i.e., n equations in n unknowns) that we can consistently solve for the parameters, we simply pre-multiply matrix equation (6.2.5) by the transpose matrix of H, that is, H^T where

$$H^T = \begin{bmatrix} h_1(x_1) & h_1(x_2) & \cdots & h_1(x_m) \\ h_2(x_1) & h_2(x_2) & \cdots & h_2(x_m) \\ \vdots & & & \\ h_n(x_1) & h_n(x_2) & \cdots & h_n(x_m) \end{bmatrix}$$

and we have from Eq. (6.2.5) that

$$H^T HP = H^T Y \tag{6.2.7}$$

where

$$H^T H = \begin{bmatrix} \sum_i^m h_1^2(x_i) & \sum_i^m h_1(x_i)h_2(x_i) & \cdots & \sum_i^m h_1(x_i)h_n(x_i) \\ \sum_i^m h_2(x_i)h_1(x_i) & \sum_i^m h_2^2(x_i) & \cdots & \sum_i^m h_2(x_i)h_n(x_i) \\ \vdots & & & \\ \sum_i^m h_n(x_i)h_1(x_i) & \sum_i^m h_n(x_i)h_2(x_i) & \cdots & \sum_i^m h_n^2(x_i) \end{bmatrix}$$

$$
H^T Y =
\begin{bmatrix}
\sum\limits_{i}^{m} h_1(x_i) y(x_i) \\[2mm]
\sum\limits_{i}^{m} h_2(x_i) y(x_i) \\[2mm]
\vdots \\[2mm]
\sum\limits_{i}^{m} h_n(x_i) y(x_i)
\end{bmatrix}
$$

Matrix equation (6.2.7) may be solved for P to obtain

$$
P = (H^T H)^{-1} H^T Y \tag{6.2.8}
$$

where $(H^T H)^{-1}$ is the inverse matrix of $(H^T H)$ and is assumed to be nonsingular. Computational effort can be reduced by observing that the matrices $H^T H$ and thus $(H^T H)^{-1}$ are symmetric if H is composed of real coefficients.

Example of Least Squares Curve Fitting.
Suppose we want to fit the linearly independent functions

$$
h_1(x) = 1 \quad \text{and} \quad h_2(x) = x^3
$$

in least squares fit to the following data:

x	0	1.0	2.0	4.0	Independent variable
$y(x)$	0	1.2	3.5	17.0	Dependent variable
$h_1(x)$	1	1	1	1	1st fit function
$h_2(x)$	0	1	8	64	2nd fit function

Recalling Eq. (6.2.6), we find for H and Y the following:

$$
H =
\begin{bmatrix}
1 & 0 \\
1 & 1 \\
1 & 8 \\
1 & 64
\end{bmatrix}
\qquad
Y =
\begin{bmatrix}
0 \\
1.2 \\
3.5 \\
17.0
\end{bmatrix}
$$

And we have for $H^T H$

$$
H^T H =
\begin{bmatrix}
1 & 1 & 1 & 1 \\
0 & 1 & 8 & 64
\end{bmatrix}
\begin{bmatrix}
1 & 0 \\
1 & 1 \\
1 & 8 \\
1 & 64
\end{bmatrix}
=
\begin{bmatrix}
4 & 73 \\
73 & 4161
\end{bmatrix}
$$

For the inverse of $H^T H$, we have

$$
(H^T H)^{-1} = \frac{1}{11,315}
\begin{bmatrix}
4161 & -73 \\
-73 & 4
\end{bmatrix}
$$

So from Eq. (6.2.8), we have for the column matrix P, which contains our least-squares parameters c_1 and c_2,

$$P = \left(H^T H\right)^{-1} H^T Y = \frac{1}{11,315} \begin{bmatrix} 4161 & -73 \\ -73 & 4 \end{bmatrix} \begin{bmatrix} 1 & 1 & 1 & 1 \\ 0 & 1 & 8 & 64 \end{bmatrix} \begin{bmatrix} 0 \\ 1.2 \\ 3.5 \\ 17.0 \end{bmatrix}$$

and finally

$$P = \begin{bmatrix} c_1 \\ c_2 \end{bmatrix} = \begin{bmatrix} 0.77 \\ 0.26 \end{bmatrix}$$

So our analytic equation of the data given is obtained through least-squares fit as follows:

$$g(x) = c_1 h_1(x) + c_2 h_2(x) = 0.77 + 0.26 x^3$$

Dimensional Analysis

Another modeling technique that requires experimental data to complete the system kernel is based on the method of dimensional analysis. Dimensional analysis has been employed with considerable success in disciplines in the physical sciences such as continuum mechanics and the thermal sciences. However, the method is quite general and can be applied to a broad range of problems. As with most modeling techniques, the critical step is in determining what parameters and variables are essential for obtaining the inputs and outputs desired. After this is done, the dimensional analysis then prescribes the general mathematical form in which the parameters appear in the system kernel. We will demonstrate dimensional analysis by two specific examples.

Business Model

Consider the following single-input, single-output model for a business firm derived using dimensional analysis. A business firm would like to estimate future income from sales produced by its traveling sales force. The firm has maintained historical data for several years on various parameters or variables believed to be of value in assessing firm operations. The parameters are as follows:

x_1 = number of salespersons employed as a function of time
x_2 = average sales income per item sold
x_3 = average number of contacts made per item sold
x_4 = average number of contacts made per salesperson
x_5 = average distance traveled per salesperson per year
x_6 = average number of contacts made per unit distance traveled

The system equation (assumed to be continuous) of interest for the firm is

$$d\$ = g(\mathbf{x})dt$$

where $d\$$ is the differential change in sales income (the system output), $g(\mathbf{x})$ is the system kernel, and the parameters are $\mathbf{x} = x_1, x_2, ..., x_6$, which have been defined previously, and dt is the differential change in time (the system input).

Now let us assume that the system kernel is dependent on the business parameters in the following manner:

$$g[x(t)] = K[x_1(t)]^{n1}[x_2(t)]^{n2}[x_3(t)]^{n3}[x_4(t)]^{n4}[x_5(t)]^{n5}[x_6(t)]^{n6}$$

where K, $n1$, ..., $n6$ are dimensionless constants to be determined as required for dimensional compatibility of the system equation. Now the independent dimensions or units associated with this system model are sales, salesperson, contacts, items sold, distance traveled, and time. If we equate the exponential powers to which each of the dimensions occurs in the system equation, we have from the system equation

$$d\$(\text{sales}) = g(x)dt(\text{time})$$

that

Sales (\$):	$1 = n2$
Salespersons:	$0 = n1 - n4 - n5$
Contacts:	$0 = n3 + n4 + n6$
Items sold:	$0 = -n2 - n3$
Distance traveled:	$0 = n5 - n6$
Time (t):	$0 = n5 + 1$

With six independent parameters and six independent linear equations, we anticipate a single unique set for the dimensionless constants, $n1$, $n2$, ..., $n6$.

Solving this system for each ni, we have $n1 = n2 = 1$, $n3 = n5 = n6 = -1$, and $n4 = 2$. Thus our system equation has the general for

$$d\$ = K\frac{x_1(t)x_2(t)[x_4(t)]^2}{x_3(t)x_5(t)x_6(t)}dt$$

If we treat the system equation as being separable, where in general $x = x(t)$ and $K = mean$ *value of* $[K(t)]$, we may integrate the equation to obtain as the general solution

$$\$(t) = K\int_0^T \frac{x_1(t)x_2(t)[x_4(t)]^2}{x_3(t)x_5(t)x_6(t)}dt$$

where we have assumed as an initial condition that sales are zero, $\$(0) = 0$, when $t = 0$. To evaluative this integral, let us assume that time-averaged values for all terms entering into the integrand over the interval $0 < t < T$ are used. Then we have, upon completion of the integration of the system equation, that

$$\$(T) = K\frac{x_1x_2x_4^2}{x_3x_5x_6}T$$

It is interesting to see how the time-averaged parameters influence the system solution for the sales income in the time interval 0 to T. Observe that sales income (\$) is proportional to the number of salesperson (x_1), the average sales income per item sold (x_2), and to the square of the number of contacts made per salesperson (x_4^2). Furthermore, the sales income is inversely proportional to the number of contacts required per item sold (x_3), the average distance traveled per salesperson per year (x_5), and the number of contacts made per distance traveled (x_6). Obviously, for the company to increase sales income, it should increase

x_1, x_2, and x_4 and decrease x_3, x_5, and x_6. Suppose from a past time period, say the prior year (viz., T^*), we know average values for each of the parameters x_1^*, x_2^*, ..., x_6^* and the resulting sales income $\$^*$, so we can use this result to evaluate K as follows:

$$K = \frac{\$^* x_3^* x_5^* x_6^*}{T^* x_1^* x_2^* (x_4^*)^2}$$

We have then for the system equation solution that

$$\frac{\$(T)}{S^*} = \left\{\frac{x_1}{x_1^*}\right\}\left\{\frac{x_2}{x_2^*}\right\}\left\{\frac{x_3^*}{x_3}\right\}\left\{\frac{x_4}{x_4^*}\right\}^2\left\{\frac{x_5^*}{x_5}\right\}\left\{\frac{x_6^*}{x_6}\right\}\frac{T}{T^*}$$

Assume that this next year the company believes it can improve these time-averaged performance parameters as follows:

$$\frac{x_1}{x_2^*} = 110\%, \frac{x_2}{x_2^*} = 110\%, \frac{x_3}{x_3^*} = 90\%, \frac{x_4}{x_4^*} = 120\%, \frac{x_5}{x_5^*} = 100\%, \frac{x_6}{x_6^*} = 95\%$$

Then the improvement in sales income is projected to increase for the next year as follows (note $T/T^* = 1$):

$$\frac{S}{S^*} = (1.1)(1.1)(1/0.9)(1.2)^2(1)(1/0.95)(1) = 2.03$$

Whether the company can actually more than double its sales income next year is dependent on how accurate the company's estimates for performance improvement, as measured by the parameters, prove to be. What is important here is that the company now has a quantitative model to predict future sales income, and the model indicates what the factors are and how they individually and collectively affect sales income.

Hydrodynamic Model

Now consider a hydrodynamic system in which we want to estimate the volume of flow of an incompressible fluid through a uniform circular pipe of variable length. Our system equation is then given by

$$dQ = g(Q, t)dt$$

where dQ/dt is the volumetric flow rate (i.e., L^3/T). It is proposed that the essential variables and parameters that influence the flow system kernel $g(Q, t)$ are

$$g = g(D, v, \rho, \mu, L) = g\left(D^{n1} v^{n2} \rho^{n3} \mu^{n4} L^{n5}\right)$$

where

L = length
M = mass
T = time
D = inside pipe diameter (dimension L)
v = flow velocity at some fixed position in pipe (dimension L/T)
ρ = mean density of fluid (dimension M/L^3)
μ = mean viscosity of fluid (dimension M/LT)
L = pipe length (dimension L)

and $n1, n2, ..., n5$ are dimensionless exponent constants to be determined. We observe that there are three independent dimensions (L, M, T) and five independent parameters and variables in our model system equation, so we anticipate that two dimensionless number groupings will appear in the final model.

Now to ensure dimensional compatibility for our system equation, we require from the system equation that

$$L^3 = (L)^{n1} \left(\frac{L}{T}\right)^{n2} \left(\frac{M}{L^3}\right)^{n3} \left(\frac{M}{LT}\right)^{n4} (L)^{n5} T$$

and equating exponent coefficients of like dimensions, we find

$$L: \quad 3 = n1 + n2 - 3n3 - n4 + n5$$
$$M: \quad 0 = n3 + n4$$
$$T: \quad 0 = -n2 - n4 + 1$$

If we solve these three equations in terms of $n4$ and $n5$, which we will treat as arbitrary, we find that

$$n1 = 2 - n4 - n5$$
$$n2 = 1 - n4$$
$$n3 = -n4$$

Expressing our system equation in terms of its grouped parameters and variables, we have

$$dQ = g\left[D^3 v \left(\frac{\mu}{vD\rho}\right)^{n4} \left(\frac{L}{D}\right)^{n5}\right] dt$$

In general, we would expect the argument of our system kernel to define an arbitrary function, perhaps of the general form (since $n4$ and $n5$ are still arbitrary and could define the integer set for an infinite power series expansion for an arbitrary function)

$$g\left[D^3 v \left(\frac{\mu}{vD\rho}\right)^{n4} \left(\frac{L}{D}\right)^{n5}\right] dt = \frac{\pi D^2}{4} v \sum_{i,j=1}^{\infty} a_{ij} \left(\frac{vD\rho}{\mu}\right)^i \left(\frac{L}{D}\right)^j$$

where the a_{ij}'s are arbitrary, dimensionless constants. We have grouped $(\pi D^2/4)$ together since it represents the cross-sectional area of the pipe, and we identify $(vD\rho/\mu)$ as the Reynolds number, Re, which plays an important role in hydrodynamics, and finally (L/D) as the pipe length ratio in equivalent diameters. For example, suppose that these two dimensionless numbers were known to have the functional form

$$\sum_{i,j=1}^{\infty} a_{ij} \, Re^i \left(\frac{L}{D}\right)^j = \frac{L}{D}[1 - \exp(Re)]$$

Then we would find that the series coefficients are defined as

$$a_{ij} = \begin{cases} 0 & \text{for all } i \text{ and } j \text{ except } j = 1 \\ (-1/i!) & \text{for } i = 1, 2, ..., n \quad \text{for } j = 1 \end{cases}$$

In general, however, unless some knowledge of the functional behavior of the dimensionless numbers is known, we must approximate the infinite power series by a single term of arbitrary power ($n1$ and $n2$) for the Reynolds number and pipe length ratio and coefficient a_o as follows:

$$dQ = a_o \left(\frac{\pi D^2}{4}\right) v \left(\frac{vD\rho}{\mu}\right)^{n1} \left(\frac{L}{D}\right)^{n2} dt$$

where we determine a_o, $n1$, and $n2$ for the best overall fit with experimental data. Now we can easily solve our system equation if we assume that the only term in the kernel of the system equation that is time-dependent is the velocity of flow $v(t)$; so we have upon integration

$$Q(T) = a_o \left(\frac{\pi D^2}{4}\right) \left(\frac{D\rho}{\mu}\right)^N \left(\frac{L}{D}\right)^M \int_0^T v^{n_1+1} dt + Q(0)$$

Furthermore, if we define a mean flow velocity over the time period $0 \leq t \leq T$ as v_m, where

$$v_m^{n_1+1} = \frac{1}{T} \int_0^T v^{n_1+1}(t) dt$$

then we have for the system solution

$$Q(T) = a_o \left(\frac{\pi D^2}{4}\right) \left(\frac{D\rho v_m}{\mu}\right)^N \left(\frac{L}{D}\right)^M v_m T + Q(0)$$

Of course, the final test of our system equation is how well it describes the behavior of the real hydrodynamic system.

Analytical Curve Fitting

Occasionally, in modeling a system, the quantitative behavior for the system kernel is reasonably well known for incremental or differential changes in each of the important input variables. Generally, such quantitative data are obtained from experimental observations or other data sources. What is now desired is to formulate an accurate mathematical expression for the model kernel from this known database. A process called "curve fitting" is another useful tool used to develop these mathematical expressions. We have seen how least-squares methods can be used to "best fit" curves with a specified mathematical form. Another technique is to visually examine and compare the system response with simplified sets of generalized mathematical functions curves that might be reasonably expected to apply to the system model under study. Obviously, the skill and breadth of experience of the system investigator are essential factors in the successful implementation of such a technique, which is heavily dependent on judgment and experience.

For example, let us restrict our attention to the change in one system output resulting from a change in one system input (holding all other system inputs and outputs constant). Suppose, as is often the case, that the observed response of the system output variable is a monotonically changing (i.e., consistently increasing or decreasing) function with a continuous change in the input variable from its minimum value to its maximum value. Let us normalize the system response for convenience to the range 0 to 1 for both the single input

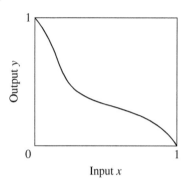

Fig. 6-8 Monotonic decreasing response.

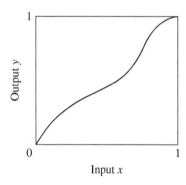

Fig. 6-9 Monotonic increasing response.

and the single output. Mathematically, for these assumptions, we have two cases to consider. In Figure 6-8, we have typified the response of the system for which the system's output change or ordinate value monotonically decreases from its maximum value (1) to its minimum value (0), as the system's input change or abscissa value increases from its minimum value (0) to its maximum value (1). Figure 6-9 typifies the system for which this response is reversed, and the system's output change monotonically increases with the input change.

Of the set of all continuous monotonic functions that possess the behavior typified by Figures 6-8 and 6-9, it seems reasonable to limit or restrict this functional set to those functional forms that are common or anticipated for natural or real systems. For example, we would expect that the system kernel would exhibit only small integral exponents $(0, \pm 1, \pm 2,$ etc.) of the system variables (inputs and outputs) and, furthermore, would exhibit a rational form (that is, a ratio of polynomial-type functions). Our basis for these assumptions is past experience with physical systems and our expectations for similar behavior. So we anticipate a system kernel with the following general mathematical form as simple power series:

$$y(x) = \frac{\displaystyle\sum_{n=0}^{N} a_n x^n}{\displaystyle\sum_{m=0}^{M} b_m x^m} \tag{6.2.9}$$

where x is the independent (input) variable and y is the dependent (output) variable. The values for N and M are finite integers. For the system whose response range is $y(0) = 1$ to $y(1) = 0$, we have from Eq. (6.2.9) that

$$y(x) = \frac{\displaystyle\sum_{n=1}^{N} a_n (x^n - 1)}{\displaystyle\sum_{m=1}^{M} b_n x^n - \sum_{n=1}^{N} a_n} \tag{6.2.10}$$

while for the response range $y(0) = 0$ to $y(1) = 1$, we have

$$y(x) = \frac{\displaystyle\sum_{n=1}^{N} a_n x^n}{\displaystyle\sum_{m=1}^{M} b_n (x^n - 1) + \sum_{n=1}^{N} a_n} \tag{6.2.11}$$

Even Eqs. (6.2.10) and (6.2.11) are more general and complex than we usually need to consider. For example, with the simple forms

$$y_1 = \frac{(1+a)x^N}{1+ax^M} \quad \text{with} \quad y_1(0) = 0, y_2(1) = 1 \tag{6.2.12}$$

and

$$y_2 = \frac{1-x^N}{1+ax^M} \quad \text{with} \quad y_2(0) = 1, y_2(1) = 0 \tag{6.2.13}$$

We can obtain a variety of functional behavior shown in Figures 6-10 through 6-13. Observe that all simple curves that exhibit monotonically increasing values of y with increasing x are defined by Eq. (6.2.12) as $y_1 = f(x)$. On the other hand, all curves that exhibit monotonically decreasing values of y with increasing x are defined by Eq. (6.2.13) as $y_2 = g(x)$. The purpose of Figures 6-10 through 6-13 is to provide quantitative trend indications as to what values of the parameters a, N, and M, as defined by Eqs. (6.2.12) and (6.2.13) might provide comparable and acceptable responses observed for a specific system. Then Eqs. (6.2.12) and (6.2.13) can be employed as analytical models for the system kernel.

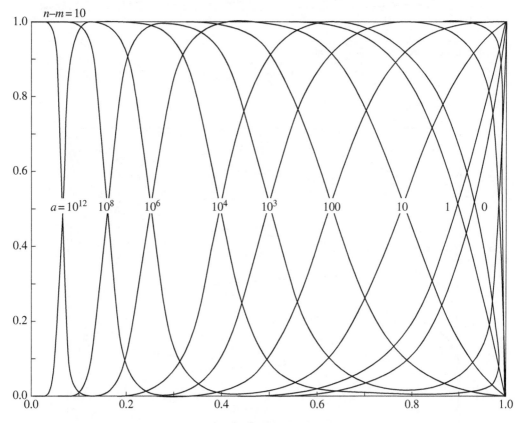

Fig. 6-10 Curves of $y = \dfrac{(1+x^{10})}{(1+ax^{10})}$ and $y = \dfrac{(1+a)x^{10}}{(1+ax^{10})}$.

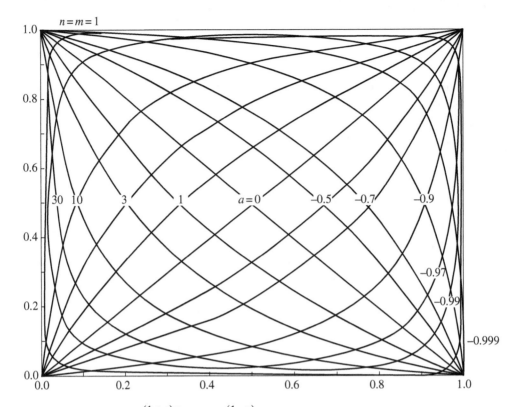

Fig. 6-11 Curves of $y = \dfrac{(1 + a)x}{(1 + ax)}$ and $y = \dfrac{(1 - x)}{(1 + ax)}$.

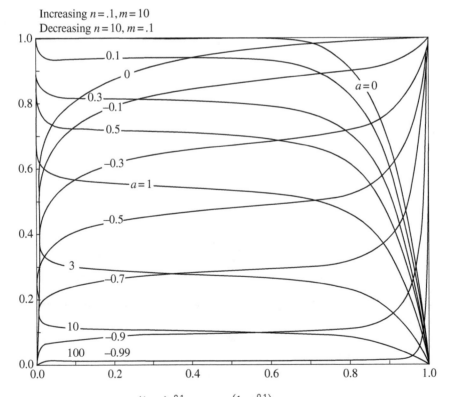

Fig. 6-12 Curves of $y = \dfrac{(1 + a)x^{0.1}}{(1 + ax^{0.1})}$ and $y = \dfrac{(1 - x^{0.1})}{(1 + ax^{0.1})}$.

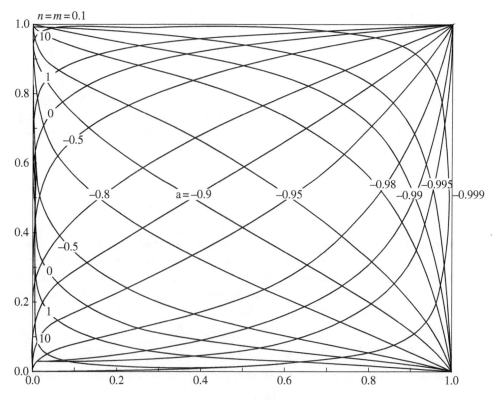

Fig. 6-13 Curves of $y = \dfrac{(1 + a)x^{0.1}}{(1 + ax^{10})}$ and $y = \dfrac{(1 - x^{10})}{(1 + ax^{0.1})}$.

6.2.5 Stochastic Modeling

Stochastic systems are those systems whose inputs and outputs may not be correctly interpreted on an effective single-event basis. The variables and parameters of such systems are subject to significant statistical variation and therefore must be interpreted on a statistical basis. Generally, stochastic systems can be modeled and analyzed employing methods from probability theory and assumptions about the distribution functions for the variables and parameters of the system. We will examine only two simple methods for stochastic analysis. From this limited exposure, we will obtain at least some insight into typical stochastic methods and procedures.

Contracting Trees

Occasionally, we encounter a situation in system investigations where we have a given output or event for which we wish to determine all possible or significant causes or inputs that might influence or affect that event. Often the single output that occupies our interest is a very important event with major consequences, and we wish to establish the important contributing causes that influenced the event. An example here of a single significant event would be the assassination of U.S. President John F. Kennedy in 1963. Because of the impact of this tragedy, great efforts were made to determine and establish the various causes that may have contributed to this single event.

Such single-output event systems often have many varied inputs, which pass through preceding compound cause and effect sequence trains. These causal trains or sequences are often called reverse or contracting trees and include the fault trees that arise in the fields of risk analysis or risk assessment. All these sequences may be depicted as tree diagrams with the output event as the top of the tree or output branch and an expanding or branching growth of inputs as we move down the broadening tree. A simple single output tree for a discrete system might appear, as shown in Figure 6-14.

Using the graphical methods for system reduction given in Section 6.1, it is easy to show that the overall system equation for the shaded system block is

$$\Delta e = g_{12}\Delta c_{12} + g_{11}g_{22}\Delta c_{22} + g_{11}g_{21}g_{32}\Delta c_{32} + g_{11}g_{21}g_{31}g_{41}\Delta c_{41}$$

which can be represented as a four-input, single-output system.

As a practical example of the single-output system, let us consider Figure 6-14 as a contracting or fault tree system. The stochastic system kernel components assume the role of probability functions, and each system block becomes a two-state system with the two kernel values equal to a probability P for one state and a probability $1 - P$ for the other state. So let us define the kernel components as follows

$$g_{11} = P_{11}, \quad g_{12} = 1 - P_{11}, \quad g_{21} = P_{21}, \quad g_{22} = 1 - P_{21}$$
$$g_{31} = P_{31}, \quad g_{32} = 1 - P_{31}, \quad g_{41} = 1 - P_4$$

Then the overall system probability equation as given by the system equation becomes

$$\Delta e = (1 - P_{11})\Delta c_{12} + P_{11}(1 - P_{21})\Delta c_{22}$$
$$+ P_{11}P_{21}(1 - P_{31})\Delta c_{32} + P_{11}P_{21}P_{31}(1 - P_{41})\Delta c_{41}$$

Now, this form of the system equation can be viewed as an equation yielding the probability of event Δe occurring, given the probabilities for four input causes

$$(\Delta c_{12}, \Delta c_{22}, \Delta c_{32}, \Delta c_{42})$$

Suppose we have the following values:

$$P_{11} = 0.10, \quad P_{11} = 0.20, \quad P_{11} = 0.15, \quad P_{11} = 0.05$$
$$\Delta c_{12} = 0.20, \quad \Delta c_{22} = 0.15, \quad \Delta c_{32} = 0.20, \quad \Delta c_{41} = 0.10$$

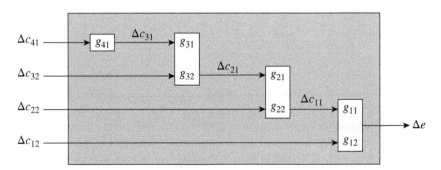

Fig. 6-14 Typical contracting tree diagram.

Then we find for Δe

$$\Delta e = 0.9(0.2) + 0.1(0.8)(0.15) + 0.1(0.2)(0.85)(0.2)$$
$$+ 0.1(0.2)(0.15)(0.95)(0.1) = 0.20$$

and thus $\Delta e = 0.20$.

The probability of the output event is 0.20.

Expanding Trees

In an opposite vein, which is in contrast to the contracting tree sequence, we may encounter situations where we have a single cause or input system for which we seek to determine all the significant consequences or outputs that can occur in an expanding causal sequence or expanding tree fashion. Often such expanding tree diagrams are called event trees or decision trees, usually depending on the systems investigator's background. An example here of a single input or cause for which we might seek the full consequences is the discovery of the neutron by James Chadwick in 1932. This significant event led to the subsequent discovery of nuclear fission, the wartime development of the atom bomb, the production of electrical power from nuclear reactors, the nuclear arms race, and other significant developments in science and technology.

A simple single-input cause for a discrete system model might have an expanding tree diagram, as shown in Figure 6-15. With standard graphical reduction methods, we have for the overall shaded system block

$$\Delta e_{31} = g_{11}g_{21}g_{31}\Delta c$$
$$\Delta e_{22} = g_{11}g_{22}\Delta c$$
$$\Delta e_{23} = g_{12}g_{23}\Delta c$$
$$\Delta e_{32} = g_{12}g_{24}g_{32}\Delta c$$
$$\Delta e_{33} = g_{12}g_{24}g_{33}\Delta c$$

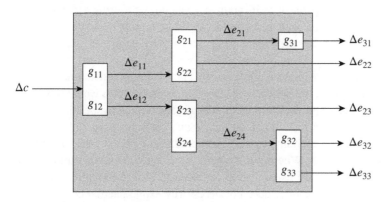

Fig. 6-15 Typical expanding tree diagram.

As we did for the fault tree system, let us assign to the kernels of this expanding tree model the following probability measures:

$$g_{11} = P_1, \quad g_{12} = 1 - P_1, \quad g_{21} = P_2, \quad g_{22} = 1 - P_2, \quad g_{32} = P_3$$
$$g_{24} = 1 - P_3, \quad g_{31} = P_4, \quad g_{32} = P_5, \quad g_{33} = 1 - P_3$$

Then we have for the stochastic system equation that

$$\Delta e_{31} = P_1 P_2 P_4 \Delta c$$
$$\Delta e_{22} = P_1 (1 - P_2) \Delta c$$
$$\Delta e_{23} = (1 - P_1) P_3 \Delta c$$
$$\Delta e_{32} = (1 - P_1)(1 - P_3) P_5 \Delta c$$
$$\Delta e_{33} = (1 - P_1)(1 - P_3)(1 - P_5) \Delta c$$

and for the probability data set

$$P_1 = 0.10, \quad P_2 = 0.20, \quad P_3 = 0.40, \quad P_4 = 0.15, \quad P_5 = 0.25, \quad \Delta c = 0.10$$

We have as probabilities for the various output events that

$$\Delta e_{31} = 0.1(0.2)(0.15)(0.1) = 0.0003$$
$$\Delta e_{22} = 0.1(0.8)(0.1) = 0.008$$
$$\Delta e_{23} = 0.9(0.4)(0.1) = 0.036$$
$$\Delta e_{32} = 0.9(0.6)(0.25)(0.1) = 0.0135$$
$$\Delta e_{33} = 0.9(0.6)(0.75)(0.1) = 0.0405$$

We observe that the probability associated with output events that result from a many sequence expanding tree model can become rather small.

6.2.6 Heuristic Modeling

Heuristic methods for system modeling constitute all those methods and techniques that, although unconfirmed, may be valuable for discovering, revealing, or suggesting the qualitative and quantitative nature and characteristics of actual systems. Such methods, although unproven or even incapable of verification, can be useful, nevertheless, in the construction of both practical and accurate system models. What is probably most distinctive about heuristic methods is that they generally lack firm theoretical bases and are, therefore, subject to intuition and opinion. Therefore, some disagreement and controversy may exist over the use of heuristic methods. For example, a possible heuristic method in establishing a given system model and ranking the significance of the inputs and outputs would be to conduct a poll among individuals who are familiar with the given system. As a specific example, suppose our system is a nation's economy, then the single significant output of interest is inflation. We might simply poll economists, bankers, workers, and politicians to determine the significant causes of inflation. Whether the ranked set of causes established by such a poll is actually representative of the real system is conjectural. We are all familiar with the fact that an opinion held by a majority does not necessarily express actual conditions or reality. Indeed, in the fourteenth century, the majority opinion was that the earth was not only flat but also the center of the universe. Nevertheless, such heuristic methods can have value, and the system investigator should be acquainted with these methods and their general use.

Comprehensive, detailed descriptions of the few heuristic methods we will briefly consider may be found in the literature. Our purpose is to acquaint the reader with some of the more common heuristic methods employed and some of the advantages and disadvantages associated with each method. Obviously, the investigator must use caution and judgment in employing any heuristic method. The possibility always exists that a given heuristic method may result in wrong assumptions about the system and an inaccurate model. Perhaps a reasonable guideline for the appropriateness of a given heuristic method for studying a particular system is past experience with similar systems. If experience has shown that a certain heuristic method, such as polling or Delphi, has been apparently successful, then probably fewer doubts exist about the appropriateness of that method applied to a similar system. Table 6-5 provides a brief description of some common heuristic methods for system

Table 6-5 Major System Modeling Methods

I. Scientific Methods

1) **Theoretical** – Use of scientific logic (usually mathematical) to construct a hypothetical model of system that can be verified and confirmed by experimental evidence.

   Advantages – Method is strongly coupled to independent experiments and measurements, and incorrect models are generally quickly identified and discarded.

   Disadvantages – Initially based on intuition and concepts of an individual or small group. Method can be abstract and complex.

2) **Experimental** – Determination and evaluation of direct experimental data and evidence to reveal physical nature of physical system.

   Advantages – Considered ultimate assessment of truth for empirical systems.

   Disadvantages – May be impractical or difficult to acquire data, and correct interpretation may be difficult to achieve.

II. Heuristic Methods

1) **Investigator intuition, opinion, judgment, and experience** – Initial source of most system models is the result of individual creative efforts to conceive descriptive models for rational systems.

   Advantages – For experienced investigators, this source of modeling can be productive and successful.

   Disadvantages – Method is subject to personal bias, emotion, and error.

2) **Delphi** – Construction of a system model based on consensus and collective experience of a large group who are familiar with the system. Group members individually and independently establish significant features of a system model.

   Advantages – This democratic polling of opinion in regard to a given system provides a collective majority view without single individuals or groups dominating the assessment of the system.

   Disadvantages – Subject to popular opinion and bias and may lack rigorous oversight.

3) **Polling of experts** – A small group that is highly knowledgeable about a given system is polled. Through interaction and feedback with group a reasonable consensus of nature of system can often be obtained.

   Advantages – Draws on "best thinking and experience" in fields relating to given system.

   Disadvantages – Often individuals or small groups can bias or dominate results.

4) **Brainstorming** – Attractive modeling process for new or complex systems for which fresh insights and concepts are sought.

   Advantages – Can often reveal previously unknown variables and parameters that affect the system's behavior.

   Disadvantages – Such spontaneous responses can be grossly in error and misleading.

---

modeling, together with the standard scientific-based methods of theoretical and experimental modeling.

6.3 Paradigm for System Modeling, Analysis, and Evaluation

As a system investigator, gains experience in system science and familiarity with system classes, the sequence and manner in which modeling, analysis, and evaluation are performed by the investigator for a given system study are a matter of individual preference and skill. However, to initiate an investigation, an overview for performing a complete system study is useful.

Table 6-6 provides a typical scheme for conducting a system study. Obviously, each study phase and effort expended is system specific.

Table 6-6 Paradigm for a Typical System Study

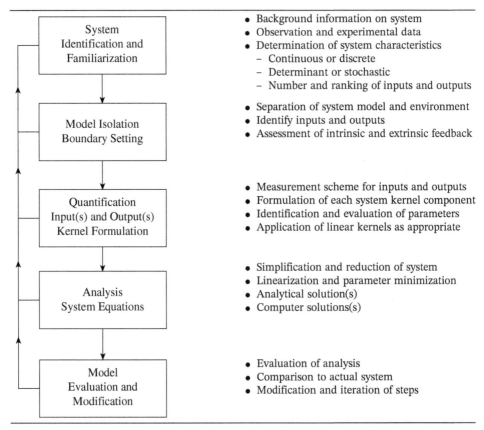

System Identification and Familiarization	• Background information on system
	• Observation and experimental data
	• Determination of system characteristics
	  – Continuous or discrete
	  – Determinant or stochastic
	  – Number and ranking of inputs and outputs

Model Isolation Boundary Setting	• Separation of system model and environment • Identify inputs and outputs • Assessment of intrinsic and extrinsic feedback
Quantification Input(s) and Output(s) Kernel Formulation	• Measurement scheme for inputs and outputs • Formulation of each system kernel component • Identification and evaluation of parameters • Application of linear kernels as appropriate
Analysis System Equations	• Simplification and reduction of system • Linearization and parameter minimization • Analytical solution(s) • Computer solutions(s)
Model Evaluation and Modification	• Evaluation of analysis • Comparison to actual system • Modification and iteration of steps

6.4 Summary

In this chapter, we have briefly examined some of the important methods and techniques generally employed in modeling systems. The graphical methods that were considered are block diagrams, signal-flow graphs, and organizational diagrams that are useful for system modeling and reduction.

Then various methods for quantification and formulation of the system inputs, outputs, and kernels were considered. The single-input, single-output system model was presented as the prospective, initial model for the study of most systems since the opportunity for obtaining an analytical solution is good. Furthermore, the single-input, single-output model serves as an excellent basis for more complex and sophisticated models with multiple inputs and outputs.

Physical science modeling was then considered, and the variables and parameters that might be encountered were discussed on a dimensional or units basis. The modeling of physical science systems was briefly contrasted when with nonphysical systems, and some of the important aspects in modeling these systems were discussed.

Three important quantitative methods that are often applied to experimental systems modeling were examined. These methods were least-squares curve fitting, dimensional analysis, and analytical curve fitting. These three methods allow the system investigator to produce quantitative system kernels from experimental and observational data. Finally, the chapter closes with a brief examination of several heuristic modeling methods for the system where other modeling methods are impractical and inappropriate.

Bibliography

ANDREWS, J. G., and R. R. MCLONE, *Mathematical Modeling*, Butterworth & Co., London, 1971. A general reference on quantitative modeling of general systems.

BLUNDELL, A. J., *Bond Graphs for Modeling Engineering Systems*, John Wiley & Sons, New York, 1982. An introductory treatment of bond graphs in system science.

BREIPOHL, A. M., *Probabilistic Systems Analysis*, John Wiley & Sons, New York, 1970. A basic treatment of the application of probability concepts and theory to systems science.

COX, D. R., and H. D. MILLER, *The Theory of Stochastic Processes*, Methuen & Co., London, 1964. A general reference for stochastic modeling and analysis.

CUNDY, H. H., and A. P. ROLLETT, *Mathematical Models*, Oxford University Press, New York, 1952. An excellent introduction to quantitative modeling.

DEO, N., *Graph Theory*, Prentice-Hall, Englewood Cliffs, NJ, 1974. An excellent book on the theory of graphs and their applications to system modeling.

DIRECTOR, S. W., and R. A. ROHRER, *Introduction to Systems Theory*, McGraw-Hill Book Co., New York, 1972. A standard introduction to system modeling and analysis.

DUNCAN, W. J., *Physical Similarity and Dimensional Analysis: An Elementary Treatise*, Edward Arnold Publishers, London, 1953. An elementary treatment of modeling by similarity and dimensional methods.

FREEMAN, H., *Discrete Time Systems*, Polytechnic Press, New York, 1965. A standard reference for discrete time systems.

GUKHMAN, A. A., *Introduction to the Theory of Similarity*, Academic Press, New York, 1965. Useful reference for similarity modeling and analysis.

LANGHAAR, H. L., *Dimensional Analysis and Theory of Models*, John Wiley & Sons, New York, 1951. An introduction to dimensional analysis.

LORENS, C. S., *Flowgraphs: For the Modeling and Analysis of Linear Systems*, McGraw-Hill Book Co., New York, 1964. A standard treatment for linear system modeling and analysis by flow graphs.

MACHOL, R. E., et al., ed., *Systems Engineering Handbook*, McGraw-Hill Book Co., New York, 1965. Handbook of useful information for the system investigator.

ROBICHAUD, L. P. A., *Signal Flow Graphs and Applications*, Prentice-Hall, Englewood Cliffs, NJ, 1962. A useful reference for the theory and application of signal-flow graphs.

SAGE, A. P., *Methodology for Large Scale Systems*, McGraw-Hill Book Co., New York, 1977. Useful reference for modeling multiple-input and output systems.

VEMURI, V., *Modeling of Complex Systems*, Academic Press, New York, 1978. An advanced treatment of multiple input and output systems.

WILSON, I. G., and M. E. WILSON, *From Idea to Working Model*, John Wiley & Co., New York, 1970. An introduction to defining and developing models for general systems.

ZADEH, L. A., and E. POLAK, eds., *Systems Theory*, McGraw-Hill Book Co., New York, 1969. A broad coverage of many areas of system science.

Problems

1 Combine and reduce the following compound block diagrams to canonical form. Find the system equation for the canonical system. Note: Doubled signal paths indicate multiple input or output signals.

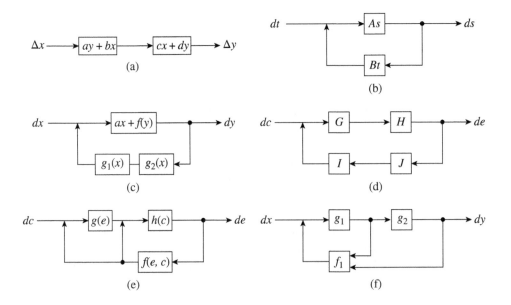

(a)

(b)

(c)

(d)

(e)

(f)

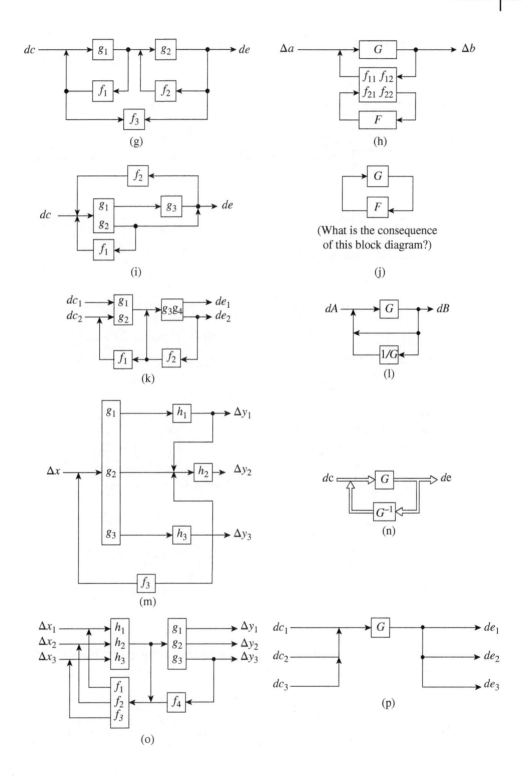

(g)

(h)

(i)

(j)

(What is the consequence
of this block diagram?)

(k)

(l)

(m)

(n)

(o)

(p)

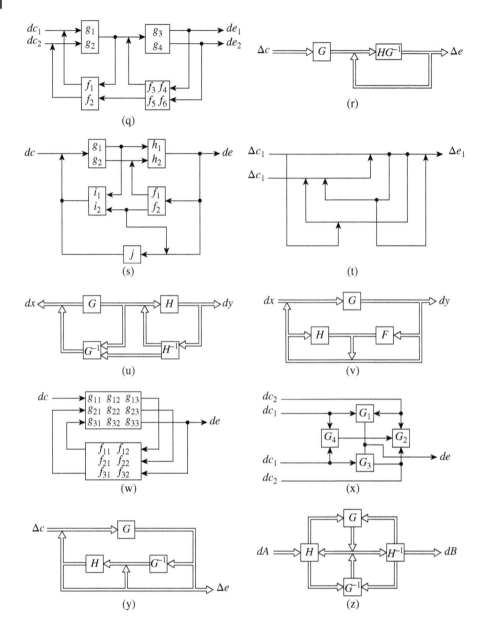

(q)

(r)

(s)

(t)

(u)

(v)

(w)

(x)

(y)

(z)

2 Determine the system equations for the following signal-flow graph systems. Note: Doubled signal paths indicate multiple input or output signals.

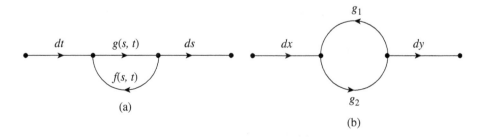

(a)

(b)

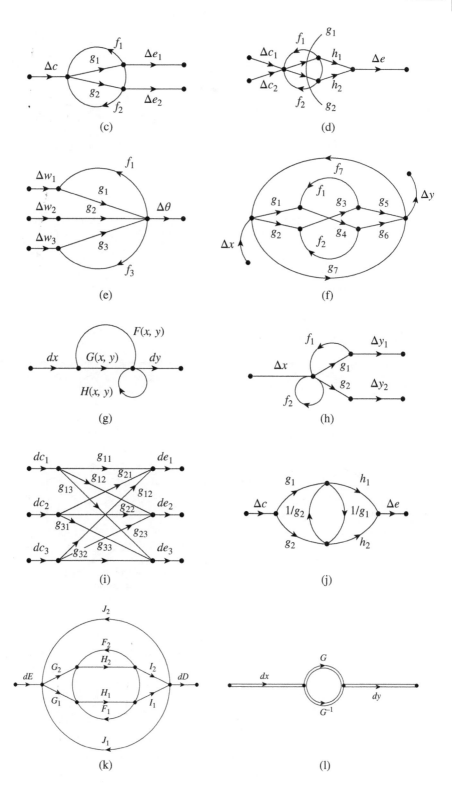

(c)

(d)

(e)

(f)

(g)

(h)

(i)

(j)

(k)

(l)

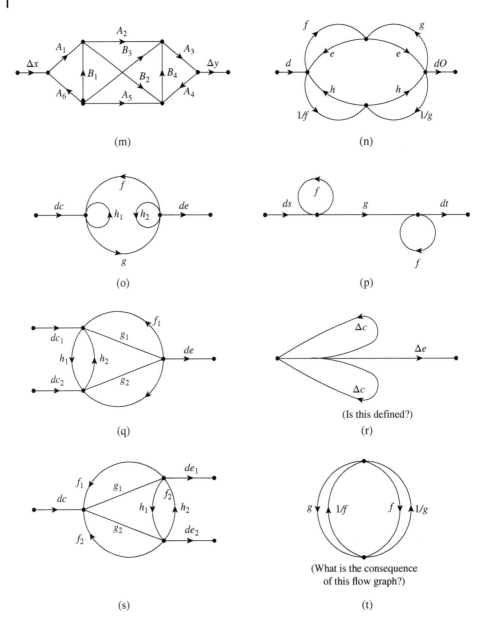

(m)

(n)

(o)

(p)

(q)

(Is this defined?)

(r)

(s)

(What is the consequence
of this flow graph?)

(t)

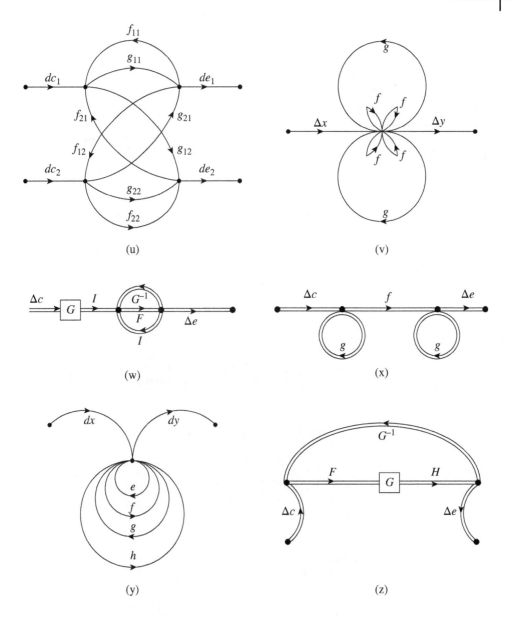

(u)

(v)

(w)

(x)

(y)

(z)

3 Organizational Diagrams: Determine the overall system equation for the following organizations. Obtain and provide the appropriate organizations diagram where necessary.

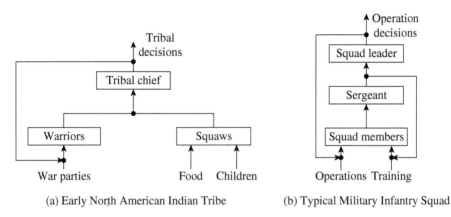

(a) Early North American Indian Tribe   (b) Typical Military Infantry Squad

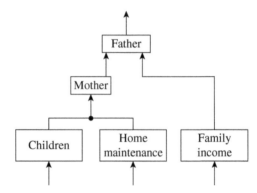

(c) Typical Family (father centered)

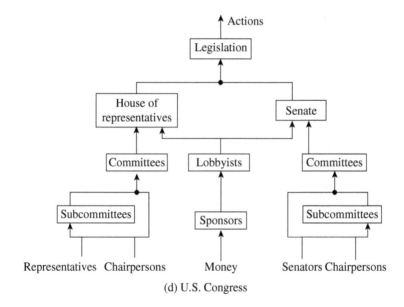

(d) U.S. Congress

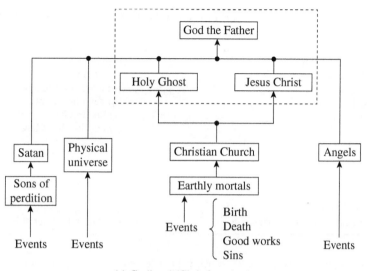

(e) Godhead (Christian view)

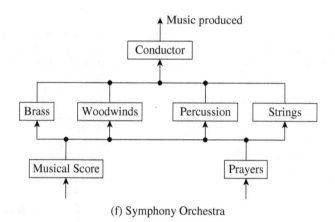

(f) Symphony Orchestra

(g) Organized crime syndicate	(q) Jewish Sanhedrin
(h) NATO alliance organization	(r) Roman Catholic Church
(i) Communist Party in China	(s) Human language classification
(j) UN Security Council	(t) Earth's geological ages
(k) British Parliament	(u) Flora and fauna classification
(l) Japanese Diet	(v) Philosophical schools of thought
(m) English Monarchy	(w) Major world rivers
(n) OPEC	(x) London Commodity Market
(o) European Common Market	(y) New York Stock Exchange
(p) Government of Roman Empire	(z) ISIS

4 Identify the system and the appropriate cause and effect variables for following physical laws and effects.

(a) Nuclear fission	(n) Cosmic rays
(b) Brownian motion	(o) Coulomb's law
(c) Ohm's law	(p) Bernoulli's theorem
(d) Radioactivity	(q) Kirchhoff's laws
(e) Stoke's law	(r) Eotvos effect
(f) Gauss's law	(s) Le Chatelier's law
(g) Zeeman effect	(t) Kepler's laws
(h) Pinch effect	(u) Newton's laws
(i) Thermoelectric effect	(v) Joule-Thompson effect
(j) Schrödinger's wave equation	(w) Hooke's law
(k) Cavitation	(x) Purkinje effect
(l) Piezoelectric effect	(y) Triboelectricity
(m) Periodic law for the elements	(z) Weber-Fechner effect

5 Identify a possible system and associated cause and effect variables for the following physical and nonphysical science systems.

(a) The atom	(n) Higher education
(b) Solar system	(o) Music
(c) A star	(p) Production of electrical power
(d) Earth's atmosphere	(q) Agriculture
(e) Earth's geological record	(r) Christianity
(f) Evolution of humans	(s) Islam
(g) Heredity	(t) The crusades
(h) Human health	(u) Communism
(i) Human behavior	(v) World War II
(j) A biological cell	(w) Logic
(k) Written language	(x) Mathematics
(l) Macroeconomics	(y) Technology
(m) International law	(z) Philosophy

6 Determine the least square fit specified function for the data set indicated in terms of the parameters A, B, and C as given.

Least Square Function	Data Set	Range
(a) Linear $y = Ax + B$	$y = x^2 - 10x$	$x = $ integers $0 \leq x \leq 10$
(b) Linear $y = Ax + B$	$\begin{array}{c\|c\|c\|c\|c\|c} x & 0 & 1 & 3 & 5 & 9 \\ \hline y & 0 & 2 & 5 & 6 & 8 \end{array}$	$x = 0, 1, 3, 5, 9$ $0 \leq x \leq 9$
(c) Quadratic $y = Ax^2 + Bx + C$	$\begin{array}{c\|c\|c\|c\|c\|c} x & -1 & -0.5 & 0 & 0.5 & 1 \\ \hline y & 4 & 1.75 & 0 & -1.25 & -2 \end{array}$	$-1 \leq x \leq 1$
(d) $y = A \exp(Bx)$ (*Hint:* Take log y)	$y = \dfrac{x}{(x+1)}$ $x = 0, 0.1, 0.2, ..., 1$	$0 \leq x \leq 1$

Least Square Function	Data Set	Range
(e) Logarithmic: $y = A + B\text{Ln}x$	$y = x^2$; $x = 1, 10/9, 10/8, 10/7, 10/6, 10/5$	$1 \leq x \leq 2$
(f) Cubic $y = Ax^3 + B$	$y = \exp(-x)$ for $x = 0, 1, 2, 3, 4, 5$	$0 \leq x \leq 5$
(g) Rational: $y = Ax + B/(x + 1)$	<table><tr><td>x</td><td>0</td><td>1</td><td>2</td><td>3</td><td>4</td></tr><tr><td>y</td><td>1</td><td>1</td><td>0.5</td><td>0.4</td><td>0.3</td></tr></table>	$0 \leq x \leq 4$
(h) General: $y = A(x)^{1/2} + B/x$	$y = \exp(-1/x)$; $x = 1, 2, 3, ..., 10$	$1 \leq x \leq 10$
(i) Linear-Log: $y = Ax + B\text{Ln}(x + C)$	<table><tr><td>x</td><td>−10</td><td>−1</td><td>0.1</td><td>0.1</td><td>1</td><td>5</td><td>20</td></tr><tr><td>y</td><td>−10</td><td>−1</td><td>−12</td><td>10</td><td>8</td><td>5</td><td>20</td></tr></table>	$10 \leq x \leq 20$
(j) Linear-Exp: $y = Ax\exp(cx)$	<table><tr><td>x</td><td>0</td><td>1</td><td>2</td><td>4</td><td>8</td><td>16</td></tr><tr><td>y</td><td>0</td><td>1</td><td>2</td><td>4</td><td>8</td><td>16</td></tr></table>	$0 \leq x \leq 16$
(k) Elliptical: $y^2 = Ax + Bx + Cx^2$	<table><tr><td>x</td><td>−1</td><td>0</td><td>1</td><td>2</td><td>3</td></tr><tr><td>y</td><td>0</td><td>1.5</td><td>2</td><td>1.5</td><td>0</td></tr></table>	$-1 \leq x \leq 3$
(l) Mixed polynomial: $y = A(1 + x)^B(1 + 1/x)^C$	$y = \exp(-x^2)$; $x = -3, -1, 0.01, 1, 3$	$0 \leq x \leq 3$
(m) Rational fraction: $\dfrac{y}{1 + y} = A + \dfrac{B}{x} + \dfrac{C}{x^2}$	<table><tr><td>x</td><td>−1</td><td>0</td><td>1</td><td>2</td><td>3</td></tr><tr><td>y</td><td>0</td><td>1.5</td><td>2</td><td>1.5</td><td>0</td></tr></table>	$0 \leq x \leq 10$
(n) Rational fraction: $y = \dfrac{(A + B/x + C/x^2)^n}{1 + x + x^2}$	$y = [1 - (x - 1)^2]^{1/2}$; $x = -2, -1, 0, 1, 2$	$0.1 \leq x \leq 2$
(o) $\sin y = Ax + Bx^2 + \dfrac{C}{x}$	<table><tr><td>x</td><td>−10</td><td>−1</td><td>−0.1</td><td>0.1</td><td>1</td><td>5</td></tr><tr><td>y</td><td>−10</td><td>−1</td><td>−12</td><td>−10</td><td>8</td><td>5</td></tr></table>	$-5 \leq x \leq 5$
(p) $y = A + \dfrac{B}{x} + \dfrac{C}{1 + 1/x}$ $\quad + \dfrac{D}{1 + x/(1 + x)}$	$y = 10\dfrac{x^3}{(1 + x)}$; $x = -2, -1, 0, 1, 2$	$-2 \leq x \leq 2$
(q) $y = A + Bx +$ $\quad C\exp\left(-x^2\right) + \dfrac{D}{x}$	$y = \text{Ln}x$; $x = 0.1, 1, 10, 100$	$0.1 \leq x \leq 100$
(r) $\sin y = A + B\sin x$ $\quad + C\cos x$	<table><tr><td>x</td><td>0</td><td>π/2</td><td>π</td><td>3π/2</td><td>π/1</td></tr><tr><td>y</td><td>−3</td><td>−1</td><td>2.5</td><td>−0.2</td><td>−2.9</td></tr></table>	$0 \leq x \leq 2\pi$

(Continued)

(Continued)

Least Square Function	Data Set	Range
$(s)\ y = A + B\sin(x + C)$ Expand Sine function	$y = 3\cos x;$ $x = 0, \dfrac{\pi}{4}, \dfrac{\pi}{2}, \dfrac{3\pi}{4}, \pi$	$0 \le x \le \pi$
$(t)\ y = Ax^B \exp(Cx + Dx^2)$	$y = 10\dfrac{x^3}{1+x}$ for $x = 0, 1, ..., 10$	$0 \le x \le 10$
$(u)\ y = A + B\mathrm{Ln}x$	$y = x$	$0.1 \le x \le 0.5$
$(v)\ y = \dfrac{A + Bx}{x + 1} + C\dfrac{x^2}{(x+1)^2}$	$y = 1 - \exp(-x)$ for $x = 0, 1, 2, ..., 10$	$0 \le x \le 10$
$(w)\ y = A + Bx + \dfrac{C}{x}$	$y = x\exp(-x);$ $x = 0.1, 1, 5, 10$	$0.1 \le x \le 10$
$(x)\ y(y + x) = A + Bx$	$x = 0, 1, 2, 3;$ $y = 0, 2, 4, 6, 8$	$0 \le x \le 4$
$(y)\ y = A + A^2 x + Bx^2$	$y = -x$	$0 \le x \le 2$
$(z)\ y + \exp(y) = A\exp(Bx)$	$y = 2x;$ $x = 0, 1, 2, 10$	$0 \le x \le 10$

7 Establish the general form of the system equation for the following system models using dimensional analysis.

  a) *Output:* Δe, where e is gross national product (GNP) per person in units of $/(yr − person).

  Input: Δc, where c is years.

  Parameters: P1 is population growth in people/year, P2 is personal income in $/person, P3 is economic growth rate in $/yr.

  b) *Output:* dn, where n is number of atoms in a sample.

  Input: dt, where t is time in seconds.

  Parameters: P1 is decay constant (atoms decayed/atom).

  Assume system equation may also depend on n and t explicitly.

  c) *Output:* dF, where F is number of nuclear fission events.

  Input: dt, where t is time in seconds.

  Parameters: P1 is microscopic fission cross section (fissions-cm^2)/(atom-neutron), P2 is number of atoms, P3 is neutron density (neutrons/cm^3), P4 system volume in cm3, P5 is neutron speed in cm/s.

  d) *Output:* dI, where I is intensity of electromagnetic radiation flux (photons/m^2-s).

  Input: dx, where x is distance traveled by radiation flux in meters through a specified material.

  Parameters: P1 is mass attenuation coefficient of material in units of m^2/kg, and P2 is material density in kg/m^3.

  e) *Output:* dE, where E is energy of a moving mass particle (kg-m^2/s^2).

  Input: dt, where t is time in seconds.

Parameters: P1 is mass of particle in kg, *P2* is particle speed in m/s, *P3* is Boltzmann's constant in kg-m²/K-s, and *P4* absolute temperature in K.

f) *Output:* Δ$, where $ is financial success of individual in units of $.

Input: Δ*W*, *W* is individual's work output in units of Watts (W).

Parameters: P1 is effective wage rate in $ per unit of work (W). *P2* is successful work-wage rate in units of successful work (SW) per $. *P3* is effective work ratio in units of SW/W.

g) *Output:* Δ*ES* is number of enemy naval ships sunk in a military engagement.

Input: Δ*S* is number of enemy ships at sea.

Parameters: P1 is average number of enemy ship contacts per locator (ships/locator), *P2* is ship locator areal density (locator/km²), *P3* is enemy area of operations (km²), *P4* is ship distribution factor (force asset/locator), and *P5* is average battle success (ships sunk/force asset).

h) *Output:* Δ$, where $ is effective product value of New York stock market adjusted Dow-Jones average in units of [product ($)].

Input: Δ*V*, where *V* is daily volume of shares traded in New York Stock Market, each trading day in units of (shares).

Parameters: P1 is time-dependent U.S. Gross National Product in units of [product ($)/person], *P2* is time-dependent Consumer Price Index in units off cost ($)], *P3* is inflationary adjusted cost index for an average share of stock in units of [share ($)/cost ($)], *P4* is time-dependent average cost of a share of stock in units of [share ($)/share], and *P5* is time-dependent U.S. population in units of (persons).

i) *Output: dD*, where *D* is probability of contracting a contagious disease per person in units of (disease/person).

Inputs: dP, where *P* is potential exposable population in persons, and *dC*, where *C* is number of disease-transferable human contacts in units of contacts.

Parameters: P1 is average per capita contact rate (contacts/person day), *P2* is infection probability in disease/contact, *P3* is population of potentially exposable persons, *P4* effective duration of disease in days.

j) *Output: dQ*, where *Q* is a quantity of heat energy in Joules (J).

Input: dT, where *T* is temperature of system in degrees Kelvin (K).

Parameters: P1 is mass of system in kilograms (kg), *P2* is specific heat of system in J/kg-K.

8 Graphically plot following system equations over output interval $0 < x < 1$ for each of the following functions. (If needed use as parameter values $a = 0, 1; n = 1, 2; m = 0$ and $p = 1; m = 1$ and $p = 0; m = 1$ and $p = 1$; and $b = 0, 1$).

a) $y = x[a(x - 1) + 1]$

b) $y = \dfrac{a}{1 + bx}$

c) $y = ax(x - 1) + 1 - x$

d) $a = \dfrac{1 + b x^m y^p}{x + y + n}$

e) $y^n = \dfrac{(1 + a)x^m}{1 + a x^p}$

f) $y = a \exp(bx + n)$

g) $\dfrac{(1+b)y^n}{1+by^n} = \dfrac{1-x^n}{1+ax^p}$

h) $y = aLnx + bx$

i) $(y+b)^2 = ax^2 + (1+a+2b)x + b^2$

j) $(y+b)^2 = ax^2 - (1+a+2b)x + (1+b)^2$

k) $y^m x = a\,x^n y$

l) $y = ax^m \exp(bx^n + 1/x^p)$

m) $y + aLny = bx + nLnx$

n) $y = ax^n \sin(bx + a/x^m)$

9 Using Figures 6-10 to 6-13, find analytical expressions of the form given by Eqs. (6.2.12) and (6.2.13) to approximately fit the curves shown in the following figure.

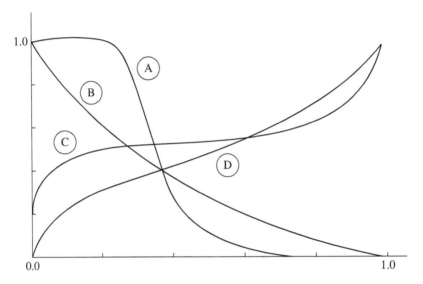

10 *Tree Diagrams:* Determine the overall system equations for each event sequence after constructing the appropriate event tree. Assume reasonable values and evaluate the expectation value or probability of the given event(s).

a) Nuclear power reactor Loss Of Coolant Accident (LOCA).

b) Probability of throwing six heads sequentially in six sequential throws of an unbiased coin. How many sequential throws are required for the probability to be less than one chance in one million?

c) Outcome of state election for governor with two gubernatorial candidates.

d) Expected mean value of the cost of oil per barrel ($/bbl.) on the world market as determined by OPEC members.

e) Development and selection process for awarding Nobel Prize for Peace.

f) Possible sequence of events in possible Iran–Israel conflict in the Middle East regarding Iranian nuclear weapon development.

g) Sequence of required events for worldwide nuclear disarmament.

11 *Heuristic Modeling Methods*

    a) For each of the systems identified in problem 5 of this chapter, propose a system modeling scheme (e.g., theoretical, experimental, intuition, Delphi, expert polling, brainstorming, etc.) and defend your choice. Outline your steps for executing your model scheme.

    b) Perform the assigned tasks in part (a) above for the systems defined in Problems 1, 4, and 10 of the problem section of Chapter 2.

7

Analysis Methods for Systems with Linear Kernels

Linear system kernels represent such a significant sector of system science that we will devote this entire chapter to the description and analysis of models that have linear kernels or can be approximated by linear kernels. As we will observe, essentially all quantitative system kernels can be approximated by linear system kernels over some appropriate range of definitions.

7.1 Background and Justification

As should be apparent by now, the simplest continuous system models are those with linear system kernels. Systems with constant kernels are simpler but much less useful and may be considered a restricted class of linear system kernels. By the classification "linear," we mean system kernels whose mathematical expressions are linear functions of the dependent variables only. For example, consider the general multiple-input, multiple-output system equation where we assign the independent vector variable to be $\mathbf{c}$ and the dependent vector variable to be $\mathbf{e}$

$$d\mathbf{e} = g(\mathbf{e}, \mathbf{c})\, d\mathbf{c} \tag{7.1.1}$$

As usual, $d\mathbf{e}$ is the $(n \times 1)$ differential output column matrix for the dependent variable $\mathbf{e}$, $g(\mathbf{e}, \mathbf{c})$ is the $(n \times m)$ system kernel matrix, and $d\mathbf{c}$ is the $(m \times 1)$ differential input column matrix for the independent variable $\mathbf{c}$. The system kernel matrix $g(\mathbf{e}, \mathbf{c})$ is linear if each component of the matrix $g(\mathbf{e}, \mathbf{c})$ for all i and j such that $i = 1,...,n$ and $j = 1,...,m$ can be expressed in the general form

$$g_{ij}(\mathbf{e}, \mathbf{c}) = G_{ij}(\mathbf{c}) + \sum_{k=1}^{n} F_{ijk}(\mathbf{c})\, e_k \tag{7.1.2}$$

where $G_{ij}(\mathbf{c})$ and $F_{ijk}(\mathbf{c})$ are arbitrary, continuous functions of the independent variables $\mathbf{c}$. It is important to observe that linearity of the system kernel is determined solely by how the dependent variables [$\mathbf{e}$ and e_k in Eq. (7.1.2)] appear in the mathematical expression for the kernel. However, we should recall from Chapter 4 that the role of the independent and dependent variables in the system equation could be exchanged in some cases to

Introduction to System Science with MATLAB, Second Edition.
Gary Marlin Sandquist and Zakary Robert Wilde.
© 2023 John Wiley & Sons Ltd. Published 2023 by John Wiley & Sons Ltd.

improve solution approaches. To exchange the role of the dependent and independent variables for the multiple-input, multiple-output system equation, let us premultiply both sides of Eq. (7.1.1) by the transpose of the system kernel, $g(\mathbf{e}, \mathbf{c})^T$ an $(m \times n)$ matrix to obtain.

$$g(\mathbf{e}, \mathbf{c})^T d\mathbf{e} = g(\mathbf{e}, \mathbf{c})^T g(\mathbf{e}, \mathbf{c}) d\mathbf{c} \tag{7.1.3}$$

Now, if the resulting matrix $g(\mathbf{e}, \mathbf{c})^T g(\mathbf{e}, \mathbf{c})$ is nonsingular [i.e., the determinant of $g(\mathbf{e}, \mathbf{c})^T$ $g(\mathbf{e}, \mathbf{c}) \neq 0$], then we can premultiply both sides of Eq. (7.1.3) by the inverse matrix, viz., $[g(\mathbf{e}, \mathbf{c})^T g(\mathbf{e}, \mathbf{c})]^{-1}$ to obtain

$$d\mathbf{c} = \left[g(\mathbf{e}, \mathbf{c})^T g(\mathbf{e}, \mathbf{c}) \right]^{-1} g(\mathbf{e}, \mathbf{c}) d\mathbf{e} = G(\mathbf{e}, \mathbf{c}) \, d\mathbf{e} \tag{7.1.4}$$

Given that $g(\mathbf{e}, \mathbf{c})^T g(\mathbf{e}, \mathbf{c})$ is nonsingular, the original independent $\mathbf{c}$ may now be treated as the dependent variable, and linearity is then determined by how the variable $\mathbf{c}$ enters the components of the matrix kernel $G(\mathbf{e}, \mathbf{c})$ and not the variable $\mathbf{e}$. However, exchanging the roles of the dependent and independent variables may not be justified by causality, and the system investigator must justify this transition and be aware of the consequences of this action.

Assuming that Eq. (7.1.4) is acceptable and the components of $G(\mathbf{e}, \mathbf{c})$ are expressible as linear expressions in $\mathbf{c}$, then the inverse matrix kernel is linear in $\mathbf{c}$, and the inverse system Eq. (7.1.4) may possibly be solved more easily than Eq. (7.1.1). Furthermore, even if Eq. (7.1.4) is not linear, it may be simpler to solve the kernel $G(\mathbf{e}, \mathbf{c})$ than for the kernel $g(\mathbf{e}, \mathbf{c})$. Note that Eq. (7.1.4) may be considered as an alternative system equation for Eq. (7.1.1).

The reasons for our great interest in linear system kernels are significant and useful. These reasons can be summarized as follows:

- The mathematical theory and methods for analyzing, solving, and evaluating linear ordinary and partial differential equations are well developed and relatively simple to apply. Since we have established the equivalence between system equations and differential equations in Chapters 4 and 5, the observations about linear differential equations also apply to linear difference equations.
- For linear systems, we know what general form the solutions must have, how the arbitrary constants of integration appear in the solutions, and when we have obtained the complete solution for the system. Nonlinear systems often contain singular and/or hidden solutions, and singularities where the solution becomes undefined; these conditions may not be apparent. Establishing the complete solution for a nonlinear system may pose challenges.
- All linear systems exhibit the property of superposition. Therefore, the sum of any finite number of linear mathematical expressions is also a linear expression. Furthermore, if we know two independent solutions of a differential equation or, equivalently, a system equation, then the sum of these two solutions, each multiplied by any finite arbitrary constant, is also a solution for the system equation.
- Through rather standard mathematical techniques (some of which will be described in this chapter), we will find that any continuous system kernel can always be approximated by a linear expression over some appropriate range of variation with an associated error

that usually can be estimated and controlled. We will refer to these approximation techniques as "linearization" techniques and will find that linearization of system kernels provides the investigator with a powerful tool for performing analytical studies of most systems because of the advantages of solving linear equations.

We are now properly motivated to propound one of the most significant propositions in system science. This proposition is fundamental to system modeling and must be understood and appreciated by the successful system investigator.

Presumption of System Linearity   In the absence of convincing evidence otherwise, all continuous system kernels should be initially considered and approximated as linear mathematical functions of the system's dependent variables (outputs) for systems that do not readily yield analytical solutions.

This proposition serves as a benchmark for us when we begin the analysis of any system and particularly those systems with nonlinear system kernels. Furthermore, recall that the presence of external feedback generally makes the system nonlinear even if the system without feedback is linear. Thus, unless we have previous knowledge or experience with the system or similar systems that dictates that the system kernel cannot be assumed to have a linear form because of unacceptable modeling consequences, we should assume that the system kernel for the model outputs is initially linear, even if nonlinearities or intrinsic system feedback are known or suspected to exist. Of course, for extrinsic system feedback, it is presumed that we know and can even control this feedback source. Thus, it can be assumed, generally, that we know or can determine adequately the mathematical expression for extrinsic feedback systems that may be present.

Thus, for example, when we begin the modeling and analysis of an unfamiliar or difficult continuous system, we can generally initially assume that the system will exhibit a linear kernel for some restricted range of differential inputs and outputs. Note that a linear system kernel does not imply that the solution for the system equation is restricted to a simple linear relationship between the input **e** and the output **c**. Indeed, the linearity of the system equation is not determined or controlled by the functional behavior of the input or independent variable **c** within the system kernel.

Furthermore, the system investigator should not feel overly constrained or limited using a linear system kernel, particularly for the initial or scoping analysis and evaluation of a given system. If the subsequent evaluation shows that a nonlinear kernel form should be considered for acceptable analysis, the previous linear modeling efforts will serve as a useful basis for comparing results and gaining an understanding of the actual system.

7.2   Linearization Methods

Unfortunately, it is not generally possible to solve analytically a continuous system equation if the system kernel contains nonlinear functions of the system output or dependent variables. Although it is only the appearance of the output variables that makes the system equations nonlinear, the presence of system feedback generally makes the system nonlinear. Furthermore, variable coefficient linear kernels pose serious analytical difficulties, particularly for multiple-input or multiple-output continuous systems. Of course, there are exceptions, when certain fortuitous, nonlinear system kernels arise, and the system equations are adaptable to appropriate solution schemes. This is particularly the case with single-input systems with few outputs where many solution schemes exist, as we have seen in Chapter 4. However, if the general system kernel is linear or can be approximated by a linear kernel, then the prospect for achieving analytical solutions is greatly improved.

Techniques for producing linear approximations of mathematical functions are often referred to as linearization techniques, and these techniques constitute a very important resource and capability in system analysis. We will examine several techniques, most of which are based on the use of the Taylor series expansion formula.

In practice, the use of linearization methods to realize analytical solutions is primarily applicable to single-input systems with one or more outputs and to multiple-input systems with a single output. Only in special cases, usually where the coefficient functions of the outputs are constant, do linearization methods produce integrable system equations for multiple-input, multiple-output systems.

For single-input systems, linearization methods will generally result in system equations that can be solved analytically if all singular points that may exist for the input variables are at least regular singular points. The singular point of a function $F(x)$ is a value of the variable, say x_0, for which $F(x_0)$ is undefined or infinite. For example, the function $1/x$ has a regular singular point or order one for $x \to 0$. However, the function $1/\sin(x)$ exhibits a singularity whenever $\sin(x)$ is zero and $x = n\pi$ for $n = 0, \pm1, \pm2, \dots$. This class of singularities is termed an essential singularity and becomes undefined repeatedly for $x = n\pi$. An essential singularity cannot be removed by a single exponent of x.

It is our purpose here to develop and demonstrate several linearization techniques that are generally applicable for system kernels that can be approximated as linear functions of the system's output (dependent) variables. Of course, the incentive for linearizing nonlinear kernels is that, in principle, linear kernels possess analytical solutions that can be realized, hopefully, by direct integration or other analytical methods. Furthermore, it is highly advisable that analytical solutions, even crude analytical approximations of the nonlinear system kernel, be obtained when possible to verify and corroborate other solution methods. This is particularly important when numerical solutions are obtained by computer analysis, and the tendency is to accept whatever results the computer provides, even incorrect data resulting from erroneous computer programs or faulty data entry or processing schemes.

7.2.1 Taylor Series Expansion

To introduce the theoretical basis for the linearization of system kernels, let us begin by considering the general single-input, single-output system kernel, $g(e, c)$. Now we wish to expand $g(e, c)$ in a power series expansion around an arbitrary output value of interest, say e_0. If we assume that $g(e, c)$ is an analytical function at the point e_0 [if not, this implies that $g(e, c)$ becomes undefined or infinite, at or near the point e_0]. Then $g(e, c)$ possesses a Taylor series expansion that is unique for some neighborhood [i.e., region for which the

Taylor series is equivalent to $g(e, c)$] around e_0 with well-defined derivatives of all orders. Thus, if $g(e, c)$ is analytical at e_0, then

$$g(e, c) = g(e_0, c) + \delta e \frac{\partial g}{\partial e}(e_0, c) + \delta e^2 \frac{\partial^2 g}{\partial e^2}(e_0\, c)$$
$$+ \cdots + \frac{\delta e^n}{n!} \frac{\partial^n g}{\partial e^n}(e_0, c) + \cdots \tag{7.2.1}$$

where we have defined for all positive integral values of k.

$$\delta e^k = (e - e_0)^k$$

Observe that the dependent variable, e, or the system output, is effectively removed from the original system kernel $g(e, c)$ by the Taylor series expansion. The system kernel is now expressed as an equivalent infinite power series expansion around e_0. Furthermore, as the value of e approaches e_0, the increasing power terms in δe become progressively smaller, and fewer terms in the Taylor series are required to provide a given accuracy for evaluating the system kernel $g(e, c)$. In particular, if δe is sufficiently small, then it is possible to approximate $g(e, c)$ as a linear function in e by terminating or truncating the infinite series of all terms containing δe to the second degree or higher. Therefore, Eq. (7.2.1) provides a linear approximation for the kernel as follows:

$$g(e, c) \approx g(e_0, c) + (e - e_0) \frac{\partial g}{\partial e}(e_0, c) \tag{7.2.2}$$

And we have effectively linearized the system kernel. We can now replace the following single-input, single-output system equation

$$de = g(e, c)\, dc$$

by the approximate linearized system equation

$$de \approx \left[g(e_0, c) + (e - e_0) \frac{\partial g}{\partial e}(e_0, c) \right] dc$$

where $g(e_0, c)$ and $(e - e_0) \frac{\partial g}{\partial e}(e_0, c)$ are functions only of the independent variable c and e_0 is an arbitrary output value around which we seek a linear system solution.

It can be shown that the error incurred in truncating the Taylor series of all nonlinear terms has the form

$$\text{Truncation error} = \frac{(e - e_0)^2}{2} \frac{\partial^2 g}{\partial e^2}(e^*, c) \tag{7.2.3}$$

where the second partial derivative of the system kernel is evaluated at some value e^* (generally unknown) such that

$$e_0 \le e^* \le e$$

What is important to observe about the truncation error is that if the function is not rapidly changing around the point e_0 so that the second derivative is small, and if e is held sufficiently close to e_0, then the truncation error is small. On the other hand, the error made by linearizing the kernel increases with large changes in e as we extend our solution range farther from the expansion point e_0.

We will now formally extend the use of the Taylor series for system linearization to multiple-output and multiple-input systems. However, the extension should be apparent to the reader. For a general system kernel with m inputs and n outputs, the linearized Taylor series approximation for kernel component g_{ij} is

$$g_{ij}(e,c) \approx g_{ij}(e_0,c) + \sum_{k=1}^{n}(e_k - e_{k0})\frac{\partial g_{ij}}{\partial ek}(e_0,c) \tag{7.2.4}$$

where $i = 1, ..., n$ and $j = 1, ..., m$.

As a generalization of Eq. (7.2.3), it can be shown that the truncation error resulting from the use of Eq. (7.2.4) is a linear approximation for the system kernel and has the form

$$\text{Truncation error} \approx \sum_{k,l}^{n}(e_k - e_{k0})(e_l - e_{l0})\frac{\partial^2 g_{ij}}{\partial e_k \partial e_l}(e^*,c)$$

where the second partial derivative of the multiple-input, multiple-output system kernel is evaluated at some multiple-output value e^* such that

$$e_0 \leq e^* \leq e$$

Example of Taylor Series Expansion for Linear Approximation
Consider the double-input, single-output nonlinear system equation

$$de = g_1\, dc_1 + g_2\, dc_2$$

where

$$g_1 = A(c_1,c_1)\,e^2 + B(c_1,c_2)\,\exp(e)$$

$$g_2 = \frac{D(c_1,c_2)}{E(c_1,c_2) + e^2}$$

To form the linear kernel approximations for this system using a Taylor series expansion, we must form the following truncated series for each kernel g_1 and g_2, where

$$g(e,c) = g(e_0,c) + (e - e_0)\frac{\partial g}{\partial e}(e_0,c)$$

So we find that

$$g_1(e,c) \approx A(c_1,c_2)\,e_0^2 + B(c_1,c_2)\exp(e_0)$$
$$+ (e - e_0)[2A(c_1,c_2)\,e_0 + B(c_1,c_2)\exp(e_0)]$$

$$g_2(e,c) \approx \frac{D(c_1,c_2)}{E(c_1,c_2) + e_0^2} - (e - e_0)\frac{2D(c_1,c_2)}{[E(c_1,c_2) + e_0^2]^2}$$

and we have determined the linearized approximation for our double-input, single-input system equation about the point e_0. To complete the solution for the system equation, we would need to specify the form for the dependent variable coefficient functions, A, B, D, and E.

7.2.2 Perturbation Methods

An alternative method for obtaining a linear approximation of the general system kernel is often referred to as a first-order perturbation approximation or perturbation method. The method is also based on the use of a Taylor series expansion around any value of a variable for which the function to be expanded is analytical. However, the method is often simpler to apply and is a common approximation tool for system investigators. Furthermore, higher-order perturbation methods can be used for higher degree approximations than linear or first degree, although we will not consider these higher-degree methods here.

Let us return to our single-input, single-output system kernel $g(e, c)$ to demonstrate the perturbation method. First, we redefine all dependent variable outputs, e, as follows:

$$e = e_0 + \delta e \tag{7.2.5}$$

where e_0 [assumed to be an analytical value for $g(e, c)$] is the value of e around which we seek our system solution and δe represents an acceptably small change or perturbation around e_0. Now we express the system kernel in terms of our new definition and perform a series expansion of the kernel in terms of δe, as follows:

$$g(e, c) = g(e_0 + \delta e, c) = g(e_0, c) + \delta e\, h(e_0, c)$$
$$+ \text{(higher order terms in } \delta e) \tag{7.2.6}$$

It is possible to show that the function $h(e_0, c)$ which serves as the coefficient of the linear or first-degree term in δe is given by

$$h(e_0, c) = \frac{\partial g}{\partial e}(e_0, c)$$

Since all power series expansions of a given analytic function about the same point are unique or identical, the Taylor series formula defines this coefficient function as given by Eq. (7.2.1). Of course, to form a linear approximation for the system kernel from Eq. (7.2.6), we simply neglect all the higher-degree terms in δe as being small compared to

$$g(e_0, c) + \delta e\, h(e_0, c)$$

which defines our linearized system kernel in terms of δe.

Example of Perturbation Methods
Consider the single-input, double-output, nonlinear system equation

$$de_1 = [A_1 e_1 + B_1 e_1 e_2 + D_1 e_{22}]\, dc = g_1 dc$$
$$de_2 = [A_2 e_2 + B_2 e_1/e_2]\, dc = g_2 dc$$

If we now let

$$e_1 = e_{10} + \delta e_1, \qquad de_1 = d\delta e_1$$
$$e_2 = e_{20} + \delta e_2, \qquad de_2 = d\delta e_2$$

then our system kernels become

$$g_1 = A_1(e_{10} + \delta e_1) + B_1(e_{10} + \delta e_1)(e_{20} + \delta e_2) + D_1(e_{20} + \delta e_2)^2$$

$$g_2 = A_2(e_{20} + \delta e_2) + \frac{B_2(e_{10} + \delta e_1)}{e_{20}}$$

$$\left\{ 1 - \frac{\delta e_2}{e_{20}} + \left(\frac{\delta e_2}{e_{20}}\right)^2 - \cdots (-1)^k \left(\frac{\delta e_2}{e_{20}}\right)^k + \cdots \right\}$$

If we retain terms only of the first degree or less in δe_1 and δe_2, then we obtain the following linear approximations for our system kernel:

$$g_1 \approx A_1 e_{10} + B_1 e_{10} e_{20} + D_1 e_{20}^2 + \delta e_1 (A_1 + B_1 e_{20}) + \delta e_2 (B_1 e_{10} + 2D_1 e_{20})$$

and

$$g_2 \approx A_2 e_{20} + \frac{B_2 e_{10}}{e_{20}} + \delta e_1 \left\{\frac{B_2}{e_{20}}\right\} + \delta e_2 \left\{A_2 - \frac{B_2 e_{10}}{e_{20}^2}\right\}$$

It is simple to show that the coefficients of the terms containing δe_1 and δe_2 are the appropriate first derivative expression defined by the Taylor series expansions for g_1 and g_2.

7.2.3 Variable Coefficient Elimination

It is possible to obtain analytical solutions for some multiple-input and output systems with variable coefficient linear kernels as was demonstrated in Chapters 4 and 5. However, it is obvious that with the exception of single-input, single-output systems, the solution methods for variable coefficient linear systems are rather complex and demanding. We want to provide an approximation method for treating such variable coefficient linear kernels that employ a generalized perturbation method for the approximation of the original variable coefficient linear kernel by a constant coefficient linear kernel. In general, constant-coefficient linear kernels are considerably simpler to solve since the powerful methods of operational mathematics are then generally applicable, as we will see in Section 7.3.

To see how we may approximate a linear variable coefficient kernel as a linear constant-coefficient kernel, consider the general linear variable coefficient system equation

$$d\mathbf{e} = g(\mathbf{e}, \mathbf{c})d\mathbf{c} \tag{7.2.7}$$

where for each component of the m input, n output matrix kernel $g(\mathbf{e}, \mathbf{c})$ we have, for $i = 1, \ldots, n$ and $j = 1, \ldots, m$, that

$$g_{ij}(\mathbf{e}, \mathbf{c}) = A_{ij}(\mathbf{c}) + \sum_{k=1}^{n} B_{ijk}(\mathbf{c}) e_k \tag{7.2.8}$$

We observe that since each coefficient term $B_{ijk}(\mathbf{c})$ is, in general, a function of the input variables $\mathbf{c}$, we have a variable coefficient linear system equation. It is significant to realize that by using the methods of kernel linearization described earlier in this chapter that any general system kernel can be approximated by a linear variable coefficient kernel as specified by Eq. (7.2.4). Of course, if the particular form of the linear system kernel is sufficiently simple or requires solution as a variable coefficient form, then the series solution methods considered in Sections 4.3 and 5.3 should be considered. However, let us assume that we seek a constant coefficient approximation of Eq. (7.2.8). We can accomplish this goal by using a modified form of the perturbation method considered in Section 7.2.2. Let us assume that we can express both the coefficient functions $B_{ijk}(c)$ and the output variables as follows

$$e_k = e_{k0} + \delta e_k \tag{7.2.9.1}$$

$$B_{ijk}(c) = B_{ijk0} + \delta B_{ijk}(c) \tag{7.2.9.2}$$

where both e_{k0} and B_{ijk0} are appropriately chosen constants, and δe_k and $\delta B_{ijk}(c)$ represent the variations of e_k and $\delta B_{ijk}(c)$ about the constant values e_{k0} and B_{ijk0}, respectively. Now, if we substitute Eqs. (7.2.9) into Eq. (7.2.8) for the linear variable coefficient, we have

$$g_{ij}(e, c) = A_{ij}(c) + \sum_{k=1}^{n} \left[B_{ijk0} + \delta B_{ijk}(c) \right] (e_{k0} + \delta e_k)$$

Expanding the sum, we find

$$g_{ij}(e, c) = A_{ij}(c) + \sum_{k=1}^{n} \left[B_{ijk0} + e_{k0}\, \delta B_{ijk}(c) \right]$$
$$+ \sum_{k=1}^{n} \left[B_{ijk0} + e_{k0}\, \delta B_{ijk}(c) \right] \delta e_k \tag{7.2.10}$$

The approximation that must be made in Eq. (7.2.10) to realize a constant coefficient linear kernel is as follows:

$$\sum_{k=1}^{n} \delta B_{ijk}(c)\, \delta e_k \ll \sum_{k=1}^{n} \left[B_{ijk0}(e_{k0} + \delta e_k) + e_{k0}\, \delta B_{ijk}(c) \right] + A_{ij}(c) \tag{7.2.11}$$

This requires that the range of variation of both δe_k and δB_{ijk} around e_{k0} and B_{ijk0} be sufficiently small so that Eq. (7.2.11) is satisfied. If this condition is satisfied, then our linear system kernel, Eq. (7.2.10), maybe approximated as follows:

$$g_{ij}(e, c) \approx A_{ij}^{m}(c) + \sum_{k=1}^{n} B_{ijk0}\, \delta e_k \tag{7.2.12}$$

where

$$A_{ij}^{m}(c) = A_{ij}(c) + \sum_{k=1}^{n} e_{k0}\, B_{ijk0}(c)$$

$$B_{ijk}(c) \approx B_{ijk0} + \delta B_{ijk0}(c)$$

We recognize that Eq. (7.2.12) for our linear system kernel is the constant coefficient linear kernel approximation desired. Observe that the nonhomogeneous term $A_{ij}(c)$ has been modified from its form in the original linear kernel, Eq. (7.2.8), to accommodate the approximation.

We must now address the issue of determining the perturbation expansions defined by Eqs. (7.2.9.1) and (7.2.9.2). Equation (7.2.9.1) represents the standard expansion process considered in Section 7.2 for expanding the output or dependent system variables around a solution point of interest. The general requirement here is that the perturbation term δe_k be small so that

$$\delta e_k \ll e_{k0}$$

or in a related manner as required by Eq. (7.2.11).

Equation (7.2.9.2) represents a perturbation expansion of the coefficient function $B_{ijk}(\mathbf{c})$ around a dominant value B_{ijk0}. This dominant value may be interpreted in many ways. We will consider two possibilities. If we know the range of the input or independent variables $\mathbf{c}(\min) \le \mathbf{c} \le \mathbf{c}(\max)$, then B_{ijk0} might be given as the maximum value assumed by $B_{ijk}(\mathbf{c})$ over the range $\mathbf{c}$; thus

$$B_{ijk0} = \max\left\{B_{ijk}(\mathbf{c})\right\}$$

for $\mathbf{c}(\min) \le \mathbf{c} \le \mathbf{c}(\min)$. If $B_{ijk0}(\mathbf{c})$ varies significantly over this range of dependent variables, then this particular perturbation expansion may not be acceptable. An alternative expansion is determined if we use the mean value of $B_{ijk0}(\mathbf{c})$ over the range of $\mathbf{c}$ defined by

$$B_{ijk0} = \frac{1}{\prod_{i=1}^{m}[c(\max) - c_i(\min)]} \int_{c_m(\min)}^{c_m(\max)} \cdots \int_{c_1(\min)}^{c_1(\max)} B_{ijk}(\mathbf{c})\, dc_1 \cdots dc_m$$

where

$$\prod_{i=1}^{m}[c(\max) - c_i(\min)] = [c_1(\max) - c_1(\min)] \cdots [c_m(\max) - c_m(\min)] \qquad (7.2.13)$$

If the variation of $B_{ijk0}(\mathbf{c})$ over the original range of $\mathbf{c}$ is too large to satisfy Eq. (7.2.11), then it is necessary to restrict this range of $\mathbf{c}$ to an interval for which Eq. (7.2.11) is satisfied.

Observe that, regardless of what procedure is used to determine B_{ijk0}, the perturbation component $\delta B_{ijk0}(\mathbf{c})$ is defined as follows for consistency:

$$\delta B_{ijk}(\mathbf{c}) = B_{ijk}(\mathbf{c}) - B_{ijk0}$$

This equation may be considered as the defining relationship for the perturbation variation of the variable coefficient. In general, the smaller the absolute value of δB_{ijk0}, the more easily is Eq. (7.2.11) satisfied, and the more accurate is the constant coefficient approximation, Eq. (7.2.12), for the general variable coefficient linear system kernel, Eq. (7.2.8).

7.3 Single-Input Linear Systems

Linear single-input systems constitute a major area of interest and impact in system science. It is an essential skill for the system investigator to be able to produce a linearized set of system equations for any continuous single-input system.

7.3.1 Single-Input, Single-Output Systems

We have established the general solution for the linear single-input, single-output system in Section 4.2.4. However, because of the insight and utility offered by the following approach, let us consider an algebraic equation that mathematically expresses the form of the general solution of a linear ordinary differential equation. Observe that we set a general linear function in $y(x)$ equal to an arbitrary constant as follows:

$$G(x) + F(x)y = K \tag{7.3.1}$$

where y, the dependent variable, is a function of x, the independent variable. Observe that if we solve Eq. (7.3.1) for y [assuming that $F(x) \neq 0$] then, alternatively, y may be considered to define a linear function in terms of the arbitrary constant K. We will show later that the arbitrary constant K is related to the integration constant or initial condition that results from solving the differential equation.

Let us differentiate Eq. (7.3.1) with respect to x, and we find

$$\frac{dG}{dx} + y\frac{dF}{dx} + F\frac{dy}{dx} = 0 \tag{7.3.2}$$

Observe that we can express this differential equation in the following form, again assuming that $F(x) \neq 0$:

$$\frac{dy}{dx} = -\frac{1}{F}\frac{dG}{dx} - \frac{y}{F}\frac{dF}{dx} \tag{7.3.3}$$

By defining the functions

$$g_1(x) = -\frac{1}{F}\frac{dG}{dx} \tag{7.3.4.1}$$

and

$$g_2(x) = -\frac{1}{F}\frac{dF}{dx} = -\frac{d\,\mathrm{Ln}\,F}{dx} \tag{7.3.4.2}$$

Equation (7.3.3) can be expressed in the form of the general linear first-order differential equation as follows:

$$\frac{dy}{dx} = g_1(x) + y\,g_2(x) \tag{7.3.5}$$

Furthermore, as we have shown in Chapter 4, Eq. (7.3.5) is equivalent to a system equation of the general form

$$dy = [g_1(x) + y\,g_2(x)]\,dx \tag{7.3.6}$$

which is the general expression for the linear single-input, single-output system equation.

Equations (7.3.4.1) and (7.3.4.2), which relate the functions $F(x)$ and $G(x)$ to our arbitrary functions $g_1(x)$ and $g_2(x)$, which define the linear differential equation, can be used to define the general solution for the first-order differential equation. If we integrate Eq. (7.3.4.2), we find that

$$\mathrm{Ln}\,F(x) = -\int g_2(x)\,dx$$

or

$$F(x) = \exp\left\{-\int g_2(x)\,dx\right\}$$

Substituting this equation for $F(x)$ into Eq. (7.3.4.1) and solving the resulting separable equation for $G(x)$, we find that

$$G(x) = - \int g_1(x) \exp \left\{ - \int g_2(x) \, dx \right\} dx$$

and so, as defined by Eq. (7.3.1), we have, for the general solution for the general linear first-order differential equation defined by Eq. (7.3.5), the standard solution with K as the arbitrary integration constant:

$$y(x) = \exp \left\{ \int g_2(x) \, dx \right\} \left\{ K + \int g_1(x) \exp \left[- \int g_2(x) dx \right] dx \right\}$$

which is, as expected, equivalent to the solution for the linear single-input, single-output system equation as determined in Section 4.2.4.

Returning to Eq. (7.3.1), observe that, if the general algebraic equation [which now can be considered to represent the solution for differential Eq. (7.3.5) or the system Eq. (7.3.6)] is linear in the arbitrary constant K as well as linear in y, then the associated differential equation used to eliminate K is also a linear differential equation in the dependent variable y. This observation applies not only to first-order ordinary differential equations but also to sets of ordinary and even partial differential equations. Thus, this general observation also applies to all multiple-input, multiple-output system equations with linear kernels, as we will learn later in this chapter.

The obvious question that now arises is, what happens if we begin with an algebraic equation in which the dependent variable or similarly the arbitrary constant appears in a nonlinear manner? For example, suppose we have the following nonlinear algebraic expression in y and K, where y, the dependent variable, is a function of x, the independent variable.

$$F(x) + \frac{G(x)}{y(x) + H(x)} = K \tag{7.3.7}$$

Differentiating this expression with respect to x, we find that

$$\frac{dF}{dx} + \frac{1}{y + H} \frac{dG}{dx} - \frac{G}{(y + H)^2} \left[\frac{dy}{dx} + \frac{dH}{dx} \right] = 0$$

Solving this expression for $\dfrac{dy}{dx}$, we have for any continuous function $F(x)$

$$\frac{dy}{dx} = g_1(x) + g_2(x)y + g_3(x)y^2 \tag{7.3.8}$$

where

$$g_1 = \frac{H^2}{G} \frac{dF}{dx} + \frac{H}{G} \frac{dG}{dx} - \frac{dH}{dx} \tag{7.3.9.1}$$

$$g_2 = 2 \frac{H}{G} \frac{dF}{dx} + \frac{1}{G} \frac{dG}{dx} \tag{7.3.9.2}$$

$$g_3 = \frac{1}{G} \frac{dF}{dx} \tag{7.3.9.3}$$

Equation (7.3.8) is a nonlinear differential equation if $g_3 \neq 0$. We observe that $g_3 = 0$ only if F is a constant, and our differential equation becomes linear. But then Eq. (7.3.7) reduces to a linear equation in y and K as expected.

Equation (7.3.8) expressed as the single-input, single-output system equation

$$dy = \left[g_1(x) + g_2(x)y + g_3(x)y^2 \right] dx$$

was encountered in Section 4.2.9 and has the general form of a Ricatti-type system kernel, which is second degree in the dependent variable y.

If we attempt to determine the general solution for the Ricatti equation as given by Eq. (7.3.7) in terms of the functions g_1, g_2, and g_3 as defined by Eqs. (7.3.9), we find after some algebraic manipulations that we must solve the following differential equation for $H(x)$ in terms of g_1, g_2, and g_3:

$$\frac{dH}{dx} = -g_1(x) + g_2(x)H + g_3(x)H^2$$

Curiously, this equation is also a Ricatti-type differential equation, and if we could solve this general equation for H, then we would also know the general solution for Eq. (7.3.8).

Thus, although Eqs. (7.3.7) and (7.3.9) do not provide us with the solution for the general Ricatti equation, Eq. (7.3.7) does show us what form the general solution for a Ricatti-type system kernel has and how the arbitrary integration constant K enters into the general solution. Unfortunately, expressing the general solutions for differential equations in terms of their arbitrary solution constants (which are called primates in formal mathematical applications) and then eliminating the constant by differentiation to obtain the differential equation associated with the initial general solution form does not usually prove useful as a technique for solving system equations. But this method of elimination of arbitrary constants or primates does indicate what general mathematical form the solutions have, and the method provides some insight into the general behavior of the system and its solutions.

For example, if we express Eq. (7.3.7) in terms of $y(x)$, we find that we have a nonlinear expression in terms of the constant K, as follows:

$$y(x) = \frac{G(x)}{K - F(x)} - H(x)$$

We also observe that $y(x)$ has a singular point [i.e., a value of the independent variable, x, for which $y(x)$ becomes infinite and undefined] when

$$F(x) = K \tag{7.3.10}$$

Since K is an arbitrary constant, the value of x for which $F(x) = K$ is also arbitrary, and we have an arbitrary singular point for the function $y(x)$. This is a critical situation for which $y(x)$ becomes undefined or infinite for any finite value of x that satisfies Eq. (7.3.10). This condition generally only occurs with nonlinear equations.

The usual general form of the solution for a first-order differential equation in terms of its arbitrary constant is

$$K = G(x, y)$$

which we previously considered in Section 4.2.2 for continuous system kernels. Eliminating K from this expression by differentiation gives the result we found previously in Section 4.2:

$$\frac{dy}{dx} = -\frac{\partial G/\partial x}{\partial G/\partial y} = g(x, y)$$

where we identify $g(x, y)$ as the continuous single-input, single-output system kernel.

7.3.2 Single-Input, Multiple-Output Linear Systems

All linear, single-input, multiple-output system equations can be solved, in principle, if the range of values of the independent variable spans only regular, isolated singular points. However, in practice, analytical solutions of the general single-input, multiple-output system equation can be difficult to realize, and numerical and symbolic solutions obtained from software such as MATLAB and MAPLE are preferred.

Constant Coefficient Linear Systems

In the entire class of linear system kernels, the simplest kernels to analyze and solve are those that have constant coefficients for each of the linear output terms that appear in the system kernel. We should recognize that it is usually possible to approximate any linear kernel by an associated constant-coefficient kernel, as was shown in Section 7.2.3.

For the single-input, multiple-output system, the linear system kernel with constant coefficients has the general form

$$g(e, c) = A\left(c\right) + B\,e$$

where we have defined the following matrices:

$$A\left(\mathbf{c}\right) = \begin{bmatrix} A_1(\mathbf{c}) \\ \vdots \\ A_n(\mathbf{c}) \end{bmatrix}, \quad B = \begin{bmatrix} B_{11} & \cdots & B_{1n} \\ & \vdots & \\ B_{n1} & & B_{nn} \end{bmatrix}, \quad \mathbf{e} = \begin{bmatrix} e_1(\mathbf{c}) \\ \vdots \\ e_n(\mathbf{c}) \end{bmatrix}$$

We note that although each of the A_i matrix elements may be functions of the input variable $\mathbf{c}$, the matrix elements, B_{ij}, are all assumed to be constant. Because the B matrix is constant, there are several analytical methods available for solving the system equation. We will review three of the standard methods and consider their relative individual advantages and disadvantages.

The first analytical solution method we will review is the classic method presented in Section 4.3.4. Referring to that section, we recall that we assumed that each dependent variable exhibited an exponential solution format for the general homogeneous solution. Substitution and reduction of these homogeneous trial solutions resulted in a characteristic equation that determined the full homogeneous solution through superposition. Refer to Section 4.3.4 for the solution details.

The other two solution methods we will briefly discuss fall within the broad class of operational methods for solving systems of differential equations. These two methods are associated with (1) the Laplace transform and (2) the Fourier transform. Although both transform methods are related to each other, preference for the use of one or the other is usually conditioned by the particular form of the constant coefficients, the linear system equations encountered, the range of the independent variable specified, and the boundary conditions imposed. Generally, in System Analysis, Laplace transform methods are usually employed for time domain problems and Fourier transform methods for frequency-domain problems. Our focus here is Laplace transform methods.

Laplace Transform Methods

Single-input, multiple-output systems with constant coefficient linear kernels that are defined for $0 \leq c \leq \infty$ for the single input and have the general form as follows can be readily subjected to Laplace transform methods:

$$d\,e = [A\,(c) + B\,e]\,d\,c \tag{7.3.11}$$

The Laplace transform method is a method in operational mathematics that employs the transform

$$L\,[F(x)] \equiv \int_0^\infty F(x)\,\exp\,(-sx)\,dx = \overline{F}(s)$$

to transform or convert functions, which satisfy very mild conditions, from one independent variable x into another complex-valued variable s. A major advantage of most transform methods is that, in general, both derivatives and integral expressions in the independent variable(s) can be easily transformed into algebraic functions in terms of the transformed variable.

For example, a consequence of the Laplace transform of particular interest to system science is that

$$L\left[\frac{dy}{dx}\right] = \int_0^\infty \frac{dy}{dx}\,\exp\,(-sx)\,dx = sL\,[y(x)] - y\,(0^+)$$

To establish this result, integrate the integrand of the integral expression by parts and let $(dy/dx)\,dx = dv$ so that $v = y$, and $u \, \exp(-s\,x)$ so $du = -s\,\exp(-s\,x)\,dx$; then

$$\int_0^\infty \frac{dy}{dx}\,\exp\,(-sx)\,dx = y(x)\exp\,(-sx)\Big|_0^\infty + s\int_0^\infty y(x)\exp\,(-sx)\,dx$$

Now, if Lim $[y(x)\exp\,(-s\,x)] = 0$ as $x \to \infty$, which is satisfied for functions, $y(x)$ that are of exponential order, then

$$\int_0^\infty \frac{dy}{dx}\,\exp\,(-sx)\,dx = s\int_0^\infty y\exp\,(-sx)\,dx - y(0^+)$$

This provides the result we sought to establish:

$$L\left[\frac{dy}{dx}\right] = sL\,[y(x)] - y\,(0^+)$$

where $y(0^+)$ is defined as Lim $y(x)$ as $x \to 0$ for $x > 0$.

If we express system Eq. (7.3.11) as a set of linear differential equations of the form

$$\frac{d\mathbf{e}}{d\mathbf{c}} = A(\mathbf{c}) + B\mathbf{e} \tag{7.3.12}$$

where $A(\mathbf{c})$ is an $(n \times 1)$ column matrix, and B is a $(n \times n)$ matrix with constant coefficients. Taking the Laplace transformation of $\mathbf{e}(\mathbf{c})$ we have, where $L[\mathbf{e}\,(\mathbf{c})] = \bar{\mathbf{e}}\,(s)$ and $L\,[A(\mathbf{c})] = \overline{A}(s)$. Observe that we use a vertical bar over the variable to indicate the Laplace Transform of that function.

$$s\bar{\mathbf{e}}(s) - \mathbf{e}(0^+) = \overline{A}(s) + B\bar{\mathbf{e}}(s)$$

and

$$(s\,I - B)\,\mathbf{e}(s) = \overline{A}(s) + \mathbf{e}(0^+)$$

where I is the unit or identity matrix whose only nonzero elements are "1" on the main diagonal.

Solving this matrix equation, assuming that the matrix $(s\,I - B)$ is not singular, gives

$$\bar{\mathbf{e}}(s) = (s\,I - B)^{-1}\left[\overline{A}\,(s) + \mathbf{e}(0^+)\right]$$

Using the Convolution Theorem (Refer to a standard Matrix Theory text), we have the column matrix solution for $\mathbf{e}(\mathbf{c})$

$$\mathbf{e}(\mathbf{c}) = \int_0^c L^{-1}\left[(s\,I - B)^{-1}\right] A(c - \lambda)\,d\lambda + \mathbf{e}(0^+)\,\mathbf{L}^{-1}\left[(s\,I - B)^{-1}\right]$$

where L^{-1} represents the inverse Laplace transform operator. This function for $\mathbf{e}(\mathbf{c})$ represents the solution for system Eq. (7.3.11) or (7.3.12).

Example System Problem Using Laplace Transforms
Consider the following constant coefficient, linear system equation for two inputs and two outputs. (Note: The same system equation is used in the example problems in Section 4.3.4, which should be referred to for a comparison of the methods employed.)

$$\begin{bmatrix} \dfrac{d e_1}{d c} \\ \dfrac{d e_2}{d c} \end{bmatrix} = \left[\begin{pmatrix} a_1 \\ a_2 \end{pmatrix} + \begin{pmatrix} b_{12} & b_{12} \\ b_{21} & b_{22} \end{pmatrix} \begin{pmatrix} e_1 \\ e_2 \end{pmatrix} \right] \tag{7.3.13}$$

Taking the Laplace transformation of this system equation provides the following matrix equation:

$$s\begin{bmatrix} \bar{e}_1 \\ \bar{e}_2 \end{bmatrix} = \left[\begin{pmatrix} a_1/s \\ a_2/s \end{pmatrix} + \begin{pmatrix} b_{11} & b_{12} \\ b_{21} & b_{22} \end{pmatrix} \begin{pmatrix} \bar{e}_1 \\ \bar{e}_2 \end{pmatrix} \right] + \begin{bmatrix} e_1(0) \\ e_2(0) \end{bmatrix}$$

where $\bar{e}_1$ and $\bar{e}_2$ denote the Laplace transforms of $e_1(c)$ and $e_2(c)$, respectively. Gathering on $\bar{e}_1$ and $\bar{e}_2$, we have as an equivalent matrix equation

$$\left[s\begin{pmatrix} 1 & 0 \\ 0 & 1 \end{pmatrix} - \begin{pmatrix} b_{11} & b_{12} \\ b_{21} & b_{22} \end{pmatrix} \begin{pmatrix} \bar{e}_1 \\ \bar{e}_2 \end{pmatrix} \right] = \begin{bmatrix} a_1/s + e_1(0) \\ a_2/s + e_2(0) \end{bmatrix}$$

This equation constitutes a linear matrix equation, which may be solved for $\bar{e}_1$ and $\bar{e}_2$ to give

$$\begin{bmatrix} \bar{e}_1 \\ \bar{e}_2 \end{bmatrix} = \frac{1}{D(s)} \begin{bmatrix} s - b_{22} & -b_{12} \\ -b_{21} & s - b_{11} \end{bmatrix} \begin{bmatrix} a_1/s + e_1(0) \\ a_2/s + e_2(0) \end{bmatrix} \tag{7.3.14}$$

where

$$D(s) = (s - b_{11})(s - b_{22}) - b_{12}b_{21} = s^2 - (b_{11} + b_{22})s + b_{11}b_{22} - b_{12}b_{21}$$

and it is recognized that $D(s)$ is the characteristic equation for the system equation. Since $D(s)$ is a quadratic equation, it possesses two roots, s_1 and s_2, that can be expressed as

$$D(s) = (s - s_1)(s - s_2)$$

We observe that we can express $1/D$ as

$$\frac{1}{D(s)} = \frac{1}{(s - s_1)(s - s_2)} = \frac{1/(s_1 - s_2)}{(s - s_1)} + \frac{1/(s_2 - s_1)}{(s - s_2)}$$

where

$$\left.\begin{matrix} s_1 \\ s_2 \end{matrix}\right\} = \frac{b_{11} + b_{22}}{2} \pm \frac{1}{2}\sqrt{(b_{11} + b_{22})^2 + 4(b_{12}b_{21} - b_{11}b_{22})}$$

If we expand matrix Eq. (7.3.14) by performing matrix multiplication of the right-hand terms of the equation, we have

$$\begin{bmatrix} \bar{e}_1 \\ \bar{e}_2 \end{bmatrix} = \frac{1}{D} \begin{bmatrix} s - b_{22})\{a_1/s + e_1(0)\} + b_{12}\{a_2/s + e_2(0)\} \\ b_{21}\{a_1/s + e_1(0)\} + (s - b_{11})\{a_2/s + e_2(0)\} \end{bmatrix}$$

The terms in the two rows of the right-hand matrix can be gathered on common terms in s to give

$$\begin{bmatrix} \bar{e}_1 \\ \bar{e}_2 \end{bmatrix} = \frac{1}{D} \begin{bmatrix} \frac{1}{s}(a_2b_{12} - a_1b_{22}) + s\,e_1(0) + a_1 - b_{22}e_1(0) + b_{12}e_2(0) \\ \frac{1}{s}(a_1b_{21} - a_2b_{11}) + s\,e_2(0) + a_2 - b_{11}e_2(0) + b_{21}e_1(0) \end{bmatrix}$$

and in expanded equation form we have

$$\bar{e}_1 = \frac{1}{sD}(a_2b_{12} - a_1b_{22}) + \frac{s}{D}e_1(0) + \frac{1}{D}[a_1 - b_{22}e_1(0) + b_{12}e_2(0)] \tag{7.3.15.1}$$

$$\bar{e}_2 = \frac{1}{sD}(a_1b_{21} - a_2b_{11}) + \frac{s}{D}e_2(0) + \frac{1}{D}[a_2 - b_{11}e_2(0) + b_{21}e_1(0)] \tag{7.3.15.2}$$

By using partial fractions, we recognize the following relationships:

$$\frac{1}{sD} = \frac{1}{s}\left(\frac{1}{s_1 s_2}\right) + \frac{1}{s - s_1}\left(\frac{1/s_1}{s_1 - s_2}\right) + \frac{1}{s - s_2}\left(\frac{1/s_2}{s_2 - s_1}\right)$$

$$\frac{1}{D} = \frac{1}{s - s_1}\left(\frac{1}{s_1 - s_2}\right) + \frac{1}{s - s_2}\left(\frac{1}{s_2 - s_1}\right)$$

$$\frac{S}{D} = \frac{1}{s - s_1}\left(\frac{s_1}{s_1 - s_2}\right) + \frac{1}{s - s_2}\left(\frac{s_2}{s_2 - s_1}\right)$$

If these partial fraction relationships are used in Eqs. (7.3.15), the equations may easily be inverted (i.e., the inverse Laplace transform may be found) to provide for the system equations for e_1 the solution

$$e_1(c) = (a_2b_{12} - a_1b_{22})\left[\frac{1}{s_1 s_2} + \frac{1/s_1}{s_1 - s_2}\exp(s_1 c) + \frac{1/s_2}{s_2 - s_1}\exp(s_2 c)\right]$$

$$+ e_1(0)\left[\frac{s_1}{s_1 - s_2}\exp(s_1 c) + \frac{s_2}{s_2 - s_1}\exp(s_2 c)\right] \qquad (7.3.16.1)$$

$$+ [a_1 - b_{22}e_1(0) + b_{12}e_2(0)]\left[\frac{1}{s_1 - s_2}\exp(s_1 c) + \frac{1}{s_2 - s_1}\exp(s_2 c)\right]$$

And for e_2 we have

$$e_2(c) = (a_1b_{21} - a_2b_{11})\left[\frac{1}{s_1 s_2} + \frac{1/s_1}{s_1 - s_2}\exp(s_1 c) + \frac{1/s_2}{s_2 - s_1}\exp(s_2 c)\right]$$

$$+ e_2(0)\left[\frac{s_1}{s_1 - s_2}\exp(s_1 c) + \frac{s_2}{s_2 - s_1}\exp(s_2 c)\right] \qquad (7.3.16.2)$$

$$+ [a_2 - b_{11}e_2(0) + b_{21}e_1(0)]\left[\frac{1}{s_1 - s_2}\exp(s_1 c) + \frac{1}{s_2 - s_1}\exp(s_2 c)\right]$$

Equations (7.3.16) may be compared with solution equations in the example problem in Section 4.3.4. The MATLAB Symbolic Toolbox also provides this solution with much less effort.

Variable Coefficient Linear Systems

In the case of linear continuous systems that have variable coefficients, the only methods that have general application are those based on series expansions. To obtain analytical solutions for these systems, the investigator assumes a trial series expansion with arbitrary coefficients for each output variable. These series are generally infinite expansions, and their form depends on the nature of the variable coefficient functions and their singularities [i.e., the values of the input variables for which the coefficient function(s) becomes undefined or infinite].

If the coefficient functions are analytical, then by definition, these functions can be represented by Taylor series or simple (positive integer) power series expansions. The solution method for system equations with analytical coefficient functions is given in Section 4.3.5 for single-input, multiple-output systems. For multiple-input, single-output, and multiple-output systems, the series expansions must also make the system equation exact as a condition of integrability, as was demonstrated in Sections 5.2.4 and 5.3.5.

If the coefficient functions of the linear system equation exhibit singularities, but all such singularities are limited to regular singular points, then these system equations may have series solutions, which may involve negative integral, fractional, or even complex-valued exponent series expansions. Again, for multiple-input system equations, these series expansions must also make the system equation exact as a condition for integration. Such series expansion solution methods are generally referred to as Method of Frobenius techniques and may be found in appropriate, advanced mathematical textbooks.

Generalized Solution Form

We demonstrated in Section 7.3.1 that the general solution for the single-input, single-output system equation could be expressed as a linear relationship in terms of the single arbitrary integration constant K, as follows:

$$e(c) = g(c) K + f(c)$$

where $e(c)$ satisfies the general single-input, single-output system equation

$$de = g(e, c)dc$$

and where

$$g(e, c) = \frac{df(c)}{dc} + [e - f(c)]\frac{\text{Ln } [g(c)]}{dc}$$

which is a linear expression for the output variable e.

Now we want to establish the general format that the general solution of the single-input, multiple-output linear system equation has in terms of the arbitrary integration constants (also called the independent solution constants). For convenience, we will demonstrate the result for a single-input, double-output system equation and then extend the result to any number of system equation inputs.

Consider the following linear expressions in terms of e_1, e_2 and k_1, k_2.

$$e_1 = k_1 g_{11}(c) + k_2 g_{12}(c) + f_1(c)$$
$$e_2 = k_1 g_{21}(c) + k_2 g_{22}(c) + f_2(c)$$

or in matrix equation form

$$\begin{bmatrix} e_1 \\ e_2 \end{bmatrix} = \begin{bmatrix} g_{11} & g_{12} \\ g_{21} & g_{22} \end{bmatrix} \begin{bmatrix} k_1 \\ k_2 \end{bmatrix} + \begin{bmatrix} f_1 \\ f_2 \end{bmatrix} \tag{7.3.17}$$

Now let us differentiate matrix Eq. (7.3.17) with respect to the input c, so we have that

$$\begin{bmatrix} \dfrac{de_1}{dc} \\ \dfrac{de_2}{dc} \end{bmatrix} = \begin{bmatrix} \dfrac{dg_{11}}{dc} & \dfrac{dg_{12}}{dc} \\ \dfrac{dg_{21}}{dc} & \dfrac{dg_{22}}{dc} \end{bmatrix} \begin{bmatrix} k_1 \\ k_2 \end{bmatrix} + \begin{bmatrix} \dfrac{df}{dc} \\ \dfrac{df}{dc} \end{bmatrix} \tag{7.3.18}$$

Observe that Eq. (7.3.17) can be written in the equivalent, simple matrix form

$$\mathbf{e} = g\,\mathbf{k} + \mathbf{f} \tag{7.3.19}$$

If we solve the linear matrix Eq. (7.3.19) for the column matrix $\mathbf{k}$, we find that

$$\mathbf{k} = g^{-1}(\mathbf{e} - \mathbf{f})$$

or, equivalently,

$$\mathbf{k} = \begin{bmatrix} k_1 \\ k_2 \end{bmatrix} = \frac{1}{D} \begin{bmatrix} g_{22} & -g_{12} \\ -g_{21} & g_{11} \end{bmatrix} \begin{bmatrix} (e_1 - f_1) \\ (e_2 - f_2) \end{bmatrix} \tag{7.3.20}$$

where the determinant D of the matrix g^{-1} (assumed nonzero) is given by

$$D = g_{11} g_{22} - g_{12} g_{21}$$

If we substitute Eq. (7.3.20) into Eq. (7.3.18) to eliminate the solution constants k_1 and k_2, we have

$$\begin{bmatrix} \dfrac{de_1}{dx} \\ \dfrac{de_2}{dx} \end{bmatrix} = \frac{1}{D} \begin{bmatrix} \dfrac{dg_{11}}{dx} & \dfrac{dg_{12}}{dx} \\ \dfrac{dg_{21}}{dx} & \dfrac{dg_{22}}{dx} \end{bmatrix} \begin{bmatrix} g_{22} & -g_{12} \\ -g_{21} & g_{11} \end{bmatrix} \begin{bmatrix} (e_1 - f_1) \\ (e_2 - f_2) \end{bmatrix} + \begin{bmatrix} \dfrac{df}{dx} \\ \dfrac{df}{dx} \end{bmatrix} \tag{7.3.21}$$

Performing the indicated matrix multiplications, Eq. (7.3.21) can be expressed as the general linear single-input, double-output system equation, as follows:

$$d \begin{bmatrix} e_1 \\ e_2 \end{bmatrix} = \left[\begin{pmatrix} G_{11} & G_{12} \\ G_{21} & G_{22} \end{pmatrix} \begin{pmatrix} e_1 \\ e_2 \end{pmatrix} + \begin{pmatrix} F_1 \\ F_2 \end{pmatrix} \right] dc \tag{7.3.22}$$

where

$$G_{11} = \frac{1}{D} \left(g_{22} \frac{dg_{11}}{dc} - g_{21} \frac{dg_{12}}{dc} \right)$$

$$G_{12} = \frac{1}{D} \left(g_{11} \frac{dg_{12}}{dc} - g_{12} \frac{dg_{11}}{dc} \right)$$

$$G_{21} = \frac{1}{D} \left(g_{22} \frac{dg_{21}}{dc} - g_{21} \frac{dg_{22}}{dc} \right)$$

$$G_{22} = \frac{1}{D} \left(g_{11} \frac{dg_{22}}{dc} - g_{12} \frac{dg_{21}}{dc} \right)$$

$$F_1 = \frac{df_1}{dc} - f_1 G_{11} - f_2 G_{12}$$

$$F_2 = \frac{df_2}{dc} - f_2 G_{21} - f_1 G_{22}$$

Thus, we have established that the general solution for the single-input, double-output linear system equation given by Eq. (7.3.22) has the form given by Eq. (7.3.17).

To demonstrate this result for the single-input, multiple-output linear system, assume that the general solution for the linear system equation has the particular form

$$\mathbf{e} = g(\mathbf{c}) \, \mathbf{k} + f(\mathbf{c}) \tag{7.3.23}$$

where

$$\mathbf{e} = \begin{bmatrix} e_1 \\ \vdots \\ e_n \end{bmatrix}, \quad g(\mathbf{c}) = \begin{bmatrix} g_{11} & \cdots & g_{1n} \\ & \vdots & \\ g_{n1} & & g_{nn} \end{bmatrix}, \quad \mathbf{k} = \begin{bmatrix} k_1 \\ \vdots \\ k_n \end{bmatrix}, \quad f(\mathbf{c}) = \begin{bmatrix} f_1 \\ \vdots \\ f_n \end{bmatrix}$$

If we differentiate this matrix equation with respect to the independent variable c, we find that

$$\frac{d\mathbf{e}}{dc} = \frac{dg(c)}{dc}\mathbf{k} + \frac{df(c)}{dc} \tag{7.3.24}$$

Solving matrix Eq. (7.3.23) for $\mathbf{k}$, we have that

$$\mathbf{k} = g^{-1}[\mathbf{e} - \mathbf{f}] \tag{7.3.25}$$

where g^{-1} is the inverse of matrix g, assuming that g is a nonsingular matrix. Substituting Eq. (7.3.25) into Eq. (7.3.24), we find that

$$d\mathbf{e} = \left[\frac{dg(c)}{dc}g^{-1}(\mathbf{e} - f) + \frac{df(c)}{dc}\right]dc$$

or, equivalently,

$$d\mathbf{e} = [F + G\mathbf{e}]dc$$

where

$$G = \frac{dg(c)}{dc}g^{-1}$$

$$F = \frac{df(c)}{dc} - \frac{dg(c)}{dc}g^{-1}f$$

which is the generalized matrix form for the single-input, multiple-output linear system equation.

7.4 Multiple-Input Linear Systems

Linear multiple-input systems provide the bases for most of the practical modeling and quantitative analysis that has been realized in system science. In the physical sciences, the basic equations for describing most physical phenomena are associated with linear partial differential equations or related equivalent integral equations. Of course there are important nonlinear equations, such as the Navier–Stokes equation and the mass continuity equation, that are recognized as basic mathematical models, but the intractable nature of nonlinear multiple-input system equations has greatly limited analysis and evaluation of such systems. One of the great promises held for the computer is that nonlinear multiple-input system equations may be routinely analyzed and evaluated by system investigators.

Nevertheless, because of the basic nature of linear system models and the analytical resources and developments that exist, knowledge and skill with modeling, analyzing, and evaluating linear system equations will continue to be essential credentials for the competent system investigator.

The general solution methods for the linear multiple-input system equations were examined in Sections 5.2.5 and 5.3.5. It was apparent from those sections that the process for obtaining analytical solutions for variable and constant-coefficient linear systems was well defined but increasingly demanding as the number of input and output variables increased. Nevertheless, the system investigator has assured a solution for any linear system whose coefficient functions could be expressed as simple power series expansions. Extended methods, such as the method of Frobenius, can be used to provide series solutions when the system equation exhibits regular singular points.

7.4.1 Multiple-Input, Single-Output

Multiple-input, single-output continuous system equations are equivalent mathematically to first-order partial differential equations, as we observed in Chapter 5. Such system equations arise when a single or dominant output is sought from a system in which multiple inputs each significantly contribute to the output.

Consider the linear multiple-input, single-output system equation

$$d e = \sum_{all \ i} [G_i(\mathbf{c}) + H_i(\mathbf{c}) \, e] \, d c_i \tag{7.4.1}$$

We learned in Section 5.2.9 that any continuous multiple-input, single-output system equation could also be expressed as an equivalent first-order partial differential equation of the form

$$P_1 \frac{\partial e}{\partial c_1} + \cdots + P_m \frac{\partial e}{\partial c_m} = Q \tag{7.4.2}$$

We also found that Eq. (7.4.2) could often be solved using the Method of Lagrange, which stated that if the related set of ordinary differential equations

$$\frac{d c_1}{P_1} = \frac{d c_2}{P_2} = \cdots = \frac{d c_m}{P_m} = \frac{d e}{Q} \tag{7.4.3}$$

could be solved to provide m solutions of the form for $i = 1, ..., m$ and an arbitrary constant k_i

$$F_i \left(\mathbf{c}, e \right) = k_i$$

then the general solution of Eq. (7.4.2) was, as follows, where ϕ is an arbitrary function:

$$\phi \left[F_i \left(\mathbf{c}, e \right), F_i \left(\mathbf{c}, e \right), ..., F_m \left(\mathbf{c}, e \right) \right] = 0$$

Since system equation (7.4.1) is linear, it is valuable to determine the particular form of the first-order partial differential equation that is associated with Eq. (7.4.1).

We know from the Method of Lagrange that the general solution for the coupled set of ordinary differential equations given by Eq. (7.4.3) is also a solution for the partial differential equation defined by Eq. (7.4.1). Let us express Eq. (7.4.3) as m differential equations of the general form for $i = 1,...,m$:

$$\frac{dc_i}{P_i} = \frac{de}{Q} \tag{7.4.4}$$

If we multiply Eq. (7.4.4) by the arbitrary, nonzero function $a_i(e, c) Q$ and add the resulting sequence of equations, we obtain the result

$$Q \sum_{i=1}^{m} \frac{a_i}{P_i} dc_i = \sum_{i=1}^{m} a_i de \tag{7.4.5}$$

Assuming that $\sum_{i=1}^{m} a_i \neq 0$, we can express Eqs. (7.4.5) and (7.4.1) as follows:

$$de = \frac{Q}{\sum\limits_{i=1}^{m} a_i} \sum_{i=1}^{m} \frac{a_i}{P_i} dc_i = \sum_{i=1}^{m} (G_i + H_i e) dc_i \qquad (7.4.6)$$

For equivalence of Eq. (7.4.6), we will require for all $i = 1, ..., m$, that

$$\frac{Q}{\sum\limits_{i=1}^{m} a_i} = \frac{P_i}{a_i} (G_i + H_i e) = \phi(e, c) \qquad (7.4.7)$$

Solving Eq. (7.4.7) for a_i provides

$$a_i = \frac{P_i}{\phi(e, c)} (G_i + H_i e)$$

Finally, substituting this expression into Eq. (7.4.6) yields the result that

$$de = \frac{Q \phi(e, c)}{\sum\limits_{i=1}^{m} P_i (G_i + H_i e)} \sum_{i=1}^{m} \frac{(G_i + H_i e) dc_i}{\phi(e, c)} \qquad (7.4.8)$$

We conclude that for Eq. (7.4.8) to be equivalent to Eq. (7.4.1), we require that

$$Q = \sum_{i=1}^{m} P_i (G_i + H_i e)$$

where the functions $P_i(e, c)$ are arbitrary functions generally chosen as integrating factors for Eq. (7.4.4).

Thus, to summarize, the multiple-input, single-output system equation

$$de = \sum_{i=1}^{m} [G_i(c) + H_i(c) e] d c_i \qquad (7.4.9)$$

is equivalent to the first-order partial differential equation

$$\sum_{i=1}^{m} \frac{1}{P_i(e, c)} \frac{\partial e}{\partial c_i} = \sum_{i=1}^{m} P_i (G_i + H_i e)$$

which can be solved by solving the following sequence of first-order differential equations

$$\frac{d c_1}{P_1} = \frac{d c_2}{P_2} = \cdots = \frac{d c_m}{P_m} = \frac{de}{\sum\limits_{i=1}^{m} P_i (G_i + H_i e)} \qquad (7.4.10)$$

The basis for determining the arbitrary functions P_i in Eq. (7.4.10) is to realize a set of integrable differential equations that will provide a solution for system Eq. (7.4.9). Let us express Eq. (7.4.10) as the following set of equivalent equations:

$$P_k dc_i = P_i dc_k \qquad (7.4.11)$$

for i and $k = 1, ..., m$, and

$$P_k \, de = \sum_{i=1}^{m} P_i (G_i + H_i e) \, dc_k \tag{7.4.12}$$

for some value of j such that $1 < j < m$.

Now we must integrate each of these equations, and to be integrable they must each be exact, as defined in Section 5.2.4. So we require for the functions P_i that

$$\frac{\partial P_i}{\partial c_i} = -\frac{\partial P_k}{\partial c_k} \tag{7.4.13}$$

for i and $k = 1, ..., m$ and for some j such that $1 < j < m$

$$\frac{\partial P_i}{\partial c_i} = -\sum_{i=1}^{m} \left[P_i H_i + (G_i + H_i e) \frac{\partial P_i}{\partial e} \right] \tag{7.4.14}$$

If we can determine a consistent set of P_i functions that satisfy Eqs. (7.4.13) and (7.4.14), then we are assured a solution for Eq. (7.4.9), the multiple-input, single-output linear system equation. Of course, if system Eq. (7.4.9) is already exact as defined in Section 5.2.4, then the equation is directly integrable, and determination of integrating factors is unnecessary.

Example of Linear Double-Input, Single-Output System Equation
Consider the following linear system kernel where the G's and H's are

$$de = (G_1 + H_1 e) \, dc_1 + (G_2 + H_2 e) \, dc_2 \tag{7.4.15}$$

Now for this double input system equation to possess a solution of the form $K = \phi(e, \mathbf{c})$ we have from Eq. (5.2.24) found in Section 5.2.4 that the system equation coefficients must satisfy the condition that

$$G_1 H_2 = G_2 H_1 \tag{7.4.16}$$

If Eq. (7.4.16) is satisfied, then effort at determining an integrating factor is justified. We will see how condition (7.4.16) arises as we seek a solution of Eq. (7.4.15) using the Method of Lagrange.

Equations (7.4.13) and (7.4.14) yield the following requirement for Eq. (7.4.15)

$$-\frac{\partial P_1}{\partial c_1} = \frac{\partial P_2}{\partial c_2} \tag{7.4.17}$$

and

$$-\frac{\partial P_1}{\partial c_1} = G_1 \frac{\partial P_1}{\partial e} + H_1 \frac{\partial P_1 e}{\partial e} + G_2 \frac{\partial P_2}{\partial e} + H_2 \frac{\partial P_2 e}{\partial e} \tag{7.4.18}$$

Now let us explore the consequences if both P_1 and P_2 are nonzero constants. Equation (7.4.17) is satisfied identically, and Eq. (7.4.18) provides

$$0 = H_1 P_1 + H_2 P_2$$

If we let $P_1 = -P_2 H_2/H_1$, then for Eqs. (7.4.11) and (7.4.12) we have

$$H_1 dc_1 = -H_2 dc_2$$
$$H_2 de = (H_2 G_1 - H_1 G_2) dc_1$$

General solutions for these differential equations are

$$K_1 = H_1 c_1 + H_2 c_2$$
$$K_2 = H_2 e + (H_1 G_2 - H_2 G_1) c_1$$

The solution as defined by the Method of Lagrange has the form $\varphi(K_1, K_2) = 0$ or equivalently

$$e = X(x) + (G_1 H_2 - G_2 H_1) c_1/H_2 \tag{7.4.19}$$

where

$$X = H_1 c_1 + H_2 c_2$$

Now to determine the appropriate form of the function $X(x)$, we recall that for a double-input system equation to be exact, we have that

$$de = \frac{\partial e}{\partial c_1} dc_1 + \frac{\partial e}{\partial c_2} dc_2$$

So from Eqs. (7.4.15) and (7.4.19), we find that

$$\frac{\partial e}{\partial c_1} = X'(x) H_1 + (G_1 H_2 - G_2 H_1)/H_2 = G_1 + H_1[X(x) + (G_1 H_2 - G_2 H_1) c_1/H_2]$$

$$\frac{\partial e}{\partial c_2} = X'(x) H_2 = G_2 + H_2[X(x) + (G_1 H_2 - G_2 H_1) c_1/H_2]$$

where X' denotes the derivative of X with respect to its single argument x. Now, these two equations provide the general solution for the system Eq. (7.4.15)

$$X(x) = e(c_1, c_2) = K \exp(H_1 c_1 + H_2 c_2) - G_2/H_2 \tag{7.4.20}$$

where K is an arbitrary constant.

It is instructive to determine the format of the general solution for the multiple-input, single-output linear system equation. The general solution of such a system equation has the form

$$e(\mathbf{c}) = f(\mathbf{c}) + g(\mathbf{c})k \tag{7.4.21}$$

where k is the independent, arbitrary integration constant. Let us form the partial derivative of e for each input function of $\mathbf{c}$ as follows:

$$\frac{\partial e}{\partial c_i} = \frac{\partial f}{\partial c_i} + \frac{\partial g}{\partial c_i} k \tag{7.4.22}$$

for $i = 1, ..., m$. If we eliminate the constant k between Eqs. (7.4.21) and (7.4.22), we have after some effort that

$$\frac{\partial e}{\partial c_i} = f \frac{\partial \mathrm{Ln}(f/g)}{\partial c_i} + e \frac{\partial \mathrm{Ln}g}{\partial c_i} \tag{7.4.23}$$

The total differential of e is given by

$$de = \sum_{i=1}^{m} \frac{\partial e}{\partial c_i} dc_i \tag{7.4.24}$$

and if we substitute Eq. (7.4.23) into Eq. (7.4.24), we obtain

$$de = \sum_{i=1}^{m} [G_i(\mathbf{c}) + H_i(\mathbf{c})\, e]\, dc_i \tag{7.4.25}$$

where

$$G_i(\mathbf{c}) = f \frac{\partial \mathrm{Ln}\,(f/g)}{\partial c_i}$$

$$H_i(\mathbf{c}) = \frac{\partial \mathrm{Ln}\,g}{\partial c_i}$$

We recognize Eq. (7.4.25) as the general form of the multiple-input, single-output system equation.

7.4.2 Multiple-Input, Multiple-Output Continuous System Equations

Multiple-input, multiple-output continuous system equations are equivalent mathematically to sets of first-order partial differential equations, as was shown in Chapter 5. Such system equations are the most general equations encountered in system science and are capable of modeling any continuous system.

The evaluation and analysis methods developed in the previous section for multiple-input, single-output linear systems may be extended to multiple-output systems. Consider the linear system equations for $\mathbf{e}$ for $i = 1, ..., n$:

$$de_i = \sum_{j=1}^{m} \left[G_{ij}(\mathbf{c}) + H_{ijk}(\mathbf{c})\, e_k \right] dc_j \tag{7.4.26}$$

The condition for the exactness of this system equation is given in Section 5.3.4. The system investigator should determine if the system equation is exact and therefore directly integrable before attempting to find integrating factors for the equation. If the system equation is not exact, the system equation should be converted to an equivalent coupled set of ordinary differential equations. Then we have that Eq. (7.4.26) is equivalent to the following set of first-order partial differential equations for $i = 1, ..., n$:

$$\sum_{j=1}^{m} P_{ij} \frac{\partial e_i}{\partial c_j} = Q_i \tag{7.4.27}$$

Using the Method of Lagrange, Eq. (7.4.27) is found to be related to the following set of coupled ordinary differential equations:

$$\frac{dc_1}{P_{i1}} = \frac{dc_2}{P_{i2}} = \cdots = \frac{dc_m}{P_{im}} = \frac{de}{Q_i} \tag{7.4.28}$$

If a set of integrating factors, that is, the P_{ij}'s can be determined so that Eq. (7.4.28) can be solved in the form

$$F_{ij}(\mathbf{c},\ e) = k_{ij} \tag{7.4.29}$$

for $i = 1, ..., n$ and $j = 1, ..., m$, then the general solution for Eq. (7.4.27) and also (7.4.26) has the form, for $i = 1, ..., n$

$$W_i\left[F_{i1}, F_{i2}, ..., F_{im}\right] = 0 \tag{7.4.30}$$

where W_i is an arbitrary function.

The essential requirement for the P_{ij}'s given in Eqs. (7.4.27) and (7.4.28) to be integrating factors is that

$$\frac{\partial P_{ij}}{\partial c_j} = -\frac{\partial P_{ik}}{\partial c_k} \tag{7.4.31}$$

and

$$\frac{\partial P_{ij}}{\partial c_j} = -\sum_{j=1}^{m}\left\{\left[G_{ij} + \sum_{k=1}^{n} H_{ijk}e_k\right]\frac{\partial P_{ij}}{\partial c_i} + P_{ij}H_{ijk}\right\} \tag{7.4.32}$$

for $i = 1, ..., n$ and $j = 1, ..., m$. Thus, if Eqs. (7.4.31) and (7.4.32) can be satisfied for some set of functions P_{ij}, then Eq. (7.4.28) can be successfully solved, and the solutions for the multiple-input, multiple-output system Eq. (7.4.26) are given by Eqs. (7.4.29) and (7.4.30).

The general solution format for the multiple-input, multiple-output linear system equation can be developed by assuming the solution for this system equation has the general form

$$\mathbf{e}\,(\mathbf{c}) = f\,(\mathbf{c}) + g(\mathbf{c})\,\mathbf{k} \tag{7.4.33}$$

where

$$\mathbf{e}\,(\mathbf{c}) = \begin{bmatrix} e_1 \\ \vdots \\ e_n \end{bmatrix},\quad f(\mathbf{c}) = \begin{bmatrix} f_1 \\ \vdots \\ f_n \end{bmatrix},\quad g(\mathbf{c}) = \begin{bmatrix} g_{11} & \cdots & g_{1n} \\ & \vdots & \\ g_{n1} & & g_{nn} \end{bmatrix},\quad \mathbf{k} = \begin{bmatrix} k_1 \\ \vdots \\ k_n \end{bmatrix}$$

The column matrix $\mathbf{k}$ forms the independent, arbitrary, integration constants associated with the system equation.

Observe that Eq. (7.4.33) can be solved for $\mathbf{k}$ as follows:

$$\mathbf{k} = g\,(\mathbf{e}, \mathbf{c})^{-1}\,[\mathbf{e}\,(\mathbf{c}) - f\,(\mathbf{c})] \tag{7.4.34}$$

where $g(\mathbf{c})$ is assumed to be a nonsingular matrix. If we form the partial derivative of the output e_i with respect to the input c_k, then from Eq. (7.4.33), we have

$$\frac{\partial e_i}{\partial c_k} = \frac{\partial f_i}{\partial c_k} + \sum_{j=1}^{n} k_j \frac{\partial g_{ij}}{\partial c_k} \tag{7.4.35}$$

If we use the appropriate component of Eq. (7.4.34) to eliminate k_j from Eq. (7.4.35), we have that

$$\frac{\partial e_i}{\partial c_k} = \frac{\partial f_i}{\partial c_k} + \sum_{j=1}^{n} \frac{\partial g_{ij}}{\partial c_k} \left[\sum_{l=1}^{n} g_{jl}^{-1} (e_l - f_l) \right] \tag{7.4.36}$$

Now we observe that the total derivative of output e_i is given by

$$d e_i = \sum_{k=1}^{m} \frac{\partial e_i}{\partial c_k} d c_k \tag{7.4.37}$$

Equations (7.4.37) and (7.4.36) can be combined to be expressed as the ith component of the general multiple-input, multiple-output linear system equation, as follows:

$$d e_i = \sum_{k=1}^{m} \left\{ \frac{\partial f_i}{\partial c_k} + \sum_{j=1}^{n} \frac{\partial g_{ij}}{\partial c_k} \left[\sum_{l=1}^{n} g_{jl}^{-1} (e_l - f_l) \right] \right\} d c$$

It is apparent that this system equation is equivalent to the general linear system equation.

7.5 Summary

This chapter has been devoted to an examination and review of linear systems. The great majority of our physical models and their associated theory is based on linear system equations because of the better likelihood of successfully obtaining analytical and numerical solutions and a greater understanding of the system under investigation. However, it is probably true that all real, physical systems are nonlinear when examined in great depth at a fundamental level. Material systems are composed of atoms and energy quanta, and the presence of more than one atom interacting with another atom results in nonlinear, Quantum Mechanical Wave Equations describing the system. Nevertheless, despite this reality, science and technology have achieved immense and crucial accomplishments, particularly in engineering and technical applications. The highly developed state of our present technology, including chemical and nuclear energy sources, the electrical power grid, the internet and world-wide-web (WWW), machines, computers, transportation, mass communications, etc., are achievements witnessed by contemporary humans over their single lifespan. Disparagingly, civil engineers claim greater overall improvement in human health through the application of sanitary engineering rather than all medical advances and applications combined. The Black Plague that impacted humans during the Middle Ages (endemic during the fourteenth and fifteenth centuries) ravaged the human population. Historians estimate that the human death toll was between a third and a half of the European population from the "Black Death" (probably Bubonic Plague). Poor hygiene and sanitation was the main vector by which the disease swept rapidly through the populations impacted. Reference and details for the Black Death are given in Wikipedia, the free encyclopedia. Furthermore, the rapid development by medical science and wide distribution of COVID-19 vaccines have undoubtedly prevented innumerable deaths and injuries since the occurrence of the Corona Virus Pandemic of 2019.

So despite the reality of nonlinearity in the physical world, we have shown that by utilizing linearization methods it is generally possible to approximate any set of continuous system equations by linear approximations over some restricted range of applicability. Furthermore, linearization methods were shown to be capable of converting linear variable coefficient system equations into approximate, constant-coefficient linear system equations.

Single-input linear system equations are the simpler class to analyze and evaluate with analytical, series, and operational methods generally applicable to such equations. Such methods were reviewed, and the general solution format for these single-input system equations was determined in terms of the integration constants for the system equations. MATLAB is particularly useful in addressing linear system equations.

Multiple-input linear system equations were then reviewed, and the important Method of Lagrange was demonstrated. By converting such multiple-input equations into mathematically equivalent sets of first-order partial differential equations, these equations can often be solved as coupled sets of ordinary differential equations by identifying appropriate integrating factors as necessary. Finally, the general solution format in terms of the system equations integration constants was demonstrated for the linear multiple-input system equation.

Bibliography

ADKINS, W. A., et al., *Ordinary Differential Equations*, Springer-Verlag, NY, 2012. A popular introductory text for differential equations.

ASELTINE, J. A., *Transform Methods in Linear System Analysis*, McGraw-Hill Book Co., NY, 1958. A useful reference for transform methods and their applications.

BATCHELDER, P. M., *An Introduction to Linear Difference Equations*, Dover Publications, 1967. Well respected treatment of linear difference equations.

BOYCE, W. E., *Elementary Differential Equations and Boundary Value Problems*, 4th edition. John Wiley & Sons, 1967

BROCKETT, R. W., *Finite Dimensional Linear Systems*, John Wiley & Sons, NY, 1970. A general reference for linear systems.

BROWN, R. G. and J. W. NILSSON, *Introduction to Linear Systems Analysis*, John Wiley & Sons, NY, 1962. An introductory treatment of linear systems.

CAMPBELL, S. L. and R. HABERMAN, *Introduction to Differential Equations with Dynamical Systems*, Princeton University Press, 2008. A concise and up-to-date textbook

DAVIS, H. T., *Introduction to Nonlinear Differential and Integral Equations*, Dover Publications, Inc., NY, 1962. An essential reference for nonlinear ordinary differential equations addressed in this chapter.

INCE, E. L., *Ordinary Differential Equations*, Dover Publications, Inc., NY, 1956. An essential reference for ordinary differential equations addressed in this chapter.

MILLER, K. S., *Linear Difference Equations*, W. A. Benjamin, 1968.

TENEBAUM, M., *Ordinary Differential Equations*, Dover Books, 1985.

ZADEH, L. A., and C. A. DESOER, *Linear System Theory*, McGraw-Hill Book Co., NY, 1963. A standard exposition in the theory and analysis of linear systems.

ZILL, D. G., *A First Course in Differential Equations*, 9th editionAcademia, Edu, 2013. A popular introduction to modelling with differential equations.

ZILL, D. G., *A First Course in Differential Equations with Modeling Applications*, Cengage Learning, 2012.

Problems

1 Linearize and solve the following system equations. Compare the linearized solution with the exact solution, which can be obtained as shown in notes.

(Note for any $F(x)$ then $F'(x) = \dfrac{dF(x)}{dx}$)

a) $z\,dz + \left(ayz^2 + \dfrac{1}{y}\right)dy = 0$ (let $z^2 = u$ and solve linear equation)

b) $x^2 \exp(y)y' = \exp(y) - x + x^2$ (let $\exp(y) = u$ and solve linear equation)

c) $dy = \dfrac{(1+y)y}{[(1+y)x]}dx$ (separable equation)

d) $xy\,dy = a(x\,dy + 2y\,dx)$ (separable equation)

e) $y' = \text{Ln}\left(\dfrac{y}{x}\right) + \left(\dfrac{y}{x}\right)$ (homogeneous equation)

f) $t\,ds = [s + (s^2 + t^2)^{1/2}]dt$ (homogeneous equation)

g) $(t + s^2)\,dt + 2s\,t\,ds = 0$ (exact equation)

h) $\dfrac{dy}{dx} + y = y^2$ (Bernoulli equation)

i) $df = \left[\dfrac{(2f + 3t - 4)}{(f - 2t + 3)}\right]^{\frac{1}{2}}dt$ (let $2f + 3t - 4 = u$ and $f - 2t + 3 = v$)

j) $de_1 = \dfrac{(A + Bc)}{e_1^2 dc}$

$de_2 = \dfrac{(D + Ec^2)}{e_2}$ (both equations are separable)

k) $\dfrac{dx}{dt} = (Hax)(Hby)$ (autonomous equations)

$\dfrac{dy}{dt} = (Hcx)(Hey)$ $\left(\text{Form } \dfrac{dy}{dx} \text{ and solve separable equation in } y \text{ and } x\right)$

l) $y' = \text{Ln}\left(\dfrac{y}{x}\right) + A\left(\dfrac{y}{x}\right)$ (homogeneous equation)

m) $de_1 = (A + Be_2^2)\,c\,dc$

$de_2 = (D + E\,e_1)\,c\,dc$ $\left(\text{form } \dfrac{de_1}{de_2} \text{ and solve separable equation}\right)$

n) $\dfrac{dy}{dx} = Ay^2 + \dfrac{By}{x}$ (Bernoulli equation)

o) $dy = \dfrac{(y - xy^2)}{(x^2 y - x)} dx$ $\left(\text{integrating factor is } \dfrac{1}{xy}\right)$

p) $de_1 = (Ae_1^2 + Be_2^2) dc$

$de_2 = (Ce_2 + De_1) e_1 dc$ $\left(\text{form } \dfrac{de_1}{de_2} \text{ and solve homogeneous equation}\right)^t$

q) $de_1 = \dfrac{(A + Bc)}{e_1^2} dc$ and $de_2 = \dfrac{(D + Ec^2)}{e_2} dc$

r) $dx = (2x - a x^2 y^n) dy$

s) $de = A e (c - e) \dfrac{dc}{c^2}$

t) $dy = \dfrac{(Ax + By + C)}{(Dx + Ey + 1)} dx$ (homogeneous equation)

u) $dx = \left[1 + \dfrac{1}{(x^2 y^2)}\right] \dfrac{dy}{y^2}$ (let $xy = u$)

v) $(x^2 + y^2 + 2x) dx = 2y \, dx$ (find an integrating factor)

w) $dx = \dfrac{[(a + bs)t]}{[(c + t)s]} dt$ (separable?)

x) $dA = \dfrac{(B^2 + Ac)}{(A^2 - Be)} dB$ (exact?)

y) $dx = \left(\dfrac{2x}{y} - a \dfrac{x^2}{y^2}\right) dy$

z) $\dfrac{dx^2}{dt^2} + A \left(\dfrac{dx}{dt}\right)^2 + f(x) = 0$

$\left(\text{let } \dfrac{dx}{dt} = v \, \& \, \dfrac{dv}{dt} = v \dfrac{dv}{dx} \, \& \text{ solve linear equation in } v^2\right)$

2 Convert the following linear system equations into constant-coefficient linear system equations and attempt to solve the resulting system equation.

a) $de_1 = (ace_1 + e_2) dc$ and $de_2 = (e_1 + bce_2) dc$

b) $dx = \dfrac{(x + y)}{(a + t)} dt$

$dy = \dfrac{(x - y)}{(b + t)} dt$

c) $de_1 = (e_1 c - 2e_2 + c^2 e_3) dc$

$de_2 = (c^2 e_1 + c e_2 - 2e_3) dc$

$de_3 = (2e_2 + c^2 e_2 + c e_3) dc$

d) $de_1 = (c_1 c_2 + e_1 + c_2) dc_1 + (1 + c_2 e_2) dc_2$

$de_2 = (c_1 e_1 + c_2 e_2) dc_1 + (c_1 + c_2) dc_2$

e) $de = (a + b c_1^2 e) dc_1 + (f + g c_2^2 e) dc_2 + h dc_3$

f) $dx = (x + y - r) dr$

$dy = (y - z + s) ds$

$dz = (z - x + t) dt$

g) $de = (ac_1^2 + c_2 e) \, dc_1 + \left(c_2^2 + \dfrac{e}{c_1}\right) dc_2$

h) $du = (u + \sqrt{v}) \, dx$

$dv = \left(x^2 u + \dfrac{v}{x}\right) dx$

i) $de = (Ac_1 + c_1 c_2) \, dc_1 + \left(Bc_2 + \dfrac{Dc_1}{c_2}\right) dc_2 + E \, dc_3$

j) $df = (A + Bx^2 g) f \, dx$

$dg = \left(\dfrac{c}{x} + D \, Ln f\right) dx$ (Make equations linear by variable change)

3   Use the Laplace Transform Toolbox provided by MATLAB to solve the following system equations:

a) $dx = (At + Bx) \, dt$ for $x(0) = 1$

b) $de_1 = (c + 2e_1 + e_2) \, dc$

$de_2 = \left(\dfrac{3}{c} + e_1 - 2e_2\right) dc$ for $e_1(0) = 1, e_2(0) = 0$

c) $du = (A_1 + B_1 v + C_1 w) dt$

$dv = (A_2 + B_2 u + C_2 w) dt$

$dw = (A_3 + B_3 u + C_3 v) dt$ for $u(0) = v(0) = w(0) = 0$

d) $dA = (t + aA + B) dt$

$dB = \left(\dfrac{1}{t} + bB - c\right) dt$

$dC = (t^2 + eA + C) \, dt$ for $A(0) = b, B(0) = a, C(0) = -e$

e) $dw = (zw - x - 2y + z) \, dt$

$dx = (w + 2x - 3y - z) \, dt$

$dy = (-w + 2x - y + 2z) \, dt$

$dz = (2w - x - y + z) \, dt$ for $z(0) = 0, w(0) = x(0) = y(0) = 1$

f) $dx = \left(A + Bt + Ct^2 + \dfrac{D}{t} + Ex\right) dt$ for $x(0) = 1$

4   Determine the linear system equations associated with the following system equation solutions by eliminating the arbitrary integration constants, the k's.

a) $Ln\left[\dfrac{(x + y)}{(x^2 - 1)}\right] = k$

b) $k = (Ax + Bx^2 + Cy)^{\frac{2}{3}}$ (Redefine k to make equation linear in y)

c) $e = k \, f(c_1, c_2) + g(c_1, c_2)$

d) $e = k \, f(c) + g(c)$

e) $e_1 = k_1 f_1(c) + k_2 g_1(c) + h_1(c)$

$e_2 = k_1 f_2(c) + k_2 g_2(c) + h_2(c)$

f) $e_1 = k_1 c_1 + k_2 c_2 + c_1 c_2$

$e_2 = k_1 c_2 - c_1 c_2 - c_1 c_2$

5 Convert the following linear system equations into related sets of first-order partial differential equations. Attempt to solve these first-order partial differential equations using the Method of Lagrange outlined in Section 7.4

a) $de = (A + e) dc_1 + (B - e) dc_2$

b) $de = (a\ c_1 + 2e) dc_1 + (b\ c_1 - 3e) dc_2$

c) $de = f_1(c_2)g(e) dc_1 + f_2(c_1)g(e) dc_2$

d) $de_1 = (A_{11} + A_{12}c_1e_1 + A_{13}c_2e_2) dc_1 + A_{14} dc_2$

   $de_2 = B_{11}dc_1 + (B_{12} + B_{13}c_1c_2 + B_{14}e_2) dc_2$

e) $de = (c_2 + 2e) dc_1 + (c_1 - 3e) dc_2$

f) $de = (A + ae) dc_1 + (B + be) dc_2 + (F + fe) dc_3$

g) $de = (c_1 + c_2e) dc_1 + (c_2 + c_1e) dc_2$

   $de_1 = (c_1c_2c_3 - e_1) dc_1 + e_2 dc_2 - e_3 dc_3$

h) $de_2 = (c_1 + e_1) dc_1 + (c_2 - e_2) dc_2 + (c_3 + e_3) dc_3$

   $de_3 = c_2dc_1 + c_1c_2e_1dc_2 + dc_3$

8

Generalized System Analysis Methods

In this chapter, we examine some of the important considerations and methods for processing and analyzing systems. These methods include reduction and modification of inputs and outputs, normalization and transformation of system variables, parameter reduction and minimization, sensitivity analysis of system parameters, treatment of systems with feedback, and computer-aided analysis of system models. The significance and utilization of each of these methods for specific systems are briefly addressed.

8.1 Simplification and Reduction of System Kernels

It is often possible and necessary to simplify the components of the system kernel or to reduce the number of inputs and outputs for a given system equation. Such simplification and reduction efforts are particularly useful if the system is very complex or an analytical solution for the system equation is required. It is essential for efficient system studies that the set of system equations that is finally prepared for analysis and solution be commensurate with the amount of time and effort available to perform the system analysis and evaluation. If the system equations as modeled for the system study cannot be subsequently analyzed and evaluated adequately because of the system equation's complexity and the number of inputs and outputs, then the system study will be improperly executed and probably unsuccessful.

Efforts for model simplification usually begin by examining the system's kernel and determining systematic means of acceptably simplifying and reducing the quantitative relationship employed for that system kernel. For example, to simplify the general continuous multiple-input, multiple-output system kernel, $g(\mathbf{e}, \mathbf{c})$, we will consider approximation methods that will allow us to replace the quantitative expressions for $g(\mathbf{e}, \mathbf{c})$ by simpler kernel expressions denoted by $g^*(\mathbf{e}^*, \mathbf{c}^*)$. The components of g^* are reduced in number or are simplified approximations of the original kernel components and are often chosen so that the system equation can be integrated and solved using one of the analytical solution methods given in Chapters 4 and 5.

Reduction of a general multiple-input, multiple-output kernel can be accomplished by converting the original kernel $g(\mathbf{e}, \mathbf{c})$ from a size $(n \times m)$ [viz., n dependent variable outputs

Introduction to System Science with MATLAB, Second Edition.
Gary Marlin Sandquist and Zakary Robert Wilde.
© 2023 John Wiley & Sons Ltd. Published 2023 by John Wiley & Sons Ltd.

and *m* independent variable inputs] to a smaller system kernel, $g^*(\mathbf{e}^*, \mathbf{c}^*)$, of size $(k \times p)$, where $k \leq n$ and $p \leq m$.

There are many methods by which such simplification and reduction can be accomplished, and familiarity with the specific system and experience with different reduction methods are the preferred basis for realizing satisfactory results. We will consider a few methods here, but the reader is cautioned against applying any of these methods indiscriminately to a given system equation. The results may produce a system kernel that is inaccurate and misleading as a model for the actual system under study.

The simplification and reduction methods considered here will be developed for application to continuous system kernels. The extended application of the methods to discrete systems will generally be obvious. Furthermore, since discrete systems are usually readily adapted to computer simulation and processing, the need for simplification and reduction of discrete variable systems is not nearly so evident and necessary.

8.1.1 Conversion of Variable System Kernels to Constant Kernels

A practical alternative in analyzing a system equation with a variable kernel, which is difficult to solve and analytically integrate, is to convert each variable component of the system kernel into a suitable constant value chosen to adequately represent the variable component over the selected range of interest of the system's inputs and outputs. Observe that such a procedure for conversion of a general system kernel to a constant kernel is a necessary, preliminary step that is required before the kernel reduction procedure given in Section 8.1.2 can be utilized. Thus, we seek practical methods for converting the general multiple-input, multiple-output system kernel $g(\mathbf{e}, \mathbf{c})$, defining the system equation

$$d\mathbf{e} = g(\mathbf{e}, \mathbf{c}) \, d\mathbf{c}$$

into a constant system kernel, as follows:

$$g(\mathbf{e}, \mathbf{c}) \Rightarrow E^*[g(\mathbf{e}^*, \mathbf{c}^*)] = \begin{bmatrix} g_{11}^* & \cdots & g_{1m}^* \\ \vdots & & \vdots \\ g_{n1}^* & \cdots & g_{nm}^* \end{bmatrix}$$

The values $\mathbf{e}^*$ and $\mathbf{c}^*$ are specified as constant parameter values of the output and input variables that effectively represent the kernel over the value range intended. The components g_{ij}^* (for $i = 1, ..., n$, and $j = 1, ..., m$) are suitable chosen constants that serve as representative values for the variable system kernel. The operator $E^*[g(\mathbf{e}^*, \mathbf{c}^*)]$ is meant to define some appropriate weighting or averaging operator that provides the expected or mean value of the variable component over its range of variation. We will provide several examples of such averaging operators for kernels.

If the variable system kernel can be successfully converted to a constant kernel, then the related approximate system equation can be easily integrated to provide the following solution:

$$\mathbf{e} = g(\mathbf{e}^*, \mathbf{c}^*) \, \mathbf{c} + \mathbf{k}$$

where $\mathbf{k}$ is the matrix integration constant.

Table 8-1   Kernel Modification Applications

General System Kernel: $g(\mathbf{e}, \mathbf{c}) = \begin{bmatrix} g_{11}(\mathbf{e}, \mathbf{c}) & \cdots & g_{1m}(\mathbf{e}, \mathbf{c}) \\ \vdots & & \\ g_{n1}(\mathbf{e}, \mathbf{c}) & \cdots & g_{nm}(\mathbf{e}, \mathbf{c}) \end{bmatrix}$

System kernels may be approximated as follows:

I) Kernel average: $g_{ij}(\mathbf{e},\ \mathbf{c}) = g_{ij}^*$ where $g_{ij}^* = \prod\limits_{l=1}^{m} \prod\limits_{k=1}^{n} \int_{(k)} \int_{(l)} g_{ij}(\mathbf{e}, \mathbf{c})\, dc_l\, de_k$

II) Averaging independent variable: $g_{ij}(\mathbf{e},\ \mathbf{c}) = g_{ij}^*(\mathbf{c})$ where $g_{ij}^*(\mathbf{c}) = \prod\limits_{k=1}^{n} \int_{(k)} g_{ij}(\mathbf{e}, \mathbf{c})\, de_k$

III) Averaging dependent variable: $g_{ij}(\mathbf{e},\ \mathbf{c}) = g_{ij}^*(\mathbf{e})$ where $g_{ij}^*(\mathbf{e}) = \prod\limits_{l=1}^{n} \int_{(l)} g_{ij}(\mathbf{e}, \mathbf{c})\, dc_l$

IV) Separable: $g_{ij}(\mathbf{e},\ \mathbf{c}) = g_i^*(e_i)\, g_j^*(c_j)$ where $g_{ij}^*(c_j) = \prod\limits_{l=1,\ l\neq j}^{m} \prod\limits_{k=1}^{n} \int_{(k)} \int_{(l)} \dfrac{g_{ij}(\mathbf{e}, \mathbf{c})}{g_i{*}(e, c)}\, dc_l\, de_k$

V) Linearized: $g_{ij}(\mathbf{e},\mathbf{c}) = g_{ij}^* + \sum\limits_{k=1}^{n} \partial\dfrac{g_{ij}^*}{\partial e_k} e_k + \sum\limits_{l=1}^{m} \partial\dfrac{g_{ij}^*}{\partial c_l} c_l$ where g_{ij}^*, $\partial g_{ij}^* / \partial e_k$, and $\partial g_{ij}^* / \partial c_l$ found
from Case I or Case VI

VI) Constant: $g_{ij}(\mathbf{e},\ \mathbf{c}) = g_{ij}^*(e, c)$ and $g_{ij}(\mathbf{e},\ \mathbf{c})$ is evaluated at any appropriate fixed value such as:

a) $\mathbf{e}^* = \mathbf{e}[\mathbf{c}(\min)],\quad \mathbf{c}^* = \mathbf{c}(\min)$
b) $\mathbf{e}^* = \{\mathbf{e}[\mathbf{c}(\max)] + \mathbf{e}[\mathbf{c}(\min)]\}/2,\ \mathbf{c}^* = \{\mathbf{c}(\max) + \mathbf{c}(\min)\}/2$
c) $\mathbf{e}^*$ and $\mathbf{c}^*$, where $g(\mathbf{e},^* \mathbf{c}^*) = 0$

Note: The definition for the product of integrals is given below

$$\prod\limits_{i=1}^{n} \int_{(i)} F(\mathbf{x})\, dx_i = \dfrac{1}{(x_{n2} - x_{n1}) \cdots (x_{12} - x_{11})} \int_{x_n} \cdots \int_{x_1} F(x_1, ..., x_n)\, dx_1 ... dx_n$$

where $x_{i2} - x_{i1} = x_i(c_{\max}) - x_i(c_{\min})$ over interval $c_{\min} < c < c_{\max}$

Table 8-1 indicates several methods for approximating system kernels. In general, the methods are straightforward and of varying difficulty to apply. Interestingly, often the most difficult aspect of these methods, as is true of many approximation methods, is attempting to quantify the errors resulting from the approximation methods employed.

Example of Replacement of System Kernel by Mean Value Constants.
Consider the linear single-input, single-output system equation

$$de = g(e, c)\, dc = (A + Be)\, dc \tag{8.1.1}$$

We want to replace the dependent variable kernel

$$g(e, c) = A + Be$$

by an appropriate mean or expectation value for the system kernel over the range of interest for e. We have for the mean value from the averaging operator that

$$E\{g\} = \dfrac{1}{c(\max) - c(\min)} \int_{c(\min)}^{c(\max)} \dfrac{1}{e(\max) - e(\min)} \int_{e(\min)}^{e(\max)} (A + Be)\, de\, dc$$

and we find for Eq. (8.1.1) that

$$E\{g\} = A + \frac{B}{2}\left[e\left(\max\right) + e\left(\min\right)\right] \tag{8.1.2}$$

If we replace the variable kernel in Eq. (8.1.1) with our expectation value for the kernel given by Eq. (8.1.2), we can easily integrate this system equation with the resulting constant kernel to give

$$e^* = \left\{A + \frac{B}{2}\left[e(\max) + e(\min)\right]\right\}\left[c - c(\min)\right] + e(\min) \tag{8.1.3}$$

where e^* is the approximate solution for Eq. (8.1.1) obtained by using a constant, expectation value kernel. For this example, which is not generally possible or practical in practice, we can easily determine the actual solution for the simple linear system Eq. (8.1.1). We find that

$$e = e(\min) \exp\left\{B\left[c - c\left(\min\right)\right]\right\} + \frac{A}{B}\left[\exp\left\{B\left[c - c(\min)\right]\right\} - 1\right] \tag{8.1.4}$$

The question now obviously arises as to how accurate an approximation is Eq. (8.1.3) for e^* to the actual system solution Eq. (8.1.4) for e? A reasonable test for the error between e^* and e is given by the mean square error, as follows:

$$\text{error } (e - e^*)^2 = \int_{c\,(\min)}^{c\,(\max)} (e - e^*)^2 \, dc$$

The indicated integration is straightforward, but the result is rather complex and not easily interpreted. A direct comparison is simpler in this case. If in Eq. (8.1.4), we assume that $B[c - c(\min)]$ is sufficiently small so that the exponential term may be approximated as

$$\exp\left\{B\left[c - c\left(\min\right)\right]\right\} \approx 1 + B\left[c - c\left(\min\right)\right]$$

then a direct comparison between Eqs. (8.1.4) and (8.1.3) shows that the normalized difference between the e and e^* is

$$e\left(\max\right) \le e(c) \le 3\, e\left(\min\right)$$

So, if we restrict the range of $e(c)$ to the interval $e(\min) \le e(c) \le 3e(\min)$ and that

$$B\left[c\left(\max\right) - c\left(\min\right)\right] < < 1$$

then $e*$ given by Eq. (8.1.3) is a good approximation for the solution e given by Eq. (8.1.4) for the example system equation.

8.1.2 Reduction of System Kernels

Consider a system composed of m inputs and n outputs with the $(n \times m)$ system kernel, as follows:

$$g(\mathbf{e}, \mathbf{c}) = \begin{bmatrix} g_{11}(\mathbf{e}, \mathbf{c}) & \cdots & g_{1m}(\mathbf{e}, \mathbf{c}) \\ \vdots & & \\ g_{n1}(\mathbf{e}, \mathbf{c}) & \cdots & g_{nm}(\mathbf{e}, \mathbf{c}) \end{bmatrix} \tag{8.1.5}$$

Now assume that by some appropriate method (such as the methods given in Section 8.1.1, we have approximated the various system kernel components by representative constants, as follows:

$$g_{ij}(\mathbf{e}, \mathbf{c}) = G_{ij} \tag{8.1.6}$$

where the approximate system kernel G is a constant matrix defined by the matrix array of G_{ij}'s, each of which is a real constant.

Assume further that relative weighting factors, $W(c_i)$, for each input c_i and weighting factors, $W(e_j)$, for each output e_j have been determined and that each of these weighting factors is a positive, real constant. The numerical value for the weighting factors could, for example, be determined by the maximum absolute value or mean absolute value anticipated for the range of values of interest for input and output variables for a given system model.

Let us now form the normalized weighting matrix, defined as follows:

$$W_{n\,m} = \frac{1}{S} \begin{bmatrix} w(c_1)\,w(e_1)\,|G_{11}| & w(c_2)\,w(e_1)\,|G_{12}| \cdots & w(c_m)\,w(e_1)\,|G_{1m}| \\ \vdots & \vdots & \vdots \\ w(c_1)\,w(e_n)\,|G_{n1}| & w(c_2)\,w(e_n)\,|G_{n2}| \cdots & w(c_m)\,w(e_n)\,|G_{nm}| \end{bmatrix} \tag{8.1.7}$$

where for Eq. (8.1.7)

$$S = \sum_{i,j=1}^{n,m} w(e_i)w(c_j)\left|G_{ij}\right| \tag{8.1.8}$$

Equation (8.1.8) is the effective weighted sum of each element of Eq. (8.1.7) and $|G_{ij}|$ denotes the absolute value of the i, j element in the constant system kernel given by Eq. (8.1.6).

Observe that the normalized weighting matrix has the property that the sum of all its individual elements is unity, since

$$\frac{1}{S} \sum_{i,j=1}^{n,m} w(e_i)\,w(c_j)\left|G_{ij}\right| = \frac{S}{S} = 1$$

Thus, each element of $W_{n\,m}\{$ i. e., $w(e_i)\,w(c_j)\,|G_{i\,j}|\,\}$ represents the normalized weighted value of the corresponding absolute value of the element $|G_{ij}|$ in the system kernel. By forming partial sums for each row and column, it is possible to order and quantitatively select a reduced system matrix. To see how this is accomplished, consider the following augmented weighting matrix, where

$$b_{ij} = w\,(c_i)\,w\,(e_j)\,\left|G_{ij}\right|$$

An augmented matrix is the modification of a given matrix by the addition of an additional column and/or row. The augmented matrix for Eq. (8.1.7) is defined to be

$$W_{n+1,m+1} = \begin{bmatrix} b_{11} & b_{12} & \cdots & b_{1m} & B_{r1} \\ b_{21} & b_{22} & \cdots & b_{2m} & B_{r2} \\ \vdots & & & & \\ b_{nm} & b_{nm} & \cdots & b_{nm} & B_{rn} \\ \hline B_{c1} & B_{c2} & \cdots & B_{cm} & S \end{bmatrix}$$

where

$$B_{cj} = \sum_{i=1}^{n} b_{ij} \qquad (8.1.9)$$

Equation (8.1.9) represents the column sum over all rows for the jth column and

$$B_{rj} = \sum_{j=1}^{m} b_{ij} \qquad (8.1.10)$$

represents the row sum over all columns for the ith row. The effective sum value S has been previously defined by Eq. (8.1.8). Now let us exchange each row in the augmented matrix $W_{n+1,\,m+1}$ as necessary with neighboring rows so that the row sums, B_{ri}, are arranged in decreasing magnitude from top to bottom. Thus, the last column of the augmented matrix decreases in value as we descend the partial sum column of the B_{ri}'s. It is important to recognize that exchanging these rows does not affect the column sums, nor does exchanging the columns affect the row sums. Indeed, we are simply rearranging the original sequencing of the system Eq. (8.1.5). This process is then repeated for the columns, and each column in the augmented matrix is exchanged with neighboring columns as necessary so that the column sums, B_{cj}, are arranged in decreasing magnitude from left to right. Thus, the last row of the augmented matrix decreases in value as we move from left to right along the partial sum row of the B_c's. Now the reordered partial sum row, the B_r's, and the partial sum column, the B_{cj}'s, are deleted, and a reordered weighting matrix, defined as RW_{nm}, of size $n \times m$ is obtained for $W_{n\,m}$, which has the property that the elements of greatest numerical value and thus greatest importance are concentrated in the upper left corner of the RW_{nm}, and the significance or numerical value of the elements decreases as we move down and to the right in the RW_{nm} matrix array.

Let us reinforce the details of this ordering and reduction procedure with an example. Consider a system equation with four inputs and three outputs, as follows:

$$\begin{bmatrix} de_1 \\ de_2 \\ de_3 \end{bmatrix} = \begin{bmatrix} g_{11}(e,c) & g_{12}(e,c) & g_{13}(e,c) & g_{14}(e,c) \\ g_{21}(e,c) & g_{22}(e,c) & g_{23}(e,c) & g_{24}(e,c) \\ g_{31}(e,c) & g_{32}(e,c) & g_{33}(e,c) & g_{34}(e,c) \end{bmatrix} \begin{bmatrix} dc_1 \\ dc_2 \\ dc_3 \\ dc_4 \end{bmatrix}$$

Now assume that we have approximated the variable matrix kernel $g(\mathbf{e}, \mathbf{c})$ by some appropriate method such as that given in Section 8.1.1. Assume that $g(\mathbf{e}, \mathbf{c})$ can be approximated as follows:

$$g(e,c) = \begin{bmatrix} 6 & 4 & 3 & -8 \\ 4 & -9 & 7 & 5 \\ -7 & 2 & -1 & 8 \end{bmatrix} = G \qquad (8.1.11)$$

where we have designated this (3×4) constant matrix kernel as G. Suppose the weighting factors determined for the inputs and outputs are as follows: For the input weighting factor $w(\mathbf{c})$, we have the diagonal matrix

$$w(\mathbf{c}) = D \begin{bmatrix} w(c_1) \\ w(c_2) \\ w(c_3) \\ w(c_4) \end{bmatrix} = \begin{bmatrix} 1 & 0 & 0 & 0 \\ 0 & 2 & 0 & 0 \\ 0 & 0 & 1 & 0 \\ 0 & 0 & 0 & 2 \end{bmatrix}$$

And for the output weighting factor $w(\mathbf{e})$ we have the diagonal matrix

$$w(\mathbf{e}) = D\left[w(e_1)\ w(e_2)\ w(e_3) \right] = \begin{bmatrix} 1 & 0 & 0 \\ 0 & 1 & 0 \\ 0 & 0 & 2 \end{bmatrix}$$

If we take the absolute values for each of the elements of the system kernel G and post-multiply each row by the input weighting factors $w(\mathbf{c})$, we have

$$\begin{bmatrix} 6 & 4 & 3 & 8 \\ 4 & 9 & 7 & 5 \\ 7 & 2 & 1 & 8 \end{bmatrix} \begin{bmatrix} 1 & 0 & 0 & 0 \\ 0 & 2 & 0 & 0 \\ 0 & 0 & 1 & 0 \\ 0 & 0 & 0 & 2 \end{bmatrix} = \begin{bmatrix} 6 & 8 & 3 & 16 \\ 4 & 18 & 7 & 10 \\ 7 & 4 & 1 & 16 \end{bmatrix}$$

And if we pre-multiply each column of the matrix in this equation by the output weighting factors $w(\mathbf{e})$, then

$$\begin{bmatrix} 1 & 0 & 0 \\ 0 & 1 & 0 \\ 0 & 0 & 2 \end{bmatrix} \begin{bmatrix} 6 & 8 & 3 & 16 \\ 4 & 18 & 7 & 10 \\ 7 & 4 & 1 & 16 \end{bmatrix} = \begin{bmatrix} 6 & 8 & 3 & 16 \\ 4 & 18 & 7 & 10 \\ 14 & 8 & 2 & 32 \end{bmatrix} \tag{8.1.12}$$

We now form the augmented matrix, which contains the partial row sums and column sums for matrix Eq. (8.1.12). So

$$W_{34} = \frac{1}{S(128)} \left[\begin{array}{cccc|c} 6 & 8 & 3 & 16 & B_{r1} = 33 \\ 4 & 18 & 7 & 10 & B_{r2} = 39 \\ 14 & 8 & 2 & 32 & B_{r3} = 56 \\ \hline B_{c1} = 24 & B_{c2} = 34 & B_{c3} = 12 & B_{c1} = 58 & S = 128 \end{array} \right]$$

where $S\,(=128)$ is the total absolute value sum of all the elements in the matrix. Reordering the rows to provide decreasing row sums from top to bottom, we have

$$\left[\begin{array}{cccc|c} 14 & 8 & 2 & 32 & B_{r3} = 56 \\ 4 & 18 & 7 & 10 & B_{r2} = 39 \\ 6 & 8 & 3 & 16 & B_{r1} = 33 \\ \hline B_{c1} = 24 & B_{c2} = 34 & B_{c3} = 12 & B_{c1} = 58 & S = 128 \end{array} \right]$$

Now we reorder the column to provide decreasing column sums from left to right:

$$\left[\begin{array}{cccc|c} 32 & 8 & 14 & 2 & B_{r3} = 56 \\ 10 & 18 & 4 & 7 & B_{r2} = 39 \\ 16 & 8 & 6 & 3 & B_{r1} = 33 \\ \hline B_{c1} = 58 & B_{c2} = 34 & B_{c3} = 24 & B_{c1} = 12 & S = 128 \end{array} \right]$$

Observe that the reordering procedure for the system matrix kernel has resulted in the reordering of the rows into the sequence e_3, e_2, e_1, for the output variables and reordering of the column into the sequence c_4, c_2, c_1, c_3 for the input variables. Now we can recollect the reordering for our system matrix kernel by replacing the corresponding row partial sum designator B_{ri} by the corresponding input changes, dc_i, and the column partial sum designator, B_{cj}, by the corresponding output change, de_j. Also, let us divide each element in the matrix by the total sum S to provide a normalized array correct to two significant figures for each element, as follows:

$$
\begin{bmatrix} d\,e_3 \\ d\,e_2 \\ d\,e_1 \end{bmatrix} = S \begin{bmatrix} 0.25 & 0.06 & -0.11 & -0.02 \\ 0.08 & 0.14 & 0.03 & 0.05 \\ -0.13 & 0.06 & 0.05 & 0.02 \end{bmatrix} \begin{bmatrix} dc_4 \\ dc_2 \\ dc_1 \\ dc_3 \end{bmatrix}
$$

This modified, weighted system matrix equation provides us with a quantitative measure given by the modified normalized system kernel for the relative contribution of each input and output.

We observe that the contribution to the change of output e_3 due to a change in the input c_4 accounts for 25% of the weighted, normalized constant system kernel's response. However, the contribution of a change in e_1 or e_3 due to a change in c_3 accounts for only 2% of the same system kernel's response.

To demonstrate how this reorder system kernel can be employed to provide a general quantitative method for deciding on how to achieve system reduction, let us define the reorder system matrix kernel as a matrix RW, where

$$
RW = \begin{bmatrix} b_{11} & b_{12} & b_{13} & b_{14} \\ b_{21} & b_{22} & b_{23} & b_{24} \\ b_{31} & b_{32} & b_{33} & b_{34} \end{bmatrix} = \begin{bmatrix} 0.25 & 0.06 & 0.11 & 0.02 \\ 0.08 & 0.14 & 0.03 & 0.05 \\ 0.13 & 0.06 & 0.05 & 0.02 \end{bmatrix}
$$

Now let us form the accumulation matrix A from RW, which is defined as

$$
A = \begin{bmatrix} b_{11} & \cdots & (b_{11} + b_{12} + \cdots + b_{1n}) \\ (b_{11} + b_{21}) & & \\ \vdots & & \\ (b_{11} + b_{21} + \cdots + b_{m1}) & \cdots & (b_{11} + b_{12} + \cdots + b_{mn}) \end{bmatrix}
$$

where

$$
A_{ij} = \sum_{l=0}^{j-1} \sum_{k=0}^{j-1} b_{i-j,k-l}
$$

Thus, the element A_{ij} represents the sum of all the b_{ij} elements in the upper-left rectangle, which contains b_{ij} as the lower-right corner element of the rectangular array. So, for our example, the accumulation matrix A is given by

$$
A = \begin{bmatrix} 0.25 & 0.31 & 0.42 & 0.44 \\ 0.33 & 0.53 & 0.67 & 0.74 \\ 0.46 & 0.72 & 0.91 & 1.00 \end{bmatrix} \tag{8.1.13}
$$

Thus, the value of any given element of the accumulation matrix, say A_{ij}, is the fraction of the system response provided by the reduced system equation if the system were truncated at the ith row and the jth column of the reordered system kernel matrix. Then the reduced system equation would have i outputs and j inputs. Note that the last element (i.e., the bottom-right sum term) of the accumulation matrix must always equal unity.

Observe that in Eq. (8.1.13), if we were satisfied with at least 50% of the system's response, we could truncate the system to two rows and two columns, where $A_{22} = 0.53$, and the reduced equation for the reordered system kernel would be, from Eq. (8.1.13)

$$
\begin{bmatrix} de_3 \\ de_2 \end{bmatrix} = \begin{bmatrix} g_{34} & g_{32} \\ g_{24} & g_{22} \end{bmatrix} \begin{bmatrix} dc_4 \\ dc_2 \end{bmatrix}
$$

Perhaps we wish to truncate the accumulation matrix at the element for which we have the greatest accumulation rate. This can be determined by constructing an accumulation rate matrix A_r, with a uniform element change.

For an $(n \times m)$ matrix, a uniform accumulation rate matrix A_r is given by

$$
A_r = \begin{bmatrix} \dfrac{1}{n\,m} & \dfrac{2}{n\,m} & \cdots & \dfrac{m}{n\,m} \\ \dfrac{2}{n\,m} & \dfrac{4}{n\,m} & & \dfrac{2m}{n\,m} \\ \vdots & & & \\ \dfrac{n}{n\,m} & \dfrac{2n}{n\,m} & \cdots & \dfrac{n\,m}{n\,m} \end{bmatrix} \tag{8.1.14}
$$

For a (3×4) matrix, using Eq. (8.1.14) for our example reordered system kernel, the corresponding uniform accumulation matrix is then given by

$$
A_r = \begin{bmatrix} 0.08 & 0.17 & 0.25 & 0.33 \\ 0.17 & 0.33 & 0.50 & 0.67 \\ 0.25 & 0.50 & 0.75 & 1.00 \end{bmatrix} \tag{8.1.15}
$$

A comparative evaluation of the accumulation rate for a given reordered system kernel can be realized by determining the difference between the corresponding accumulation rate matrix coefficients and the uniform accumulation rate matrix coefficients. Thus, from Eqs. (8.1.13) and (8.1.15), we can show the following:

$$
A - A_r = \begin{bmatrix} 0.17 & 0.14 & 0.17 & 0.11 \\ 0.16 & 0.20 & 0.17 & 0.07 \\ 0.21 & 0.22 & 0.16 & 0.00 \end{bmatrix} \tag{8.1.16}
$$

The element of greatest absolute value, 0.22, indicates that the best choice for this system reduction technique is to truncate the reordered system kernel at the third row and second column. Thus, based on accumulation rate, the best reduced system kernel associated with the reordered system kernel determined by Eq. (8.1.16) is as follows:

$$
\begin{bmatrix} de_3 \\ de_2 \\ de_1 \end{bmatrix} = \begin{bmatrix} 8 & 2 \\ 5 & -9 \\ -8 & 4 \end{bmatrix} \begin{bmatrix} dc_4 \\ dc_2 \end{bmatrix}
$$

and this reduced system in terms of e_1, e_2, e_3, c_2 and c_4 provides approximately 72% of the response of the original system, which had four inputs and three outputs. If appropriate numerical values had been assigned to the original system kernel elements, g_{ij}, to permit ordering by this technique, it would be appropriate then to return to the original elements and then attempt an analytical solution. Thus, for example, we might attempt then to solve the following reduced original system equation:

$$\begin{bmatrix} de_3 \\ de_2 \\ de_1 \end{bmatrix} = \begin{bmatrix} g_{34} & g_{32} \\ g_{24} & g_{22} \\ g_{14} & g_{12} \end{bmatrix} \begin{bmatrix} dc_4 \\ dc_2 \end{bmatrix}$$

Other similar techniques are available for comparing and reducing the system kernel. Furthermore, the system investigator may simply wish to restrict analysis efforts for the system equation to a specified matrix size [e.g., (2×2)]. These comparative techniques can be used to assess the relative error resulting from such a decision.

8.2 System Normalization and Parameter Reduction

After the vital task of formulating the system kernel and quantifying the system equation is done, the system investigator begins the challenging task of solving the system equation. This is usually accomplished by analytical methods or computer analysis. In either situation, it is often desirable to normalize the system's input and output variables and minimize the number of system equation parameters for reasons that include the following:

- Simplification of the system equation for analytic analysis and adoption for specific analysis methods, such as linearization and other approximation methods.
- Preparation of the system equation for computer analysis by variable normalization and reduction of the number of parameters that must be subjected to parametric studies.
- Assistance in quantitative modeling and analysis by placing variables into a similar range of variation.
- Improved insight into the behavior and importance of each variable and kernel component and efficiency in the reduction of the number of system variables as necessary.

8.2.1 System Variable Normalization

In initial attempts to model a rational system, we commonly seek to identify and quantify the familiar and hopefully most significant inputs and outputs associated with the system. Since this is generally a difficult but very important task in accurate modeling, any method of or insight into greater system awareness and understanding that assists us in better representing the significant cause and effect relationships should be employed. At this stage of the modeling effort for the given system, the investigator may and should ignore efforts to combine, synthesize, or transform system variables, since this action may result in the loss or oversight of some significant system variables, parameters, or other data source. It is usually advisable to be as complete and accurate in setting up the initial quantitative model as is possible and to delay, until this stage of the model study, efforts to normalize and transform the system variables to enhance mathematical conciseness and efficiency of the system equations. In other words, common wisdom and experience in system studies imply that system studies are generally most effectively accomplished in a series of small steps that entail rather specific modeling operations. We will consider in this section the step of system normalization and transformation that is generally useful in making the system equations more efficient and useful for subsequent system studies. However, this step of normalization and transformation should not be imposed, in general, on the system equation until the investigator believes that the system equations are reasonably complete and adequate for the system study requirements. The basis for this caution is that the mathematical transformations that will be imposed on the system equation will often tend to combine and rearrange variables and parameters into forms that will not be as familiar or recognizable to the system investigator, although the influence of these variables and parameters on the system equation will not be altered.

To demonstrate the typical normalization and transformation procedures that can be imposed on system equations, consider the general multiple-input, multiple-output continuous system equation with feedback

$$d\mathbf{e} = [I - g(\mathbf{e}, \mathbf{c}) f(\mathbf{e}, \mathbf{c})]^{-1} g(\mathbf{e}, \mathbf{c}) \, d\mathbf{c} \tag{8.2.1}$$

where $[I - g(\mathbf{e}, \mathbf{c}) f(\mathbf{e}, \mathbf{c})]^{-1}$ is the inverse ($n \times n$) matrix that accounts for feedback arising from the feedback kernel, $f(\mathbf{e}, \mathbf{c})$ is a ($m \times n$) matrix. Also, $g(\mathbf{e}, \mathbf{c})$ is the system kernel without feedback, a ($m \times n$) matrix, $d\mathbf{e}$ is the ($n \times 1$) differential output matrix, and $d\mathbf{c}$ is the ($m \times 1$) differential input matrix.

In general, we seek solutions for system Eq. (8.2.1) for some range of the input variable $\mathbf{c}$ such that

$$\mathbf{c}(\min) < \mathbf{c} < \mathbf{c}(\max)$$

and this input variable range defines a range for the output variable $\mathbf{e}$ determined by the system's kernel (e.g., $[I - gf]^{-1}g$) such that

$$\mathbf{e}(\min) < \mathbf{e} < \mathbf{e}(\max)$$

Within these ranges for the input (or independent) variables [$\mathbf{c}(\min)$ to $\mathbf{c}(\max)$] and the output (or dependent) variables [$\mathbf{e}(\min)$ to $\mathbf{e}(\max)$], there are often certain sets of values, say $\mathbf{c}^*$ and $\mathbf{e}^*$, that are of particular interest to the system investigator. This is particularly the case for the output or dependent variables, where $\mathbf{e}^*$ may represent a value of vital concern, such as an equilibrium or quasi-stability value or an initial or final asymptotic value. Let us consider some of the important values of the variables and the procedures by which we may accomplish variable normalization.

Table 8-2 provides a sample tabulation of critical values for the input and output variables that may be of interest for a given system. Depending on the nature of the critical value and its effect on the system equation, the output response of the system may vary from zero to infinity.

Table 8-3 provides a sample of possible variable normalization definitions that are useful in modifying system equations to improve analysis and evaluation efforts. Observe that if the multiple-input, multiple-output system equation

Table 8-2 Critical Values for System Equations

$(I - g f)\, de = g\, d\mathbf{c}$

Critical Value $(\mathbf{e}, \mathbf{c} \rightarrow \mathbf{e}^*, \mathbf{c}^*)$	System Designation for Critical Value	Consequences for System Equation
$g(\mathbf{e}^*, \mathbf{c}^*) = 0$	Zero gain (steady-state if $\mathbf{c}$ is time or zero gradient if $\mathbf{c}$ is position)	$d\mathbf{e} = 0$ or $d\mathbf{c} \rightarrow \pm \infty$
$g(\mathbf{e}^*, \mathbf{c}^*)\, f(\mathbf{e}^*, \mathbf{c}^*) \rightarrow 1$	Infinite gain from feedback	$d\mathbf{c} = 0$ or $d\mathbf{e} \rightarrow \pm \infty$
$g(\mathbf{e}^*, \mathbf{c}^*) \rightarrow \pm \infty$	Mixed gain	$d\mathbf{c} = 0$ or $[I - gf]\, d\mathbf{e} \rightarrow \pm \infty$
$g(\mathbf{e}^*, \mathbf{c}^*)\, f(\mathbf{e}^*, \mathbf{c}^*) \rightarrow \pm \infty$	Mixed gain	$d\mathbf{e} = 0$ or $g\, d\mathbf{c} \rightarrow \pm \infty$
$g(\mathbf{e}^*, \mathbf{c}^*) = G$ where $0 < |G| < \infty$	Specific gain	Various
$f(\mathbf{e}^*, \mathbf{c}^*) = |G|$ where $0 < |G| < \infty$	Specific gain	Various

Table 8-3 Various Variable Normalization Definitions

Variable Transformation	Designation	Range
$y = \dfrac{e - e\,(\min)}{e\,(\max) - e\,(\min)}$	Fractional difference ratio	$0 < y < 1$
$y = e/e^*$ (for $e^* \neq 0$)	Simple ratio	$e(\min)/e^* < y < e(\max)/e^*$
$y = e - e(\min)$ or $e = e - e^*$	Simple difference	$0 < s < e(\max) - e(\min)$ $e(\min) - e^* < y < e(\max) - e^*$
$\delta y = (e - e^*)/e^*$ (for $e^* \neq 0$)	Perturbation	$[e(\min) - e^*]/e^* < \delta y < [e(\max) - e^*]/e^*$
$y = ae/e^* + b$	General linear form	$ae(\min) + b < y < ae(\max) + b$

$$d\mathbf{e} = g(\mathbf{e}, \mathbf{c})\, d\mathbf{c}$$

is transformed into the equivalent system equation

$$d\mathbf{y} = G(\mathbf{y}, \mathbf{x})\, d\mathbf{x} \qquad (8.2.2)$$

where $\mathbf{y}$ and $\mathbf{x}$ are variable transformations of the fractional difference ratio, simple ratio, perturbation, or general linear form; then both $\mathbf{y}$ and $\mathbf{x}$ are dimensionless since the dimensions carried by the original variables $\mathbf{e}$ and $\mathbf{c}$ are eliminated through normalization. Furthermore, if $\mathbf{y}$ and $\mathbf{x}$ are dimensionless in the system Eq. (8.2.2), then the normalized kernel, $G(\mathbf{y}, \mathbf{x})$, must also be dimensionless.

Example for System Variable Normalization.
Consider the following nonlinear system equation with feedback as a model for the change in the population P of a sample with time t.

$$dP = \frac{g\,(P, T)}{1 - g(P, T) f(P, T)}\, dT = g_f(P, T)\, dT \qquad (8.2.3)$$

where

$$g(P, T) = a P$$

$$f(P, T) = b (P - P_0)$$

We will assume that $P(T = 0) = P_0$, where $P_0 \neq 0$. So, we have for the overall system kernel with feedback that

$$g_f (P, T) = \frac{a P}{1 - a b P (P - P_0)} \tag{8.2.4}$$

Now let us define as normalized variables

$$p = P/P_0, \quad d P = P_0 \, d p \quad \text{(note that } d P_0 \neq 0\text{)}$$

$$t = a T, \quad d t = a \, d T$$

So, we have for our system equation in terms of the normalized variables p and t that

$$d p = \frac{p \, d t}{1 + a b P_0^2 p (1 - p)}$$

This equation is separable and can be expressed in the separated form (see Section 4.2.6) as follows:

$$d p/p + a b P_0^2 p (1 - p) \, d p = d t \tag{8.2.5}$$

Integrating Eq. (8.2.5), we have

$$\text{Ln} \, p + a b P_0^2 \, p (1 - p/2) = t + \text{Ln} \, K$$

or, equivalently,

$$p = K \exp \left[t - a b P_0^2 p (1 - p/2) \right] \tag{8.2.6}$$

where K is the integration constant. If we impose on Eq. (8.2.6) the initial condition that $P(t = 0) = P_0$ or, equivalently, $p(t = 0) = 1$, then

$$p = \exp \left[t + \lambda (p - 1)^2 \right] \tag{8.2.7}$$

where

$$\lambda = (1/2) \, a b P_0^2$$

We observe in this example that a single parameter λ controls the response of the normalized system solution given by Eq. (8.2.7). A plot of Eq. (8.2.7) for various values of the parameter λ, shown in Figure 8-1, indicates that generalized responses the system can yield as the system parameter varies.

The case when $\lambda = 0$ (which results in no feedback) indicates typical exponential growth of the sample, while for the case when λ becomes arbitrarily large and negative, the system maintains the initial population without change in time. We observe that negative feedback (i.e., $a \, b < 0$) is essential in this model to bound the population growth for the system described by Eq. (8.2.3). However, the objective of this example is to demonstrate that

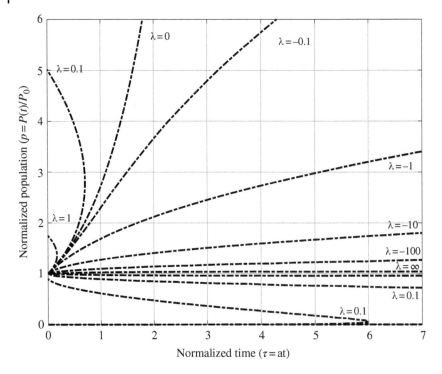

Fig. 8-1 General model for population change with time.

the system model defined by Eq. (8.2.3), which contained three apparently independent parameters (i.e., P_0, a, b), there is only one independent parameter

$$\lambda = (1/2)a\,b\,P_0^2$$

that defines and controls the nature of the system's behavior. Thus, the response of the system is invariant with dimensionless time, t, if $a\,b\,P_0^2$ remains constant

8.2.2 Parameter Reduction and Minimization

In the initial formulation and quantification stages of modeling a system, the system investigator often tends to generously endow the system kernel with many parameters (i.e., constants that are not functions of the inputs or outputs but do change as the properties or characteristics of the system are varied) so that the model will satisfy or fit the observed or anticipated behavior of the system. However, after a reasonably satisfactory system kernel has been modeled, the issue then arises as to the minimum number of parameters required that still uniquely define the system's kernel. The answer to this question is very useful for several reasons.

- For every system model, there exists a certain minimum set of parameters that, together with the quantitative form of the system kernel, define and control the essential behavior of the system model.
- Reduction of a system model to its minimum parameter set assists in the presentation of the system data and graphical analysis in a generalized or universal format.

- Comparable systems that have similar quantitative kernel formats may appear unrelated because of the occurrence of different parameter sets. Reduction of these parameter sets to minimum sets will generally reveal the similarity in these system models and permit useful inter-comparisons and insight to be made about otherwise apparently unrelated systems.
- The reduction of the system kernel model to a minimum parameter set will often improve the efficiency of the solution and evaluation efforts.
- Finally, the casting of a system model into an equivalent minimum parameter set model can provide basic insight and greater understanding of the fundamental nature of the model and its associated rational system.

We will limit our presentation on parameter reduction and minimization to the use of linear transformation techniques because of their simplicity and broad application. Nonlinear transformations and some heuristic methods can also be useful in parameter reduction and kernel simplification, but they require careful judgment and considerable skill. Such methods are most appropriately learned through individual efforts and broad experience in system science.

Linear transformation methods require that we transform all system input and output variables as linear functions of new variables. For example, consider the general multiple-input, multiple-output system equation

$$d\mathbf{e} = g(\mathbf{e}, \mathbf{c})\, d\mathbf{c} \tag{8.2.8}$$

where the multiple output $\mathbf{e}$ is, as usual, an $(n \times 1)$ column matrix, the multiple input $\mathbf{c}$ is an $(m \times 1)$ column matrix, and $g(\mathbf{e}, \mathbf{c})$ is an $(n \times m)$ matrix kernel. The linear transform we will impose upon the inputs and outputs is

$$\mathbf{c} = P\mathbf{u} + Q \quad \text{so} \quad d\mathbf{c} = P d\mathbf{u} \tag{8.2.9}$$

and

$$\mathbf{e} = R\mathbf{v} + S \quad \text{so} \quad d\mathbf{e} = R d\mathbf{v} \tag{8.2.10}$$

where

$$P = \begin{bmatrix} p_{11} & \cdots & p_{1m} \\ \vdots & & \\ p_{m1} & \cdots & p_{mm} \end{bmatrix}, \quad \det|P| \neq 0, \quad R = \begin{bmatrix} r_{11} & \cdots & r_{1m} \\ \vdots & & \\ r_{n1} & \cdots & r_{nm} \end{bmatrix}, \quad \det|R| \neq 0,$$

$$Q = \begin{bmatrix} q_1 \\ \vdots \\ q_m \end{bmatrix} \quad S = \begin{bmatrix} s_1 \\ \vdots \\ s_n \end{bmatrix}$$

and the new variable set is $\mathbf{u}$ and $\mathbf{v}$

$$\mathbf{u} = \begin{bmatrix} u_1 \\ \vdots \\ u_m \end{bmatrix} \quad \text{and} \quad \mathbf{v} = \begin{bmatrix} v_1 \\ \vdots \\ v_n \end{bmatrix}$$

The elements of the matrices P, Q, R, and S are arbitrary constants, which will be assigned to minimize the system model parameters. However, we must recognize that the determinants of P or R cannot vanish; otherwise, we will have the result that components of the inputs $\mathbf{c}$ or outputs $\mathbf{e}$ will similarly vanish and fail to constitute a linearly independent variable.

If we substitute the transformed input and output variables as defined by Eqs. (8.2.9) and (8.2.10) into the system Eq. (8.2.8), we have

$$R\,d\mathbf{v} = g(R\mathbf{v} + S, P\mathbf{u} + Q)\,P\,d\mathbf{u} \tag{8.2.11}$$

Let us now define the transformed system kernel, which may be expressed as a function of the transformed input and output variables $\mathbf{c}$ and $\mathbf{e}$, respectively, as (note that R^{-1} exists if the determinant of $R \neq 0$)

$$G\,(\mathbf{u}, \mathbf{v}) = R^{-1}\,g(R\mathbf{u} + S, P\mathbf{v} + Q)\,P \tag{8.2.12}$$

Then system Eq. (8.2.11) becomes

$$d\mathbf{v} = G(\mathbf{v}, \mathbf{u})\,d\mathbf{u} \tag{8.2.13}$$

which is the transformed system equation related to system Eq. (8.2.8) through the transformations given by Eqs. (8.2.9), (8.2.10), and (8.2.12). Our objective now is to determine a consistent set of matrix constants for P, Q, R, and S such that the number of essential parameters contained in Eq. (8.2.13) is a minimum or at least reduced as much as practical.

To demonstrate the parameter reduction technique, let us begin with a single-input, single-output system that has the following nonlinear kernel (observe, however, that the system equation is linear for c as the dependent variable):

$$de = \frac{A + Be}{Dc + e}\,dc \tag{8.2.14}$$

We observe that there are three parameters, A, B, and D, that define this system equation. The variable transformations defined by Eqs. (8.2.9) and (8.2.10) reduce to the following scalar equations:

$$c = pu + q \quad \text{and} \quad dc = p\,du$$
$$e = rv + s \quad \text{and} \quad de = r\,dv$$

Substitution of these transforms into Eq. (8.2.14) provide

$$dv = G(v, u)\,du \tag{8.2.15}$$

where

$$G(v, u) = \left(\frac{p}{r}\right) \frac{A + Bs + Brv}{Dq + s + DPu + rv} \tag{8.2.16}$$

We are free to choose the transform parameters, p, q, r, and s, as we wish, but we are constrained not to let $p = 0$ or $r = 0$; otherwise, our kernel $G(\mathbf{v}, \mathbf{u})$ vanishes or is undefined. The application of linear transform techniques to nonlinear systems such as Eq. (8.2.14) is somewhat of an art, and various approaches can be employed. In the case of system Eq. (8.2.15), our major consideration, besides parameter reduction, is, if possible, to express the kernel given by Eq. (8.2.16) in a form that permits analytical solution. In this case, we observe that if the terms $A + Bs$ in the numerator and $Dq + s$ in the denominator both vanish, then the kernel becomes homogeneous, as defined in Section 4.2.7. So, we will require that $A + Bs = 0$, and we have that

$$s = -A/B$$

and, similarly, that $D\,q + s = 0$, so

$$q = -s/D = A/(BD)$$

For the other transform parameters, let us set $r = D$ and $p = 1$. Then the system kernel, Eq. (8.2.16), becomes

$$G(v, u) = \lambda v/(v + u) \tag{8.2.17}$$

where the single parameter $\lambda = B/D$ constitutes the minimum parameter set.

The general solution for system Eq. (8.2.15) with the kernel given by Eq. (8.2.17) is found by using the method for homogeneous equations outlined in Section 4.2.7. The system equation solution is found to be

$$u = \left[K v^{\frac{1}{\lambda - 1}} + \frac{1}{\lambda - 1} \right] v \tag{8.2.18}$$

where K is the integration constant. Observe that two constants, $\lambda = B/D$ and the integration constant K, uniquely define the characteristics of system Eq. (8.2.15). It is interesting to express the system solution, Eq. (8.2.18), for the transformed system Eq. (8.2.15), in terms of the original system equation variables as defined by Eq. (8.2.14). The result is

$$c = \frac{A}{BD} + \left[K \frac{e}{D} + \frac{A}{BD} \right]^{(D-B)/B} + \frac{D}{B-D} \left(\frac{e}{D} + \frac{A}{BD} \right) \tag{8.2.19}$$

Observe that, whereas in Eq. (8.2.19) for the system solution, we apparently have three independent system parameters, besides the constant of integration K, which appears in both Eqs. (8.2.18) and (8.2.19), in the transformed system solution given by Eq. (8.2.18), we have reduced the system parameters from three to one, where $\lambda = B/D$. Thus, the original system equation, as defined by Eq. (8.2.14), can be completely characterized by only one essential parameter.

Establishing the reduced or minimum parameter set for nonlinear kernels requires a variety of techniques that depend on the form of the kernel. For linear system kernels, the technique is specific and straightforward for parameter minimization, as we will now demonstrate.

To demonstrate parameter reduction for a multiple-input, multiple-output linear system, consider the single-input, double-output system equation that has a linear, constant-coefficient kernel as follows:

$$de = (A + Be)dc \tag{8.2.20}$$

where

$$\mathbf{e} = \begin{bmatrix} e_1 \\ e_2 \end{bmatrix}, \quad A = \begin{bmatrix} a_1 \\ a_2 \end{bmatrix}, \quad B = \begin{bmatrix} b_{11} & b_{12} \\ b_{21} & b_{22} \end{bmatrix}, \quad \det|B| \neq 0$$

We observe that this system kernel has six parameters, a_1, a_2, b_{11}, b_{12}, b_{21}, and b_{22}, which control the response of the single-input, double-output linear system equation with constant coefficients. Imposition of the linear variable transformations as prescribed by Eqs. (8.2.9) and (8.2.10) gives

$$c = p\,u \quad \text{so} \quad dc = p\,du \ (p \neq 0)$$

and

$$\mathbf{e} = R\mathbf{v} + S \quad \text{so} \quad d\mathbf{e} = R\,d\mathbf{v}, \quad \text{where} \quad \det|R| \neq 0$$

With these transformations substituted into Eq. (8.2.20), we have

$$R\,d\mathbf{v} = [A + B(R\mathbf{v} + S)]p\,du \tag{8.2.21}$$

We require that determinant $R \neq 0$ so that we are assured that R^{-1} exists. Pre-multiplying Eq. (8.2.21) by R^{-1}, we have

$$\frac{d\mathbf{v}}{du} = pR^{-1}(A + BS) + pR^{-1}BR\mathbf{v} \tag{8.2.22}$$

The greatest simplification in Eq. (8.2.22) results if we require that

$$PR^{-1}(A + BS) = 0 \tag{8.2.23}$$

which implies that $S = -B^{-1}A$. Also, we choose R such that

$$pR^{-1}BR = pD = p\begin{bmatrix} \lambda_1 & 0 \\ 0 & \lambda_2 \end{bmatrix} \tag{8.2.24}$$

where λ_1 and λ_2 are the eigenvalues of the matrix B found from

$$\det |B - I| = 0$$

which in expanded form is

$$\lambda^2 - \lambda\,(b_{11} + b_{22}) + b_{11}\,b_{22} - b_{12}\,b_{21} = 0 \tag{8.2.25}$$

Equation (8.2.25) is easily solved for the eigenvalues λ_1 and λ_2 to give

$$\left.\begin{matrix} \lambda_1 \\ \lambda_2 \end{matrix}\right\} = \frac{b_{11} + b_{22}}{2} \pm \frac{1}{2}\sqrt{b_{11}{}^2 - 2\,b_{11}\,b_{22} + 4\,b_{12}\,b_{21} + b_{22}{}^2}$$

With Eqs. (8.2.23) and (8.2.24) defining our variable transformations, the solution for the transformed system Eq. (8.2.22), is found to be

$$v_1 = v_{10}\,\exp(u)$$

$$v_2 = v_{20}\,\exp(\lambda_2 u / \lambda_1)$$

where v_{10} and v_{20} are integration constants, and we have chosen $p = 1/\lambda_1$. Observe that our set of six parameters provided in Eq. (8.2.20) have been reduced to two essential parameters (i.e., λ_1 and λ_2) together with initial conditions for our transformed solutions, v_1 and v_2.

Table 8-4 presents a sampling of general solutions for full parameter single-input, single-output system equations together with their associated minimum parameter system equation, a particular solution, and its graphical depiction. Observe that the technique for

Table 8-4 Minimum Parameter System Equations

General Form/Solution	Reduced Form/Solution	Graph of Reduced Solution
$dy = a\,dx$ $y = ax + K$	$ds = dt$ $s = t$	
$dy = (2ax + b)\,dx$ $y = ax^2 + bx + K$	$ds = 2t\,dt$ $s = t^2$	
$dy = (3ax^2 + 2bx + c)\,dx$ $y = ax^3 + bx^2 + cx + K$	$ds = (3t^2 + 2t)\,dt$ $s = t^2\,(t + 1)$	
$dy = (4ax^3 + 3bx^2 + 2cx + e)\,dx$ $y = ax^4 + bx^3 + cx^2 + ex + K$	$ds = (4t^3 + 3t^2 + 2pt)\,dt$ $s = t^2(t^2 + t + p)$	$(p=0)$
$dy = \left[\dfrac{a}{2y + b}\right] dx$ $y^2 + by = ax + K$	$ds = \dfrac{1}{2\sqrt{t}}\,dt$ $s^2 = t$	
$ds = \dfrac{(t + 1/2)}{\sqrt{t(t + 1)}}\,dt$ $s^2 = t(t + 1)$	$ds = \dfrac{(t + 1/2)}{\sqrt{t(t + 1)}}\,dt$ $s^2 = t(t + 1)$	
$dy = \left[\dfrac{3ax^2 + 2bx + c}{2y + c}\right] dx$ $y^2 + ey = ax^3 + bx^2 + cx + K$	$ds = \dfrac{t\,(3t + 2)}{2\sqrt{t^2\,(t + 1)} + p}$ $s^2 = t^2(t + 1) + p$	$(p=0)$

parameter reduction and minimization using linear variables transforms is useful for simplifying both linear and nonlinear system equations and in characterizing the behavior of their general solutions. Table 8-4 presents the system equation solution in its simplest reduced form, which is the system solution that is usually sought.

8.2.3 Sensitivity Analysis of System Parameters

The system equation used to quantitatively model a typical system is usually endowed by the system investigator with a least one or more somewhat arbitrary constants or parameters. These parameters are subsequently assigned values to uniquely specify the system under study. However, although the parameters may vary for a particular model characterization, the parameters are usually assumed to remain constant over some appropriate range of variation of the system's inputs and outputs.

Generally, as we have learned in Section 8.2.2, these parameter sets for a given system equation can be reduced in number by appropriate combinations to smaller and even minimum, essential parameter sets. Furthermore, for both the original and reduced parameter sets, it is apparent that some parameters have greater significance and influence upon the system's behavior and characteristics than other parameters contained in the system equation. The quantitative assessment of individual system parameter influence upon system outputs is the basis and goal of system sensitivity analysis, which we will consider next.

Consider the multiple-input, multiple-output continuous system equation

$$d\mathbf{e} = g(\mathbf{e}, \mathbf{c}, \mathbf{p})\, d\mathbf{c} \tag{8.2.26}$$

where we assume that the system kernel g is a function of the $(m \times 1)$ input matrix $\mathbf{c}$, the $(n \times 1)$ output matrix $\mathbf{e}$, and k independent parameters represented by the $(k \times 1)$ parameter matrix $\mathbf{p}$. Now let us define the $(n\,k \times 1)$ sensitivity matrix $\mathbf{s}$ for the system parameters as

$$\mathbf{s} = \frac{\partial \mathbf{e}}{\partial \mathbf{p}} \tag{8.2.27}$$

Observe that $\mathbf{s}$ is defined as the differential change in output per unit change of the parameter for continuous systems. Thus, the sensitivity of given system output to a given parameter is measured by $\mathbf{s}$. If we form the partial derivatives of the system Eq. (8.2.26) with respect to each parameter $\mathbf{p}$, we find that

$$d\left[\frac{\partial \mathbf{e}}{\partial \mathbf{p}}\right] = \left[\frac{\partial g}{\partial \mathbf{e}} \frac{\partial \mathbf{e}}{\partial \mathbf{p}} + \frac{\partial g}{\partial \mathbf{c}} \frac{\partial \mathbf{c}}{\partial \mathbf{p}} + \frac{\partial g}{\partial \mathbf{p}}\right] d\mathbf{e} + g\, d\left[\frac{\partial \mathbf{c}}{\partial \mathbf{p}}\right] \tag{8.2.28}$$

Now we observe that the inputs $\mathbf{c}$ constitute the independent variables, which are assumed to be independent of changes in the parameters, so for each value of i and j such that $1 \le i \le m$ and $1 \le j \le k$, we have that

$$\frac{\partial c_i}{\partial p_j} = 0 \tag{8.2.29}$$

Using Eq. (8.2.27), which defines the sensitivity matrix $\mathbf{s}$ and Eq. (8.2.29) then Eq. (8.2.28) can be expressed as

$$d\mathbf{s} = \left(\frac{\partial g}{\partial \mathbf{s}}\mathbf{s} + \frac{\partial g}{\partial \mathbf{p}}\right) d\mathbf{c} = G(\mathbf{e}, \mathbf{c}, \mathbf{p}, \mathbf{s})\, d\mathbf{c} \tag{8.2.30}$$

Observe that Eq. (8.2.30) has the general form of a multiple input, $\mathbf{c}$; multiple output, $\mathbf{s}$; system equation. Equations (8.2.26) and (8.2.30) together constitute a coupled set of system equations for the $(m \times 1)$ matrix input $\mathbf{c}$, the $(n \times 1)$ matrix output $\mathbf{e}$, and the $(n\,k \times 1)$ parameter sensitivity matrix $\mathbf{s}$. In general, this coupled set of expanded system equations is very difficult to solve because of the increased number of outputs, viz., $\mathbf{e}$ and $\mathbf{s}$. However, under certain conditions, it is possible to realize approximate solutions. Now the system solution sought for Eq. (8.2.30) is the dependency of the sensitivity matrix $\mathbf{s}$ upon the system outputs $\mathbf{e}$ and the parameters $\mathbf{p}$ as

$$\mathbf{s} = \frac{\partial \mathbf{e}}{\partial \mathbf{p}} = F(\mathbf{e}, \mathbf{p})$$

Considerable simplification in sensitivity analysis occurs if the kernels g in Eq. (8.2.26) and G in Eq. (8.2.30) are not explicit functions of the inputs $\mathbf{c}$ or can be approximated as such. If this is true, then the coupled system equations are autonomous and elimination of the differential input $d\mathbf{c}$ between Eqs. (8.2.26) and (8.2.30) provides the result

$$d\mathbf{s} = \left[\frac{\partial g}{\partial \mathbf{e}} \mathbf{s} + \frac{\partial g}{\partial \mathbf{p}} \right] g^{-1}(\mathbf{e}, \mathbf{p}) \, d\mathbf{e} \tag{8.2.31}$$

Equation (8.2.31) is a linear differential equation in $\mathbf{s}$ if we hold the matrix parameter matrix $\mathbf{p}$ constant. Assuming that this approximation is justified, we can employ the methods developed in Chapters 4, 5, and 7 to obtain the general solution of Eq. (8.2.31). The solution of Eq. (8.2.31) provides a useful approximation of the sensitivity matrix $\mathbf{s}$ and its dependency upon the system outputs. Thus, the sensitivity or quantitative influence of each system parameter upon the system's output is known.

Example of Analysis for System Parameter Sensitivity.
Consider the single-input, single-output system equation

$$de = \left(\frac{e + p}{e} \right) dc \tag{8.2.32}$$

where e is the output, c is the input, and p is the system equation parameter, which may assume different values around some expectation value p_0. Now let us partially differentiate the system equation with respect to p to determine the sensitivity variable $\partial c / \partial p = 0$, so we find that

$$d\left(\frac{\partial e}{\partial p} \right) = ds = \left(\frac{e - p s}{e^2} \right) dc \tag{8.2.33}$$

recognizing that $\partial c / \partial p = 0$. Our objective is to obtain the system solution for s as a function of the output e and parameter c that satisfies Eqs. (8.2.32) and (8.2.33). Recognizing that the system kernels for Eqs. (8.2.32) and (8.2.33) are not explicit functions of the system input c, we can eliminate the differential dc between these equations to obtain the following reduced system equation for s and e

$$de = \frac{e(e + p)}{e - ps} ds \tag{8.2.34}$$

Equation (8.2.34) can be expressed as the following linear differential equation in s by setting $p - p_0$ as a fixed value

$$\frac{ds}{de} + \left(\frac{1}{e} - \frac{1}{e + p_0} \right) s = \frac{1}{e + p_0} \tag{8.2.35}$$

With determination of the integrating factor [i.e., $e/(e + p_0)$, see Section 4.2.5 for details], Eq. (8.2.35) may be easily integrated to provide the following solution for the parameter sensitivity variable

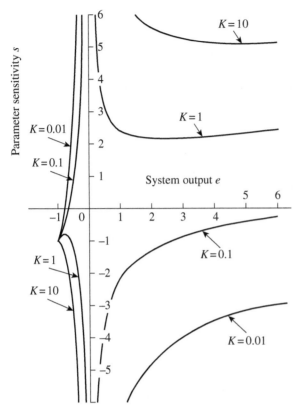

Fig. 8-2 Graph of parameter sensitivity vs. system output.

$$s = \left(1 + \frac{p_0}{e}\right) \text{Ln}\left[K\left(e/p_0 + 1\right)\right] + \frac{p_0}{e} \tag{8.2.36}$$

Figure 8-2 provides a graphical depiction of Eq. (8.2.36) for various $e/p_0 \rightarrow 0$, and this system Eq. (8.2.32) becomes increasingly sensitive to the parameter p as $e/p_0 \rightarrow 0$ and $p_0 \rightarrow \infty$

8.3 Systems with Feedback

Perhaps one of the most significant and fascinating phenomena exhibited by most systems is that of system feedback. As we recall from Chapter 2, system feedback occurs when one or more of the system's outputs modifies one or more of the inputs to the system. This process of feedback permits the system to become "goal seeking" in the sense that some operational trait, tendency, or condition is sought by or for the system through the mechanism of intrinsic feedback or by an external control agent through extrinsic feedback. In practice, the outputs from the system may be modified by either element within the system (intrinsic feedback) or by elements outside the system (extrinsic feedback via the system's environment) or a combination of to modify the subsequent input to the system to influence the system's operation.

8.3.1 System Kernel Feedback Gain

A principal aspect of system feedback is that by appropriate modification of a system output, it is possible to significantly modify (e.g., increase or decrease) the subsequent output of the system through input modification. This modification or amplification process is usually referred to as feedback gain, which may be positive, negative, or zero and of arbitrary magnitude. A quantitative description of the system gain that is possible with feedback is perhaps most easily demonstrated by consideration of the simplest system capable of exhibiting feedback, a single-input, single-output system with feedback.

Recall that the general form of the single-input, single-output system equation with feedback is

$$de = \frac{g(e,c)\,dc}{1 - g(e,c)\,f(e,c)} = g_f(e,c)\,dc \tag{8.3.1}$$

where we defined the kernel g_f with feedback as

$$g_f(e,c) = \frac{g(e,c)}{1 - g(e,c)\,f(e,c)} \tag{8.3.2}$$

When no feedback exists (i.e., $f = 0$), the system kernel is $g(e, c)$.

Since feedback amplification or gain is usually assessed without regard to sign value, let us take the absolute values of Eq. (8.3.2) so that we have

$$\left| g_f(e,c) \right| = \left| \frac{g(e,c)}{1 - g(e,c)\,f(e,c)} \right| = \frac{|g(e,c)|}{|1 - g(e,c)\,f(e,c)|} \tag{8.3.3}$$

We observe that if the feedback kernel f assumes values such that

$$|1 - g(e,c)\,f(e,c)| < 1$$

then we have from Eq. (8.3.3) that

$$\left| g_f(e,c) \right| > |g(e,c)|$$

and the system is said to exhibit positive feedback gain (or an absolute gain greater than unity), since the output response of the system with feedback present is now amplified by a factor greater than $|g(e, c)|$ (i.e., the system kernel without feedback present).

If the feedback kernel f assumes a value such that

$$|1 - g(e,c)\,f(e,c)| > 1$$

then we have from Eq. (8.3.3) that

$$\left| g_f(e,c) \right| < |g(e,c)|$$

and now the system is said to exhibit negative feedback gain (or an absolute gain less than unity), since the output response of the system with feedback present is amplified by a factor less than $|g(e, c)|$.

When the feedback kernel $f(e, c)$ satisfies the condition

$$|1 - g(e,c)\,f(e,c)| = 1$$

the system exhibits no feedback gain or a neutral feedback gain (since we have an absolute gain of unity), and we have that

$$\left| g_f(e,c) \right| = \left| g(e,c) \right|$$

A particularly important value of the feedback kernel occurs when

$$\left| 1 - g(e,c) f(e,c) \right| \to 0$$

since in this critical situation the absolute feedback gain becomes arbitrarily large, and we have that

$$\left| g_f(e,c) \right| \to \infty$$

which can occasionally be desired or sought as a system response, but generally will prove to be stressful or even destructive to the system and produce unacceptable outputs.

Figure 8-3 provides a graphical description of these various feedback gain conditions that can occur for the normalized system kernel, g_f/g, as a function of the normalized feedback kernel, g_f. Figure 8-3 employs Eq. (8.3.3) expressed in terms of the normalized system kernels as follows:

$$\frac{g_f(e,c)}{g(e,c)} = \frac{1}{1 - g(e,c) f(e,c)}$$

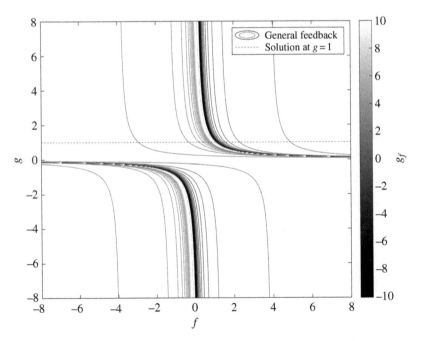

Fig. 8-3 Graph of the effect of feedback on the general single-input, single-output system equation.

Table 8-5 Feedback Kernel and Effect on System Gain

Feedback (as $\varepsilon^2 \to 0$)	System Kernel g_f	Comment
$gf < 0$	$g_f / g < 1$	Negative feedback gain
$gf = 0$	$g_f = g$	Neutral feedback gain
$0 < gf < 1$	$g_f / g > 1$	Positive feedback gain
$gf \to 1 - \varepsilon^2$	$g_f \to \infty$	Infinite feedback gain
$gf \to 1 + \varepsilon^2$	$g_f \to -\infty$	Infinite feedback gain
$1 < gf < 2$	$g_f / g < -1$	Positive feedback gain
$gf = 2$	$g_f = -g$	Neutral feedback gain
$gf > 2$	$g_f / g > -1$	Negative feedback gain

A tabulated sequence of values for the feedback kernel and the consequences on the system gain are given in Table 8-5.

Let us consider now the single-input, multiple-output and the multiple-input, single-output system equations with feedback to assess how feedback gain influences these systems. The general system equations are

$$d\mathbf{e} = \frac{g(\mathbf{e}, c)}{1 - \sum\limits_{i=1}^{n} g_i f_i} \, dc = g_f dc$$

as given by Eq. (3.2.12) for the single-input, multiple-output system, and for $d\mathbf{e} = g(e, \mathbf{c})$ $d\mathbf{c} = g_f \, d\mathbf{c}$

$$de = \frac{g(e, \mathbf{c})}{1 - \sum\limits_{i=1}^{m} g_i f_i} \, d\mathbf{c} = g_f \, d\mathbf{c}$$

as given by Eq. (3.2.18) for the multiple (m) input, single-output system. We observe that both system equations with feedback have the same general form for their denominators, where the feedback functions (i.e., the f's) appear in the overall system kernel. Observe that for

$$0 < \sum\limits_{i=1}^{n \text{ or } m} g_i f_i < 2$$

then $|g_f| > |g|$, and we obtain positive gain. If we have that

$$\sum\limits_{i=1}^{n \text{ or } m} g_i f_i = \begin{cases} > 2 \\ < 0 \end{cases}$$

then $|g_f| < |g|$ and negative gain results. Furthermore, the critical condition for positive feedback gain occurs when

$$\sum\limits_{i=1}^{n \text{ or } m} g_i f_i = 1$$

since $|g_f| \to \infty$

Finally, for the general multiple-input, multiple-output system equation with feedback, we have from Eq. (3.2.24) that

$$d\mathbf{e} = [I - g(\mathbf{e}, \mathbf{c})f(\mathbf{e}, \mathbf{c})]^{-1}g(\mathbf{e}, \mathbf{c})\,d\mathbf{c} = g_f(\mathbf{e}, \mathbf{c})\,d\mathbf{c}$$

and if

$$-1 < \det|I - g(\mathbf{e}, \mathbf{c})f(\mathbf{e}, \mathbf{c})| < 1$$

then we obtain positive gain and find that

$$\left|g_f\right| > |g|$$

For the condition that

$$\det|I - g(\mathbf{e}, \mathbf{c})f(\mathbf{e}, \mathbf{c})| > 1$$

we find that

$$\left|g_f\right| < |g|$$

and negative feedback gain occurs. Finally, if the det $|I - gf| \to 0$, then the matrix $[I - gf]$ becomes singular, and we witness infinite system gain since

$$\left|g_f\right| \to \infty$$

8.3.2 Effect of Feedback on Linear Kernels

In general, we will find that a system without feedback that has a linear kernel will not preserve the property of linearity with the addition of feedback. To establish this result, consider the general single-input, single-output system equation with feedback:

$$de = \frac{g(e,\ c)}{1 - g(e,c)f(e,c)}\,dc = g_f(e,c)\,dc \tag{8.3.4}$$

where

$$g_f(e,c) = \frac{g(e,\ c)}{1 - g(e,c)f(e,c)} \tag{8.3.5}$$

With no feedback present, $f = 0$, then Eq. (8.3.5) becomes

$$g_f(e,c) \Rightarrow g(e,c) \tag{8.3.6}$$

and if this system kernel without feedback is a linear function of the output variable e, then Eq. (8.3.6) has the form

$$g_f(e,c) \Rightarrow g(e,c) = A(c) + B(c)e \tag{8.3.7}$$

where $A(c)$ and $B(c)$ are arbitrary functions of the input variable c. If we now employ Eq. (8.3.7) as the defining relationship for the kernel function $g(e, c)$, then in the presence of system feedback as defined by Eq. (8.3.5), we have

$$g_f = \frac{A(c) + B(c)\,e}{1 - [A(c) + B(c)\,e]\,f(e,c)}$$

It is obvious that the system kernel g_f with feedback present will not be linear unless f assumes one of the special values

$$f(e,c) = \begin{cases} 0 & (8.3.8) \\ \dfrac{h(c)}{g(e,c)} = \dfrac{h(c)}{A(c) + B(c)\,e} & (8.3.9) \end{cases}$$

Of course, if $f = 0$ as given by Eq. (8.3.8), then there is no feedback present. For Eq. (8.3.9), where $h(c)$ is an arbitrary function of the input, then Eq. (8.3.7) becomes

$$g_f(e,c) = \frac{1}{1 - h(c)}\,[A(c) + B(c)\,e] = A_1(c) + B_1(c)\,e$$

which is also a linear function of the output variable e.

The observation that the addition of feedback to system kernels, which are linear in the output variables, does not preserve kernel linearity except for very special forms for the feedback kernel also applies to multiple-input and multiple-output systems. This result perhaps explains to some degree why most models for systems frequently omit feedback, at least for preliminary studies of systems. However, it is important to recognize that feedback generally exists for most real systems, and the presence of feedback can significantly affect the behavior (and, unfortunately, complicate the analysis) of these systems.

Recall that, for the special case in which we have a system with an equal number of input and output variables, kernel linearity can be interpreted either in terms of the output variables or the input variables with equal validity for analysis and solution purposes. Generally, the system investigator is advised to assume a point of view for such systems that are symmetric in the input and output mode (i.e., have the same number of inputs and outputs) that may result in an easier system equation to solve.

For example, for the single-input, single-output system equation and kernel as defined by Eqs. (8.3.4) and (8.3.5), assume that the system equation without feedback is a linear function of the input variable c, and so the system kernel can be expressed as

$$g_f(e,c) \Rightarrow g(e,c) = \frac{1}{D(e) + E(e)\,c}$$

Observe that this form for the kernel results in a linear system equation when expressed as follows:

$$d\,c = [D(e) + E(e)\,c]\,d\,e$$

In the presence of feedback, as defined by Eq. (8.3.5), the system kernel $g_f(e,c)$ becomes

$$g_f(e,c) = \frac{1}{D(e) + E(e)\,c - f(e,c)}$$

and it is apparent that if the feedback kernel $f(e,c)$ vanishes or is also a linear function of the input variable, that is,

$$f(e,c) = F(e) + G(e)\,c$$

then g_f preserves its linear form, and we have that 1

$$g_f(e,c) = \frac{1}{[D(e) - F(e)] + [E(e)\,c - G(e)]\,c}$$

The linear system kernel for the system equation then has the form

$$dc = \{\,[D(e) - F(e)] + [E(e)\,c - G(e)]\,c\}\,de$$

Thus, we have established the interesting and important result that the single-input, single-output system kernel that is a linear function of the input variable c without feedback will preserve linearity in the presence of a feedback kernel that is also a linear function of the input.

8.3.3 Single-Input, Single-Output Systems with Feedback

Even for the simplest continuous system with feedback, which is the single-input, single-output system, it is not possible to determine analytical solutions for all system models. Indeed, it was not possible to find solutions for all single-input, single-output systems even when no feedback was present. The occurrence of feedback, however, greatly complicates (and, therefore, enriches and expands the types of possible solutions) attempts to find analytical solutions. As we saw in Chapter 8, Section 8.3.2, even the single-input, single-output system that has a linear kernel without feedback becomes nonlinear in general in the presence of feedback, and therefore an analytical solution usually cannot be found.

Therefore, since no single method will be generally applicable to analytically solving continuous system equations with feedback present, we must be satisfied with recognizing specialized forms for system kernels with feedback that can be solved. Of course, all the techniques that we applied to single-input, single-output system kernels given in Chapter 4, Section 4.2 are applicable here if the system kernel with feedback present satisfies the necessary conditions required to use the technique. In this section, we wish to explicitly examine the feedback kernel $f(e, c)$ and its functional form and identify those forms that can be solved analytically.

The general method we will employ in attempting to determine some of those classes of system equations with feedback that can be analytically solved is the requirement that the resulting system equation (or, equivalently, the differential equation in differential form) be exact, as we have seen in Sections 4.2.5, 4.3.6, 5.2.4, and 5.3.4. All differential equations that can be directly integrated must exist in exact or total differential form. For example, the general first-order differential equation in the differential form

$$M(x,y)\,dx + N(x,y)\,dy = 0 \tag{8.3.10}$$

may be directly integrated as follows with y held constant in the first integral

$$\int_a^x M(x,y)\,dx \; + \; \int_o^y N(x,y)\,dy = \text{constant}$$

if and only if we satisfy the general requirement for exactness:

$$\frac{\partial M}{\partial y} = \frac{\partial N}{\partial x} \tag{8.3.11}$$

Equation (8.3.11) assumes that M and N in Eq. (8.3.10) are continuous functions with continuous derivations, but this condition is satisfied for most system equations. To transform or convert a given single-input, single-output system equation into a form of Eq. (8.3.10) for which the requirement for exactness, Eq. (8.3.11), is satisfied may require, in general, considerable manipulation, transformation, and multiplication by certain functions. Such a process is succinctly referred to as finding the integrating factor (i.e., finding the combination of functional changes that makes the resulting equation exact) for a given differential equation.

The single-input, single-output continuous system equation with feedback has the general form

$$de = \frac{g}{1-gf}\, dc \tag{8.3.12}$$

There are unlimited ways of modifying this equation to achieve an exact differential equation form. Suppose we multiply Eq. (8.3.12) by

$$g^m f^n\,(1-gf)$$

and, after arranging the equation into standard form, we have

$$g^m f^n\,(gf-1)\,de + g^{m+1}f^n dc = 0 \tag{8.3.13}$$

For this differential equation to be exact, we require that

$$\frac{\partial}{\partial c}\,[\,g^m f^n\,(gf-1)\,] = \frac{\partial}{\partial e}\,[\,g^{m+1}f^n\,]$$

or

$$[(n+1)\,g^{m+1}f^n - n\,g^m f^{n-1}]\,\frac{\partial f}{\partial c} + [\,(m+1)\,g^m f^{n+1} - m\,g^{m-1}f^n\,]\,\frac{\partial g}{\partial c}$$
$$= n\,g^{m+1}f^{n-1}\,\frac{\partial f}{\partial e} + (m+1)\,g^m f^n\,\frac{\partial g}{\partial e} \tag{8.3.14}$$

Dividing Eq. (8.3.14) by $g^m f^n$ and collecting on common partial derivatives, we have that

$$\left[(n+1)\,g - \frac{n}{f}\right]\frac{\partial f}{\partial c} + \left[(m+1)f - \frac{m}{g}\right]\frac{\partial g}{\partial c} = n\,\frac{g}{f}\,\frac{\partial f}{\partial e} + (m+1)\,\frac{\partial g}{\partial e} \tag{8.3.15}$$

This is one equation that $g(e, c)$ and $f(e, c)$ may satisfy for Eq. (8.3.13) to be exact and, therefore, directly integrable.

As an example of the use of Eq. (8.3.15), let $n = m = -1$; then we have

$$\frac{1}{f}\,\frac{\partial f}{\partial c} + \frac{1}{g}\,\frac{\partial g}{\partial c} = -\frac{g}{f}\,\frac{\partial f}{\partial e} \tag{8.3.16}$$

A simple set of functions for f and g that satisfy Eq. (8.3.16) is (where K is an arbitrary constant)

$$g\,(e,c) = G(e)\,H(c) \tag{8.3.17.1}$$

$$f\,(e,c) = \frac{K}{H(c)} \tag{8.3.17.2}$$

Substitution of Eqs. (8.3.17.1) and (8.3.17.2) into Eq. (8.3.12) provides the following system equation:

$$de = \frac{G(e)\,H(c)}{1 - K\,G(e)}\,dc$$

which is separable and can be integrated to give as a system equation solution

$$\int \frac{de}{G(e)} - K\,G(e) = \int H(c)\,dc + \text{constant}$$

Other examples of the use of Eq. (8.3.15) and related equations will be considered in the problems given at the end of this chapter.

8.3.4 Inversion of System Kernels with Feedback

An interesting alternative method for expressing and hopefully solving system equations both with and without feedback is the inverse kernel method. To demonstrate the method, consider the continuous single-input, single-output system equation with feedback

$$de = \frac{g\,(e, c)}{1 - g\,(e, c)\,f\,(e, c)}\,dc \tag{8.3.18}$$

Let us express the system equation as

$$[1 - gf]\,de = g\,dc \tag{8.3.19}$$

Now let us assume that both the input and output variables can be expressed solely as inverse functions of the kernels g and f, as follows:

$$e = G\,(g, f) \quad \text{so} \quad de = \frac{\partial G}{\partial g}\,dg + \frac{\partial G}{\partial f}\,df \tag{8.3.20}$$

and

$$c = H\,(g, f) \quad \text{so} \quad dc = \frac{\partial H}{\partial g}\,dg + \frac{\partial H}{\partial f}\,df \tag{8.3.21}$$

Although obtaining such inverse expressions for e and c as explicit functions of f and g may be difficult, the implicit function theorem provides a theoretical guarantee for the existence of such inverse functional relationships under certain conditions. Now let us employ Eqs. (8.3.20) and (8.3.21) to eliminate the differential input dc and output de from Eq. (8.3.19), as follows:

$$g\left(\frac{\partial H}{\partial g}\,dg + \frac{\partial H}{\partial f}\,df\right) = (1 - gf)\left(\frac{\partial G}{\partial g}\,dg + \frac{\partial G}{\partial f}\,df\right)$$

If we now gather terms on the differentials df and dg, we have

$$\left[g\frac{\partial H}{\partial f} + (gf - 1)\frac{\partial G}{\partial f}\right]df + \left[g\frac{\partial H}{\partial g} + (gf - 1)\frac{\partial G}{\partial g}\right]dg = 0 \tag{8.3.22}$$

Equation (8.3.22) appears formidable and unpromising. However, let us examine what conditions must exist for Eq. (8.3.22) to constitute an exact differential of the following form for some function $\varphi(g, f)$

$$d\varphi(g,f) = 0 = \frac{\partial \varphi}{\partial g} dg + \frac{\partial \varphi}{\partial f} df \tag{8.3.23}$$

A necessary and sufficient condition for an equation of the form of Eq. (8.3.23), which is continuous, to be exact (recall Sections 4.2.5, 4.3.6, 5.2.4, 5.3.4), is that

$$\frac{\partial}{\partial g} \left(\frac{\partial \varphi}{\partial f} \right) = \frac{\partial}{\partial f} \left(\frac{\partial \varphi}{\partial g} \right) \tag{8.3.24}$$

If we associate the appropriate partial derivative coefficients in Eq. (8.3.23) with those coefficients in Eq. (8.3.22), then the requirement for exactness as defined by Eq. (8.3.24) is

$$g \frac{\partial^2 H}{\partial g \partial f} + \frac{\partial H}{\partial f} + f \frac{\partial G}{\partial f} - (1 - gf) \frac{\partial^2 G}{\partial g \partial f}$$
$$= g \frac{\partial^2 H}{\partial f \partial g} + g \frac{\partial G}{\partial g} - (1 - gf) \frac{\partial^2 G}{\partial f \partial g} \tag{8.3.25}$$

If we assume that both functions H and G are continuous functions with continuous derivatives, then

$$\frac{\partial^2 H}{\partial g \partial f} = \frac{\partial^2 H}{\partial f \partial g} \quad \text{and} \quad \frac{\partial^2 G}{\partial g \partial f} = \frac{\partial^2 G}{\partial f \partial g} \tag{8.3.26}$$

Application of Eqs. (8.3.26) to Eq. (8.3.25) reduces the equation to

$$\frac{\partial H}{\partial f} = g \frac{\partial G}{\partial g} - f \frac{\partial G}{\partial f} \tag{8.3.27}$$

or equivalently

$$\frac{\partial c}{\partial f} = g \frac{\partial e}{\partial g} - f \frac{\partial e}{\partial f}$$

Thus, Eq. (8.3.27) is the requirement on the inverse functions G and H as defined by Eqs. (8.3.20) and (8.3.21), respectively, for the system equation with feedback [i.e., Eq. (8.3.18)] to be exact and thus integrable. As an example in using Eq. (8.3.27), let us assume that

$$\frac{\partial H}{\partial f} = K \quad \text{where } K \text{ is a constant} \tag{8.3.28}$$

Integrating Eq. (8.3.28), we have

$$H = Kf + h_1(g) \tag{8.3.29}$$

where $h_1(g)$ is an arbitrary function of g. Equation (8.3.28) also provides the following requirement from Eq. (8.3.27):

$$g \frac{\partial G}{\partial g} - f \frac{\partial G}{\partial f} = K \tag{8.3.30}$$

Using the Method of Lagrange discussed in Section 5.2.9 to solve Eq. (8.3.30), we have that

$$\frac{dg}{g} = -\frac{df}{f} = \frac{dG}{K} \tag{8.3.31}$$

From Eq. (8.3.31), we obtain as a solution between g and f that

$$\text{Ln } g = -\text{Ln} f + K_1$$

where K_1 is an integration constant. Also, we can integrate the differential equation relating df and dG to give

$$G = -K_2 \text{Ln} f + K_2$$

where K_2 is also an integration constant. The method of Lagrange provides that the general solution of Eq. (8.3.30) then has the form where φ is an arbitrary function

$$\varphi(K_1, K_2) = 0$$

or equivalently,

$$K_2 = G + K_1 \text{Ln} f = h_2 (\text{Ln } g + \text{Ln} f) \tag{8.3.32}$$

where h_2 is an arbitrary function of the variable $(\text{Ln } g + \text{Ln} f)$. So, we have that the single-input, single-output system equation with feedback as given by Eq. (8.3.18) is exact and therefore integrable if the system kernels B and f can be expressed in terms of the input and output variables c and e from Eq. (8.3.29) as

$$H(g, f) = c = Kf + h_1(g) \tag{8.3.33}$$

and for Eq. (8.3.32) as

$$G(g, f) = e = h_2 \text{Ln}(fg) - K\text{Ln} f \tag{8.3.34}$$

where h_1 and h_2 are arbitrary functions and K is an arbitrary constant. Substituting Eqs. (8.3.33) and (8.3.34) into Eq. (8.3.27) provides verification that the resulting system equation is exact.

8.4 Computer-Aided Analysis of Systems

We have considered some of the many useful methods for modifying, reducing, and approximating difficult system equations so that practical methods could be employed to realize analytical solutions of the system equations. It is important that the investigator recognize the desirability and utility of acquiring such analytical solutions whenever possible, even in the presence of somewhat restrictive, limiting assumptions about the system equation. These analytical solutions provide an initial basis for assessing and evaluating the competence and applicability of the system model with the actual system. Furthermore, such analytical solutions reveal the quantitative form of the system solution and how model parameters enter and affect the solution responses. Furthermore, these approximate analytical solutions serve as a primary basis for judging the validity of other system solutions, such as those obtained by computer-based methods.

Nevertheless, since system equations, particularly those with multiple inputs, multiple outputs, or nonlinear kernels, often cannot be analytically solved, analyzed, or evaluated, the system investigator must resort to other methods to study such difficult systems. Undoubtedly, the primary method now readily available to the system investigator is the use of computers to numerically determine or simulate the solutions for both discrete and continuous systems. As we learned in Chapters 4 and 5, the computer is ideally suited for determining the solution of any discrete system equation, regardless of the quantitative form for the kernel or number of inputs and outputs.

Furthermore, for continuous system equations, there exists a great variety of numerically based procedures for solving continuous systems equations, which are mathematically equivalent to differential equations. We will restrict our attention to just two major numerical analysis methods for solving system equations. These methods are classified as (1) Runge-Kutta methods and (2) Predictor-Corrector methods. Each of these standard numerical algorithms for solving sets of differential equations has advantages and disadvantages, and the experienced system investigator should be familiar with both methods as well as other numerical algorithms to obtain satisfactory system equation solutions efficiently and accurately.

The mathematical foundation for most computer algorithms for solving continuous system equations begins with the general multiple-input, multiple-output system equation

$$d\mathbf{e} = g(\mathbf{e}, \mathbf{c}) \, d\mathbf{c} \tag{8.4.1}$$

If we integrate Eq. (8.4.1) over the arbitrary integration interval $c_k \leq c \leq c_{k+1}$ we obtain the following integral equation:

$$\mathbf{e}(\mathbf{c}_{k+1}) = \mathbf{e}(\mathbf{c}_k) + \int_{e_k}^{e_{k+1}} g(\mathbf{e}, \mathbf{c}) \, d\mathbf{c} \tag{8.4.2}$$

Section 5.3.9 treats this integral equation form of the system equation in detail.

Generally, most numerical procedures for estimating the system output assume that the integral term in Eq. (8.4.2) can be suitably approximated by an estimate of the integrated kernel based on the mean value theorem for integrals; that is,

$$\int_{e_k}^{e_{k+1}} g(\mathbf{e}, \mathbf{c}) \, d\mathbf{e} = \langle g(\mathbf{e}, \mathbf{c}) \rangle_k (\mathbf{c}_{k+1} - \mathbf{c}_k)$$

where $\langle g \rangle_k$ denotes the appropriate mean value for the kernel over the interval $\mathbf{c}_k \leq \mathbf{c} \leq \mathbf{c}_{k+1}$. Since $\mathbf{e}$ is not known over this range, we can only approximate the function $\langle g \rangle_k$. This is usually accomplished with expressions of the form

$$\langle g(\mathbf{e}, \mathbf{c}) \rangle_k \approx \sum_{i=1}^{I} G_i(\mathbf{e}_k + \Delta \mathbf{e}_{k\,i}, \mathbf{c}_k + \Delta \mathbf{c}_{k\,i}) \tag{8.4.3}$$

Detailed development and description of the various methods that can be used to approximate Eq. (8.4.3) can be found in the extensive literature on numerical analysis. The system investigator should become familiar with the standard methods so that existing algorithms and specific computer software can be used effectively for analyzing and modeling systems.

We will restrict our discussion here to a brief comparison of Runge-Kutta and Predictor-Corrector methods. Runge-Kutta methods, as broadly interpreted here, include all those numerical integration methods that are fully self-starting and do not require past system equation data from previous integration steps to proceed. The general advantages and disadvantages associated with Runge-Kutta methods include the following:

- Completely self-starting at any point in the integration range.
- Relatively simple to program on the computer.
- Inherent stability characteristics.
- Integration step size is easily changed at any step.
- However, no simple way to determine the best integration step size for minimum error without performing numerous repeated runs with different step sizes to sample errors.
- Error estimates are difficult to obtain.
- May require more computation than Predictor-Corrector methods for comparable accuracy.

Predictor-Corrector methods, also broadly interpreted here, include all those numerical integration routines that employ one or more prediction calculations and corrector calculations to execute an integration step. These methods usually employ a strategy to modify the integration step to maximize computational efficiency and minimize error. Some of the general advantages and disadvantages here include the following:

- Appropriate integration step can usually be determined to maximize procedure performance.
- Generally, less computations are required than for Runge-Kutta methods for equivalent accuracy.
- Error estimates are readily obtained.
- However, instability characteristics may arise.
- Method is not self-starting and requires special initiation procedures.
- Generally, requires a Runge-Kutta starting procedure, and programming is more complex.

Although the general multiple-input, multiple-output system equation can be numerically integrated for all the input or independent variables simultaneously with a multiple independent variable algorithm, this is seldom done because of the programming complexity and computational intensity imposed on the computer. Furthermore, the computer data output for a many-input, many-output system equation would be voluminous and difficult to assess and evaluate. Thus, standard computer software (coded algorithms) for solving sets of differential equations is usually limited to a single independent variable.

The general procedure for treating multiple-input systems is to selectively span the range of the independent variables using a single input variable, as was demonstrated in Section 5.3.8 for reducing multiple-input systems. Using this reduction procedure, multiple-input system equations can be solved quite effectively with standard computer software, including the computer integration programs given in this section.

MATLAB offers an array of subroutines for solving differential equations that employ various methods for solving initial value problems. The MATLAB IVP (Initial Value Problem) solvers, ode23 is based on a pair of second and third-order Runge-Kutta methods, and ode45 uses a mix of fourth and fifth-order methods. The solver, ode45, is usually more accurate but generally requires more computer time to address the problem. The use of both methods for a comparison of the solutions from each is usually worthwhile to provide some degree of confidence in the final solutions used for the model. Also, the MATLAB Symbolic Math Toolbox provides a software package termed "dsolve" that allows the solution of systems of first-order differential equations to be analyzed with results provided in terms of analytical functions.

8.5 Summary

In this chapter, we have examined some of the most important aspects of revising and analyzing system models. The simplification and reduction of many variable and complex system kernels are often essential for the practical analysis and evaluation of such systems. In this chapter, we examined simple, quantitative methods for simplifying kernels by the replacement of variable kernel components by representative constants. Furthermore, methods for ordering and ranking these constant kernel matrices so that reduction procedures could be made on a rational, quantitative basis were discussed.

The techniques for normalization and minimization of system equation variables and parameters were demonstrated. These techniques are particularly useful in demonstrating the similarity between diverse systems that possess related system equations and in reducing the effort necessary in parameter characterization and data gathering for system models.

The essential characteristics and influence exerted by system feedback on system models were demonstrated, and the ubiquitous and complicating effects were examined. Some simple schemes for analyzing certain system kernels with feedback were considered.

Finally, the chapter closed with a brief discussion of one of the most powerful tools available to the system investigator for the analysis and evaluation of system models, particularly those models that cannot be adequately studied with analytical methods. The powerful tools offered by computer-supported analysis software are essential possessions for the competent system investigator. Some simple but very useful examples are addressed in the chapter for plotting, examining, and solving system equations.

Bibliography

CARNABAN, B., et al., *Applied Numerical Analysis*, John Wiley & Sons, New York, 1969. A valuable resource of methods and FORTRAN programs for model analysis.

DALKEY, C., *"Delphi Research: Experiments and Prospects,"* Proceedings: *International Seminar on Trends in Mathematical Modeling*, Venice, 1971. State-of-the-art applications of Delphi modeling.

JENSEN, R., and C. TOMES, *Software Engineering*, Prentice-Hall, Englewood Cliffs, NJ, 1979. A general introduction to computer algorithms for model analysis.

GILAT, A., *MATLAB, An Introduction with Applications*, John Wiley & Sons, Inc., NY, 2004. A very useful introductory exposition for using MATLAB for the applications encountered in this text.

GRIGORIU, M., *Linear Dynamical Systems*, Springer, 2021.

MATLAB, *The Language of Technical Computing*, The MathWorks, Inc., MA. Various titles and volumes are provided with toolboxes and software packages from MATLAB.

MELSA, J. L., and S. K. JONES, *Computer Programs for Computational Assistance in the Study of Linear Control Theory*, McGraw-Hill Book Co., NY, 1973. An excellent reference on computer applications of linear systems.

MESAROVIC, M. D., et al., *Theory of Hierarchical, Multilevel Systems*, Academic Press, NY, 1970. A standard work on hierarchical systems.

OSBORN, A., *Applied Imagination*, Charles Scribner's Sons, NY, 1979. A provocative insight into imagination in practice and modeling.

PEACOCK, J., Early cosmology textbook for undergraduate and graduate students *Cosmological Physics*, Cambridge University Press, 1999.

POLYA, G., *How to Solve It*, Princeton University Press, Princeton, NJ, 1945. An elementary introduction to heuristic methods.

POLYA, G., *Mathematics of Plausible Reasoning, Volume I., Induction and Analogy in Mathematics, Volume II, Patterns of Plausible Inference*, Princeton University Press, Princeton, NJ, 1954. A useful reference set for applying mathematical models to systems.

RUBINSTEIN, M. F., *Patterns of Problem Solving*, Prentice-Hall, Englewood Cliffs, NJ, 1975. Insights into the process of quantitative thinking and modeling.

SANDQUIST, G. M., and Z. R. WILDE, *Introduction to System Science with MATLAB*, 2nd Ed., John Wiley & Sons, NY, 2023.

WEINBERG, G., *An Introduction to General Systems Thinking*, John Wiley & Sons, NY, 1975. A very useful and interesting reference on system modeling.

Problems

1   Convert or approximate the following system kernels as requested from Table 8-1 for the variable ranges given. Solve the resulting system equation if possible.

a) $de = (A + Bc + De)\, dc$ (convert to average value kernel)

   $(0 < c < 1$ and $1 < e < e_{max})$

b) $dy = \dfrac{(1 + xy)}{(x + y)}\, dx$ (constant kernel) $(0 \le x \le 1$ and $y \ge 1)$

c) $de_1 = (Ae_1 + e_2)\, dc$ (separable kernel)

   $de_2 = (Be_1\, e_2 + 1)\, dc$

   $(0 \le c \le 10, 1 \le e_1 \le 5,$ and $1 \le e_2 \le 8)$

d) $dr = (r^2 - s^2)\, ds$ (linearize kernel)

   $(0 \le s \le 1$ and $0 \le r \le 1)$

e) $de = (Ae + c_1)\, dc_1 + (Be + c_1)\, dc_2$ (eliminate independent variables)

   $(0 \le c_1, c_2 \le 1,$ and $1 \le e \le 2)$

f) $de_1 = (A\, e_1\, c + e_2 + B)\, dc$ (eliminate dependent variable)

   $de_2 = (D\, e_2\, c + e_1 + E)\, dc$

   $(0 \le c \le 1, 1 \le e_1,$ and $e_2 \le 2)$

2 The following constant system kernel matrices g are in standard form (i.e., the rows correspond to the system outputs and the columns to the system inputs both in sequential order). Employ the weighting factors for the inputs [i.e., W, (inputs)] and the outputs [i.e., W_0 (outputs)] as shown and determine the weighted, reordered system kernel as demonstrated in Section 8.1.2. Also determine the accumulation matrix for the system kernel. What ordering should the inputs and outputs assume to satisfy the weighted ordered system kernel?

Note that $D = D[a \ b] = D\begin{bmatrix} a \\ b \end{bmatrix} = \begin{bmatrix} a & 0 \\ 0 & b \end{bmatrix}$ as appropriate

a) $g = \begin{bmatrix} 1 & 8 \\ 15 & -7 \end{bmatrix}$, $\quad W_i = D\begin{bmatrix} 2 \\ 1 \end{bmatrix}$, $\quad W_0 = D\,[3\ 4]$

b) $g = [1\ 4\ 7\ 8\ -5]^T$, $\quad W_0 = D\,[1\ 2\ 1\ 1\ 2]$

c) $g = [4\ -3\ 2\ 7\ -10]$, $\quad W_i = D\,[1\ 2\ 1\ 1\ 2]^T$

d) $g = \begin{bmatrix} 1 & 4 & 8 & -2 \\ 4 & 3 & 2 & 3 \\ 8 & 2 & -7 & -6 \\ -2 & 3 & -6 & 8 \end{bmatrix}$, $\quad W_i = D\,[1\ 1\ 1\ 1]^T$, $\quad W_0 = D\,[2\ 2\ 2\ 2]$

Note: This matrix system kernel is symmetric (i.e., coefficient elements reflected across the main diagonal of the system kernel are identical). What effect does this have on the weighted reordered system kernel and the accumulation matrix?

e) $g = \begin{bmatrix} 0 & 1 & 3 \\ 4 & 0 & 2 \\ 3 & 8 & 0 \end{bmatrix}$, $\quad W_i = D\,[1\ 2\ 3]^T$, $\quad W_0 = D\,[3\ 2\ 1]$

f) $g = \begin{bmatrix} 0 & -2 & -1 \\ -2 & -1 & 3 \\ -1 & 3 & 2 \end{bmatrix}$, $\quad W_i = D\,[1\ 2\ 1]^T$, $\quad W_0 = D\,[2\ 1\ 2]$

Note: This system kernel is singular. What consequence does this fact have on kernel reduction efforts and the system equations themselves? What conclusion can be made about the system equations if the system kernel is singular?

g) $g = \begin{bmatrix} 1 & 3 & 7 \\ 2 & -8 & 5 \\ 4 & 0 & -6 \\ -3 & 5 & 9 \end{bmatrix}$, $\quad W_i = D\,[2\ 1\ 2\ ]^T$, $\quad W_0 = D\,[1\ 1\ 2\ 2]$

h) $g = \begin{bmatrix} 1 & 3 & 7 \\ 9 & 0 & 11 \\ -5 & 3 & -8 \\ 6 & 0 & -9 \\ 8 & 2 & -4 \end{bmatrix}$, $\quad W_i = D\,[1\ 2\ 1]^T$, $\quad W_0 = D\,[1\ 1\ 1\ 1\ 1]$

i) $g = \begin{bmatrix} 1 & 2 & 3 & 4 & 5 \\ -5 & -3 & -1 & -2 & -4 \\ 5 & 4 & -3 & -1 & 2 \\ 4 & -5 & 1 & 2 & -3 \\ 3 & -1 & 4 & 5 & -2 \end{bmatrix}$, $\quad W_i = D[1\,5\,2\,3\,4]^T \quad W_o = D[5\,1\,4\,3\,2]$

j) $g = \begin{bmatrix} 0.1 & 100 & -0.001 \\ 1000 & -10 & 0.01 \\ -10 & 0.1 & 1 \end{bmatrix}$, $\quad W_i = D\,[1\ 10\ 0.1]^T, \quad W_o = D\,[1\ 0.1\ 10]$

k) $g = \begin{bmatrix} 10^2 & -10^3 & 10^{-1} & -10^2 \\ 10 & 10^0 & 10 & 10^2 \\ -10^3 & 10^3 & 10^2 & -10^{-2} \\ 10^{-2} & -10^2 & 10^{-2} & 10 \end{bmatrix}$, $\quad W_i = D\,[1\,1\,10\,1]^T, \quad W_o = D\,[1\ 10\ 1\ 1]$

l) $g = \begin{bmatrix} 1 & 2 \\ -3 & 2 \\ 4 & 1 \\ 2 & 4 \\ 3 & 1 \end{bmatrix}$, $\quad W_i = D\,[2\ 1]^T, \quad W_o = D\,[1\,2\,3\,2\,1]$

m) $g = \begin{bmatrix} 1 & 0 & 0 & 0 \\ 0 & 3 & 0 & 0 \\ 0 & 0 & 4 & 0 \\ 0 & 0 & 0 & 2 \end{bmatrix}$, $\quad W_i = D\,[1\,2\,1\,2]^T, \quad W_o = D\,[2\,1\,1\,2]$

3   Perform a linear variable transformation of each of the following system equations around the normalization values indicated. Establish the normalized, dimensionless system equation and the minimum parameter set.

a) $de = (Ae + Bc + D)\,dc$
   [for $e(c_0) = e_0$]

b) $dy = (A + By)\,y\,dx$
   [for y_0 when $dy = 0$]

c) $dr = \left[\dfrac{(A + Br)}{(1 + Cr^2)} \right] s\,ds$
   [for r_0 when $dr = 0$]

d) $de_1 = (A_1 + B_1\,e_1 + D_1\,e_2)\,dc_1$
   $de_2 = (A_2 + B_2\,e_1 + D_2\,e_2)\,dc_2$
   [for $e_1\,(c_0) = e_{10}$ and $e_2\,(c_0) = e_{20}$]

e) $de = (A_1c_1 + A_2c_1c_2 + A_3e)dc_1 + \left(B_1\dfrac{c_1}{c_2} + B_2e^2 \right)dc_2$
   [for $e(c_0) = e_0$]

f) $de_1 = (A_{11} + B_{11}\,e_1 + B_{12}\,e_2 + B_{13}\,e_3)\,dc_1 + (A_{21} + D_{11}\,e_1 + D_{12}\,e_2 + D_{13}\,e_3)\,dc_2$
   $de_2 = (A_{12} + B_{21}\,e_1 + B_{22}\,e_2 + B_{23}\,e_3)\,dc_1 + (A_{22} + D_{21}\,e_1 + D_{22}\,e_2 + D_{23}\,e_3)\,dc_2$
   $de_3 = (A_{13} + B_{31}\,e_1 + B_{32}\,e_2 + B_{33}\,e_3)\,dc_1 + (A_{23} + D_{31}\,e_1 + D_{32}\,e_2 + D_{33}\,e_3)\,dc_2$
   [for $e_{10},\ e_{20},\ e_{30}$ all arbitrary]

g) $dy = \left(A + By + C \operatorname{Ln} y + \frac{D}{y}\right) dx$

[for $y(1) = 1$]

h) $dy = \dfrac{(A + By)}{y} dx$

[for $y_0 \neq 0$ when $dy = 0$]

i) $dr = \left[\dfrac{(A + B \operatorname{Ln} r)}{(1 + Cr^2)}\right] s \, ds$

[for $r_0 \neq 0$ when $dr = 0$]

j) $de_1 = \left(A_1 + \dfrac{B_1}{e_1} + \dfrac{D_1}{e_2}\right) dc_1$

$de_2 = \left(A_2 + \dfrac{B_2}{e_1} + \dfrac{D_2}{e_2}\right) dc_2$

[for $e_1\,(c_0) = e_{10}$ and $e_2\,(c_0) = e_{20}$]

4 Solve the following system equations using MATLAB and the appropriate Toolbox (e.g., ode23, ode45 in the Runge-Kutta Toolbox or dsolve in the MATLAB Symbolic Math Toolbox).

a) $dy = -2\,x\,y\,dx$ $\left[\text{for } 0 \leq x \leq 10 \text{ and } y(0) = \left(\frac{\pi}{2}\right)^{\frac{1}{2}}\right]$

Compare your result with the exact solution $y = \left(\frac{\pi}{2}\right)^{\frac{1}{2}} \exp\left(-x^2\right)$

b) $de = (e^2 - 2e\,c + c^2)\,dc$ [for $0 \leq c \leq 1$ and $e(0) = 1$]

Compare your result with the exact solution Hint: let $x = e - c$

c) $ds = (s^2 - 2s + 3)\,dt$ [for $0 \leq t \leq 1$ and $s(0) = 1$]

Compare your result with the exact solution

d) $df = \operatorname{Ln}\,(f + g)\,dg$ [for $1 \leq g \leq 2$ and $f\,(1) = 0$]

e) $dw = \dfrac{(w + x + 2wx)}{(w - x)}\,dx$ [for $0 \leq x \leq 1$ and $w(0) = 1$]

f) $de_1 = (e_1 - e_2)\,c\,dc$ [for $1 \leq c \leq 2$, $e_1\,(1) = e_2\,(1) = 1$]

$de_2 = (e_1 + e_2)\,c\,dc$

Compare with exact solution.

g) $dx = (x + z\,y)\,dz$ [for $0 \leq z \leq 1$, $x(0) = y(0) = 0$]

$dy = (y - z\,x)\,dz$

h) $de_1 = \left(ce_1 - \dfrac{e_2}{e_1}\right) dc$ [for $0 \leq c \leq 10$, $e_1\,(0) = e_2\,(0) = 1$]

$de_2 = (e_1\,e_2 - c^2)\,dc$

i) $de = (1 + e - c_1 - c_2)\,dc_1 + (1 - e + c_1 + c_2)\,dc_2$

[for $0 \leq c_1,\ c_2 \leq 1$ and $e(0, 0) = 1$]

Hint: See Section 6.2 and let $(dc_1, dc_2) = (\tau_1\,\tau_2)\,dt$.

Note: Exact solution is $e = k \exp\,(c_1 - c_2) + c_1 + c_2$.

j) $de = e\,c_2\,c_3\,dc_1 + e\,c_1\,c_3\,dc_2 + e\,c_1\,c_2\,dc_3$

[for $0 \leq c_1,\ c_2,\ c_3 \leq 1$ and $e\,(0, 0, 0) = 1$]

Compare with exact solution.

k) $de_1 = c_1 e_1 \, dc_1 + c_2 e_2 \, dc_2$ [for $1 \le c_1$, $c_2 \le 2$, $e_1 (1, 1) = e_2 (1, 1) = 1$]

$de_2 = c_1 e_2 \, dc_1 + c_2 e_1 \, dc_2$

l) $de_1 = (e_1 + e_2 \, e_3 - e_4) \, dc$, $\quad de_2 = (e_1 \, e_2 - e_3 \, e_4) \, dc$

$de_3 = (e_1 \, e_2 \, e_3 - e_4) \, dc$, $\quad de_4 = (e_1 - e_2)/c \, dc$

[for $1 \le c \le 2$, and $e_1 (1) = e_2 (1) = e_3 (1) = e_4 (1) = 1$]

m) $de_1 = \left(\dfrac{e_1}{e_2} - e_3 c \right) dc$, $\quad de_2 = \left(e_1 c + \dfrac{e_2}{e_3} \right) dc$, $\quad de_3 = \dfrac{(e_1 + e_2)}{c} dc$

[for $1 \le c \le 2$, and $e_1 (1) = e_2 (1) = e_3 (1) = 1$]

n) $de_1 = (e_1 + c \operatorname{Ln} e_2 - e_1 \, e_2) dc$, $de_2 = \left(e_1 c + \dfrac{e_2}{e_3} \right) dc$

[for $1 \le c \le 2$, and $e_1 (1) = e_2 (1) = e_3 (1) = 1$]

5 Attempt to solve the following single-input, single-output system equation with feedback.

$$g(e, c) de = \frac{g(e, c)}{[1 - g(e, c) f(e, c)]} dc$$

a) If $g(e, c) = g_0$ (constant) and $f(e, c) = f(e)$

b) If $g(e, c) = g(e)$ and $f(e, c) = f_0$ (constant)

c) If $g(e, c) = g_0$ (constant) and $f(e, c) = f(c)$

d) If $g(e, c) = g(c)$ and $f(e, c) = f_0$ (constant)

e) If $g(e, c) = [A(e) + B(e)c + D(e) \, c^n]^{-1}$ and $f(e, c) = f(e)$

f) If $g(e, c) = g(c) \, G(e)$ and $f(e, c) = \dfrac{F(e)}{g(c)}$

g) If $g(e, c) = g(c) \, G(e)$ and $f(e, c) = \dfrac{F(c)}{G(e)}$

h) If $g(e, c) = g\left(\dfrac{e}{c} \right)$ and $f(e, c) = f\left(\dfrac{e}{c} \right)$

i) If $g(e, c) = g\left[\dfrac{(Ae + Bc + D)}{(ae + bc + d)} \right]$ and $f(e, c) = f\left[\dfrac{(Ae + Bc + D)}{(ae + bc + d)} \right]$

j) If $g(e, c) = g(e, c)$ and $f(e, c) = \dfrac{1}{g(e, c)} - G(e) F(c)$

k) If $g(e, c) = g(e, c)$ and $f(e, c) = \dfrac{[1 - G(e) F(e)]}{g(e, c)}$

l) If $g(e, c) = [f(e, c) + G(e)]^{-1}$ and $f(e, c) = f(e, c)$

m) If $g(e, c) = g(c)$ and $f(e, c) = \dfrac{[1 - f(e)]}{g(c)}$

n) If $g(e, c) = \dfrac{g(c)}{f(e)}$ and $f(e, c) = f(e)$

6 Show that the single-input, single-output system equation with feedback

$$de = \frac{g(e, c)}{[1 - g(e, c) f(e, c)]} dc$$

is exact (see Section 4.2.5) and therefore integrable if the following conditions are satisfied.

a) $\partial \dfrac{\left[\dfrac{g}{(1-gf)}\right]}{\partial c}$

b) $f\dfrac{\partial g}{\partial c} + g\dfrac{\partial f}{\partial c} = \dfrac{\partial g}{\partial e}$

Hint: Express equation as $(1 - g\,f)\,de = g\,dc$ and impose exactness condition.

c) $\partial \dfrac{\left[\dfrac{g}{(1-gf)}\right]}{\partial c} = 0$

d) $\partial \dfrac{[1-gf)]}{\partial c} = -\dfrac{\partial g}{\partial e}$

e) $g\dfrac{\partial f}{\partial e} + \dfrac{\partial f}{\partial c} + \partial \ln \dfrac{g}{\partial c} = 0$

f) $f\partial \ln \dfrac{(fg)}{\partial c} = \partial \ln \dfrac{g}{\partial e}$

7 Attempt to solve the following multiple-input, multiple-output system equations with feedback. Attempt to obtain a manual analytical solution, and if unsuccessful, try the MATLAB Symbolic Toolbox.

a) $de = \dfrac{(A\,dc_1 + B\,dc_2)}{[1 - Af_1(e) - Bf_2(e)]}$

b) $de = \dfrac{(c_2\,dc_1 + c_1\,dc_2)}{[1 - (A+B)c_1 c_2]}$

c) $de_1 = \dfrac{(A\,e_1 e_2)}{[1 - (Af_1 + Bf_2)e_1 e_2]}\,dc, \quad de_2 = \dfrac{(B\,e_1 e_2)}{[1 - (Af_1 + Bf_2)e_1 e_2]}\,dc$

d) $de_1 = \dfrac{(-A\,e_1)}{(1 + AB_1 c e_1 + AB_2 c e_2)}\,dc, \quad de_2 = \dfrac{(A\,e_2)}{(1 + AB_1 c e_1 - AB_2 c e_2)}\,dc$

e) $de_1 = g_1 \dfrac{(e_1)}{\{1 - c[A_1 g_1(e_1) + A_2 g_2(e_2) + A_3 g_3(e_3)]\}}\,dc$

$de_2 = g_2 \dfrac{(e_2)}{\{1 - c[A_1 g_1(e_1) + A_2 g_2(e_2) + A_3 g_3(e_3)]\}}\,dc$

$de_3 = g_3 \dfrac{(e_3)}{\{1 - c[A_1 g_1(e_1) + A_2 g_2(e_2) + A_3 g_3(e_3)]\}}\,dc$

f) Show that if

$$\sum_1^n \dfrac{\partial}{\partial c}(g_i f_i) = \dfrac{\partial g_1}{\partial e_1} = \dfrac{\partial g_2}{\partial e_2} = \cdots = \dfrac{\partial g_n}{\partial e_n}$$

then the following system equations for $i = 1, ..., n$ are exact.

$$de_i = \dfrac{g_i\,dc}{1 - \sum\limits_{i=1}^{n} g_i f_i}$$

8 Refer to Section 8.3.3 for single-input, single-output system equations with feedback. Obtain the general solution for the system equation when the following conditions are satisfied in Eq. (8.3.15).

a) $n = m = -1$, $f = f(e)$, $g = \dfrac{H(e)}{G(c)}$ where $d\dfrac{G(c)}{dc} = $ constant

b) $n = m = -1$, $g = g(e)$, $f = F(e)\,H(c)$ where $d\operatorname{Ln}\dfrac{H(c)}{dc} = $ constant

c) $n = m = 0$, $g = g(e)$, $\dfrac{\partial f}{\partial c} = \partial\operatorname{Ln}\dfrac{g}{\partial e}$

d) $n = m = 0$, $f = f(e)$, $\dfrac{\partial f}{\partial c} = \partial\operatorname{Ln}\dfrac{g}{\partial e}$

e) $n = 0$, $m = -1$, $\partial\dfrac{\left(f - \dfrac{1}{g}\right)}{\partial c} = 0$

f) $n = -1$, $m = 0$, $f = f_0$ (constant), $f_0\dfrac{\partial g}{\partial c} = \dfrac{\partial g}{\partial e}$

9 Refer to Section 8.3.4 for inversion of system kernels with feedback. Obtain the general solution for the single-input, single-output system equations defined by Eqs. (8.3.20), (8.3.21), and (8.3.27).

a) $G = \operatorname{Ln}(f^a\,g^b)$,	$H = \operatorname{Ln}(f^m\,g^n)$
b) $G = \operatorname{Ln}(f^a\,g^b)$,	$H = A\,f + B$
c) $G = A\,g + 2\,B\,f + D$,	$H = E\,g\,f - B\,f^2 + F$
d) $G = $ constant,	$H = A + F(g)$

9

System Science Applications

In this chapter, we will examine a variety of systems from many fields of human knowledge and the environment. These systems provide a broad sampling of simple but practical quantitative models in which to apply the various methods of modeling, analysis, and evaluation activities provided in this text. However, since the detailed examination, analysis, and evaluation of any of the systems considered here would be an unreasonable venture in time and effort. Our purposes will be primarily to initiate a model development and briefly analyze the systems posed in this chapter for their pedagogical and demonstrative value rather than for detailed knowledge and understanding of the specific systems considered.

We introduce the chapter by providing a comprehensive but simple classification of human knowledge into categories of practical interest for system science. Then we devote the remainder of the chapter to specific system studies and attempt to demonstrate many of the modeling and analysis concepts, methods, and schemes we have developed previously. Selected MATLAB software tools, especially the Symbolic Toolbox, are employed to manipulate, model, analyze, graph, and otherwise assist in the evaluation of many of the system applications presented in this chapter. However, detailed, and comprehensive processing and evaluation of these system models using MATLAB and its various Toolboxes are deferred to the reader. It is assumed that the reader is already competent with MATLAB and its Toolboxes or can become so.

Finally, an important requirement for any rational system model is the correctness and consistency of the units or dimensions associated with the inputs, outputs, and system parameters. Generally, in each of the system models we examine in this chapter, we will also assign units associated with the variables and constants appearing in the model and are generally shown in parentheses (units). The system investigator should ensure consistency of the units and dimensions used in the modeling and analysis. Performing a consistency check frequently during the development of the model will indicate errors in the modeling effort and permit early correction.

9.1 Classification of System Science by Topics

There are so many distinct systems within the recognition and interest of the human intellect that it is prudent to establish some basis for classifying and cataloging systems. The range of identifiable systems within the discipline of system science is as broad and varied

Introduction to System Science with MATLAB, Second Edition.
Gary Marlin Sandquist and Zakary Robert Wilde.
© 2023 John Wiley & Sons Ltd. Published 2023 by John Wiley & Sons Ltd.

as the full imagination and accumulated knowledge of humankind. Indeed, the system models and the system equations of state by which we comprehend these systems constitute the rational basis by which we understand our universe and ourselves.

The *Encyclopaedia Britannica* has attempted the ambitious task of outlining and classifying all rational fields of human knowledge. We will use the *Britannica's* Classification of Knowledge as a useful guide for sorting and identifying system science topics. Table 9-1 is an adaptation of the Table of Contents of the 15th Edition of the *Encyclopaedia Britannica's* Outline of Knowledge (Propaedia), which classifies human interests and knowledge into 10 major classifications or parts as follows:

- Part 1 addresses physical science concepts of space, time, matter, and energy.
- Part 2 deals with physical science related to Earth Sciences.
- Part 3 treats all life forms (flora and fauna) other than humans that are found on the Earth.
- Parts 4 through 9 address human life, society, art, technology, religion, and history, respectively.
- Part 10 classifies branches of formal human knowledge.

Table 9-1 Classification of System Science Topics

1.0 Physical Science (space, time, matter, energy)

    1.1 Atoms: atomic nuclei and elementary particles

        1. Atomic nucleus, elementary particles

        2. Structure and properties of atoms

        3. Standard Model of Particle Physics (SMPP)

        4. Quantum Mechanics

        5. Dark Matter and Dark Energy

        6. Theory of Everything (TOE)

    1.2 Energy, radiation, states, and transformations of matter

        1. Chemical elements

        2. Chemical compounds

        3. Chemical and nuclear reactions

        4. Heat, thermodynamics, nonsolid states of matter

        5. Solid-state of matter

        6. Mechanics of rigid and deformable bodies

        7. Electricity and magnetism

        8. Waves and wave motion

    1.3 Physical universe

        1. Nature of space and time (spacetime)

        2. Cosmology

        3. Galaxies and stars

        4. Milky Way

        5. Solar system

        6. Exoplanets

2.0 Earth (air, water, land, and climate)

    2.1 Earth's properties, structure, composition

        1. Planet

        2. Physical properties

        3. Structure and composition of interior

        4. Constituent minerals and rocks

Table 9-1 (Continued)

2.2 Earth's atmosphere and hydrosphere
1. Atmosphere
2. Hydrosphere
3. Sea Motions, tides, currents
4. Weather and climate

2.3 Earth's surface features
1. Physical surface features
2. Features produced by geomorphic processes

2.4 Earth's history
1. Origin and development
2. Interpretation of geological record
3. Eras and periods of geological time

3.0 Earth Life (flora and fauna)

3.1 Nature and diversity of living things
1. Characteristics of living things
2. Origin of life and evolution of living things
3. Classification of living things

3.2 Molecular basis of biological processes (Genomics, DNA, RNA)
1. Chemical and vital biological processes
2. Metabolism: bioenergetics and biosynthesis
3. Biological processes at molecular level

3.3 Structures and functions of organisms
1. Cellular basis of form and function
2. Relation of form and function in organisms
3. Regulation and integration of vital processes
4. Covering and support
5. Procurement and processing of nutrients
6. Gas exchange, internal transport, elimination
7. Reproduction and sexual function
8. Development: growth, differentiation, morphogenesis
9. Heredity

3.4 Behavioral responses of organisms
1. Nature and patterns
2. Development and range of individual and group behavior

3.5 Biosphere
1. Basic chemical and physical features
2. Community of interacting populations
3. Hazards of life, disruptions, death
4. Patterns of life
5. Humans in biosphere

4.0 Human life (evolution, health, diseases, behavior, communications)

4.1 Stages in development of human life on Earth
1. Order of primates
2. Quaternary hominidae
3. Evolution of humanity
4. Races of humanity

4.2 Human health and diseases
1. Human health

Table 9-1 (Continued)

    2. Structures and functions of human body

    3. Manifestations, recognition, treatment of human diseases

    4. Diseases and disorders of human body

    5. Practice of medicine and health care

  4.3 Human behavior and experience

    1. General theories

    2. Conditions and developmental processes

    3. Influence of environment

    4. States affecting behavior and conscious experience

    5. Capacities to integrate human behavior and conscious experience

    6. Development of teaming and thinking

    7. Guide for organizational behavior and conscious experience
    8. Personality and self

  4.4 Communication and language

    1. Methods and forms
    2. Language (written, oral, etc.)

  4.5 Aspects of human daily life

    1. Means of subsistence
    2. Leisure and recreation

5.0 Human society (culture, society, wealth, politics, education)

  5.1 Culture

    1. Development of human culture

    2. Variety of sociocultural forms

  5.2 Social organization and social change

    1. Structure and change

    2. Group structure of society

    3. Social status

    4. Collective behavior and mass society

    5. Urban and rural communities

  5.3 Production, distribution, utilization of wealth

    1. Economic concepts, issues, systems

    2. Human wants and their economic expression

    3. Markets, pricing, distribution of goods

    4. Organization of production and distribution

    5. Distribution or income and wealth

    6. Macroeconomics

    7. Economic growth and planning

  5.4 Politics and government

    1. Political theory and practice

    2. Structure, branches, offices of government

    3. International relations, peace, and war

  5.5 Social organization and social change

    1. Philosophies and systems of law

    2. Branches of public law, substantive and procedural

    3. Professions and practice of law

  5.6 Education

    1. Aims and methods of education

    2. Education around world

    3. Organization and delivery of education

    4. Computer-based delivery of education

6.0 Art (literature, theater, music, dance, sculpture, movies)

  6.1 Art in general

    1. Theory and classification of arts

    2. Experience and criticism of works of art

    3. Non-esthetic contexts of art

Table 9-1 (Continued)

6.2 Particular arts
1. Literature
2. Theater
3. Motion pictures
4. Music
5. Dance
6. Architecture, landscape, urban design
7. Sculpture
8. Drawing, painting, printmaking, photography
9. Arts of decoration and functional design

7.0 Technology (tools, machines, manufacturing, construction, etc.)

7.1 Nature and development of technology
1. Technology's scope and history
2. Organization of human work

7.2 Elements of technology
1. Technology of energy conversion and utilization
2. Technology of tools and machines
3. Technology of measurement, observation, control
4. Extraction and conversion of industrial raw materials
5. Technology of industrial production processes

7.3 Major Fields of technology
1. Agriculture and food production
2. Technology of major industries
3. Construction technology
4. Transportation technology
5. Information processing and communication systems
6. Military technology
7. Technology of urban community
8. Technology of Earth and space exploration

8.0 Religion (Hinduism, Buddhism, Judaism, Christianity, Islam, etc.)

8.1 Religion in general
1. Diverse religious views
2. Religious experience and phenomenology

8.2 Particular religions
1. Prehistoric and primitive religions
2. Religions of ancient peoples
3. Hinduism and other religions of India
4. Buddhism
5. Religions of East Asia, China, Korea, Japan
6. Judaism
7. Christianity
8. Islam
9. Other religions and their movements

9.0 History of humanity (ancient, European, Eastern, Western)

9.1 Civilizations of ancient Southwest Asia, North Africa, Europe
1. Southwest Asia and Egypt, Aegean, North Africa
2. Europe and Mediterranean world to 395

9.2 Civilizations of medieval Europe, North Africa, Southwest Asia
1. Western Europe, Byzantine Empire, Eastern Europe (395–1050)
2. Empire of Caliphate and its successor states to 1050

(Continued)

Table 9-1 (Continued)

3. Western Christendom in Middle Ages (1050–1500)

4. Crusades, Islamic states of Southwest Asia, North Africa, Europe, Eastern Christendom (1050–1480)

9.3 Civilizations of East, Central, South, Southeast Asia

1. China to late T'ang (755)

2. China from late T'ang (755)

3. Inner (Central and Northeast) Asia to 1750

4. Japan to Meiji Restoration (1868) and Korea to 1910

5. Indian subcontinent and Ceylon to 1200

6. Indian subcontinent (1200–1761) and Ceylon (1200–1505)

7. Southeast Asia to 1600

9.4 Civilizations of sub-Saharan Africa to 1885

1. West Africa

2. Nilotic Sudan and Ethiopia from 550

3. East Africa and Madagascar

4. Central Africa

5. Southern Africa

9.5 Civilizations of pre-Columbian America

1. Andean civilization to 1540

2. Meso-American civilization to 1540

9.6 World development up to 1920

1. Western Europe (1500–1789)

2. Eastern Europe, Southwest Asia, North Africa (1480–1800).

3. Europe (1789–1920)

4. European colonies in Americas (1492–1790)

5. United States and Canada (1763–1920)

6. Latin America and Caribbean nations to 1920

7. Australia and Oceania to 1920

8. South Asia under European imperialism (1500–1920)

9. Southeast Asia under European imperialism to 1920

10. China 1839 until 1911 revolution

11. Japan from Meiji Restoration to 1910

12. Southwest Asia and North Africa (1800–1920)

13. Sub-Saharan Africa (1885–1920).

9.7 Modern World since 1920

1. International movements, diplomacy, wars, terrorism

2. Europe

3. United States and Canada

4. Latin American and Caribbean nations

5. East Asia: China revolution, Japanese hegemony

6. South and Southeast Asia: late colonial period and new nations

7. Australia and Oceania

8. Southwest Asia and Africa

10.0 Branches of formal knowledge

10.1 Logic

1. History and philosophy of logic

2. Formal logic, metalogic, applied logic

10.2 Mathematics

1. History and foundations of mathematics

2. Branches of mathematics

3. Applications of mathematics

Table 9-1 (Continued)

10.3 Science
    1. History and philosophical basis of science
    2. Physical sciences
    3. Material sciences
    4. Earth sciences
    5. Biological sciences
    6. System sciences
    7. Medical and affiliated disciplines
    8. Social sciences, psychology, and ethics
    9. Technological sciences
10.4 History and humanities
    1. Historiography and study of history
    2. Humanities and humanistic scholarship
10.5 Philosophy
    1. Nature and divisions of philosophy
    2. History of philosophy
    3. Philosophical schools and doctrines

Source: Adapted from *Encyclopaedia Britannica's* Outline of Knowledge, 15th Edition.

Obviously, our sampling and treatment for instructive and exemplary system models from this comprehensive listing of possible system topics must be limited and brief. However, it should be apparent to the system investigator that there are numerous, unexamined, and major areas of human knowledge that can be beneficially subjected to the mathematical rigor of system science inquiry. The result of such erudite efforts would be an improved understanding of humanity and our universe. With enhanced understanding, a better quality and appreciation of life for humans would result. Improved and quantitative understanding of many areas in society, such as health, poverty, employment, religious intolerance, politics, and especially internecine terrorism, would be of great value in promoting peace, understanding, and prosperity.

In this chapter, we will select some simple but useful systems for modeling and analysis that demonstrate various methods of modeling, analysis, and evaluation including the implementation of computer software such as MATLAB and its toolboxes including the Symbolic Toolbox to assist in the quantitative modeling of a system. However, the illustrative system models addressed, and analysis efforts imposed are not meant to provide comprehensive models of any of the system examples considered. A complete analysis of even a relatively simple system with multiple inputs and outputs represents a significant investment of time and effort in modeling, analysis, and evaluation. Furthermore, the investigator must devote considerable background study and effort to become familiar and competent with the system under investigation so that the time devoted to modeling, analysis, and evaluation will be effective and successful. An excellent introductory background for a given system under study can be obtained from a comprehensive encyclopedia such as the Encyclopaedia Britannica, Google (or Google Search), and Wikipedia, the free Internet encyclopedia, coupled with the published literature in the particular field. It is estimated that Google Search performs over 3.5 billion searches daily and is the dominant, worldwide website and database. Wikipedia also has a very large, open database and is a useful starting source for system science exploration that we have employed in this text. However, data found in Wikipedia must be confirmed and used with due diligence because the database is open to users and subject to error and bias by its contributors and users.

Cataloged information sources at major libraries are also useful and obvious sources for data associated with a given system. The information available using the World Wide Web ("www") is another useful source of current and rapid information gathering regarding almost any field of interest. Using appropriate "Uniform Resource Locators" or URLs for subjects, the investigator can rapidly access pertinent and current information using the Internet. For example, computer search engines including "www.google.com, www.yahoo.com, www.aol.com, and www.msn.com, wikipedia.com" can provide extensive lists of information sources usually directly accessible through the web. The development of Internet search engines and their expanding coverage of subjects relative to system science are increasingly evident and valuable.

Finally, it is assumed that the reader is familiar with the International System of Units provided in Table 6-3 in Chapter 6 of this text. This text generally employs these units in developing and evaluating the system models found in this Textbook.

9.2 System Science Applications to Space, Time, Matter, and Energy in Physical Science

The fundamental constituents of existence, space, time, matter, and energy are important and exciting topics in system science. Comprehensive knowledge of the nature and behavior of these fundamental entities is essential to the understanding of all other physical systems found in the universe. Interestingly, some of the most profound and valuable models in the entire field of system science reside within the basic domain of space, time, matter, and energy. These models include the topics of special and general relativity, four-dimensional spacetime, entropy, quantum mechanics, dark matter and dark energy, multiple and parallel universes or multiverses and super string or M-Theory that assumes the existence of ten spatial dimensions. The Standard Model of Particle Physics (SMPP) and the Higgs Boson, whose existence was demonstrated in 2014 utilizing the Large Hadron Collider (LHC) near Geneva, Switzerland, and the "Big Bang" model for the origin of our universe, ~13.8 billion years ago, are major developments in the last few decades. We consider a few important models in this very basic and dynamic field of system science.

9.2.1 First Law of Thermodynamics

Often we become so familiar with certain basic system models, particularly those we consider as fundamental statements or "laws of nature," that we are inclined to forget that these laws are simply well-established system equations for models of important physical phenomena that have been subjected to intensive testing and repeated verification. Because of the extensive testing, verification, and acceptance of these models, we assert that these system models express experimental truth or fundamental laws of physical science. Such an example is the "First Law of Thermodynamics" for a closed thermodynamic system with continuous properties, which is defined by the following differential system equation.

$$dQ = dU + PdV \tag{9.2.1}$$

where

dQ = differential change in thermal or heat energy transiting system (J)
dU = differential change in internal energy within system (J)
dV = differential change in volume of system (m^3)
P = mean absolute pressure in system (N/m^2)
PdV = mechanical work performed on system (J or N m)

Observe that throughout our examination of systems from many fields of human knowledge and the environment, we will employ units or dimensions for the variables and parameters found in the system models. Generally, we utilize SI (*System International*) units, although other standard terminology may be used. For Eq. (9.2.1), we have used J for the Joule energy unit, N for Newton, and m for meter. Refer to Chapter 6 for a detailed evaluation of scientific units.

Returning to Eq. (9.2.1), if we consider dQ as the differential system output and dU and dV as differential inputs, then we have a double-input, single-output system with the following system diagram:

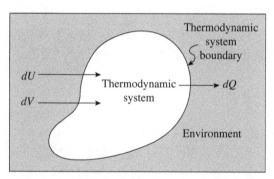

We observe that the boundary for the thermodynamic system model effectively isolates the system from the surrounding environment. This boundary prevents the exchange of any mass or energy between the system and its environment except for those exchanges that can occur through the differential inputs dU and dV and the output dQ. The first law of thermodynamics, Eq. (9.2.1), has as a possible system block diagram the following representation:

$$
\begin{array}{c}
dU \longrightarrow \\
dV \longrightarrow
\end{array}
\boxed{
\begin{array}{c}
g_1 = 1 \\
g_2 = P
\end{array}
}
\longrightarrow dQ
$$

Observe that we could consider any one or two of the differential variables dQ, dU, or dV as outputs and the remaining differentials as inputs without loss of accuracy and generality.

If the system pressure P is a function of V only, then the system equation, the first law or Eq. (9.2.1), may be integrated to give

$$Q = U + \int P(V)dV + K \tag{9.2.2}$$

where K is the integration constant. In general, P depends on V, Q, and U, and the integral in Eq. (9.2.2) cannot be evaluated unless the functional dependence of each of these variables in P is known. Therefore, we have that

$$P = P(Q, U, V)$$

The fact that our system equation requires knowledge of the dependence of P on the other input and output variables, that is Q, U, and V, is a consequence of the experimental observation that thermal energy differential dQ and the mechanical work differential PdV are thermodynamic processes that are path-dependent functions. Thus, from Eq. (9.2.2) for the solution of the system equation, the value of Q obtained depends on how P varies over the integration range of V. However, the system's internal energy, U, may be regarded as an input to the system whose integration is not dependent on the thermodynamic process taken by the system between two equilibrium states. Mathematically, U is an exact differential that depends only upon its ends points for evaluation of the change in U and does not depend upon the thermodynamic processes that brought the system to its equilibrium state. Note that end values assume that the system is at equilibrium throughout the system.

Furthermore, we from Chapter 5, Section 5.2.4 [see Eq. (5.2.21) for a double input, single-output system] that system Eq. (9.2.1) can be integrated, and it has a solution of the form

$$\phi(Q, U, V) = K$$

where $\phi(Q, U, V)$ is an arbitrary, continuous function of Q, U, V and K is the integration constant. Integrability of Eq. (9.2.2) requires that $P = P(Q, U, V)$ and satisfies the requirement that

$$\frac{\partial P}{\partial U} = -\frac{\partial P}{\partial Q} \tag{9.2.3}$$

A convenient example of a general function for $P(Q, U, V)$ that satisfies Eq. (9.2.3) is the following function where $G(V)$ are $H(V)$ are continuous functions that possess the derivatives displayed.

$$P = G(V) - (Q - U)\frac{1}{H(V)}\frac{\partial P}{\partial Q} \tag{9.2.4}$$

It is easily shown that this functional form for P leads to an integrable system equation for Eq. (9.2.1). If we substitute Eq. (9.2.4) into system Eq. (9.2.1) and multiply by $H(V)$ (which is the integrating factor for the equation), we obtain Eq.(9.2.1).

$$H(V)dQ + Q\frac{dH(V)}{dV}dV = H(V)dU + U\frac{dH(V)}{dV}dV + G(V)H(V)dV \tag{9.2.5}$$

We observe that the left side of Eq. (9.2.5) is equivalent to

$$H(V)\,dQ + Q\frac{dH}{dV}dV = d\left[H(V)Q\right]$$

and the first two terms on the right side of Eq. (9.2.5) become

$$H(V)dU + U\frac{dH}{dV}dV = d[H(V)U]$$

Therefore, Eq. (9.2.5) possesses the exact, integrable form

$$d[H(V)Q] = d[H(V)U] + G(V)H(V)dV$$

which can be integrated to provide the following solution for the system equation

$$Q = U + \frac{1}{H(V)} \int G(V) \, H(V) dV + K \qquad (9.2.6)$$

where K is the integration constant.

For an example of a specific, thermodynamic process that exhibits a pressure function (or system kernel) defined by Eq. (9.2.4), assume that the system is composed of an isolated perfect gas that is maintained at constant temperature. All thermodynamic processes exhibited by this system are isothermal and no changes in temperature occur within the perfect gas during the thermodynamic process. Thus, for an isothermal, perfect gas system where N is the number of moles of ideal gas and R is the gas constant for the given gas, we have

$$PV = N R T = \text{constant} = P_0 V_0$$

So

$$P = \frac{P_0 V_0}{V}$$

Thus in Eq. (9.2.4), we find that

$$H(V) = \text{constant so } \frac{dH}{dV} = 0$$

and

$$G(V) = P = \frac{P_0 V_0}{V}$$

We now find as the solution for the system equation, from Eq. (9.2.6), that

$$Q = U + P_0 V_0 \, \text{Ln} \, V + K$$

This is the standard solution for the heat exchanged by a closed, isothermal thermodynamic system composed of a perfect gas.

9.2.2 Particle Diffusion Model

The basic system equations for the one-dimensional diffusion of small particles within a diffusing medium in the absence of any chemical reactions between the particles are given by

$$dn = -\frac{J}{Dv} dx \qquad (9.2.7.1)$$

$$dJ = \left(s - \frac{nv}{\lambda} \right) dx \qquad (9.2.7.2)$$

where

$n = $ particle density at position x (particles/cm^3)
$J = $ particle current at position x (particles/cm^2-s)
$D = $ particle diffusion coefficient (cm)
$v = $ particle speed (cm/s)
$\lambda = $ particle removal mean free path (particles-cm/particles removed)

s = volumetric production of particles (particles produced/cm^3-s)

x = particle position along the diffusing dimension (cm)

If D, v, S, and λ are all constants with respect to x, then upon eliminating dx in Eq. (2.7.1) and Eq. (2.7.2), we obtain a separable, first-order differential equation as follows:

$$\frac{dn}{dJ} = -\frac{1}{Dv}\frac{J}{s-nv/\lambda}$$

which can be separated and integrated as follows

$$\int \left(s - n\frac{v}{\lambda}\right) dn = -\frac{1}{Dv}\int J\, dJ + K$$

to give

$$sn - \frac{n^2 v}{2\lambda} = -\frac{1}{Dv}\frac{J^2}{2} + K \tag{9.2.8}$$

where K is an arbitrary constant. Equation (9.2.8) can also be expressed as

$$\frac{J^2}{K_0} - \frac{(n - \lambda s/v)^2}{\lambda K_0/Dv^2} = 1 \tag{9.2.9}$$

This is the general equation for a hyperbola with center at $J = 0$ and $n = \lambda s/v$. K_0 is the adjusted integrating constant. If we solve Eq. (9.2.9) for J and substitute this expression into Eq. (9.2.7.1), we have

$$dn = -\frac{1}{Dv}\sqrt{K_0}\left[1 + \frac{Dv^2}{\lambda K_0}\left(n - \frac{\lambda s}{v}\right)^2\right]^{1/2} dx$$

This equation is separable and can be integrated to give the following with the additional integration constant K_1.

$$\int \left[1 + \frac{Dv^2}{\lambda K_0}(n - \lambda s/v)^2\right]^{-1/2} dn = -\frac{1}{Dv}\sqrt{K_0}\, x + K_1$$

Completing the integration and reducing the resulting expression, we find that we can express the solution for the particle density $n(x)$ as a function of x

$$n(x) = \frac{s\lambda}{v} + K_2 \exp(-ax) + K_3 \exp(ax)$$

where K_2 and K_3 are new arbitrary constants and $a = (D\lambda)^{-1/2}$. Furthermore, from system Eq. (9.2.7.1), we have for $J(x)$ that

$$J(x) = -Dv\frac{dn}{dx} = v\sqrt{\frac{D}{\lambda}}\,[K_2 \exp(-ax) - K_3 \exp(ax)]$$

Now, suppose as boundary conditions we have that $n(0) = n(L) = 0$. Then we find that

$$n(x) = \frac{s\lambda}{v}\left[1 - (1 - e^{-aL})\frac{\sinh ax}{\sinh aL}\right] \tag{9.2.10}$$

and

$$J(x) = \frac{s}{a}\left[(1-e^{-aL})\frac{\cosh ax}{\sinh aL}\right] \tag{9.2.11}$$

If we eliminate the explicit dependence of Eqs. (9.2.10) and (9.2.11) on the position x, as was done in Eq. (9.2.9), we obtain the following equation as the explicit, hyperbolic form of the solution for the particle density $n(x)$ and particle current $J(x)$.

$$[J/B]^2 - \left[\left(n - \frac{s\lambda}{v}\right)/A\right]^2 = 1 \tag{9.2.12}$$

where

$$A^2 = \left(\frac{2s\lambda}{v}\right)^2 C^2$$

$$B^2 = D\lambda s^2 C^2$$

$$C^2 = \frac{[1 - \exp(-a\lambda)]^2}{\cosh^2(a\lambda)}$$

$$a = (D\lambda)^{-1/2}$$

A generalized plot of Eq. (9.2.12) has the form of a phase plane, which depicts graphically the dependence between the particle density J and the current n. How the density n and current J change with position are shown by the following MATLAB graph.

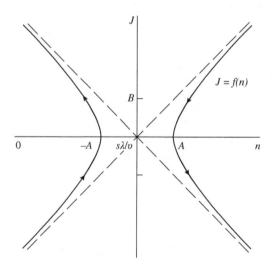

9.2.3 Relativistic Mechanics Model

One of the most significant developments to occur in physics in the early 1900s was the recognition that Newtonian mechanics did not correctly describe the behavior of physical particles whose mass was very small or zero (e.g., atomic electrons, protons, neutrons, photons, etc.). In this situation, quantum mechanics is required for a more accurate description of the physical system. Furthermore, for particles with mass whose speed approaches the speed of light, relativistic mechanics is necessary to account for the experimentally observed behavior of the particles. Furthermore, the system model for the dynamical behavior of matter based upon Newtonian mechanics is inadequate to describe the full range of particle sizes, masses, and motion that is experimentally observed. The coupling of space and time (spacetime) recognized by Einstein has now replaced Newtonian mechanics with more accurate mechanics based on Einstein's Theory of "Space-Time." Unfortunately, however, Einstein's space-time equations and quantum mechanics are presently irreconcilable. The correct mechanics model that describes the complete range of behavior for all measurement values of matter and energy is presently not known. Hopefully, as the full nature of the fundamental forces (strong, weak, electromagnetic, and especially gravitational force) is better understood, these fundamental forces can be unified into a single, comprehensive system model. Such a model, if it can be developed has been colloquially depicted as the "Theory of Everything" or TOE model. However, such a TOE model waits to be discovered, confirmed, and implemented.

Let us examine two basic equations from relativistic mechanics in the Special Theory of Relativity. The total energy E possessed by any material particle in a prescribed reference coordinate is

$$E = m(v)c^2 \tag{9.2.13}$$

where the relativistic mass m of the particle in the reference coordinate system is determined by the Lorentz equation for the velocity dependence of mass.

$$m(v) = \frac{m_0}{\sqrt{1 - v^2/c^2}} \tag{9.2.14}$$

Equation parameters are defined as follows

E = total energy of particle of mass m (J)
m_0 = rest mass [stationary particle mass in reference coordinates (kg)]
v = particle's speed with respect to reference coordinate (m/s)
c = speed of light in a vacuum as a fundamental constant (m/s)

The system equations associated with Eqs. (9.2.13) and (9.2.14) are easily obtained by differentiation. The results are found from Eq. (9.2.13) as follows

$$dE = c^2 dm \tag{9.2.15}$$

and from Eq. (9.2.14), noting that $p = mv$ is the particle's scalar momentum (kg m/s), we have

$$c^2 m\, dm = mv\, dmv = p\, dp$$

which can be expressed as

$$dp = \frac{E}{p} dm \tag{9.2.16}$$

Setting up a block diagram for Eqs. (9.2.15) and (9.2.16) with dm as the differential input and dE and dp as the differential outputs, we have the system model shown in Figure 9-1.

Observe that if we eliminate dm between Eqs. (9.2.15) and (9.2.16), we find that

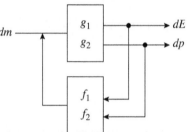

Fig. 9-1 Relativistic system model with no feedback.

$$EdE = c^2 pdp$$

This equation is easily integrated to give

$$E^2 = E_0^2 + c^2 p^2 \qquad (9.2.17)$$

where we have used the initial conditions $E = E_0$ when $p = mv = 0$.

It is instructive to solve Eq. (9.2.17) for E, and we obtain the interesting result

$$E = \pm \left(E_0^2 + c^2 p^2\right)^{1/2}$$

This equation indicates that energy E can have a positive or negative value and thus positive and negative energy states exist together with positive (ordinary matter) and negative (antimatter) mass states. Observe that both m and m_0 could also be either positive or negative in Eq. (9.2.14) with no inconsistency.

Returning to our block diagram, it is interesting to speculate whether intrinsic or internal feedback is present within the two system kernels g_e and g_p shown in Figure 9-1. To investigate our speculation, let us consider a related model block diagram to Figure 9-1, which has dm as input and dE and dp as outputs and exhibits full feedback:

We know from Chapter 3 (Section 3.2.2) that the system equations for this block diagram are

$$dE = \frac{g_1}{1 - g_1 f_1 - g_2 f_2} dm$$

$$dp = \frac{g_2}{1 - g_1 f_1 - g_2 f_2} dm$$

Our objective is an attempt to find a set of reasonable kernel functions g_1, g_2, f_1, and f_2 such that

$$\frac{g_1}{1 - g_1 f_1 - g_2 f_2} = g_e = c^2 \qquad (9.2.18.1)$$

$$\frac{g_2}{1 - g_1 f_1 - g_2 f_2} = g_p = \frac{E}{p} \qquad (9.2.18.2)$$

We have two equations containing four arbitrary unknowns. Thus, there are an infinite number of possibilities for g_1, g_2, f_1, and f_2 that will satisfy Eq. (9.2.18). However, because of factors such as symmetry, form, and limiting values, it is possible to identify practical candidate kernels for consideration.

If we divide Eq. (9.2.18.1) by Eq. (9.2.18.2), we can eliminate the feedback kernels and obtain a ratio for the kernels g_1 and g_2 independent of f_1 and f_2

$$\frac{g_1}{g_2} = c^2 \frac{p}{E} \tag{9.2.19}$$

Furthermore, let us require that $f_1 = f_1(E)$ be a function of E only and $f_2 = f_2(p)$ be a function of p only so that these feedback kernels only reference their direct output differentials dE and dp, respectively. Then Eq. (9.2.18.2) becomes, using Eq. (9.2.19) for g_1/g_2 and rearranging the resulting equation

$$\frac{1}{E} + \frac{c^2}{E} f_1(E) = \frac{1}{p} \left[\frac{1}{g_2} - f_2(p) \right] \tag{9.2.20}$$

Observe that if we assume that $g_2 = g_2(p)$ only, then we have separated Eq. (9.2.20) into a function of E on the left side of the equation and a function of p on the right side. Now, if we equate Eq. (9.2.20) to an arbitrary constant K_1, we have that

$$f_1(E) = \frac{K_1 E - 1}{c^2}$$

and

$$f_2(p) = \frac{1}{g_2(p)} - K_1 p$$

Furthermore, let us require that $g_1 = g_1(E)$ only [recall that $g_2 = g_2(p)$], so from Eq. (9.2.19) we can again separate functions to give

$$g_1(E) \frac{E}{c^2} = p\, g_2(p) = K_2$$

where K_2 is a second arbitrary constant.

Thus, identifying the results of our assumptions, we find that

$$g_1(E) = \frac{K_2 c^2}{E}, \quad g_2(p) = \frac{K_2}{p}$$

and

$$f_1(E) = \frac{1}{c^2}(K_1 E - 1), \quad f_1(p) = p\,(1/K_2 - K_1)$$

where K_1 and K_2 are arbitrary constants. We observe for dimensional compatibility that K_1 must have dimensions of reciprocal energy, so we will let $K_1 = a/E_0$, and K_2 has dimensions of energy, so let $K_2 = b\, E_0$, where E_0 is the rest mass energy and parameters a and b are dimensionless. Furthermore, from a check for consistency of these kernel functions; g_1,

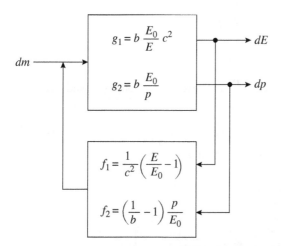

Fig. 9-2 Relativistic energy system model with feedback.

g_2, f_1, and f_2 with Eqs. (9.2.18.1) and (9.2.18.2); we find that we require that $a = 1$, while b is completely arbitrary (i.e., b could be a complex number with any finite numerical value). With these assumptions in place, a block diagram for our system model with feedback is as shown in Figure 9-2.

We note the symmetry of each of the kernels. Observe that g_1 and g_2 are inversely proportional to their respective outputs E and p, while the feedback kernels f_1 and f_2 are directly proportional to their respective inputs E and p. Since there is generally no way to observe the existence of inherent feedback in a closed system, this block diagram could be an alternative model for the open-loop (i.e., no feedback) system shown in Figure 9-1.

Observe, however, that as $E \to E_0$ and $p \to 0$ (which implies that the speed, $v \to 0$, in our reference coordinate), then both feedback kernels f_1 and f_2 approach zero. The limiting response for the open-loop system with $f_1 = f_1 = 0$ is given by

$$dE = b\,\frac{E_0}{E}c^2 dm$$

$$dp = b\,\frac{E_0}{p}\,dm$$

If these asymptotic system equations are solved, we find that

$$E^2 = 2b\,E_0 c^2 m + K_e \tag{9.2.21.1}$$

and

$$p^2 = 2b\,E_0 m + K_p \tag{9.2.21.2}$$

where K_e and K_p are arbitrary integration constants. Remembering that Eqs. (9.2.21) are valid in the limit as $E \to E_0$ for $p \to 0$, we require that

$$E_0^2 = 2b\,E_0\,c^2 m_0 + K_e$$

and

$$p = 2b\,E_0 m_0 + K_p$$

Thus, using these results to determine K_e and K_p, Eqs. (9.2.21) become

$$E^2 = 2b\, E_0\, c^2 m + E_0^2(1 - 2b) \tag{9.2.22.1}$$

$$p^2 = 2b\, E_0(m - m_0) \tag{9.2.22.2}$$

Now let us eliminate the parameter b from these two equations, and we find (Recall: $E_0 = m_0\, c^2$)

$$E^2 = E_0^2 + c^2 p^2 \tag{9.2.23}$$

which is identical with Eq. (9.2.17) as expected. Observe that for $p < m_0\, c$, we have from Eq. (9.2.23) that the expansion of the radical provides the familiar result from the Special Theory of Relativity for the kinetic energy of objects traveling at speeds less than the velocity of light, c.

$$E = \pm E_0 \left[1 + \frac{1}{2}\left\{\frac{p}{m_0 c}\right\}^2 - \frac{1}{8}\left\{\frac{p}{m_0 c}\right\}^4 + \frac{1}{16}\left\{\frac{p}{m_0 c}\right\}^6 - \cdots \right]$$

It is possible to place a constraint on b by recalling Eq. (9.2.16), so we have

$$p\,dp = E\,dm = mc^2\,dm$$

Integrating this equation and observing that $m = m_0$ when $p = 0$,

$$p^2 = c^2\left(m^2 - m_0^2\right) = c^2(m + m_0)\,(m - m_0) \tag{9.2.24}$$

Now observe that if we compare Eq. (9.2.22.2) with Eq. (9.2.24), we have

$$\mathrm{Lim}_{m \to m_0}\left(1 + \frac{m}{m_0}\right) = 2 = 2b$$

and we find that $b = 1$ is required to satisfy this limiting condition. Finally, setting $b = 1$ in our system equations with feedback, we have that

$$g_1 = \frac{E_0}{E}\, c^2 \quad \text{and} \quad g_2 = \frac{E_0}{p}$$

and

$$f_1 = \frac{1}{c^2}\left(\frac{E}{E_0} - 1\right) \quad \text{and} \quad f_2 = 0$$

and our system block diagram, Figure 9-2, with feedback assumes the following form when $b = 1$:

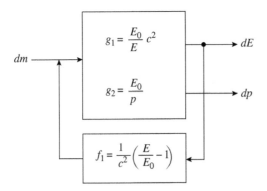

With b set to unity, intrinsic feedback from the momentum differential output has vanished and only energy feedback remains. These system equations for the above block diagram for our relativistic mechanics model are still consistent with the standard Eq. (9.2.15) and Eq. (9.2.16), but now a possible candidate for unobservable, internal system feedback has been developed. It is left to the reader to speculate on the validity and value of this system model with intrinsic feedback as an "alternative" description of relativistic mechanics now with the internal, but unobservable, feedback of energy.

9.2.4 System Problems for Matter, Energy, Space, and Time

Fusion Reaction Kinetics. A promising future energy resource presently under development, which could offer a nearly inexhaustible supply of fuel, is nuclear fusion. The fusion of hydrogen is the basic reaction that provides the enormous energy released from stars. Our own sun, which is an intermediate size star, converts hydrogen into helium with a large energy release that sustains all life and biological processes on earth. The basic fusion reactions under consideration presently for terrestrial applications may employ the two heavy isotopes of hydrogen, deuterium (D) and tritium (T). However, in the fusion reaction occurring between two deuterium nuclei, a helium-3 nucleus (He^3) may be produced, which can also undergo fusion with the hydrogen isotopes. In addition, neutrons (n), protons (H), and alphas (He^4) are produced in the fusion reaction.

The basic fusion reactions that are under investigation are as follows:

$$D + D \rightarrow He^3 + n + (3.27 \text{ MeV})$$
$$D + D \rightarrow T + H + (4.04 \text{ MeV})$$
$$D + T \rightarrow He^4 + n + (17.6 \text{ MeV})$$
$$D + He^3 \rightarrow He^4 + H + (18.3 \text{ MeV})$$

Our objective is to determine the time behavior of a fusion plasma (a high temperature, ionized gas mixture) composed of highly ionized atoms, primarily ions and electrons that collectively form a high kinetic energy, neutral electrical charged mass that is capable of exhibiting fusion of its colliding ion specie. Assuming that the two reactions for the $D + D$ reaction occur with equal probability, then an ion balance over time provides the following set of system equations for the ion densities of deuterium, n_D, tritium, n_T, and helium-3,

$$d \begin{bmatrix} n_D \\ n_T \\ n_{He} \end{bmatrix} = \begin{bmatrix} s_D - \{\sigma v\}_{DD}\, n_D{}^2 - \{\sigma v\}_{DT} n_D n_T - \{\sigma v\}_{DHe}\, n_D n_{He} \\ s_T + \dfrac{1}{4}\{\sigma v\}_{DD}\, n_D{}^2 - \{\sigma v\}_{DT} n_D n_T \\ s_{He} + \dfrac{1}{4}\{\sigma v\}_{DD}\, n_D{}^2 - \{\sigma v\}_{DT} n_D n_{He} \end{bmatrix} dt$$

where s_i is the externally supplied source, if present, of the particular ion specie denoted by the ith subscript and the parameters $\{\sigma v\}$ denote the mean product of the microscopic cross section and the center of mass interaction speed for each specific particle reaction.

We observe that our set of kinetics equations for fusion of these isotopes is nonlinear in each of the dependent variables for deuterium, n_D, tritium, n_T, and helium-3, n_{He}. No analytical solutions for these equations are known. However, to make some progress with analytical modeling, we will consider the following useful strategy.

For these system equations, perform the following:

a) Determine the equilibrium values for the ion densities, n_{D0}, n_{T0}, n_{He0}, which occur when

$$\frac{dn_D}{dt} = \frac{dn_T}{dt} = \frac{dn_{He}}{dt} = 0$$

b) Normalize the time-dependent ion density variables as follows:

$$N_D = \frac{n_D}{n_{D0}}, \quad N_T = \frac{n_T}{n_{T0}}, \quad N_{He} = \frac{n_{He}}{n_{He0}}$$

and express the resulting system equations to achieve a dimensionless form for each term in the equations. Note that the dimensionless form will also require normalization of the input variable, time.

c) Linearize the system equations by perturbation methods (see Chapter 7, Section 7.2.2) by letting

$$N_D = 1 + \delta N_D$$
$$N_T = 1 + \delta N_T$$
$$N_{He} = 1 + \delta N_{He}$$

and imposing linearization approximations as necessary

d) Obtain an analytical solution for the linear system equations obtained in part (c) using the symbolic differential equation solver from MATLAB.

e) Also with the symbolic differential equation solver from MATLAB, attempt to solve the normalized set of nonlinear system equations obtained in part (b). If the symbolic MATLAB solver is not successful (very likely), then numerically solve the system equations for a selected set of parameter values and initial conditions and provide a sample graphical output for N_D, N_T, and N_{He} as functions of time around the equilibrium values.

f) Compare the analytical solutions obtained from the linear system equations of part (d) and from the computer simulation of part (e) for the same parameter values and initial conditions. How well did the linearized system equations perform in describing the system model? In particular, over what range of time do the linearized and nonlinear solutions reasonably agree?

g) Attempt to eliminate the time dependence for these system equations and obtain a set of system equations in terms of N_D, N_T, and N_{He}. Plot the three-dimensional phase plane around the equilibrium value of operation. Use the Symbolic Toolbox in MATLAB for plotting the phase plane.

Energy State Model for the Atomic Nucleus. To a good approximation, ignoring Quantum Chromic Dynamics, there are only three discrete mass inputs, the number of protons, the number of neutrons, and the kinetic energy input, that determine the binding energy of a given atomic nucleus. Let Q represent the total binding energy state of a nucleus; then we assume for a system model

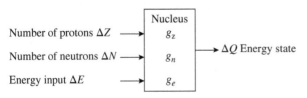

We have as the system equation for the model

$$\Delta Q = g_Z \, \Delta Z + g_N \, \Delta N + g_E \, \Delta E$$

Now we recognize that Z *and* N change only by unitary, discrete values (Z and N change by integer values) and E by the appropriate, discrete, quantum energy change.

a) Assume that g_Z, g_N, and g_E are all constants, and obtain the expected result that the total ground state energy (or binding energy) is given by

$$Q = c^2 \left[M_Z \, Z + M_N N \right] + E$$

where

Q = total energy state of nucleus (MeV)
M_Z = proton mass (amu)
M_N = neutron mass (amu)
c = speed of light (Note: $c^2 = 931$ MeV/amu)
E = total kinetic input energy to nucleus (MeV)
Z = integer number of protons in nucleus
N = integer number of neutrons in nucleus

b) Assume intrinsic feedback exists and evaluate an example for constant feedback for this model if each f_Z, f_N, and f_E is a constant.

Friedmann Equation. A fundamental equation modeling the "big bang" expansion of our Milky Way Universe in General Relativity is given by the Friedmann equation approximation. This equation is a limiting case of Einstein's equations applied to a homogeneous and isotropic universe in General Relativity. This equation relates the Hubble parameter H, to the mean mass density of the universe ρ, the spatial curvature of space K, and Newton's gravitational constant G, as shown below.

$$\frac{8\,\pi}{3} G\rho = H^2 + K$$

where

G = Universal gravitational constant (($N - m^2$)/kg^2)
H = Hubble's Constant (1/s)
K = Curvature of space in general relativity (1/s^2)
ρ = Total mean mass density of universe (kg/m^3)

and

$$K = \begin{cases} <0 & \text{if universe is spatially hyperbolic and infinite} \\ =0 & \text{if universe is spatially flat} \\ >0 & \text{if universe is spatially spherical, finite and closed} \end{cases}$$

a) Normalize the Friedmann equation, assuming that

$$r = \frac{8\pi}{3} G\rho, \quad h = H \quad \text{with} \quad h_0 = H_0 \quad \text{and} \quad k = K$$

so we have

$$dr = 2h\,dh + dk = dh^2 + dk$$

with the resulting solution form

$$r - r_0 = h^2 - h_0^2 + k - k_0$$

b) If $r \to 2r_0$ what changes in dh and dk could produce this result?

Early cosmological models assumed that our universe was contracting under the influence of gravity. However, observational evidence from the Hubble Telescope and other independent observations in deep space now confirm that our universe is actually expanding, and the expansion is accelerating and increasing with time. We still believe that our universe is spatially flat and not curved, so $K = 0$. Furthermore, we have good estimates for Hubble's constant, so we are forced to conclude that ρ, the mean mass density, is much larger than what we observe. In fact, the observable value for ρ provides only about ~5% of the total value of matter and energy needed to explain the observed expansion and behavior of the universe. This deficiency in total mass and energy has been hypothesized by cosmologists as due to "dark matter" accounting for ~27% of the observed density deficiency and "dark energy" accounting for the remaining ~68% energy deficiency. Dark matter is transparent and displays no or very little visible electromagnetic image and is thus, transparent (or non-Baryonic matter). The challenge for cosmologists is to find rational explanations for dark matter and dark energy and demonstrate their existence. Because of dark matter's unknown but significant impact on cosmology, there are numerous active national and international research programs directed at resolving the nature and composition of these dark entities. Of course, one explanation is there is no dark matter, and our present theory of gravity and other models for cosmological behavior are in error. However, this is an undesirable conclusion for most cosmologists since the evidence for the observed evidence dark matter and dark energy is broad and compelling. The true solution to this conundrum will probably merit Nobel Award (s) for its final resolution.

Relativistic Particle Model. Consider a system model composed of a relativistic particle with rest mass m_0. The model input and output differential variables of interest are time, dt, position, ds, mass, dm, momentum, dp, and energy, dE. The parameters for the model include the external force, F, on the particle and the speed of light, c, together with the particle rest mass, m_0. There are a total of 10 possible system equations relating any 2 of the differential variables defined previously. From the relativistic equations for nucleons (neutrons and protons) in physics, determine all 10 of the possible system equations and identify the system kernel for each equation. As an example, the system equation relating the differential energy and mass is

$$dE = g(v)\,dm = c^2 dm$$

and thus the system kernel is $g(v) = c^2$ for dm as the input and dE as the output.

Newton's Second Law. Newton's Second Law for the motion of a single point particle of mass m for one-dimensional motion is given by

$$F = \frac{d(mv)}{dt}$$

where F is the sum of all external forces acting on the particle along the direction of motion, v is the particle's instantaneous speed, and t is time. Newton's second law can be written as a system equation with time and speed changes as inputs and the mass change as output as follows:

$$dm = \frac{F}{v}dt - \frac{m}{v}dv \qquad\qquad (9.2.25)$$

a) Using the results of Section 5.2.4, show that the requirement on F for Eq. (9.2.25) to have a solution of the form, constant $= F(m, t, v)$, is

$$v\frac{\partial F}{\partial v} - m\frac{\partial F}{\partial m} = 0$$

b) If the force function is given by

$$F = K(mv)^k$$

where K and k are constants, show that condition (a) is satisfied for this force function and obtain a solution for Eq. (9.2.25).

Satellite Entry into the Atmosphere. A satellite returning from deep space enters the earth's atmosphere and falls to the earth along a descending path coincident with an extended radial line r passing through the center of the earth, as depicted in the figure.

Use Newton's law of gravity for the force attraction, F_g, between the earth and the satellite:

$$F_g = \frac{GMm}{r^2}$$

and for the frictional drag force, F_d, between the satellite and atmosphere, assume that

$$F_d = cv^2 \exp\left[a\left(R - r\right)\right]$$

where

G = universal gravitational constant $((\text{N-m}^2)/\text{kg}^2)$
M = earth mass (kg)
m = satellite mass (kg)
v = satellite speed (m/s)
R = earth's mean radius (m)

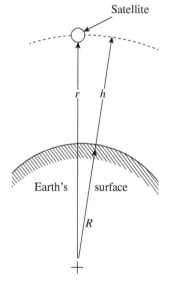

Observe that c and a are constants with dimensions of c $((\text{N-s}^2)/\text{m}^2)$ and a $(1/\text{m})$, respectively

a) Show that Newton's second law gives, as the system equation for the satellite's speed (output) as a function of time, the following:

$$dv = \left\{ \frac{g R^2}{r^2} + \frac{cv^2}{m} \exp\left[-a\,(r-R) \right] \right\} dt \qquad (9.2.26)$$

where g is the mean acceleration of gravity at the earth's surface and m is assumed to be constant.

b) Show that system Eq. (9.2.26) can also be expressed as follows by observing that $r = R + h$ and $dt = (1/v)\,dh$

$$dv^2 = -\left[\frac{2g}{(1 + h/R)^2} + 2\,\frac{c}{m}\,v^2 \exp\left(-ah \right) \right] dh \qquad (9.2.27)$$

c) Show that the solution for system Eq. (9.2.27) for the speed v in terms of the vertical height h is

$$v = \left\{ \exp\left[-f(h) \right] \int g\,(h)\, \exp\left[f(h) \right] dh \right\}^{1/2} + K$$

where

$$f(h) = -2\,\frac{c}{a\,m}\,\exp\left(-ah \right)$$

$$g(h) = -\frac{2g}{(1 + h/R)^2}$$

and K is an arbitrary constant related to the initial speed of the satellite as it enters the earth's atmosphere.

d) Expand the terms in $f(h)$ and $g(h)$ as shown

$$\exp\left(-ah \right) = 1 - ah + (\text{higher order terms})$$

$$\frac{1}{(1 + h/R)^2} = 1 - \frac{2h}{R} + (\text{higher order terms})$$

retain only $(1 + a\,h)$ and $(1 + 2h/R)$ respectively and obtain a solution for $v\,(h)$ free of quadratures. Assume that $v(h_0) = v_0$ (i.e., the satellite's entrance speed). At what speed does the satellite strike the surface of the earth?

Schrödinger Wave Equation. The space and time-dependent Schrödinger Wave Equation for a single particle is

$$i\hbar \frac{\partial \psi}{\partial t} = -\frac{\hbar^2}{2m} \nabla^2 \psi + V(r,t) \psi = H\psi \qquad (9.2.28)$$

where

$$\hbar = h/2\pi, \ h \text{ is Planck's constant (J-s)}$$
$$\nabla^2 = \text{Laplacian spatial operator } (1/m^2)$$
$$i = \text{imaginary number}$$
$$\psi(r,t) = \text{particle wave function (dimensionless)}$$
$$V = \text{potential energy (J)},$$
$$H = \text{Hamiltonian energy operator (J)}$$
$$r = \text{radial position vector (m)}$$
$$t = \text{time (s)}$$

Detailed analysis and applications using Eq. (9.2.28), the Schrödinger Wave Equation is found in standard Quantum Mechanics (QM) texts, so we will only focus on the time behavior of a single particle subjected to a potential energy field $H(t)$.

Let us form Eq. (9.2.28) into the following (QM) System equation.

$$\frac{d\psi}{\psi} = d(\text{Ln } \psi) = -\frac{iH(t)}{\hbar} dt = H(t) \, d\left(-\frac{it}{\hbar}\right) \qquad (9.2.29)$$

and we define new input and output variables as

$$\text{Ln } \psi = \varphi(\tau) \quad \text{and} \quad \tau = -it/\hbar$$

Then system Eq. (9.2.29) becomes the following, where we have expanded the potential function, $H(\varphi, \tau)$, to address the condition that the potential energy is also dependent upon the wave function

$$d\varphi = H(\varphi, \tau) \, d\tau \qquad (9.2.30)$$

a) Obtain the solution for Eq. (9.2.30) if $H(\varphi, \tau) = A(\tau) + B(\tau) \varphi$ and A and B are functions of τ. Multiply Eq. (9.2.30) by an integrating factor $I(\varphi, \tau)$ and determine the requirements on I so that Eq. (9.2.30) is exact and can be integrated in terms of elementary functions (see Section 5.2.4)

b) Express the solution for Eq. (9.2.30) in terms of the wave function $\psi(r, t)$ for the time, t, the parameters, i, and Planck's constant h. Display your results using the graphical resources in MATLAB.

Simple Pendulum. A simple pendulum is shown schematically, where m is the mass connected to a rigid massless rod of length L.

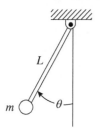

a) Show that the equation of motion for the mass is

$$\frac{d^2\theta}{dt^2} + \frac{g}{L} \sin\theta = 0$$

b) Express the equation in part (a) as a system equation with θ as the input (independent variable) and $w = d\theta/dt$ as the output (dependent variable).

c) Solve the system equation found in part (b) and generate MATLAB plots for the general responses of w versus θ for several different initial conditions.

d) Express the solution for $\theta(t)$ as a function of time in terms of a quadrature for $\theta(t)$. Solutions of this quadrature involve Elliptic Integrals.

e) Linearize the system equation given in part (a) and determine the general solution for the angular displacement $\theta(t)$.

Second Law of Thermodynamics. For a closed thermodynamic system undergoing a reversible process, we have from the Clausius Inequality for an isentropic system the following

$$dQ = TdS \tag{9.2.31}$$

where dQ is the heat transferred from the system, T is the mean absolute temperature of the system, and dS is the change in entropy within the system. Recalling the first law of thermodynamics for our system, we have that

$$dQ = dH - VdP \tag{9.2.32}$$

where dH is the change in enthalpy within the system, dP is the change in system pressure, and V is the mean absolute volume within the system. Eliminating dQ between Eqs. (9.2.31) and (9.2.32) and treating dH as the system output, we have

$$dH = T\,dS + V\,dP \tag{9.2.33}$$

a) For system Eq. (9.2.33), determine the necessary conditions for exactness of the system equation (see Section 5.2.4).

b) Determine nontrivial functions for $T(H, S, V)$ and $P(H, S, V)$ that satisfy condition (a) and then solve the system equation.

c) Repeat this analysis for a closed system using for the first law in the form

$$dQ = dU + PdV$$

where U is the internal energy of the system.

Continuity of Mass Model. The basic system equation describing the physical model for mass conservation in time in any closed system where no significant mass-energy exchange reactions are occurring is

$$d\rho = (S - \nabla \cdot \rho\mathbf{u})\,dt$$

where

ρ = mass density (kg/m^3)
$\mathbf{u}$ = mass flow vector velocity (m/s)
t = time (s)
S = net input of mass into system (kg/(s-m^3))
∇ = gradient operator (1/m)

a) Assume a specific system that exhibits one-dimensional flow in the x-direction, where

$$\mathbf{u} = \mathbf{i}\, v\, [1 + \cos(2\pi x/L)]$$
$u_0 = $ scalar speed of mass flow at $t = 0$ (m/s)
$S = $ constant

$$\nabla = \mathbf{i}\, \frac{\partial}{\partial x}$$

$\mathbf{i} = $ unit vector in x direction

So

$$\nabla \cdot \rho\mathbf{u} = \mathbf{i}\, \frac{\partial}{\partial x} \cdot \mathbf{i}\rho v_0\, [(1 + \cos(2\pi x/L)] = \rho v_0\, \frac{\partial}{\partial x}\, [1 + \cos(2\pi x/L)]$$

What is the system equation and general solution for the mass density $\rho(t)$?

Heat Conduction Model. It is observed experimentally that both the temperature T and heat flow q change with length along with a heat-conducting medium. Thus, the system equations for this model have the form

$$dT = g_1\, dx$$

$$dq = g_2\, dx$$

where g_1 and g_2 are the respective system kernels. Experimental observation has shown that g_1 is proportional to q and inversely proportional to a function of T, that is $k(T)$ called the thermal conductivity.

a) Assume that g_2 is a constant, and find solutions for q as a function of T and then q and T as functions of the heat flow path x.
b) If g_2 is a function of T and q, what functional form should g_2 have for the differential equation relating dT and dq to be exact?
c) Select a model for g_2 from part (b) and solve the system equations.

Model for Stellar Equilibrium. Consider a star of radius R and constant mass M composed of a high-temperature ionized gas (plasma) of uniform density ρ, which is in hydrostatic equilibrium (i.e., the expansive pressure forces resulting from the nuclear fusion reaction are equal at all times to the gravitational forces containing the plasma). For an isotropic plasma (i.e., uniform properties in all directions for a fixed radius), we seek to determine the output changes in the radial variation of pressure P, energy production from fusion E, and kinetic temperature T. The block diagram for our system is

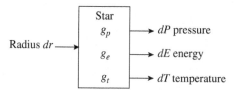

The system equations are found to be

$$dP = -\frac{GM\rho}{r^2} dr \text{ (pressure and gravity force equilibrium)}$$

$$dE = 4\pi r^2 Q\rho dr \text{ (fusion energy production)}$$

$$d(aT^4) = -3\frac{Kr}{c}\frac{Q}{4\pi r^2} dr \text{ (radiation equilibrium)}$$

or equivalently for radiation equilibrium, dT gives

$$dT = -\frac{3K\rho}{4acT^3}\frac{Q}{4\pi r^2} dr$$

We also have the following auxiliary equations coupling system variables and parameters:

$$M = 4\pi \int_0^R \rho\, r^2 dr \text{ (total stellar mass)}$$

$$P = \rho\frac{RT}{m} + \frac{aT^4}{3} \text{ (gas and radiation pressure)}$$

$$Q = Q_0\, T^b\rho \text{ (fusion power density)}$$

For the above equations, we identify the constants as G (universal gravitational constant), K (radiation opacity), m (molecular weight), a (Stefan–Boltzmann constant), c (speed of light), and Q_0 and b as specific stellar parameters.

a) Express the system equations in terms of output and input variables only.
b) By assuming that we have quasi-adiabatic equilibrium (i.e., there is no significant radiation loss from the system) so that P is proportional to p^k where k is the ratio of specific heats and assumed to be a constant, obtain a solution for the system equations for P as a function of T and P, T and E as functions of r.
c) Determine the average temperature and pressure throughout the volume of the plasma and find the total energy output for the star.
d) Plot the 3-D behavior of the star's fusion power density Q as a function the star's mass M and radiation pressure P.

Planck's Radiation Model. In 1900, Max Planck postulated the first quantum mechanics system model to establish the photonic theory of radiation emission for a black body. The differential input to the system (a black body is a perfect absorber and emitter of radiation at all frequencies of electromagnetic radiation) is the change in electromagnetic wavelength $d\lambda$, and the output is the change in power per unit emission dW. A block diagram for the black body system is given by

and the system equation for Planck's model is

$$dW = \frac{2\pi c^2 h}{\lambda^5}\frac{1}{\exp(hc/KT\lambda) - 1} d\lambda$$

It is convenient to define normalized input and output variables for the system equation as follows:

$$w = \frac{h^3 c^2}{2\pi K^4 T^4} W$$

$$x = \frac{hc}{KT} \frac{1}{\lambda}$$

Then the normalized system equation can be written as

$$dw = \frac{x^3}{1 - \exp(x)} dx = \frac{x^3 \exp(-x)}{\exp(-x) - 1} dx \tag{9.2.34}$$

a) Show that the total power emitted per unit area for all wavelengths $(0 < \lambda < \infty)$ is given by

$$W = \frac{2\pi^5 K^4}{15 c^2 h^3} T^4$$

This equation is known as the Stefan–Boltzmann equation for electromagnetic radiation emission. Observe that we have used the result that

$$\int_0^\infty \frac{x^3}{e^x - 1} dx = \frac{\pi^4}{15}$$

b) Express Eq. (9.2.34) in terms of the general single-input, single-output system equation with feedback:

$$dw = \frac{g}{1 - gf} dx$$

Identify the possible forms for the kernels g and f. Draw the block diagrams associated with each of these system equations with feedback and comment on their physical reality.

c) Assume that x is small in system Eq. (9.2.34) (i.e., $\lambda \gg hc/KT$) and obtain the following result with K as the integration constant.

$$W = -\frac{2}{3\lambda^3} \pi K T c + K$$

Van der Waals Model for Gases. Consider the following system equation for a gas:

$$d\tau = g_1 dp + g_2 dn + g_3 dv$$

where

$\tau = \text{Ln } T$ (T is absolute gas temperature in Kelvin degrees – K)
$p = \text{Ln } P$ (P is absolute gas pressure in Pascals – Pa)
$n = \text{Ln } N$ (N is number of moles of gas in moles)
$v = \text{Ln } V$ (V is volume of gas in m³)

and

$$g_1 = \frac{1}{1 + a \exp[2(n-v) - p]}$$

$$g_2 = \frac{2a}{a + \exp[2(v-n) + p]} + \frac{b}{b - a \exp(v-n)} - 1$$

$$g_3 = \frac{1}{1 - b \exp(n-v)} - \frac{2a}{a + \exp[2(v-n) + p]}$$

and a and b are constants.

a) Show that a solution for the system equation is given by the van der Waals equation for the state of a gas:

$$\left(P + \frac{a N^2}{V^2}\right)(V - bN) = NRT$$

where the arbitrary integration constant R is identified as the universal gas constant.

b) Show that if a and $b \to 0$ in the system equations, then the solution becomes the familiar, simple perfect gas law

$$PV = NRT$$

c) By assuming that $2(n-v) - p$ and $n - v$ are small, approximate each exponential term in the system kernels and obtain the approximate equation kernels,

$$g_1 = \frac{1}{1 + 2a(n-v) - a(p-1)}$$

$$g_2 = \frac{2a}{a + 1 + 2(v-n) + p} + \frac{b}{b - v + n - 1}$$

$$g_3 = \frac{1}{1 - b(n-v)} - \frac{2a}{a + 1 + 2(v-n) + p}$$

and attempt to find a solution for this approximate system equation using the MATLAB Toolbox. If successful, plot your result for P as a function of T and V.

$$P = F(T, V)$$

Maxwell's Relations. An important model for thermodynamic systems that are closed (no mass exchange) and reversible and where only PdV type of work occurs, is defined by Maxwell's relations as

$$dE = T\,dS - P\,dV$$
$$dH = T\,dS + V\,dP$$
$$dG = -S\,dT + V\,dP$$
$$dA = -S\,dT - P\,dV$$

where

E = internal energy (J)
H = enthalpy (J)
G = Gibbs free energy (J)
A = Helmholtz free energy (J)
T = absolute temperature (K)
S = entropy (J/K)
P = absolute pressure (J/m^3)
V = volume (m^3)

Observe that system outputs assumed here are E, H, G, and A, while system inputs are T, S, P, and V. Obtain the necessary conditions for the system to possess an integrating factor as provided in Section 5.3.4. If successful, use the MATLAB Toolbox for a possible solution and output plots.

Gibbs Free Energy, Equilibrium State Model. A useful model to determine if a thermodynamic system is in its maximum local entropy state is found from the Gibbs Free Energy equation as follows

$$dG = dH - T dS$$

where

G = Gibbs free energy (J)
H = enthalpy (J)
T = absolute temperature (K)
S = entropy (J/K)

Suppose for a given thermodynamic system that the temperature T is given by the following function of entropy S where a, b, c, d, and e are constants

$$T(S) = \frac{a + bS + cS^2}{1 + dS + eS^2}$$

Determine the solution for the Gibbs free energy G for the equilibrium state of this system as a function of the enthalpy and entropy. Determine the value of the system enthalpy when the Gibbs free energy is 0.

Additional Modeling Problems. For each of the systems given in Table 9-2, define an appropriate model and its pertinent elements. Express quantitatively variables for each of the model inputs and outputs and develop a simple analytical expression for the system model equation. Determine analytical solutions for the models where possible, and interpret and evaluate your result. Implement the MATLAB tools where practical.

Table 9-2 Modeling Problems from the Physical Sciences

	System	Input	Output
(a)	Binary chemical reaction rate $A + B \rightarrow AB$	Chemical Reactants A & B	Chemical Product AB
(b)	Tachyons (particles whose velocity exceed light speed)	Time	Mass (m) and momentum (mv)
(c)	Elastic solid materials	Strain	Stress
(d)	Expanding universe (assume total mass and energy fixed)	Space	Mass and Energy density
(e)	Uniform solid	Sudden stress input	Sound propagation
(f)	Relativistic time dilation	Speed	Time difference
(g)	Lagrange's equations	Time	Generalized coordinates
(h)	Particle transport	Time and displacement	Particle energy and momentum
(i)	Basic force fields (gravity, electromagnetic, weak, strong)	Displacement	Force
(j)	Wave motion	Time and displacement	Wave energy and momentum
(k)	Thermodynamic system	Enthalpy, kinetic and potential energy	Mechanical work and heat
(l)	Conservation of mass	Time and displacement	Mass density
(m)	M-Theory	Open and closed strings	Electron behavior
(n)	Cosmology	Membrane (or Brane)	Parallel Universes
(o)	Higgs Boson	Higgs Field Theory	Other Bosons?
(p)	Parallel Universe	Milky Way Galaxy	Changes in SMPP
(q)	Dark Matter	Dark Energy	Interactions of Dark entities
(r)	Supper String Theory	Quantum Mechanics	General Relativity
(s)	M-Theory	11 dimensions	Experimental verification
(t)	DNA on the Earth	Life forms elsewhere	Hydrocarbon basis
(u)	Genetic Code	Evolution	Intelligent Design
(v)	Nucleotides	Codons	Proteins
(w)	Base Pairs: C – G, A – T	Human Genome	Binary Code Basis (0, 1)
(x)	Thermodynamic System	Temperature and Entropy	Minimum entropy states
(y)	Enrico "Fermi's Paradox"	Discovery of alien life	Impact on human earth life
(z)	Accelerating expansion of Milky Way due to dark energy	Multiverses	Interactions between Milky Way and parallel universes

9.3 Earth Science Applications of System Science

Of all physical planetary bodies in our universe, the planet Earth, which serves as the cradle of humanity and is the only known planet supporting life, is of primary interest to mankind. The life support system provided for flora and fauna inhabiting the earth is apparently quite unique, as well as inspiring. Although evidence of habitable exoplanets and extraterrestrial life is being actively sought, no indisputable discovery of extraterrestrial life is yet forthcoming. Thus, a broad, accurate knowledge of mankind's terrestrial spaceship is essential not only to understand life itself but also to appreciate and protect the earth and its resources for their intrinsic, unique value as the wellspring of life and the repository of human knowledge and development.

9.3.1 Atmospheric Model

The earth's atmosphere consists mainly of nitrogen (78% by volume) and oxygen (21% by volume), with the remainder comprised by argon (0.9%), carbon dioxide (0.03%), and traces of other gases and air particulates. Water vapor in the atmosphere varies from 0% to about 4%, depending on local atmospheric conditions. Although CO_2 constitutes only about 1 part in 3000 of the atmosphere, this combustion product from animal and fuel oxidation reactions plays an important role along with water vapor in the production of carbohydrates and oxygen in the important biological process of photosynthesis. Also, atmospheric CO_2 and water vapor absorb long-wavelength infrared radiation emitted from the earth. This absorbed energy is transferred to other atmospheric gases, resulting in a global heating of the atmosphere. Finally, CO_2 is highly soluble in water, so the earth's water sources serve as absorbers or sinks for gaseous CO_2 in the atmosphere.

We seek a simple model for the change in mean atmospheric CO_2 content as a function of the change in the earth's animal population, the combustion or oxidation of fossil fuels that produce CO_2, and the CO_2 reduction process (i.e., water absorption, photosynthesis, etc.), which utilize or absorb atmospheric CO_2. The inventory of plants on the earth should appear in the model as an intrinsic feedback loop to the system reflecting the conversion of CO_2 to O_2 and carbohydrates through photosynthesis. A simple block diagram for the model is as follows:

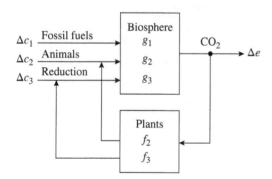

The overall system equation in canonical form is

$$\Delta e = \frac{g_1 \, \Delta c_1 + g_2 \, \Delta c_2 + g_3 \, \Delta c_3}{1 - g_2 f_2 - g_3 f_3}$$

Physical observations indicate that

$$g_1 > 0, \quad g_2 > 0, \quad f_2 > 0, \quad f_3 > 0, \quad \text{but} \quad g_3 < 0$$

The kernel g_3 is negative to reflect the fact that water (oceans, lakes, and atmospheric moisture) absorbs CO_2, while plants produce O_2 by photosynthesis with CO_2. So the effect on Δe from input Δc_3 combined with feedback f_3 is to reduce CO_2

Let us assume a very simple quantitative system model by making all the model kernels constant. So let

$$g_1 = B_f, \quad g_2 = B_a, \quad g_3 = -B_r, \quad f_2 = P_a, \quad f_3 = P_r$$

where all B's and P's are positive constants. We have as the solution for this system equation (assumed to be continuous)

$$e_f = e_0 + \frac{\left[B_f(c_1 - c_{10}) + B_a(c_2 - c_{20}) - B_r(c_3 - c_{30})\right]}{(1 - P_a B_a + P_r B_r)}$$

where we have used the integration constant to satisfy the initial conditions that

$$e = e_0 \text{ and } c_1 = c_{10}, \quad c_2 = c_{20} \text{ and } c_3 = c_{30}$$

Observe from the system equation solution that

$$\frac{\partial e}{\partial c_1} > 0, \quad \frac{\partial e}{\partial c_2} > 0 \text{ and } \frac{\partial e}{\partial c_3} < 0 \text{ if } 1 + P_r B_r > P_a B_a.$$

Thus, for this condition on the feedback kernels, increases in animal population and combustion of fossil fuels increase atmospheric CO_2, while an increase in plant coverage reduces CO_2. In view of the concern over the continual increase of CO_2 observed in the Earth's atmosphere since the Industrial Revolution, as shown in Figure 9-3, it is interesting to speculate on how a reduction or even a reversal of this buildup could be realized using our model as a basis for quantitative decision making.

Because of the observed increase of CO_2, it is evident for our model that

$$B_f c_{10} + B_a c_{20} > B_r c_{30}$$

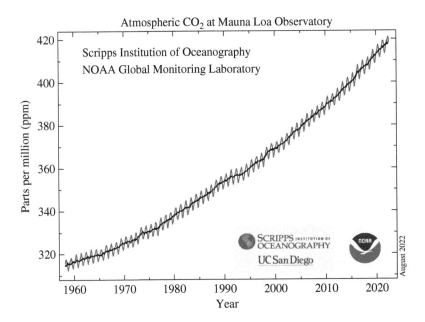

Atmospheric CO_2 at Mauna Loa Observatory

Fig. 9-3 Increase of carbon dioxide in atmosphere.

So one possible management policy to reduce CO_2 levels in the atmosphere would be to change the following system parameters:

Decrease	Increase
c_{10} and c_{20}	c_{30}
B_f, B_a, and P_a	B_r and P_r

Since $B_a c_{20} \ll B_f c_{10}$, these parameter and variable constraints imply the need for an international policy of limiting the combustion of fossil fuels and maintaining or even expanding plant coverage on the earth. Whether such a policy is possible in light of growing energy needs and societal urbanization remains to be seen. World environmental groups are actively promoting legislation for atmospheric restrictions for climate control.

9.3.2 Geothermal Model

It is a well-known geological fact that the temperature of the earth's crust (lithosphere) increases with depth into the crust. Although geologists believe that several independent factors may contribute to this internal geothermal heating, we will assume that the primary source of the heating and the observed positive temperature gradient with depth is the result of decay of naturally occurring radioactive materials found in the earth's crust. There are several very long-lived radioactive materials found in the lithosphere, including uranium (U-234, U-235, U-238), thorium (Th-232), potassium (as K-40), and other radioactive nuclides.

Since we seek to determine the temperature and heat flow or current as a function of depth, let us consider a model as portrayed by Figure 9-4.

The heat flow rate and temperature distribution will be assumed to be isotropic through the earth's volume. To obtain the system kernels, we appeal to the energy conservation law of thermodynamics and the Fourier heat conduction equation from heat transfer for heat diffusion. Thus

$$dq = g_1 \, dr = 4\pi r^2 Q(r) \, dr \text{ (energy equation)} \tag{9.3.1}$$

$$dT = g_2 \, dr = -\frac{q(r)}{4\pi r^2 k} \, dr \text{ (Fourier heat equation)} \tag{9.3.2}$$

where

$Q(r)$ = volumetric heat source (W/km^3)

$q(r)$ = radial heat flow outward at position r (W)

$k(r)$ = thermal conductivity at position r ($W/(km\text{-}K)$)

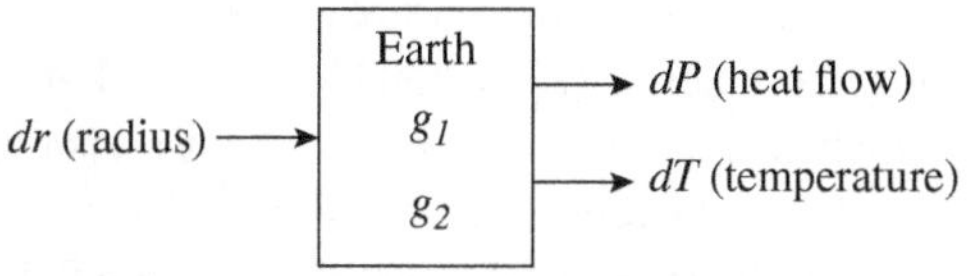

Fig. 9-4 Block diagram and model for geothermal energy system.

r = radial position within earth (km)

$T(r)$ = temperature at position r (K)

On examination of our system equations, we observe that we should solve Eq. (9.3.1) first to obtain a solution for $q(r)$, which can then be employed to solve Eq. (9.3.2). So, integrating Eq. (9.3.1), we have

$$q(r) = 4\pi \int_0^r Q(r) \, r_2 \, dr \tag{9.3.3}$$

where $q(0) = 0$. It is known from the analysis of earthquake waves transmitted through the earth that its interior is not uniform, so the conductivity k and the volumetric heat source $Q(r)$ both vary with radial position in the earth. Although we do not know accurately how $Q(r)$ varies with position, we will assume that a mean value, say Q_0, can be found, and we then have from Eq. (9.3.3) (using the mean value theorem for integrals) that

$$q(r) = \frac{4}{3} \pi \, Q_0 \, r^3$$

With $q(r)$ now known, we can integrate Eq. (9.3.2) to give

$$T(r) = -\frac{1}{3} Q_0 \int_0^r \frac{r}{k(r)} \, dr + T_0 \tag{9.3.4}$$

Again, we do not know how the conductivity $k(r)$ varies throughout the earth's interior, so we assume a mean value, k_0, for the conductivity and evaluate the integral in Eq. (9.3.4) to obtain

$$T(r) = -Q_0 \frac{r^2}{6 \, k_0} + T_0 \tag{9.3.5}$$

where $T(0) = T_0$. Some preliminary values of the parameters estimated from near-surface measurements on the earth's crust are

$$k = 700 \text{ K/km} \qquad \frac{dT}{dr} = -30 \text{ K/km}$$

$$T(R) = 300 \text{ K} (\sim 60\degree \text{F}) \qquad R = 6400 \text{ km}$$

With these parameters and boundary conditions, we find that

$$Q_0 = 9.8 \text{ W/km}^3, \qquad T_0 = 96,000 \text{ K}$$

It is apparent that T_0 is unreasonably high, and it is believed that the earth's interior becomes fluid at temperatures exceeding about 1500 K. If we solve Eq. (9.3.5) for that value of $r = R_0$ where $T(R_0) = 1500$ K, we find that the solid crustal thickness is only about 50 km, which is in reasonable agreement with seismic data. These very high internal temperatures imply that the inner region of the earth's volume is a dense, convective, molten fluid, which apparently produces the earth's magnetic field and serves to partially shield radiation emanations from the sun and outer space. This shielding is essential for life to exist and survive on the Earth.

It is also interesting to determine the amount of radioactive material (in ppm) necessary to produce the volumetric heat source we derived (i.e., $Q_0 = 9.8$ W/km) and to compare the number with measured values found in crustal material. This mass calculation is based solely on the properties of U-238 since it constitutes over 99% of naturally occurring uranium. We find that

$$\frac{g(\text{U-238})}{g\,\text{crust}} = 9.8\,\frac{\text{W}}{\text{km}^3} \times 10^{-15}\,\frac{\text{km}^3}{\text{cm}^3} \times \frac{\text{cm}^3}{2.8\,g\,\text{crust}} \times 6.3 \times 10^{12}\,\frac{\text{MeV}}{\text{W-s}}$$

$$\times \frac{\text{U atom decay}}{4\,\text{MeV}} \times 4.5 \times 10^9\,\frac{\text{yr-U atom}}{\text{U atom decay}} \times 3.1 \times 10^7\,\frac{\text{s}}{\text{yr}}$$

$$\times \frac{g\,\text{mole}}{6.0 \times 10^{23}\,\text{U atoms}} \times 238\,\frac{g(\text{U-238})}{g\,\text{mole}}$$

$$= 0.3\,\text{ppm}\,\frac{g\,(\text{U-238})}{g\,\text{crust}}$$

Interestingly, the earth's surface soil is found to contain about 2 ppm of U-238. However, the oceans carry only about 0.003 ppm of dissolved uranium, so assuming that only about 25% of the earth's surface is soil and the remainder is water implies a possible mean crustal density of uranium of about 0.3 ppm, which agrees with our current estimated value.

9.3.3 Terrestrial Water Balance Model

A feature of the earth that is presently unique among the known planets is the existence of large volumes of surface water as a liquid. The oceans, seas, lakes, and rivers of the earth have played an important role in initiating and sustaining life and in influencing the history of mankind. The presence of large masses of liquid water is probably critical for designating an exoplanet (extraterrestrial) as capable of initiating and sustaining life. The earth's water cycle, or inventory and movement of water as liquid, vapor, and ice through the geohydrological cycle, is driven primarily by energy intercepted from the sun and to a small degree by geothermal energy arising from radioactive decay within the earth's crust.

Figure 9-5 indicates the principal inventories and movement rates of terrestrial water identified within specific water cycle compartments. Water mass values in each inventory component and movement rates have been normalized to the water mass held in the atmosphere (which is set to unity for comparison purposes). It is estimated that the mean annual mass of water exchanged in the atmosphere is about 4.2×10^{14} metric tonnes (or 420,000 km^3 of liquid volume). It is apparent from the figure that the bulk of the earth's water, about 97%, is held in the oceans and ground water, while only about 0.001% is found in the atmosphere and 0.63% is found on the land surfaces.

Because of the significance of the movement of water between the oceans, the atmosphere, and the earth's continental surface water, let us treat this cycle as a closed system (ignoring the exchange between groundwater and surface water by assuming this mass exchange is in equilibrium as shown in Figure 9-5).

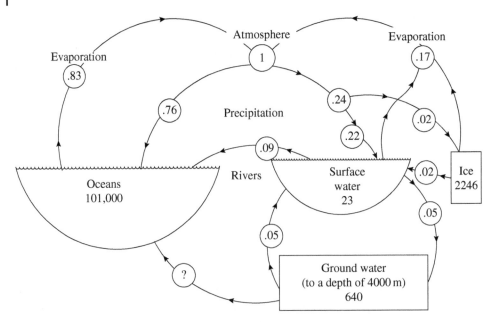

Fig. 9-5 Inventories and flow of terrestrial water. *Source:* U.S. Geological Survey, 1967.

We will assume that the total mass of water in the atmosphere, m_a, the water mass on the land surface, m_s, and the ocean water mass, m_0, satisfy the following time-dependent system equations:

Atmosphere: $\dfrac{dm_a}{dt} = k_{0a}m_0 + k_{sa}m_s - (k_{a0} + k_{as})\,m_a$ (9.3.6.1)

Surface: $\dfrac{dm_s}{dt} = k_{as}m_a - (k_{sa} + k_{as0})\,m_s$ (9.3.6.2)

Ocean: $\dfrac{dm_0}{dt} = k_{s0}m_s + k_{a0}m_a - k_{0a}m_0$ (9.3.6.3)

Our system equations are based on a first-order rate model, as depicted in Figure 9-6. If we simply add the system equations [Eqs. (9.3.6)] together, we find that

$$dm_a + dm_s + dm_0 = 0$$

and integrating we have for any value of time that

$$m_a(t) + m_s(t) + m_0(t) = M$$

where M is the total mass of terrestrial water (assumed to be constant) available in the cycle model.

To simplify our model, we observe that since the inventory of water in the oceans m_0 for any time t is so large compared to the surface and atmospheric masses m_s and m_a, we will assume that $m_0(t) = M$ and therefore set $dm_0 = 0$. However, the ocean water input to the atmosphere, $k_{0a}\,m_0$, will serve as a major control term, where the parameter k_{0a} may be dependent on time. For example, time-dependent changes in the solar energy incident

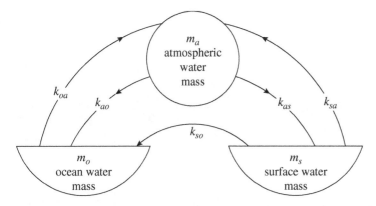

Fig. 9-6 Terrestrial water exchange model.

on the ocean surface might arise because of the greenhouse effect (CO_2) atmospheric dust, or solar cycles. These changes would make k_{0a} a time-dependent parameter. Then, for our reduced system equations, we have

$$d \begin{bmatrix} m_a \\ m_s \end{bmatrix} = \left[\begin{pmatrix} -k_a & k_{sa} \\ k_{as} & -k_s \end{pmatrix} \begin{pmatrix} m_a \\ m_s \end{pmatrix} + \begin{pmatrix} S \\ 0 \end{pmatrix} \right] dt \qquad (9.3.7)$$

where we have defined for convenience

$$k_a = k_{0a} + k_{as}, \quad k_s = k_{sa} + k_{s0}, \quad S = k_{0a} m_0$$

and we will consider $S(t)$ as our time-dependent input control function. We will also assume that k_a, k_s, k_{sa}, and k_{as} are constant in time.

System Eqs. (9.3.7) can be solved by Laplace transform methods as given in Section 7.3.2 or through the characteristic equation process as outlined in Section 4.3.4. Also, MATLAB M files can be employed to obtain the Laplace Transforms for the surface and atmospheric masses m_s and m_a in Eq. (9.3.7)

The time-dependent solutions for the system equations are found to be

$$m_s(t) = c_{11} \exp(s_1 t) + c_{12} \exp(s_2 t)$$
$$+ c_{13} \int_0^t [\sinh \{k_2(t-x)\} + \cosh \{k_2(t-x)\}] \exp \{-k_1(t-x)\} S(x) \, dx \qquad (9.3.8)$$

and

$$m_a(t) = c_{21} \exp(s_1 t) + c_{22} \exp(s_2 t)$$
$$+ c_{23} \int_0^t \sinh \{k_2(t-x)\} \exp \{-k_1(t-x)\} S(x) \, dx \qquad (9.3.9)$$

where

$$c_{11} = \frac{1}{2k_1} \frac{[(s_1 + k_s)m_{a0} + k_{sa} m_{s0}]}{(s_2 - s_1)}$$

$$c_{12} = -\frac{1}{2k_1} \frac{[(s_2 + k_s) m_{a0} + k_{sa} m_{s0}]}{(s_1 - s_2)}$$

$$c_{13} = \frac{1}{k_2} (k_s - k_1)$$

$$c_{21} = \frac{1}{2k_1} \frac{[(s_1 + k_a) m_{s0} + k_{as} m_{a0}]}{(s_1 - s_2)}$$

$$c_{22} = -\frac{1}{2k_1} \frac{[(s_2 + k_s)m_{a0} + k_{sa} m_{s0}]}{(s_2 - s_1)}$$

$$c_{23} = \frac{k_{as}}{k_2}$$

$$k_1 = \frac{1}{2}(k_a + k_s)$$

$$k_2 = \frac{1}{2} \left[k_a^2 + k_s^2 + 4k_{as}k_{sa} - 2k_s k_a\right]^{1/2}$$

$$s_1 = -k_1 + k_2$$

$$s_2 = -k_1 - k_2$$

The parameters m_{a0} and m_{s0} are the initial values of the atmospheric water mass and surface water mass, respectively.

If the time-dependent input control function $S(t)$ is specified, then the surface water mass $m_s(t)$ can be found from Eq. (9.3.8), and the atmospheric water mass $m_a(t)$ can be found from Eq. (9.3.9). Using MATLAB, plot a typical time history for this model for terrestrial water exchange.

9.3.4 Topical System Applications in the Earth Sciences

Earth's Magnetic Field. The earth's magnetic field is believed to arise from the thermally induced electric current loops within the earth's molten liquid core. Develop a simple model for the earth's vector magnetic field **B** based on a one-turn current loop centered a distance R from the earth's center core contained in a plane tilted about 11° from the earth's equatorial plane. The model output desired is the earth's magnetic field **B**, so determine those inputs and parameters required to realize this specific model. Note that a typical value of the earth's magnetic field at sea level is about 3 microTesla.

Solar Energy Input to the Earth. Solar energy normally incident at the mean radius of the earth's thermosphere is partitioned as follows. About 34% is reflected back into space, 19% is absorbed in the upper atmosphere, and the remaining 47% is absorbed by the earth's surface. Develop a model for the earth's atmospheric envelope that has as an input the solar energy incident at the thermosphere (i.e., the atmospheric layer where solar energy is reflected back into space or transmitted through the atmosphere) and that predicts the observed distribution of solar energy to the atmosphere and the earth's surface. Assume that solar energy arrives from the sun as a uniform beam of power $(1.4\,\text{kW/m}^2)$, where it interacts and is isotropically scattered by the thermosphere at a mean altitude of 80 km above the earth's surface.

Atmospheric Precipitation. Spherical droplets of water are falling through the atmosphere. Assume that water from the droplets evaporates at a time rate that is proportional to the surface area of the droplet and proportional to the atmospheric temperature above the ambient dew point. Develop a model for this system of droplets in which time is the input and the droplet radius is the output. It is experimentally observed that a 0.1 cm droplet evaporates completely in approximately 25 seconds at an atmospheric temperature of 10 K above the ambient dew point.

a) Determine the system equations.
b) Determine the solution for the system equations that provide the mass of droplets as a function of time and temperature above the dew point.
c) Estimate the lifetime (time before evaporation) of 0.25 cm droplets in air at 15 K above the dew point.

Terrestrial Radio-wave Spectrum. The radio-wave spectrum imposed on the earth from the atmosphere and extraterrestrial sources is noisy and complex. Figure 9-7 shows an approximate sampling of radio spectra sources. The 3 K curve is the remnant background noise from the "big bang" origin of the universe estimated to have occurred about 13.8 billion years ago. Other radio wave spectra sources include electromagnetic noise from our Milky Way Galaxy arising mainly from synchrotron radiation, an exponentially increasing noise source associated with all electromagnetic radiation sources, and the earth's own

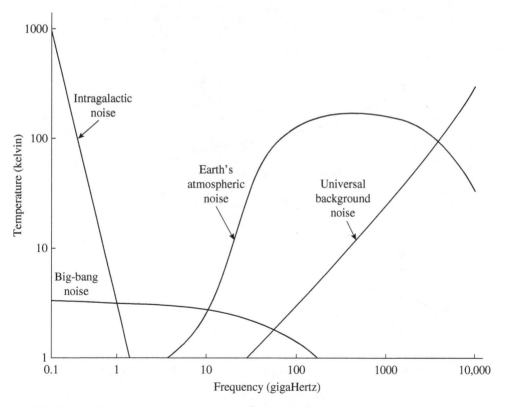

Fig. 9-7 Terrestrial radio-wave spectrum. *Source:* "Search for Extraterrestrial Intelligence," Scientific American, May 1975, p. 82.

atmospheric noise due to molecular absorption and re-radiation of water and oxygen molecules in the atmosphere.

Using Figure 9-7, determine the total quantitative temperature equivalent radio spectrum from the noise sources indicated as a function of frequency in the range from 0.1 to 10,000 GHz. Find all relative maximum and minimum spectra values within the same frequency region.

Free Electrons in the Earth's Atmosphere. The earth's ionosphere is composed of a layer of electrons that peaks in density at approximately 300 km above the earth's surface. Obviously, there are competing reactions between these atmospheric electrons and surrounding ions. However, the solar energy and cosmic ray interactions in this upper atmosphere maintain a daily equilibrium cycle. This layer of electrons plays a very important role in radio-wave communications introduced into the atmosphere by humans. If the density or temperature distribution of these electrons is perturbed, the impact on radio communications can be significant. Solar storms can dramatically disturb the atmospheric ionosphere.

Figure 9-8 gives an approximate distribution of the density and temperature of electrons with altitude in the earth's atmosphere. Observe that the electron density has a

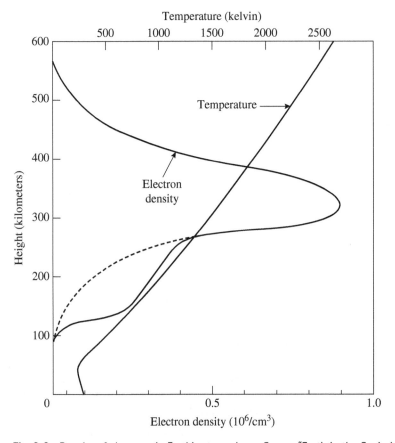

Fig. 9-8 Density of electrons in Earth's atmosphere. *Source:* "Earth in the Sun's Atmosphere," Scientific American, October 1959, p. 66.

quasi-Gaussian (or normal) distribution with an anomalous shoulder between about 100–250 km. The temperature distribution increases approximately linearly with altitude over the bulk of the electron distribution.

a) Assume that the electron density distribution is Gaussian (ignore the shoulder) and the temperature distribution is linear. Establish the system equations for this model with atmospheric altitude as the input and electron temperature and density as outputs.
b) Using the system equations, determine the total quantity of electrons in the atmosphere and the mean temperature of these electrons.
c) Assume that a 10-megaton (MT) nuclear explosive is detonated at an altitude of 300 km above the earth's surface and imparts energy uniformly to the electrons within a spherical volume of radius 100 km between the altitude ranges of 200 and 400 km. (Note: 1 MT $= 2.6 \times 10^{34}$ eV and Boltzmann's constant is 8.6×10^{-5} eV/K.) Determine the electron energy distribution with altitude, assuming that this energy input is uniformly distributed within the spherical envelope of radius 100 km between 200 and 400 km and that the electron density distribution is not significantly disturbed. What might be the effect on radio-wave communications from such a high-altitude nuclear explosion? Can such a nuclear event precede an international attack upon the affected country?

Atmospheric Models for the Planets. With the space probes sent to our neighboring planets, Mariner IV (US) to Mars and Mariner V (US), and Venera 4 (former USSR) to Venus, some experimental data on the characteristics of these planetary atmospheres are now available. Figures 9-9 and 9-10 indicate approximate atmospheric pressure and

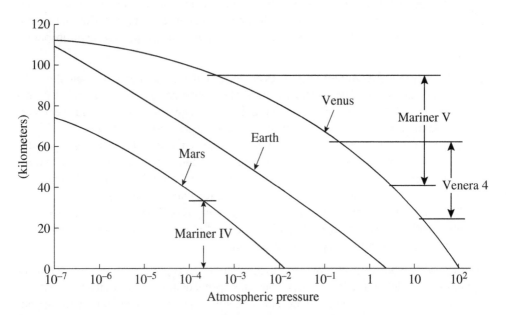

Fig. 9-9 Atmospheric pressure for Mars, Earth, and Venus. *Source:* "Atmosphere of Mars and Venus," Scientific American, March 1969, p. 85.

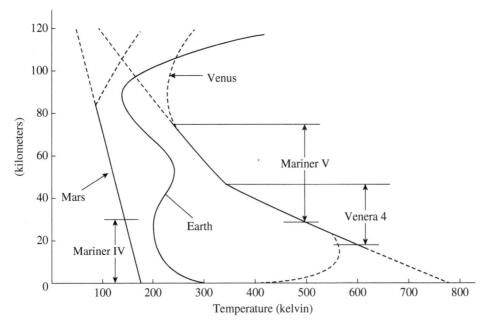

Fig. 9-10 Atmospheric temperature for Mars, Earth, and Venus. *Source:* "Atmosphere of Mars and Venus," Scientific American, March 1969, p. 85.

temperature data for Mars, Venus, and the Earth for model reference. Using these data, develop quantitative system equations that predict the atmospheric pressure and temperature (outputs) as a function of altitude (input) for each planet. Compare these equations among Milky Way planets and compare and contrast prevailing atmospheric conditions.

Solar Energy Output Model. There is clear evidence that solar energy reaching the earth from the sun exhibits periodic variations in time. The mean solar constant (about $1.35\,\mathrm{kW/m^2}$) varies about $0.02\,\mathrm{kW/m^2}$ approximately sinusoidally with a quasi-mean period of 11 years corresponding to the sunspot cycle. Furthermore, a longer cycle variation with a period of about 38 years and variation of about $0.04\,\mathrm{kW/m^2}$ is also observed. Establish a system equation for the solar energy incident on the earth's atmosphere as a function of the input variable time. Assume, arbitrarily, that a maximum for the sunspot cycle and longer period cycle both occurred in July 2020 (assumption!). Determine the total and average solar energy density in $\mathrm{kW/m^2}$ received by the earth from 2000 to the year 2020. During the time period 2020 to 2040, predict the year(s) of relative maximum and minimum solar energy. How might the earth's annual weather vary during those years?

Additional Modeling Problems. For each of the systems given in Table 9-3, define an appropriate model and its pertinent elements. Express quantitative variables for each of the model inputs and outputs and develop a simple analytical expression for the system model equation. Determine analytical or other solution forms for the system equation and interpret and evaluate your result.

Table 9-3 Modeling Problems for the Earth Science Systems

	System	Input	Output
(a)	Continental plates	Magma flow	Continental drift
(b)	Atmosphere	Solar energy	Weather
(c)	Biosphere	Solar activity	Climate
(d)	Earth	Geological eras	Earth's history
(e)	Lithosphere	Material, pressure, temperature	Constituent material radial zones
(f)	Earth's oceans	Moon and sun	Tides
(g)	Mountain building	Tectonic plate motion	Average mountain height
(h)	Solar system planets	Mean distance from sun. Normalized to Earth distance	Revolution time around sun Mean surface temperature Mass and gravity of planet
(i)	Atmosphere	Altitude	Atmospheric regions (ionosphere, stratosphere, troposphere, etc.)
(j)	Earth	Tidal and atmospheric friction	Rotational energy
(k)	Oceans	Ocean mining techniques	Mineral recovery
(l)	Volcanoes	Heat and pressure	Eruption and lava flow
(m)	Ground water	Recharge mechanisms	Extraction
(n)	Geological history	Time	Geological periods
(o)	Continental drift	North American Plate	African Plate
(p)	Chernobyl Nuclear Accident	Reactor operations	Reactor's physical responses
(q)	U.S. Clean Air Act	Regulatory requirements	Atmospheric responses
(r)	DDT	User applications	Biological impacts
(s)	U.S. EPA	U.S. Congressional actions	Economic consequences
(t)	Hurricanes	U.S. Occurrence	Predictions by scientists
(u)	Kyoto Protocol	International Participants	Country responses
(v)	Tornadoes	Katrina (U.S. 2005)	Loss of human life and property
(w)	Tropical Forests	Brazil Tropical Forest	Habitat
(x)	Tsunamis	Fukushima	Environmental Impacts
(y)	Volcanoes	Mount Saint Helena eruption	Worldwide impact
(z)	Recycling	Commercial plastic debris	Plastic recycle and reutilization

9.4 Life Systems Applications of System Science

The most significant finding humans encounter in the study of the universe is the existence of life, life in multitudinous forms, at diverse, evolutionary levels, and with abundant species widely distributed. Although restricted to a narrow range of environmental conditions, the vigor and resiliency of these life forms on the earth is impressive. Increasing our understanding of some important aspects of life on earth is the purpose of this section.

9.4.1 Continuous and Discrete Growth Models

The simple exponential growth equation for the population density N_c, for a continuous system model is given by

$$dN_c = r N_c dt \tag{9.4.1}$$

and the associated growth equation for the population density N_d for a discrete system has the form

$$\Delta N_d = r N_d \Delta t \tag{9.4.2}$$

where for both system equations the parameter r is the growth rate and t is the time. Equation (9.4.1) is separable, and the solution for the continuous system if the growth rate r is a constant in time is given by

$$N_c = N_0 \exp{(rt)} \tag{9.4.3}$$

where N_0 is the population at $t = 0$.

To solve Eq. (9.4.2) for the discrete system, we observe for our first increment that

$$\Delta N_d\{t(0)\} = N_d\{t(1)\} - N_d\{t(0)\} = rN_d\{t(0)\}\Delta t(0) = rN_d\{t(0)\}[t(1) - t(0)]$$

and solving for $N_d\{t(1)\} = N_d(\Delta t)$, we have

$$N_d(\Delta t) = N_d(0)(1 + r \Delta t) \tag{9.4.4}$$

where $\Delta t = t(1)$, $t(0) = 0$, and $N_d\{t(0)\} = N_d(0)$. We will assume that the time step Δt is uniform for each step change in N_d. For the next incremental change in N_d, we have

$$\Delta N_d(\Delta t) = N_d(2 \Delta t) - N_d(\Delta t) = N_d(\Delta t) r \Delta t \tag{9.4.5}$$

and recalling Eq. (9.4.4), we have from Eq. (9.4.5) that

$$N_d(2\Delta t) = N_d(\Delta t)(1 + r\Delta t) = N_d(0)(1 + r\Delta t)^2$$

Repeating this process, we find from the $\Delta N_d[(k-1)\Delta t]$ difference that

$$N_d(k\Delta t) = N_d(0)(1 + r\Delta t)^k \tag{9.4.6}$$

It is possible to show that under certain conditions the discrete population density N_d given by Eq. (9.4.6) approaches the continuous population density N_c, given by Eq. (9.4.3). We observe that $t = k \Delta t$, and if we let the step size $\Delta t \to 0$, observing thus that $k \to \infty$, then we have (note: $r \Delta t = r t/k$)

$$\text{Lim}_{\substack{\Delta t \to 0 \\ k \to \infty}} N_d(k\,\Delta t) = N_d(t) = \text{Lim}_{\substack{\Delta t \to 0 \\ k \to \infty}} N_d(0)\,(1 + rt/k)^k$$

Completing the limiting operation, we have

$$N_d(t) = N_d(0)\,\text{Lim}_{k \to \infty}(1 + r\,\Delta t/k)^k = N_d(0)\,\exp\,(rt) \tag{9.4.7}$$

since by definition

$$\exp\,(x) = \text{Lim}_{h \to \infty}(1 + x/h)^h$$

We see on comparing Eqs. (9.4.3) and (9.4.7) that they are equivalent in the limit for small Δt, and for appropriate conditions on the size of the input change in the discrete system model, the two models are equivalent.

Returning to Eq. (9.4.1), it is possible to obtain analytic solutions for those growth models in which the growth rate r is a simple power series in terms of the population density N_c. Determine the solution for continuous population density for $r\,(N_c) = r_0 + r_1 N_c + r_2 N_c^2$ where r_0, r_1, and r_2 are constants.

9.4.2 The Mammalian Lung Model

The pulmonary system in higher animals is an elaborately structured system of air sacs and blood vessels that serves to exchange gases, primarily oxygen and carbon dioxide, between the atmosphere and the circulating blood of the animal. Lower forms of life can satisfy their gas exchange needs by simple molecular diffusion for cellular dimensions. However, for multi-celled organisms such as fish, birds, reptiles, and humans, the exchange of vital gases via diffusion alone is insufficient and must be accomplished by other transport mechanisms, such as blood flowing through the pulmonary system.

We wish to develop a simple quantitative model of the pulmonary system that expresses the oxygen content of the blood $O_x(t)$ in the lungs as a function of time t and breathing rate r. A simple schematic for our model and block diagram is shown in Figure 9-11.

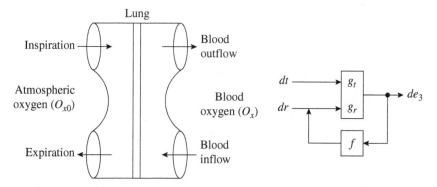

Fig. 9-11   System model for lungs.

The standard system equation deduced from the block diagram is found to be (see Chapter 6 for details)

$$dO_x = \frac{g_t \, dt + g_r \, dr}{1 - g_r f} \tag{9.4.8}$$

where g_t is the system kernel that relates the time dependence for changes of oxygen dissolved in the blood. While g_r is the system kernel that determines the dependency between blood oxygen and respiratory rate r, and k is the feedback kernel that controls the respiratory rate. For simplification, let us quantify our system kernels as follows:

$$g_r = a \left[1 - O_x/O_{x0}\right] \quad \text{for } a > 0$$

$$g_t = A \left(O_{x0} - O_x\right) \cos \left(rt\right)$$

$$k = k \text{ (a constant)}$$

The basis for this selection of functions is the assumption of simple diffusion for oxygen exchange and a sinusoidal respiratory cycle.

With these expressions for the system kernels in place, system Eq. (9.4.8) becomes

$$dO_x(r, t) = \frac{A(O_{x0} - O_x) \cos \left(rt\right) dt + a(1 - O_x/O_{x0})}{1 - ak \left(1 - O_x/O_{x0}\right)} dr \tag{9.4.9}$$

To simplify this system equation, let us define new variables and parameters as follows:

$$Y = 1 - O_x/O_{x0}, \quad dO_x = -O_{x0} \, dY, \quad b = a/O_{x0}$$

System Eq. (9.4.9) can now be expressed as

$$(af - 1/Y) \, dY = A \cos \left(rt\right) dt + b \, dr$$

This equation is exact (see Section 5.2.4 for the basis of this observation) and can be directly integrated to give

$$akY - \text{Ln } Y = \frac{A}{r} \sin \left(rt\right) + br + K \tag{9.4.10}$$

where K is the integration constant. Now suppose as an initial condition we want for $r = t = 0$ that $O_x (0, 0) = 0$ or $Y = 1$. Then we have that $K = a \, k$. Eq. (9.4.10) for the solution of our system equation can then be expressed in terms of O_x as follows:

$$O_x(r, t) = O_{x0}\left[1 - \exp\left(-\frac{A}{r} \sin \left(rt\right) - b \left(r + k \, Q_x\right)\right)\right] \tag{9.4.11}$$

We observe the following interesting consequences of our model:

1. As the respiratory rate in our model becomes very large $(r \rightarrow \infty)$, the blood oxygen level approaches that in the atmosphere $(O_x \rightarrow O_{x0})$
2. As the respiratory rate approaches zero $(r \rightarrow 0)$, the blood oxygen level approaches

$$O_x \rightarrow O_{x0}[1 - \exp\left(-A\,t - b\,k\,O_x\right)]$$

For a given value of time, the blood oxygen level may be multiple valued. That is $O_x = 0$ and $O_x \rightarrow 0$ as $r \rightarrow 0$, depending on the parameters $A, b, k,$ and time t.

9.4.3 Topical System Problems in Life Science

Cell Diffusion Model. The change in diffusion concentration $d(c_0 - c)$ of material passing through a membrane wall is controlled by the change in the osmotic pressure $d(P_0 - P)$ and the mean system temperature dT. The general system equation has the form

$$d(c_0 - c) = g_1\, d(P_0 - P) + g_2\, d\, T$$

where c and P are the concentration and pressure on one side of the membrane and c_0 and P_0 are the concentration and pressure on the other side of the membrane.

a) For a given cell diffusion model, assume that the system kernels are as follows

$$g_1 = a\, \frac{(c_0 - c)}{P_0 - P}$$

$$g_2 = -b\, \frac{(c_0 - c)}{T}$$

Show that the system equation is the following with a and b constant.

$$c = c_0 - K\, \frac{(P_0 - P)^a}{T^b}$$

b) If the parameters a and b in part (a) are functions of c, P, and T, show that the requirement for exactness and integrability is

$$\frac{\partial a}{\partial \operatorname{Ln} T} + a\, \frac{\partial b}{\partial \operatorname{Ln}(c_0 - c)} = b\, \frac{\partial a}{\partial \operatorname{Ln}(c_0 - c)} - \frac{\partial b}{\partial \operatorname{Ln}(P_0 - P)}$$

c) Given exactness of the functions in part (b), determine the integral form for the solution of the cell diffusion model.

Cell Population Model. Consider a time-varying reproductive cell population $p(t)$ in the presence of a regulated supply $s(t)$ of nutrient $n(t)$. Assume that the rate at which the cells are produced is proportional to $p(a + n)$, and the rate at which cells are destroyed is proportional to $p(p + b)$.

a) Show that the system equations governing the change in cell population and nutriment are given by

$$dp = [A\,p\,(a + n) - B\,p\,(p + b)]dt$$
$$dn = [s\,(t) - C\,p\,(a + n)]dt$$

where A, B, C, a, and b are all constants.
b) Determine the equilibrium values of p and n, that is, p_0 and n_0.
c) Normalize the system outputs in terms of the equilibrium values

$$N(t) = n(t)/n_0 \quad \text{and} \quad P(t) = p(t)/p_0$$

and minimize the number of system parameters. What are the system equations now?
d) Linearize the system equations and determine solutions for the system outputs $N(t)$ and $P(t)$.

e) Attempt to find nonlinear solutions for the system using the symbolic MATLAB operator "dsolve" described in Chapter 3.

f) If MATLAB does not provide an analytic solution, obtain numerical solutions using the MATLAB Numerical Differential Equation Solver for a selected set for the constants A, B, C, a, and b.

Life Cycle Model. An essential life cycle that supports all animal life on the earth is shown in the block diagram.

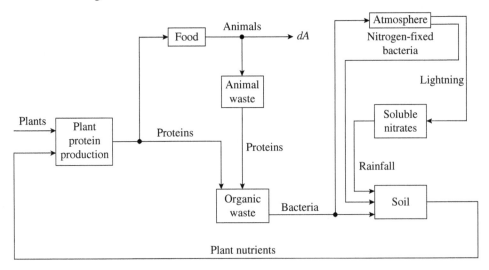

Consider the plant population change dP as the model input and animal production change dA as the output.

a) Define kernels for each model block.
b) Collapse the life cycle model into a single equivalent, canonical system form and write the overall system equation.
c) Combine and quantify the system kernels as appropriate.
d) Determine the system equation solution for a linear system kernel.
e) Discuss the implications of this system model.

Population Growth Model. The growth in time of a population for $N(t)$ members at time t is described by the system equation

$$dN(t) = r(t) N (1 - N/N^*) dt$$

where $r(t)$ is the growth rate in members per member per unit time and N^* is a constant related to the equilibrium population.

a) Obtain the solution for the system equation in terms of $\int_0^t r(t)\, dt$
b) For the mean value, $\int_0^t r(t)\, dt = \langle r \rangle t$, plot the response of $N(t)/N_0$ as a function of normalized time (i.e., $\langle r \rangle t$ for the following initial values:

$$N_0 = 0.1N^*, \quad N^*, \quad \text{and } 10N^*$$

c) Suppose that $\int_0^t r(t)\,dt = r_0\,u\,(t-t_0)\,t$, where $u(t)$ is the unit step function defined as follows: $u(t-t_0) = 0$ if $t < t_0$ and 1 if $t \geq t_0$.

Plot $N(t)/N_0$ as a function of time in units $r_o\,t$.

Classical Predator-Prey Model. The classic model for the time behavior for the Predator-Prey system is the Lotka and Volterra Model, which has the following nonlinear form for the system behavior

$$dH = H(h - aP)\,dt$$
$$dP = P(bH - p)\,dt$$

where $H(t)$ is the prey (the hunted) population and $P(t)$ is the predator (the hunter) population at time t. The parameter h is related to the birth rate of the prey, p to the death rate of the predator, and a and b are interaction parameters.

a) Show that the non-zero, equilibrium solutions P_0 and H_0 for the model are given by

$$H_0 = p/b \quad \text{and} \quad P_0 = h/aH_0 = p/b \quad \text{and} \quad P_0 = h/a$$

b) Show that the time-independent solution for the nonlinear system is given as

$$Z = h \ln P - aP + p \ln H - bH$$

where Z is the arbitrary integration constant.

c) Show the graphical behavior of the normalized pray population, P/P_0 (0 to 3) versus normalized predator population, H/H_0 (0 to 3) for various values of the Prey population, Z are shown below

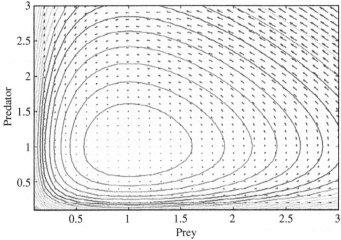

Prey-Predator dynamics described for values of Z. Arrows depict the velocity and direction of solutions in time. Data used are $a = b = p = h = 1$ and non-zero equilibria values are $P/P_0 = H/H_0 = 1$. *Source:* Wikipedia.

Generalized Predator-Prey Model. The following generalized model associated with the classical Lotka and Volterra Predator-Prey Model has broader applications for diverse fields in addition to Life Systems. By appropriate interpretation of the model outputs, e_i (for $i = 1$, 2, etc.) and parameters, this model can be applied to system models including disease epidemics, ecology, economics, competitive industries, and even hostile relations between nations.

The system equations for this model have the following non-linear form as functions of time for two interacting outputs e_1 and e_2.

$$de_1 = r_1 e_1 \left[1 - \frac{(e_1 + a_{12}e_2)}{b_1} \right] dt \tag{9.4.11.1}$$

$$de_2 = r_2 e_2 \left[1 - \frac{(e_2 + a_{21}e_1)}{b_2} \right] dt \tag{9.4.11.2}$$

where

$e_1 = $ First competing output time behavior (e_1 units at time t)
$e_2 = $ Second competing output time behavior (e_2 units at time t)
$r_1 = $ First output intrinsic growth unit rate (change in e_1 per time change)
$r_2 = $ Second output intrinsic growth unit rate (change in e_2 per time change)
$a_{12} = $ Competition coefficient of first output versus second output
    (e_1 unit change per e_2 units)
$a_{21} = $ Competition coefficient of second output versus first output
    (e_2 unit change per e_2 unit)
$b_1 = $ First output carrying capacity (change in e_1 at carrying capacity)
$b_2 = $ Second output carrying capacity (change in e_2 at carrying capacity)

Note that model output variables will have appropriate dimensions for each term, such as pure numbers, dollars, production units, human deaths, etc. for the given system

a) Determine the equilibrium values for e_1 and e_2 for all equilibrium conditions (viz., $de_i/dt = 0$).
b) Normalize the output variables in terms of the nonzero steady-state values found in part a) as

$$e(t) = \{e(\text{non} - \text{zero equilibrium value}) \, x \, E(t)\}$$

where $E(0) = 1$ at equilibrium.
   With this step the dependent variables are now dimensionless numbers for any model application.
c) Attempt to find a solution for the nonlinear normalized equations obtained in (b) both manually and using the MATLAB Symbolic Toolbox. Comment on the results of this effort.
d) Make a plot of $E_1(t)$ versus $E_2(t)$ by forming the ratio dE_1/dE_2, which eliminates the time dependence, and use the Symbolic Toolbox for a selected set of model parameter values.
e) Linearize the system equations around equilibrium output values as follows by ignoring product terms $\delta E_1 \, \delta E_2$.

$$E_i(t) = 1 + \delta E_i(t)$$

f) Obtain time-dependent solutions for the linearized system equations using the MATLAB Symbolic Toolbox.

g) Plot the time behavior of $\delta E_1(t)$ and $\delta E_2(t)$ using the MATLAB Symbolic Toolbox for a selected set of parameter values.

Hodgkin–Huxley Biophysics Model. The Hodgkin–Huxley model employs a set of four nonlinear, time-dependent differential equations that describe how voltage potentials are produced and transmitted within biological neurons. Alan Hodgkin and Andrew Huxley developed their model in 1952, and they received the Nobel Prize in Physiology or Medicine in 1963 for their great achievement in biophysics. The model is regarded as a major accomplishment in twentieth-century biophysics.

Since the Hodgkin–Huxley model is comprised of a complex, nonlinear set of differential equations that cannot be solved analytically, we will simplify the original model equations and employ these simplified equations below as an approximate, single input (time as the independent variable), multiple output (four dependent variables e_1, e_2, e_3, e_4) system for the Hodgkin–Huxley model.

$$c\frac{dV}{dt} = \left[I - c_k n^4(V - V_K) - c_{Na}m^3h\,(V - V_{Na}) - c_L(V - V_L)\right]$$

$$\frac{dn}{dt} = V\left[a_n(1-n) - b_n n\right]$$

$$\frac{dm}{dt} = V\left[a_m(1-m) - b_m m\right]$$

$$\frac{dh}{dt} = V\left[a_h(1-h) - b_h h\right]$$

The original Hodgkin–Huxley model parameters a_i's, b_i's, and c_i's are functions of the membrane voltage, V. These equations required numerical methods for analysis. Therefore, to simplify the equations, we will assume that $V(t)$ as shown and the parameters a_i's, b_i's, and c_i's are all now constants. Then single input and multiple output system variables and parameters are

$$de_1 = \frac{1}{c}\left[I - c_k e_2{}^4(e_1 - V_K) - c_{Na}e_3{}^3 e_4(e_1 - V_{Na}) - c_L(e_1 - V_L)\right]dt \qquad (9.4.12.1)$$

$$de_2 = e_1\left[a_n(1 - e_2) - b_n e_2\right]dt \qquad (9.4.12.2)$$

$$de_3 = e_1\left[a_m(1 - e_3) - b_m e_3\right]dt \qquad (9.4.12.3)$$

$$de_4 = e_1\left[a_h(1 - e_4) - b_h e_4\right]dt \qquad (9.4.12.4)$$

where

$e_1(t) = V(t)$ membrane voltage (millivolts)
$e_2(t) = m(t)$ potassium channel activation (dimensionless)
$e_3(t) = n(t)$ sodium channel activation (dimensionless)
$e_4(t) = h(t)$ sodium channel inactivation (dimensionless)

Furthermore, m, n, and h are dimensionless, output variables with a range $0{\rightarrow}1$. Also, C_m, V_k, V_{Na}, V_L, a's, b's, and c's are all constant, time independent parameters. $I(t)$ electric current (nano-amps) is measured to test the model.

Our strategy in analyzing these complex system equations, Eqs. (9.4.16), is to eliminate e_1 from Eqs. (9.4.12.2), (9.4.12.3), and (9.4.12.4) and find solutions for e_2, and e_3 in terms of e_4. We then substitute the solution for e_4 into Eq. (9.4.12.1) and obtain two equations for $e_1(t)$ and $e_4(t)$. We can then eliminate the membrane voltage, e_1, by forming the ratio of Eq. (9.4.12.2) with Eq. (9.4.12.4) to provide

$$\frac{de_2}{[a_n(1-e_2)-b_ne_2]} = \frac{de_4}{[a_h(1-e_4)-b_he_4]}$$

This equation is easily integrated as follows with K_1 as the integration constant

$$\frac{1}{a_n+b_n}\,\mathrm{Ln}\,[a_n-e_2(a_n+b_n)] = \frac{1}{a_h+b_h}\,\mathrm{Ln}\,[a_h-e_4(a_h+b_h)] + K_1$$

Solving for e_2 in terms of e_4 we have with K_2 as integration constant

$$e_2 = \frac{a_n}{a_n+b_n} + K_2\,[a_h-e_4(a_h+b_h)]^{\frac{a_n+b_n}{a_h+b_h}} \tag{9.4.13}$$

Performing the same steps, we obtain e_3 in terms of e_4 with the integration constant K_3

$$e_3 = \frac{a_m}{a_m+b_m} + K_3\,[a_h-e_4(a_h+b_h)]^{\frac{a_m+b_m}{a_h+b_h}} \tag{9.4.14}$$

Next, we form the ratio using Eq. (9.4.16a) and Eqs. (9.4.14) which provides the response of e_1 in terms of e_4

$$\frac{de_1}{de_4} = c\,\frac{[I(t)-c_ke_2^4(e_1-V_K)-c_{Na}e_3^3e_4(e_1-V_{Na})-c_L(e_1-V_L)]}{e_1[a_h(1-e_4)-b_he_4]} \tag{9.4.15}$$

where e_2 and e_3 are functions of e_4 as given by Eq. (9.4.13) and Eqs. (9.4.14) respectively. Observe the $I(t)$ is also a function of t, but we will assume it is constant for each selected set of phase plane plots for e_1, e_2, e_3, and e_4.

It is apparent that Eq. (9.4.15) poses a formidable integration task. However, using the graphical plotting programs in the Symbolic Toolbox of MATLAB, phase plane plots for Eq. (9.4.15) can be developed for selected values of the parameters I, a_i's, b_i's, c_i's, and V's and the behavior of e_1 and e_4 in the phase plane can then be examined.

Finally, various approximate analytical solutions for e_1 in terms of e_4 can be found, and the resulting solution for e_1 can be used in Eq. (9.4.15) to obtain the time-dependent solutions for e_1 and subsequently the time-dependent solutions or e_2, e_3, and e_4. For a selected time range, determine the responses for

$e_1(t) = V(t)$ membrane voltage (millivolts)
$e_2(t) = m(t)$ potassium channel activation (dimensionless)
$e_3(t) = n(t)$ sodium channel activation (dimensionless)
$e_4(t) = h(t)$ sodium channel inactivation (dimensionless)

Dissolved Oxygen in Water. Dissolved oxygen is essential for supporting life in water bodies. Yet competing factors serve to deplete and replace the dissolved oxygen in a water body that supports animal life. Consider as a model for a body of water that supports an aquatic population, the following system equations:

$$dA = g_a(A, W, T)\, dt$$
$$dW = g_w(A, W, T)\, dt$$
$$dT = g_t\,(A, W, T)\, dt$$

where

$$g_a = C\,[A_0 - A\,(t)] - k\,D\,[T(t) - T_0]\,W(t)$$
$$g_w = kR[T(t) - T_0] - D\,W(t)$$
$$g_t = EW(t)$$

for variable and parameters definitions as follows:

C = coefficient of water turbulence (1/s)
A_o = water body's maximum dissolved oxygen content [(kg O_2) /m^3]
$A\,(t)$ = dissolved oxygen content in water at time t [(kg O_2)/m^3)]
k = waste decomposition temperature coefficient [(kg O_2)/(K-kg)]
D = waste decomposition coefficient (1/s)
$T\,(t)$ = water temperature at time t (K)
T_0 = water temperature at time $t = 0$ (K)
E = solar energy absorption coefficient for waste [(m^3-K)/(s-kg)]
R = waste generation temperature coefficient [kg /(m^3-s-kg^2)]
$W(t)$ = waste density (kg/m)

Perform the following

a) Attempt to solve the nonlinear system equations. (Hint. Refer to Section 4.2.7 and treat the equation given by dT/dW first.)
b) Determine equilibrium values for each system output.
c) Assume that fish require a minimum oxygen content of 5.0 mg O_2/liter at $T_0 = 0$ °C. Consider as typical values for other parameters the following: Data *Source:* Wikipedia

$A = 15.0$ mg O_2/liter	$D = 0.5\,h^{-1}$
$E = 1.0$ cc – K/h-mg waste	$k = 1.0\,K^{-1}$
$R = R_0$ mg waste/h	$C = 1.0\,h^{-1}$ (large lake)

If $A = 10$ mg O_2/liter and $W = 0$ when $T = 0$ °C, what maximum value of R_0 (mg waste /m^3-mg O_2) is allowable without sustaining major fish kill over time?

Animal Growth Model. We seek to develop a simple model for the increase in weight and growth in size (e.g., mean height of animals from conception to maturity). Our model input is the differential change in time, and outputs are changes in weight dW and size dh. The block diagram for our system is as follows:

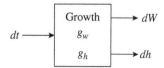

Assume that g_w is proportional to $(W_0 - W)$ and g_h is proportional to $(h_0 - h)$, where W_0 and h_0 are the mature weight and size, respectively,

a) Form the system equations and determine their solutions.
b) Assume that the animal's weight is related to size by the expression

$$W = CDh^3$$

where C is a constant and D is the body density. Can the two growth equations be made equivalent with this assumption?

Solar Radiation and Plant Radiation Absorption Model. Light, as electromagnetic (EM radiation) energy from the sun in the narrow frequency range between about 380 and 1100 nm ($1\ nm = 10^{-9}\ m$), is responsible for providing the energy input for all photosynthesis processes. Recall that photosynthesis is the essential photo-biological process in which light energy is employed to produce organic molecules by the reduction of carbon dioxide to carbohydrates. About half of the solar energy incident upon the earth's surface and, remarkably, all animal vision and other photobiological processes are enclosed within this same narrow frequency band of the full EM spectrum. Furthermore, there is compelling evidence that life found anywhere in the universe would require and be restricted to this same frequency band for all photo biological processes. Indeed, as evidence of this essential limitation on radiant energy input to biological processes, it is the relatively thin ozone layer in the earth's atmosphere found between 22 and 25 km above the earth's surface that protects all terrestrial life from extraterrestrial short-wavelength (i.e., less than about 300 nm) biocidal radiation. The ozone layer begins absorbing the sun's radiation and other extraterrestrial radiation sources at about 320 nm and becomes virtually opaque at EM wavelengths of less than 290 nm.

Chlorophylls are the biological pigments that are employed in the photosynthesis processes for absorbing solar radiation. In Figure 9-12, the frequency distribution of average

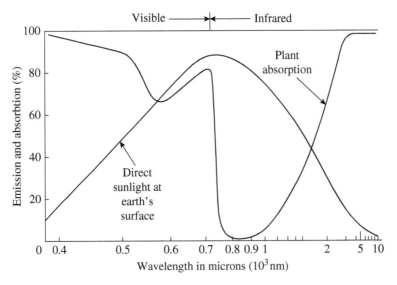

Fig. 9-12 Frequency distribution of sunlight and plant leaf absorption. *Source: Wikipedia.*

sunlight on the earth's surface is shown to peak at a broad maximum at about 500 nm (i.e., the blue-green region of the visual spectrum). Also shown in the figure is the frequency-dependent radiant energy absorption fraction for the leaves of typical plants. Observe the apparent evolutionary paradox that plant radiation energy absorption is at a near minimum value. In other words, where chlorophyll energy absorption and utilization is most efficient, sunlight exhibits a minimum output. Quantitatively model the net radiant energy absorbed by chlorophyll and possibly contribute to this continuing controversy in photobiological evolution.

Specifically, develop a simple set of analytical functions (i.e., a small number of algebraic expressions, each for an appropriate sequential bandwidth as necessary) that describe (1) the frequency-dependent fractional energy absorption by chlorophyll and (2) the frequency-dependent fractional energy content of sunlight. Both sets of functions should span the frequency range from about 400 to 900 nm. You may employ reasonable "smoothing" procedures for the graphical data presented in the figure.

a) At what frequency is the chlorophyll-solar energy utilization function (i.e., the product of the chlorophyll absorption distribution function and the solar energy emission distribution) a maximum?

b) What is the overall efficiency of the utilization function overall frequencies between 400 and 900 nm? (In other words, the ratio of the integral of the utilization function divided by the integral of the solar emission; where both integrals are defined over the frequency range of 400–900 nm.)

c) Can you explain on a biological basis why this evolutionary paradox might have occurred?

Extraterrestrial Life Model (SETI Search for Extraterrestrial Intelligence) An unresolved question that has challenged terrestrial humans for ages is the following: "Is there intelligent life beyond the earth?" The acronym for the scientific investigation addressing this question is SETI or the "Search for Extra-Terrestrial Intelligence." Presently, with recent discoveries in cosmology and mounting experimental data from International Space Programs, we have a defensible physical model for explaining the birth and development of our visible Universe. We have confirmed the "big bang" model that conceived our universe 13.8 billion years ago, discovered black holes including the black hole, Sagittarius A*, at the center of our Milky Way Galaxy and measured the accelerating expansion of space-time in the observable universe.

However, the open and unanswered question of extraterrestrial life remains. This issue has spawned the active field of astrobiology, and major financial investments and scientific efforts are evident in the international scientific community. The obscure spatial depths of our universe are now unfolding to NASA's Hubble Space Telescope, the Keppler Space Telescope and the European Southern Observatory (ESO) SPHERE Telescope along with the International Space Station (ISS). Furthermore, this unresolved, astrobiological search should be further clarified by the United States, multibillion-dollar, James Webb Space Telescope (JWST) Program for infrared imaging that went into service in 2022. With a 100-fold greater sensitivity than the Hubble Space Telescope, JWST promises to enhance exploration of the life cycle of galaxies and their planetary systems for critical properties and evidence of possible extraterrestrial life.

Interestingly, SETI may have arisen from a U.S. astronomical meeting in 1961, when a young radio astronomer named Frank Drake presented a quantitative model known as the "Drake equation" that was formulated to estimate the probability of monitoring alien life forms in the Milky Way Galaxy using radio astronomy. We will formulate an equation similar to the Drake equation but now incorporate the entire observable universe with its more than 100 billion ($\sim 10^{11}$) galaxies and the several 100 billion stars and planets within each galaxy.

Our mathematical model to investigate the existence of extraterrestrial life will employ probability occurrence factors, P_i defined as follows:

$$P = \prod_{i=1}^{7} P_i = P_1 P_2 P_3 P_4 P_5 P_6 P_7 \qquad (9.4.16)$$

where

P = occurrence of extraterrestrial intelligent life/observable universe (viz., Big Bang Universe).
P_1 = galaxies/observable universe
P_2 = stars/galaxy
P_3 = exoplanets/star
P_4 = life hospitable exoplanets/exoplanet
P_5 = liquid water in life hospital planet/life hospitable exoplanet
P_6 = DNA in planets with liquid water/liquid water on hospital planet
P_7 = extraterrestrial intelligent life/DNA in planets with liquid water

Observe that dimensionally we have captured the desired model output as follows with these estimated occurrence factors or probabilities:

$$P = P_1 P_2 P_3 P_4 P_5 P_6 P_7 = \frac{\text{extraterrestrial intelligent life}}{\text{observable universe}} \qquad (9.4.17)$$

Consider the following order of magnitude estimates for the model exponents since greater accuracy is presently unwarranted.

$$P_1 = 10^{11}, P_2 = 10^{11}, P_3 = 10^{n3}, P_4 = 10^{n4}, P_5 = 10^{n5}, P_6 = 10^{n6}, P_7 = 10^{n7} \qquad (9.4.18)$$

where we have utilized current estimates from international spatial probe data for P_1 and P_2.

Furthermore, we can provide an absolute lower bound for our model by recognizing that intelligent life is found on the Earth, so we have as the lower bound for our extraterrestrial model the following value

$$P \text{ (with earth)} = 1 = 10^0$$

We have for the exponents provided in Eqs. (9.4.17) and (9.4.18) the following

$$0 = 11 + 11 + n3 + n4 + n5 + n6 + n7$$

We assume conservatively that $P_3(n3)$ through $P_7(n7)$ are each less than unity. Note, however, that $P_3 > 1$ (a value of 9) for our Solar System, so we are conservative with our estimates for $P_3(n3)$ through $P_7(n7)$.

$$n3 + n4 + n5 + n6 + n7 \geq -22$$

If we assume that $n3$ through $n7$ are all approximately equal to a single mean value estimate $\langle n \rangle$, then we have

$$\langle n \rangle = -22/5 = -4.4$$

and

$$P_3 \approx P_4 \approx P_5 \approx P_6 \approx P_7 \approx 4.0 \times 10^{-5}$$

for the mean occurrence for each factor. If we assume that $n3 = -1$ assuming that 10% of galaxy stars have a least one orbiting planet, then we have

$$\langle n \rangle = -21/4 = -5.25$$

and for equal occurrence factors we have

$$P_4 \approx P_5 \approx P_6 \approx P_7 \approx 5.6 \times 10^{-6}$$

We conclude from a large number of galaxies and stars within each galaxy that the possibility for finding intelligent life in the universe is highly favorable.

Now let us now assume the following twofold data set for more realistic occurrence factors and investigate the model's predictions. We shall consider two boundaries for the data set for each occurrence factor. However, assume that P_1 and P_2 are the same for both boundaries and other occurrence factors are as follows. Note: [Data set P_{min} (earth)$\rightarrow P_{obs}$ (observable galaxy)] as follows:

$P_1 = $ galaxies/observable universe, same $[10^{11} \rightarrow 10^{11}]$
$P_2 = $ stars/galaxy, same $[10^{11} \rightarrow 10^{11}]$
$P_3 = $ exoplanets/star $[10^{-1} \rightarrow 10^{0}]$
$P_4 = $ life hospitable exoplanets/exoplanet $[10^{-3} \rightarrow 10^{-2}]$
$P_5 = $ liquid water in life hospital planet/life hospitable exoplanet $[10^{-4} \rightarrow 10^{-3}]$
$P_6 = $ DNA in planet liquid water/liquid water on hospital planet $[10^{-6} \rightarrow 10^{-4}]$
$P_7 = $ occurrence extraterrestrial intelligent life/DNA in planet liquid water $[10^{-8} \rightarrow$ to $10^{-6}]$

Employing these values for our earth only and observable galaxy estimates, we have the following as found for the earth only basis

$$P \text{ (earth)} = P_1\, P_2\, P_3\, P_4\, P_5\, P_6\, P_7 = 1$$

And for the observable galaxy estimate

$$P \text{ (observable galaxy)} = P_1\, P_2\, P_3\, P_4\, P_5\, P_6\, P_7 \approx 10^7 \approx 10 \text{ million}$$

We may disagree with the implications of this result, but the estimate is about what many astrophysicists actually anticipate, namely, over a million separate occurrences of extraterrestrial intelligent life in our observable galaxy. In fact, in the 1940s during Manhattan

Table 9-4 Modeling Problems for the Life Sciences

	System	Input	Output
(a)	Bioluminescence	Reacting chemicals	Light frequency and intensity
(b)	Cell	Ionizing radiation level	Radiation survival probability
(c)	DNA	Adenine (A) – Thymine (T) Cytosine (C) – Guanine (G)	Codon assembly in DNA A – T, C – G
(d)	Cell	Enzymes	Cell functions
(e)	Macromolecules	Water solutes	Hydrocarbon input
(f)	Mammalian eye	Target image	Central nervous system, cilia muscle, retinal image
(g)	Mammalian bodies	Viruses	Cancer cell
(h)	Cellular energy product	ADP	ATP
(i)	Protein production	DNA	RNA
(j)	Proteins	DNA	Protein configuration
(k)	Proteins	mRNA (messenger RNA)	tRNA(transfer RNA)
(l)	Nervous system	Nerve signals	Cellular response
(m)	Ecosystems	Energy and chemical inputs	Organic system response
(n)	Eukaryotic cell	RNA	Evolutionary events
(o)	Carbon atoms	RNA structure	DNA structure
(p)	Mutation	Original DNA	Heritable variations
(q)	Plant form and structure	Viruses	Cancer cell
(r)	Cellular energy product	ADP	ATP
(s)	Mitosis	Chromosome input	Daughter cell production
(t)	Organism	Enzyme	Protein production
(u)	Cell	Viruses	Cancer cell
(v)	Human species model	E. coli	Test Mice
(w)	Gibbs free energy	Enthalpy	Entropy
(x)	Amino acids	Peptide bonds	Ligands
(y)	DNA	RNA	Virus
(z)	DNA cell invasion	SARS-Cov-2	COVID-19 Pandemic

District (U.S. development of the "atom bomb"), the Nobel Laureate physicist, Enrico Fermi arrived at a similar estimate for the existence of extraterrestrials. Fermi reputedly posed the question to his scientific colleagues, "where are they?" (i.e., why is there no direct evidence of extraterrestrials despite their high expected occurrence for existence). The question has become known as the "Fermi Paradox" and still remains unanswered. There is an active, well-funded program searching for exoplanets in our observable universe. (Reference: Wikipedia – the free encyclopedia).

It is also useful to consider the uncertainty regarding each of the occurrence factors, which can be estimated by taking the natural log of Eq. (9.4.17) and forming the discrete differences for each occurrence as follows:

$$\frac{\Delta P}{P} = \sum_1^7 \frac{\Delta P_i}{P_i}$$

If we retain the error in each of our occurrence estimates by

$$\text{Maximum} \left(\frac{\Delta P_i}{P_i} \right) \approx 10^{-2}$$

Then make your own estimate for these occurrence factors for Eq. (9.4.17) and respond to the Fermi Paradox.

Additional Modeling Problems. For each of the systems given in Table 9-4, define an appropriate system model and the essential elements of the system. Express quantitative variables for each of the model inputs and outputs and develop a simple analytical expression for the model's system equation. Determine analytical or other solution forms for the system equation and interpret and evaluate your results.

9.5 Applications of System Science to Human Life

Of all life forms encountered on the earth, the most advanced and challenging in system science is human life. The marvelously complex, intricate life form we technically classify as homo sapiens (the present evolved human being) is an essential object for careful study and accrued knowledge. Indeed, the balance and intimacy that humans share with all life on earth and the life-support processes of the earth itself are a challenging study in system interaction and feedback.

9.5.1 Hemodynamic Circulatory System

One of the most marvelous, biological systems, which is of obvious interest to humans, is the mammalian circulatory system for blood flow. These highly adaptive and essentially similar systems provide essential circulation of the life fluids with essential suspended elements that support for all basic cell functions in mammals. Each circulatory system is a marvel of simplicity in purpose and complexity in function.

Our desire here is to present a simplified, quantitative model for the human circulatory system. The following assumptions and approximations are imposed to develop the model.

The objective for the analytical model of the circulatory system developed here will be primarily to provide a correct quantitative description of the behavior and interaction of blood flow through the major subsystems and compartments of the circulatory system. Furthermore, the model is directed at providing a description of behavior around steady-state operating conditions for which parameter values for the circulatory system can be determined.

The primary goal here in developing a mathematical model for the circulatory system is to provide a practical and reasonably accurate description of the mean hydrodynamic characteristics of the circulating blood within the major subcomponents of the circulatory system. The following assumptions and approximations are imposed to realize such a circulatory flow model:

- The circulatory system is reduced to three major subsystems, each containing two components. The subsystems are (1) the heart or blood pumps containing a left and right heart component, (2) the systemic circulation with an arterial and venous component, and (3) the pulmonary system with an arterial and venous component. The capillary beds of the systemic and pulmonary subsystems are assumed to be associated with their respective venous component.
- Within the subsystems, chemical, biological, shunting, and electrical conditions that might affect blood circulation are neglected.
- Within each major component of the circulatory system only mean, quasi-steady-state values for the component's variables and parameters are considered.
- The blood flowing through the circulatory model can be characterized as an isothermal Newtonian fluid with a constant viscosity within each subsystem component.
- In those subsystem components where transient pulsatile blood flow is significant, time-averaged (over a few heartbeats) mean values of pertinent variables and parameters (e.g., flow rate, pressure, mass, speed, viscosity, and vessel parameters) will be employed.

Figure 9-13 provides an overall schematic for the circulatory system model. Blood is assumed to flow clockwise around the system. Beginning with the pulmonary vein, blood at a mean pressure of P_{pv} flows into the left heart. This pump supplies a pumping power of W_{lh} to increase the mean blood pressure by ΔP_{lh} and provide blood flow at a mean stroke volume of Q_{lh} to the aorta in the systemic circulation system at a mean flow rate of q_{lh} and a mean pressure of P_{AO}. In the systemic arterial system, blood with a mean volume of Q_{sa} flows at a mean rate of q_{lh} against a resistance of R_{sa} and drops in pressure by ΔP_{sa}. Blood then flows into the systemic venous system at a mean flow rate of q_{sv} with a mean volume of Q_{sv} and a flow resistance of R_{sv} that reduces the pressure by ΔP_{sv}. Blood departs the systemic venous system at a mean pressure P_{vc} and flows into the right heart. The right heart provides a pumping power of W_{rh} to increase the mean blood pressure by ΔP_{rh} and delivers blood at a mean flow of q_{rh} into the pulmonary artery at a mean pressure of P_{pa} and a mean stroke volume of Q_{rh}. Blood with a mean volume Q_{pa}, flows at a mean rate q_{pa} through the pulmonary arterial system, which has a resistance of R_{pa}. and drops in pressure by ΔP_{pa}. From the pulmonary arterial system, blood then flows at a mean flow of q_{pv} into the pulmonary venous system, where the blood occupies a mean volume of Q_{pv} and experiences a flow resistance of R_{pv} that reduces the pressure by ΔP_{pv}. From the pulmonary venous system,

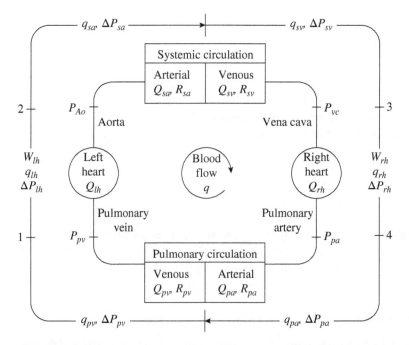

Fig. 9-13 Circulatory system model schematic. *Source:* Annals Biomedical Engineering (G. Sandquist), Vol.10, pp. 1–33, 1982.

blood then flows into the pulmonary vein at a mean pressure of P_{pv} where the blood flow repeats this circuit.

By employing the basic mathematical statements for conservation of energy, mass, and momentum and using Figure 9-13 as a schematic representation of the model, it is possible to derive from these physical system models, conservation equations for blood volume, blood flow, power delivered to the heart, and blood vessel compliance (or elasticity).

A full derivation of the complete circulatory equations will not be given (see reference noted in Figure 9-13), but the systemic circulatory system model for blood flow in a human can be expressed as follows

Blood volume:

$$Q = Q_s + \frac{q}{BPM} \tag{9.5.1.1}$$

Vessel compliance:

$$Q_s = C_{sa}P_{A0} + C_{sv}P_{vc} \tag{9.5.1.2}$$

Blood flow:

$$q = \frac{a\left(P_{A0} - P_{vc}\right) P_{A0}^2 P_{vc}^2}{b\, P_{vc}^2 + P_{A0}^2} \tag{9.5.1.3}$$

Heart pumping power:

$$W = (P_{A0} - P_{vc}) \frac{q}{e} \qquad (9.5.1.4)$$

where variables are

Q_s = blood volume in systemic system (liters)
q = blood flow in systemic system (liters/min)
BPM = heart pumping rate (beats/min)
W = power delivered to left heart (Watts)
P_{A0} = mean aortic pressure (mm Hg)
P_{vc} = mean central venous pressure (mm Hg)

and parameters are

Q = total blood volume in body (liters) (about 5 liters in humans)
C_{sa} = systemic arterial compliance (liters/mm Hg)
C_{sv} = systemic venous compliance (liters/mm Hg)
a = volume rate parameter [liters/min-(mm Hg)^3]
b = dimensionless parameter
e = effective left ventricle pumping efficiency (liters-mm Hg/W-min)

If we consider as the output variables for the system model q, W, Q_s, and BPM, and express each of these outputs as functions of the variables P_{A0} and P_{vc} (chosen as the independent variables for the system), then Eqs. (9.5.1) can be expressed as follows to produce a system equation of state for the human systemic circulatory model with two inputs and four outputs:

$$q = \frac{a(P_{A0} - P_{vc}) P_{A0}^2 P_{vc}^2}{b P_{vc}^2 + P_{A0}^2} \qquad (9.5.2.1)$$

$$W = \frac{a(P_{A0} - P_{vc})^2 P_{A0}^2 P_{vc}^2}{e(b P_{vc}^2 + P_{A0}^2)} = (P_{A0} - P_{vc}) \frac{q}{e} \qquad (9.5.2.2)$$

$$Q_s = C_{sa} P_{A0} + C_{sv} P_{vc} \qquad (9.5.2.3)$$

$$BPM = \frac{a(P_{A0} - P_{vc}) P_{A0}^2 P_{vc}^2}{(Q - C_{sa} P_{A0} - C_{sv} P_{vc})(b P_{vc}^2 + P_{A0}^2)} = \frac{q}{(Q - Q_s)} \qquad (9.5.2.4)$$

Equations (9.5.2) constitute the solutions for the systemic human circulatory model system equation when the outputs q, W, Q_s, and BPM are expressed as functions of the input pressures P_{A0} and P_{vc}. Three-dimensional graphs for q, W, Q_s, and BPM are readily obtained with MATLAB and shown in the following figures for these nominal parameter values.

Q = 5 liters (total blood volume in body)
Q_s = 4 liters (total systemic blood volume)
C_{sa} = 0.01 liters/mm Hg (systemic arterial compliance)
C_{sv} = 0.6 liters/mm Hg (systemic venous compliance)
a = 0.0001 liters/min (mm Hg)^3 (volume rate parameter)
b = 1.0 (dimensionless parameter)
e = 0.5 liters-mmHg/W-min (left ventricle pumping efficiency)

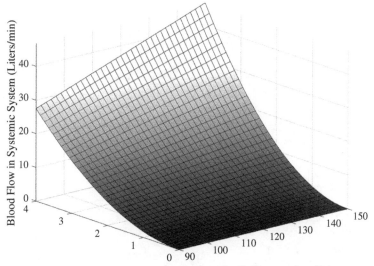

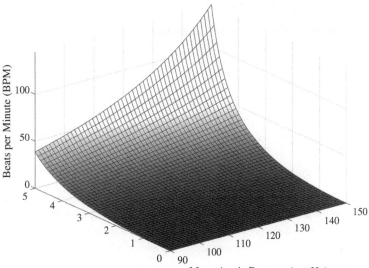

The system equations for the systemic circulatory model are easily found by differentiation the system equations of state as follows:

$$
\begin{bmatrix} dq \\ dW \\ dQ_s \\ dBPM \end{bmatrix} = \begin{bmatrix} g_{11} & g_{12} \\ g_{21} & g_{22} \\ g_{31} & g_{32} \\ g_{41} & g_{42} \end{bmatrix} \begin{bmatrix} dP_{A0} \\ dP_{vc} \end{bmatrix}
$$

where

$$g_{11} = \frac{\partial q}{\partial P_{A0}} = \left(\frac{1}{P_{A0} - P_{vc}} + \frac{2}{P_{A0}} - \frac{2P_{A0}}{bP_{vc}^2 + P_{A0}^2} \right) q$$

$$g_{12} = \frac{\partial q}{\partial P_{vc}} = \left(-\frac{1}{P_{A0} - P_{vc}} + \frac{2}{P_{vc}} - \frac{2bP_{vc}}{bP_{vc}^2 + P_{A0}^2} \right) q$$

$$g_{21} = \frac{\partial W}{\partial P_{A0}} = \left(\frac{1}{P_{A0} - P_{vc}} + \frac{1}{P_{A0}} - \frac{P_{A0}}{bP_{vc}^2 + P_{A0}^2} \right) 2W$$

$$g_{22} = \frac{\partial W}{\partial P_{vc}} = \left(-\frac{1}{P_{A0} - P_{vc}} + \frac{1}{P_{vc}} - \frac{bP_{vc}}{bP_{vc}^2 + P_{A0}^2} \right) 2W$$

$$g_{31} = \frac{\partial Q_s}{\partial P_{A0}} = C_{sa}, \quad g_{32} = \frac{\partial Q_s}{\partial P_{vc}} = C_{sv}$$

$$g_{41} = \frac{\partial BPM}{\partial P_{A0}} = \left(\frac{1}{P_{A0} - P_{vc}} + \frac{2}{P_{A0}} + \frac{C_{sa}}{Q - C_{sa}P_{A0} - C_{sv}P_{vc}} - \frac{2P_{A0}}{bP_{vc}^2 + P_{A0}^2} \right) BPM$$

$$g_{42} = \frac{\partial BPM}{\partial P_{vc}} = \left(-\frac{1}{P_{A0} - P_{vc}} + \frac{2}{P_{vc}} + \frac{C_{sv}}{Q - C_{sa}P_{A0} - C_{sv}P_{vc}} - \frac{2bP_{vc}}{bP_{vc}^2 + P_{A0}^2} \right) BPM$$

We have as a block diagram for the human systemic circulatory system the following two-input, four-output model:

For additional analysis of this circulatory model, perform the following:

a) Using the 3-D graphing functions in MATLAB, plot the normalized behavior of the circulatory outputs defined by Eqs. (9.5.2) as functions of the input pressures P_{A0} and P_{vc} for adult male human circulatory values. To generalize the graphs, normalize the outputs by plotting as outputs, the following expressions with respect to the input pressures:

q/a with $b = 1$	for Eq. (9.5.2.1)
$W(e/q)$	for Eq. (9.5.2.2)
Q_s/C_{sa} with $C_{sv}/C_{sa} = 10, 50, 100$	for Eq. (9.5.2.3)
BPM/a	for Eq. (9.5.2.4)

b) Review current research papers with quantitative data on mammalian circulatory systems and make modifications, extensions, and improvements to the system kernels to accommodate the racial specie considered.

9.5.2 Model for Medical Diagnosis Using Radioactive Nuclides

In the medical diagnosis of certain diseases, special compounds that have been labeled with selected radioactive isotopes are administered to a patient to assess the uptake/removal rate of these special compounds by particular organs or tissues. A predominate radionuclide used for such medical tests is the short-lived daughter nuclide of Mo-99, Technetium-99m (Tc-99m), which has a half-life of 6 hours and reduces to ~24E-6 of its original activity in 72 hours after administration. Thus, the activity of Tc-99m is greatly reduced in three days and poses no radiation risk to others. An administration on Friday morning would be undetectable on the following Monday. Assume that at some initial time, the patient is administered intravenously a single, instantaneous dose of a radioactive-labeled compound m_b into the bloodstream. This compound is eliminated from the bloodstream by both natural elimination processes (e.g., removal by the kidneys and the intestinal tract) and by radioactive decay.

Then the system equation material balance for compound m_b in the bloodstream is given by

$$dm_b = [S(t) - (\lambda_b + \lambda_r)m_b]dt \qquad (9.5.3)$$

where

$$S(t) = S_0\delta(t)$$

$\delta(t) = 1$ for $t = 0$ (is the delta or impulse function)

$\delta(t) = 0$ for $t \neq 0$

S_0 = initial radioactivity injected into bloodstream (atoms/s)

λ_b = biological removal constant (atoms removed/atom-s)

λ_r = radioactive decay constant (atoms decayed/atom-s)

m_b = radioactive atoms present in bloodstream at time t (atoms)

Now the particular organ or tissue under study exhibits a material balance for the number of radioactive-labeled compound atoms m_t within the organ. The system equation here is given by

$$dm_t = [c(m_b - m_t) - \lambda_r m_t]\, dt \qquad (9.5.4)$$

where

m_t = number of radioactive atoms in organ or tissue at time t (atoms)

c = uptake constant between bloodstream and organ or tissue (1/s)

Observe that the exchange of the radioactive compound between the blood and the organ or tissue is a diffusion process and blood flow is proportional to the mean concentration difference between the two media.

What is sought is a practical method for measuring c, the organ or tissue uptake constant. The solution of system Eq. (9.5.3) is found to be

$$m_b(t) = S_0 \exp[-(\lambda_b + \lambda_r)\, t] \qquad (9.5.5)$$

and using Eq. (9.5.5), the solution for system Eq. (9.5.4) is found to be

$$m_t(t) = S_0 \frac{c}{c - \lambda_b}(\exp[-(\lambda_r + \lambda_b)\, t] - \exp[-(\lambda_r + c)\, t]) \qquad (9.5.6)$$

By proper shielding and focusing of the radiation detector, the radioactivity of the organ or tissue {which is given by $A(t) = k_r\, m_r(t)$} alone can be measured, so from Eqs. (9.5.5) and (9.5.6) we can develop the following iterative relationship for c.

$$c_{n+1} = -\frac{1}{t} \, \text{Ln}\left\{ (\lambda_b/c_n - 1)\frac{A(t)}{\lambda_r S_0} + \exp\left[-(\lambda_r + \lambda_b)\,t\right]\right\} - \lambda_r \tag{9.5.7}$$

Although Eq. (9.5.7) can be used to determine c by sampling the activity $A(t)$ at specific times, a practical alternative approach is to integrate Eq. (9.5.6) over an arbitrary counting period $(t_1 \le t \le t_2)$ that is short compared to the mean lifetime of the radioactive isotope. Then an approximate estimate for the uptake constant is found to be

$$c = 2\,\frac{A_c}{\lambda_r \, S_0(t_1 + t_2)}$$

where A_c is the mean value of the radioactivity of the organ or tissue measured over the time interval $t_1 \le t \le t_2$.

Utilize the previous data for a micro curie (3.9E4 dis/s) administration of Tc-99m for the following data to find the value for c (uptake constant between bloodstream and organ or tissue).

$$A_c = 2246 \text{ disintegrations}$$
$$S_0 = 3.9E4 \text{ dis/s}$$
$$\lambda_r = 1.92E\text{-}3 \ (1/s)$$
$$t_1 + t_2 = 600 \text{ seconds}$$

Using the 3D plotting capability of MATLAB, obtain the following plot of c versus $A_c(1{,}000 < A_c < 10{,}000)$ and $(t_1 + t_2)[100 < (t_1 + t_2) < 1{,}000]$.

9.5.3 Quantitative Model for Stress

Whenever we experience changes in our normal life pattern, we can become subject to stress. Many of these stresses are productive and beneficial and are responsible for growth, learning, and improvement in our mastery of living and adjusting to changes. However, psychologists also recognize that too high a stress level within a short time period can lead to mental or physical ailments. Several studies have attempted to quantify the stress posed by common human experiences. Table 9-5 provides an abbreviated ranking and valuation of the stress posed by the events depicted. Also, given in the right-hand column is the time constant or mean time or duration that the particular stress event exists for the typical person. Using this table, a total or accumulated stress state for a typical individual can be assessed for any time period. The system equation solution for the accumulated stress $s(t)$ at time t is given by the following equation where the sum is performed over each stress event i, which has a stress value of W_i and which occurred at time t_i and has a time constant of T_i.

$$s(t) = \sum_{i=1}^{n} W_i \, u(t - t_i) \, \exp\left[\frac{-(t - t_i)}{T_i}\right]$$

Observe that this stress function is controlled by the unit step function $u(t)$ that has the property that $u(t - t_i) = 0$ if $t < t_i$ and $u(t - t_i) = 1$ if $t \ge t_i$

To exemplify an application of the model, suppose a 24-year-old female has the following stress-related history. See Table 9-5.1.

Table 9-5   Stress Values for Selected Events

Stress Event	Values (W_i)	Time Constant (T_i – yr)
Death of spouse	100	3.0
Divorce	73	2.2
Marital separation	65	2.0
Death of family member	63	1.0
Major personal injury or illness	53	1.8
Marriage	50	2.0
Loss of job	47	0.5
Family injury or illness	44	0.5
Pregnancy	40	2.2
Change of employment	36	1.0
Child leaves home	29	1.0
Spouse's employment change	26	0.5
Change of residence	20	1.2
Vacation	13	0.1
Minor law violation	11	0.05

Source: Adapted from Future Facts by Stephen Rosen, p. 225. Copyright 1976. Simon & Schuster.

Table 9-5.1

	Events	Age (yr)	Stress W_i	Ti (yr)
a.	Marriage	20	50	2.0
b.	Change of residence	21	20	1.2
c.	Pregnancy	22	40	2.2
d.	Death of parent	23	63	1.0
e.	Divorce	24	73	2.2

We want to quantitatively assess the stress state for this female at age 24 and determine the individual contribution of each past stress event to her present emotional and physical condition. From our system equation, we have for her stress state at age 24

$$s(24) = (a)\ 50\exp(-4/2) + (b)\ 20\exp(-3/1.2)$$
$$+ (c)\ 40\exp(-2/2.2) + (d)\ 63\exp(-1/1) + (e)\ 73\exp(-0)$$

and evaluating each event (a) through (e), we have for a 24-year-old woman

$$s(24) = 6.8 + 1.6 + 16.1 + 23.2 + 73.0 = 121$$

Generally, a stress value of 100 is considered to have significant potential emotional and physical impact, as shown by Table 9-6. So counseling and therapy are advisable for our female subject.

The stress state can be normalized, and each event evaluated for its contribution to the total stress state of the above female subject. For our example, this is easily accomplished as follows:

Table 9-6 Potential Impacts of Stress Levels

Total Stress State Level	Potential Impact
<50	Nominal
100	Significant effects
150	Serious disturbances
>200	Physical and/or mental breakdown

$$s(24) = \frac{121}{100\%}[6\%(\text{marriage}) + 1\%(\text{residence}) + 13\%(\text{pregnancy}) + 19\%(\text{death})$$
$$+ 61\%(\text{divorce})]$$

So we conclude that the female's present stress level at age 24 is comprised of 6% from her marriage, 1% from her residence change, 13% from her pregnancy, 19% from the death of her parent, and 61% from her divorce. The divorce and parent death events are the major factors (~80%) contributing to her present stress state and should be addressed first by a counselor or therapist.

9.5.4 Topical System Problems Associated with Human Life

Human Health Care Costs. A major economic issue facing the U.S. public is the rapidly escalating costs associated with health care. In 1960, health care costs were $26.9 billion and accounted for 5.2% of the U.S. Gross Domestic Product (GDP). By 2005 these costs had risen 73 times to $1973 billion, which represented 15.3% of the GDP. Table 9-7 provides a brief medical cost history for the United States.

a) Using the data given in this table, develop a growth model that will estimate U.S. medical costs over the range from 1960 to 2020. The kernel for the system equation should be chosen to fit the data in the table in a "least square fashion" for a standard exponential growth model.
b) Determine the total per capita funds that have been spent for a U.S. resident living from 1960 to 2020. Determine the rate of change of growth in %GDP. Is it positive, zero, or negative?
c) For what year will health costs equal 25% of U.S. GNP? Is this growth in health costs asymptotic to some final, fixed percent of the U.S. GNP or does the growth approach the total GDP (unrealistic)?

U. S. Population Growth Model. The logistic model for human population growth can be reasonably applied to the U.S. population. The system equation for the logistic model is

$$dp = p(a - bp)\, dt$$

where p is the population in millions of residents, t is the time in years, and a (1/years) and b (1/person-years) are parameters.

a) Show that the solution for the system equation is

$$p(t) = \frac{a}{b\,[1 + K \exp(-at)]}$$

where K is the integration constant.

Table 9-7 U.S. Health Care Costs Chained to 2010 Dollars

Year	Cost ($billions)	Per Capita ($)	Population (million)	%GDP
1960	27	145	186	5.2
1965	42	210	200	5.9
1970	75	356	210	7.2
1975	133	604	220	8.1
1980	253	1100	230	9.1
1985	439	1810	242	10.4
1990	714	2813	254	11.8
1995	1017	3783	269	13.7
2000	1354	4780	283	14.8
2005	1723	5800	297	15.9
2010*	2100	6780	310	17.0
2015*	2530	7900	320	18.1
2020*	2930	8880	330	19.2

Source: Adapted with *estimates: U.S. Dept. of Commerce, Bureau of Economic Analysis.

b) For the following data set, determine the best values for a, b, and K for "least squares" fit for the bi-decade census data for (2000–2040).

Year	1840	1860	1880	1900	1920	1940	1960	1980	2000	2020*	2040*
Population	17	31	50	76	106	132	186	230	283	335	389

*estimated (millions)

c) What is the equilibrium value of the U.S. population as predicted by the model?
d) If the parameters a and b in the system equation are functions of time, then the system equation becomes a Ricatti equation as defined in Section 4.2.9. Convert this Ricatti equation to a linear, second-order differential equation and find solutions that exist for various specific functions of $a(t)$ and $b(t)$. See Section 4.3.10 for assistance.
e) Explore the application of MATLAB for attempting solutions for both the linear, second-order differential equation and the Ricatti equation prescribed in (d)

Chemotherapy Treatment Model. In a hypothetical chemotherapy treatment procedure, two drugs (inputs) are administered to a cancer patient on a quasi-continuous, serial basis. The first drug, D_1, is an interferon inhibitor that must be administered jointly with drug, D_2, as an interferon promoter. Biopsies taken from the patient are then measured for cancer cell count C (the output) to assess the treatment efficacy in destroying cancer cells.

Assume that a quantitative system model is established from experimental data that has the form

$$\frac{\partial C}{\partial D_1} = C\left[\frac{1}{\exp(D_1 - aD_2) - 1} - \frac{1}{D_1}\right]$$

and

$$\frac{\partial C}{\partial D_2} = - \frac{aC}{\exp(D_1 - aD_2) - 1}$$

then do the following:

a) Establish the system equation for dC as a function of dD_1 and dD_2.
b) Attempt to obtain an analytical solution for the system equation.
c) Attempt a solution for the system equations using the symbolic equation solver in MATLAB.
d) Using either part (a) or (b), obtain an analytical solution for the original system equations. What happens to the solution when $a = D_1/D_2$?

Decision Tree for Medication. The decision to undertake a given action to realize a proposed consequence or goal is a system that can be quantitatively modeled as a decision tree. A practical example is the decision of a patient to receive a prescribed medication presumed to reduce the risk or impact of a particular health effect. The decision tree and its logical consequences are shown in Figure 9-14.

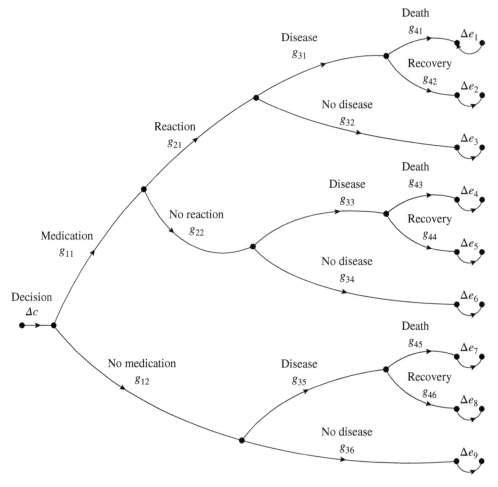

Fig. 9-14 Decision tree for medication.

a) Write the complete system equation for the model.
b) Assume that each kernel expresses the probability associated with the event given. Write the system equation that expresses the total probability for the events (1) no disease, (2) recovery, and (3) death.
c) Assume the following numerical values for the model kernels:

$$g_{12} = 0.5000, \quad g_{21} = 0.0200, \quad g_{32} = 0.9925, \quad g_{34} = 0.9935$$
$$g_{36} = 0.9750, \quad g_{42} = 0.9998, \quad g_{44} = 0.9997, \quad g_{46} = 0.9995$$

assume all remaining numerical values are 0.1
What are the overall probabilities for the events "no disease," "recovery," and "death"?

Model of System for Regulating Human Body Temperature. Consider the following block diagram for a model for the regulation of human body temperature.

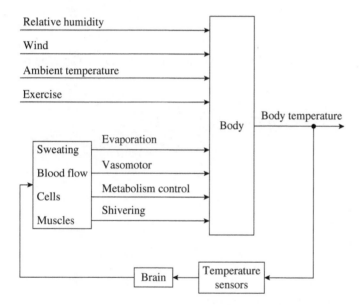

a) Develop the system equation in a general symbolic functional form for the model, including intermediate input and output variables.
b) Collapse the system equation to the equivalent canonical form using MATLAB and obtain the canonical system equation with feedback.
c) Develop a reasonable set of parameter values for each of the model inputs and determine the body temperature output. Is your output body temperature reasonable? Adjust parameters until you achieve a normal body temperature of 37 °C (98.6 °F)

Model for Blood Analysis. With advances available in rapid, high resolution, multi-molecular blood compound analysis, a diagnostic tool routinely used is multichannel (up to 50 or more compounds) analysis of a patient's blood. Such high-level, computerized quantitative analysis is very valuable for medical diagnosis and projections of a patient's state of health. Because of extensive background data available on normal (or mean values c_{i0}) values of various compounds found in human blood and their anticipated variation (or standard deviation s_{ri}), it is possible to develop an accurate quantitative model for the state of the blood chemistry of a patient. Consider as a model for such a blood analysis system the following block diagram, where each differential input dc_i represents the change in concentration for a given molecular compound found in human blood, and de represents the change in the overall state of the blood.

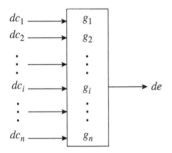

Assume that each kernel g_i has the quantitative form (which will produce a Gaussian or normal distribution for the model)

$$de = \frac{1}{\sum_{i=1}^{n} G_i} \sum_{i=1}^{n} W_i \sigma_i^{-5/2} (c_{i0} - c_i) \exp\left[\frac{-(c_{i0} - c_i)^2}{2\sigma_i^2}\right] dc_i$$

where

$$G_i = W_i \exp\left[\frac{-(c_{i0} - c_i)^2/2\sigma_i^2}{\sqrt{2\pi}\,\sigma_i}\right]$$

and

$\quad W_i =$ weighting factor for ith input
$\quad e =$ model output for blood's total quality index
$\quad c_{i0} =$ mean or expected value for the concentration of ith input
$\quad \sigma_i^2 =$ variance expected for ith input

a) Show that the solution for the blood quality index from the system equation is given by

$$e = \frac{1}{\sqrt{2\pi}} \sum_{i=1}^{n} \frac{W_i}{i} \left\{ \exp\left[\frac{-(c_{i0} - c_i)^2}{2\sigma_i^2}\right] - 1 \right\}$$

where $e\,(c_{10}, c_{20}, ..., c_{n0}) = e\,(c_0) = 0$.

b) Using the data given in Table 9-8 for 12 sample inputs, develop the quantitative expression for $e(c)$ for this data.

c) Assume that $W_i = 1/12$ for $i = 1,..., 12$; then what is $e\,(c^*)$ from part (b) if $c^* = 1.1\,c_0$ (i.e., a value 10% greater than the mean for each input)?

Table 9-8   Typical Human Blood Chemistry

Sample Inputs	Normal/Mean (c_{i0})	Standard Deviation (σ_i)
Calcium ion	9.5 mg%	1.0 mg%
Inorganic phosphate	3.4 m%	1.0 mg%
Glucose	80.0 mg%	22.5 mg%
BUN (blood urea nitrogen)	15.0 mg%	5.0 mg%
Uric acid	6.3 mg%	2.5 mg%
Cholinesterase	230.0 mg%	75.0 mg%
Total Proteins	7.4 mg%	0.8 mg%
Albumin	4.4 mg%	0.7 mg%
Total bilirubin	0.7 mg%	0.4 mg%
Alkaline phosphate	2.6 total units	1.6 total units
LDH (lactate dehydrogenase)	70.0 total units	50.0 total units
SGOT (aspartate aminotransferase)	25.0 total units	15.0 total units

Source: Adapted from Wikipedia (see definitions for medical and chemical compounds.

Heart Disease Risk Model. Diseases of the human heart are a major medical problem in the United States and in most industrialized nations. As an effort to inform people of the factors that are believed to contribute to heart disease and to provide a mechanism for indications of early warnings that might indicate the incipience of heart disease, several indexes have been proposed and used for quantifying these factors associated with heart disease. Such risk models are employed by life insurance companies to assess applicants for policies.

Consider the following heart risk index I and its contributing factors:

$$
\begin{aligned}
I = \ & W_{bp}\left[A_{bp}(BP_s - 120) + B_{bp}(BP_d - 80)\right] + W_s A_{bs} S \\
& + W_{sc}\left[SC - SC_0\right] + W_{sx} SEX \\
& + W_w[A_w(WT - 75)\,SEX + B_{bw}(WT - 60)\,(1 - SEX)] \\
& + W_a[A_a(AGE - 40)\,SEX + B_a(AGE - 45)\,(1 - SEX)]
\end{aligned}
$$

where

BP_s = mean systolic blood pressure in mm Hg
BP_d = mean diastolic blood pressure in mm Hg
S = level of cigarette smoking in cigarettes per day
SC = serum cholesterol level in the blood in mg%
SEX = gender (0 for females and 1 for males)
WT = body weight in kilograms
AGE = age in years

and the W's, A's, and B's are parameters and negative values are set to zero.

Consider this heart risk index as the quantitative model for this heart disease risk system

a) Define the inputs and outputs for the model.
b) Draw the block diagram for the model and label all inputs, outputs, and kernels. Determine the quantitative expressions for each kernel.

c) Use the following data

$$A_{bp} = B_{bp} = 0.1, \quad A_w = B_w = 0.025, \quad A_a = B_a = 0.04$$
$$A_s = 0.03, \quad SC_0 = 30mg\%, \quad W_{bp} = 45\%, \quad W_s = 25\%$$
$$W_{sc} = 7\%, \quad W_{sx} = 3\%, \quad W_w = 15\%, \quad W_a = 5\%$$

to determine the risk indexes for the following individuals:

Female: $BP_s = 120$ mm Hg, $\quad BP_d = 85$ mm Hg, $\quad\quad S = 0$,
$\quad\quad\quad SC = 30$ mg%, $\quad\quad\quad WT = 50$ kg, $\quad\quad\quad\quad GE = 60$ years
Male: $\quad BP_s = 160$ mm Hg, $\quad BP_d = 100$ mm Hg, $\quad S = 10$ cigarettes/day
$\quad\quad\quad SC = 50$ mg%, $\quad\quad\quad WT = 95$ kg, $\quad\quad\quad\quad AGE = 40$ years

d) Using the heart risk index as base, how might a health insurance company index the premium charged these two individuals? Quantify this difference in premiums for the male and female by determining the variation of the premium from the average value of premium for both.

e) As the deviations of particular indicators such as blood pressure, smoking, and weight become large, the risk of heart disease increases rapidly. Modify the index by assuming that each risk variable (except for age) varies as the square of the variation around the normal value. For example, for serum cholesterol, let that term be

$$W_{sc}(SC - SC_0)^2$$

What index values now might be used for the individuals described in part (c)?

Medical Diagnosis Model. A model is sought that is capable of assessing the state of health of an individual, recognizing existing illnesses and abnormalities, diagnosing these conditions and prescribing treatment, requesting further clinical tests or other data if required, and evaluating preconditions of latent diseases and health conditions that might occur without medical intervention or corrective action. Finally, the model should assess the itemized costs for such services.

The generalized model will have the graphical depiction shown:

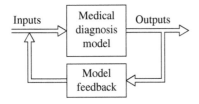

Consider as possible inputs for the model

- Patient medical history
  - Past illness and diseases (physical and mental, hospitalization, etc.)
  - Immediate family member medical history (age of parents' death)
  - Lifestyle and habits (employment, diet, exercise, smoking, stress, etc.)
  - Present and past medications, drugs, etc.

- Physical background
  - Age, sex, weight, height, build, race, physical handicaps
- Biological and physiological data

  - Temperature, blood pressure, blood chemistry, x-rays, urinalysis, electrocardiographs, electroencephalographs, radioimmunoassays, CT-scans, MRI, acoustic analysis, proctoscopies, biopsies, other tests
  - Other specific diagnostic tests as determined by the physician or requested through feedback from the computer assessment.

As model outputs, consider detailed quantitative statements with data on the following:

- Overall state of health.
- Existing diseases or abnormal conditions.
- Requests for further tests or data.
- Recommended treatment for diseases and abnormal conditions found.
- Assessment of latent conditions or trends toward abnormal conditions and recommended remedial actions.
- Itemized costs for services given and recommended treatment costs.

Human Endocrine System Model The human endocrine system is composed of about nine major, distinct glandular tissues that secrete hormones into the bloodstream to regulate other cells and regions in the body. Chemically, these hormones may be classified as proteins, glycoproteins, polypeptides (large chains of amino acids), and steroids. Some hormones are capable of providing positive feedback, while others provide negative feedback for control of certain biological functions. There appear to be two mechanisms by which hormones control cellular function. One is the regulation of protein synthesis (control of DNA replication), and the other is through the formation of the nucleotide cyclic adenosine-3, 5-monophosphate (i.e., cyclic AMP). Table 9-9 summarizes selected human endocrine tissues with the hormones secreted and their chemical nature and presumed function.

Table 9-9 Human Endocrine Types

Tissue or Gland	Hormone	Chemistry	Function
Testes	Testosterone	Steroid	Sexuality (male)
Ovaries	Estradiol	Steroid	Sexuality (female)
Thyroid	Thyroxine	Amino acid	Stimulation of metabolism
Thymus	Thymosin	Polypeptide	Cellular immunity

a) Become familiar with human endocrine system, particularly with the four glands given in Table 9-9 and their hormonal secretion and function.
b) Establish a block diagram for these hormones and their function, together with possible feedback mechanisms.
c) Create a simple system equation for the endocrine model developed and assess the output control for protein synthesis and cyclic AMP (adenine monophosphate) production.

Enzymatic Reaction Rate Model. In 1913, two German biochemists, L. Michaelis and M. L. Menten developed the following mathematical expression to describe the rate for enzymatic reactions as a function of their substrate (compound on which the enzyme acts) concentration.

$$V = \frac{V_m s}{K + s}$$

where

$V(s) =$ enzymatic reaction rate (enzyme/s)
$V_m =$ maximum enzymatic reaction rate (enzyme/s)
$s =$ substrate concentration (substrate cm^3/cm^3)
$K =$ a constant (defined as the Michaelis constant (substrate cm^3/cm^3)

a) Show that the system equation for the model can be expressed as the following separable differential equation by normalizing the reaction rate.

$$dv = (1 - v)\, v\, \frac{ds}{s}$$

where $v = V/V_m$ is the fractional reaction rate.

b) Define new input and output variables so that the system equation has the general form

$$dy = -dx$$

c) Give physical interpretations to these variables. What role does K (Michaelis constant) perform for this system equation transformation?

d) The solution for the transformed system equation can be expressed as $y = c/x$, where c is a constant. Relate c to K, y to v, and x to s

e) Plot the 3-D figure for $V(s)$ versus s and the parameter K for a reasonable range of values using MATLAB.

National Level of Health Model.

System:	Develop a system model that provides a numerical value (0–10 or 0–100%) for the level of human health for a given nation's population based on a model that employs the following important contributing factors (i.e., inputs) to human health. The numerical value could be obtained as the nation's health variation around the mean value for each input for total world values of these inputs. Access U.S. NIH (National Institute of Health) and Wikipedia for U.S. human health data for a selected year.
Inputs:	Consider and rank (%) the following factors: Life expectancy Infant death rate Mortality rates Morbidity rates Potable water quality Air quality Dietary adequacy Worker safety and environment Mental health
Outputs:	Overall level or assessment of nation's level of human health.

Model for Epidemics/Pandemics. Around the year 1347, the Black Death, probably bubonic plague, swept through Europe, spreading from Mediterranean ports, through Sicily (1347), North Africa, into Europe through Italy, Spain, France (1348), Austria, Hungary, Switzerland, Germany, the Low Countries and then across the channel to England (1349). In 1350 the Black Death then moved into the Scandinavian and Baltic countries. Recurrences of the plague arose in the years 1360–1363, 1369–1371, 1374–1375, 1390, and 1400.

It is estimated that over one-third of the world's total population in the fourteenth century died as a consequence of this epidemic. The impact of this depopulation was to alter national economics, reduce social stratification, and deemphasize the influence of the Catholic Church throughout Europe.

In 2019, the SARS-CoV-2, virus termed as COVID-19 imposed a similar worldwide health pandemic that inflicted both immense health and economic issues upon the world. The pandemic reportedly began in Wuhan, China, and spread rapidly throughout each continent in the world. The actual source and final extent of the impact and consequences are still to be fully determined and evaluated.

Although such an international, complex disease transmission system as an epidemic/pandemic has many inputs, outputs, and interactive parameters, it is possible to develop a simple but useful model for epidemics as follows.

Let us assume that a human population that can be affected by a given epidemic/pandemic (or contagious disease vector) is composed of two groups. Those individuals who are infected and can transmit the disease, whom we will define as infectives or $I(t)$ at time t; and those individuals who do not have the disease at time t but are susceptible to the infection, whom we will designate as susceptible, $S(t)$, at time t.

A reasonable set of nonlinear system equation that exhibits much of the behavior of such actual epidemic cycles is given by

$$dI = (AS - B)I dt \tag{9.5.8.1}$$
$$dS = (C - AIS) dt \tag{9.5.8.2}$$

where

$I(t) =$ number of infectives in population at time t
$S(t) =$ number of susceptibles in population at time t
$A, B,$ and C are positive constants

a) Set $C = 0$ and eliminate dt between Eqs. (9.5.8.1) and (9.5.8.2). Then determine S as a function of I

b) Plot the phase plane solution found in part (a) for S versus I for various values of the integration constant.

c) Determine the behavior in time for both $S(t)$ and $I(t)$ as found from parts (a) and (b).

d) Return to system Eqs. (9.5.8) with $C \neq 0$ and determine a linearized solution for the system equations in time.

e) Compare the results of parts (c) and (d).

f) Execute a MATLAB solution of the system equation found in (c) before linearization for $I(0) = 1$ and $S(0) = 10$ and determine the asymptotic result (i.e., $I(t)$ and $S(t)$ as time t goes to infinity)

g) Comment on use of this model for the Spanish Flue pandemic (1918's) and the COVID-19 pandemic (started 2019) and seek relative data for the modelling effort.

Additional Modeling Problems. For each of the systems given in Table 9-10, define an appropriate system model and the essential elements of the system. Express quantitative variables for each of the model inputs and outputs and develop a simple analytical expression for the model's system equation. Determine analytical or other solution forms for the system equation and interpret and evaluate your results.

Table 9-10 Modeling Problems for Human Biology

	System	Inputs	Outputs
(a)	Kidneys	Blood	Urine
(b)	Cell	Ionizing radiation	Cell damage
(c)	Brain	Sensory inputs	Experiences
(d)	Reproduction	Ovum and sperm	Zygote
(e)	Human life	Physical condition, education, occupation, mental outlook	Longevity Good health
(f)	Humans	Bacteria, viruses, fungi, parasites	Diseases
(g)	Quaternary hominidae	Homo erectus, Homo sapien, Neanderthal, Cro-Magnon	Temporal appearance Each population type
(h)	Human cell`	Viruses	Cancer incidence
(i)	Human population	Accidents, cancer, heart disease, other causes	Mortality rate
(j)	Human health	Diet, exercise, sleep, heredity, mental conditioning	Quality of life
(k)	Blood	Plasma, red cells, white cells, platelets	Properties and function
(l)	Psychological system	Intelligence level	Psychological health
(m)	Early Earth Oceans	Chemical reactions	H, C, N, and O roles
(n)	Protein structures	Amino acids	Polymerization
(o)	Nucleic acids	DNA structure	RNA structure
(p)	Carbohydrates	Sugars	Polysaccharides
(q)	Cellular respiration	Krebs cycle	Metabolic pathways
(r)	Meiosis	Effects of genetic errors	Source of genetic errors
(s)	Natural selection	Homology and analogy	Time behavior
(t)	Bacteria and archaea	Biochemical lineages	Spirochetes
(u)	COVID-19	Genetic material	mRNA (SARS-CoV-2)
(v)	Plant reproduction	Pollination and refertilization	Reproductive structures
(w)	Animal nutrition	Digestion and absorption	Glucose role
(x)	Animal reproduction	Male/female roles	Role of sex hormones
(y)	Biodiversity	Quantifying diversity	Human threats
(z)	Population ecology	Population growth	Endangered species

9.6 Applications of System Science to Human Society

It has been asserted that the major challenges and unresolved issues confronting humanity today are not technical but social, not philosophical but moral, not legal but ethical. The basis for this assertion is undoubtedly conditioned by the major concerns of our times, which include peace, poverty, racial intolerance, understanding, morality, and self-control. Humanity's scientific and technical progress and knowledge have provided great societal benefits but, unfortunately, similar progress and mutual understanding have not occurred in the area of social understanding and compassionate behavior of humans, where economic turmoil, national wars, terrorism, ethnic cleansing, disease, and famine still plague the planet. The COVID-19 Pandemic (the onset of this major disaster health disaster that occurred in 2019 in Wuhan, China) resulted in millions of human causalities from morbidity and mortality and severe economic and geopolitical ramifications. The full impact of this pandemic upon the earth's nations and their people is unknown. The critical issue that arises is whether a better understanding of the causes and effects of such critical social manifestations can result in improved control and amelioration of such human exigencies?

9.6.1 World Cultural and Economic Regions

Although the world with its approximate ~7 billion human inhabitants represents a vast, complex, interactive system of human culture and actions and physical resources, it is possible and, indeed, productive to simplify and reduce this real-world system to a simple model. The model should be representative of world humanity but still be amenable to analysis.

Let us represent the world's nations and their people in terms of ten interactive subsystems or regions. The important variables and parameters that might be used to describe each region of the world model (i.e., political, economic, physical, technical, etc.) are assumed to be described by single, mean, or effective values averaged over the region in some way.

Thus, it is meaningful to talk about a mean human population, political type, economic output, technical level, and other characteristics for each region. We will limit our attention here to population distributions among the ten major regions of the world model as shown in Table 9-11. This table identifies the ten regions and their respective populations in 1950, 1975, and 2000. Also shown with each region's population figure are the percentages that the region's population contributes to the total world population for that year.

The usual system equation employed for the modeling of a human population count, $P(t)$, in time is given by

$$dP = r\,P\,dt \tag{9.6.1}$$

where

P = number of living persons in designated group at time t
t = input or independent variable time (usually in years by calendar year)
r = effective growth or change constant with dimensions of reciprocal time (usually 1/yr)

Table 9-11 Existing Regional Population Model

	Region	Population (millions)					
		1950	%	1975	%	2000	%
1	North America	166	6.5	262	6.4	334	5.5
2	Western Europe	322	12.6	450	11.0	565	9.3
3	Japan	82	3.2	119	2.9	164	2.7
4	Australia and South Africa	26	1.0	49	1.2	73	1.2
5	Russia and Eastern Europe	273	10.7	392	9.6	511	8.4
6	Latin America	176	6.9	331	8.1	541	8.9
7	North Africa and Middle East	79	3.1	143	3.5	243	4.0
8	Tropical Africa	166	6.5	274	6.7	420	6.9
9	South and Southeast Asia	682	26.6	1186	29.0	1885	31.0
10	China	585	22.9	887	21.7	1344	22.1
	World	2557	100.0	4093	100.0	6080	100.0

Source: Data adapted from UN Population Division, World Population Prospects.

System Eq. (9.6.1) is separable and can be integrated to give

$$\text{Ln } P = rt + \text{Ln } k \tag{9.6.2}$$

where k is the integration constant. Equation (9.6.2) may be arranged into the following growth equation:

$$P = P_0 \exp (rt)$$

where $P = P_0$ when $t = 0$.

The growth or change constant can also be expressed as

$$r = \text{Ln } (1 + i) \tag{9.6.3}$$

where i is annual or equivalent growth interest or percent growth. System Eq. (9.6.3) can be expressed in terms of i as

$$P = P_0(1 + i)^t \tag{9.6.4}$$

which is the standard equation for compound interest growth. The annual growth percentages given in Table 9-12 are defined as the parameter i expressed as an annual percentage for the preceding 25-year growth periods from 1975 and 2000. Using the growth percentages indicated for each region in 2000, the population for the year 2025 is predicted using Eq. (9.6.4) and indicated in Table 9-12. Also shown is the percentage of the world population, the region population, will bear in the year 2025. It is interesting to observe the shifts in population percentage predicted by the model. Of course, whether such shifts will actually occur is dependent on the accuracy of our database.

Table 9-12   Projected Regional Population Model

Region		1950	%	1975	%	2000	%	2025*	%
		**Population (millions)**							
1	North America	166	6.5	262	6.4	334	5.5	443	5.9
2	Western Europe	322	12.6	450	11.0	565	9.3	609	8.1
3	Japan	82	3.2	119	2.9	164	2.7	178	2.4
4	Australia and South Africa	26	1.0	49	1.2	73	1.2	84	1.1
5	Russia and Eastern Europe	273	10.7	392	9.6	511	8.4	701	9.3
6	Latin America	176	6.9	331	8.1	541	8.9	742	9.9
7	North Africa and Middle East	79	3.1	143	3.5	243	4.0	322	4.3
8	Tropical Africa	166	6.5	274	6.7	420	6.9	693	9.2
9	South and Southeast Asia	682	26.6	1186	29.0	1885	31.0	2219	29.0
10	China	585	22.9	887	21.7	1344	22.1	1560	21.0
	World	2557	100.0	4093	100.0	6080	100.0	7551	100.0

Source: Data adapted from UN Population Division, World Population Prospects, *projections

a) Using the data in Table 9-12, determine the predicted years when the world population reaches 10 billion, 11 billion, and 15 billion.
b) Various estimates have been made for the maximum population carrying capacity for Earth. How might the parameter r or i become functions of time to reflect this limiting capacity for population growth?
c) Assume a simple power series expansion for $r(t)$ as follows

$$r(t) = \sum_{k=0}^{n} r_k t^k$$

and solve the resulting system given by Eq. (9.6.1). Propose various conditions on the predicted future populations to determine the power series expansion coefficients, r_k, used in your expansion.

9.6.2   Solow Model for Economic Growth

The 1987 Nobel Prize in Economics was awarded to Robert Solow for his quantitative model for economic growth. The Solow Model is regarded by economists as a major tool for understanding the economic growth of an economic system. The Solow Model has served as the basis for the development of numerous quantitative macroeconomic studies since its inception in 1987. The Solow Model assumes a continuous behavior in the absence of government or international controls for a single produced product. We shall briefly investigate a few of these models (reference: Wikipedia).

The basic Solow Model expresses the total economic production function, $Y(t)$, as a generalized function $F(t)$ of the following economic input variables:

$$Y(t) = F\left[K(t), A(t)\, L(t)\right] \tag{9.6.5}$$

where

$Y(t) = $ Total production function
$K(t) = $ Capital influence on total production Y
$A(t) = $ Augmentation technology that influences labor L
$L(t) = $ Labor's direct input to total production Y

The Solow Model is not explicitly time-dependent but is controlled temporally by the implicit time dependence of $K(t)$, $A(t)$, and $L(t)$. Observe that the contribution of labor on the total production function generally enters as the product of $A(t)$ and $L(t)$, namely $[A(t)\, L(t)]$, which is called the technically augmented labor input. Various model extensions to the basic Solow Model (including the Solow–Swan Model) exist with the specific assumptions made for the analytical forms for $K(t)$, $A(t)$, and $L(t)$.

A simple extension of the Solow Model is the Cobb–Douglas Model for constant total production returns, which employs the mathematical form

$$Y(t) = F[K(t), A(t)\, L(t)] = [K(t)]^{m}\, [A(t)\, L(t)]^{1-m} \tag{9.6.6}$$

where m is a dimensionless parameter over the range $0 \le m \le 1$.

This range for m satisfies the Inada[*] condition for "elasticity" of total production that ensures the sum of the exponential components of capital and effective labor are equal to one.

The simple growth for the effective labor terms in the Cobb–Douglas Model is assumed to be

$$A(t) = A(0)\exp\,(at) \tag{9.6.7.1}$$

$$L(t)\ = L(0)\,\exp\,(bt) \tag{9.6.7.2}$$

and the capital contribution, $K(t)$, to total production is found as the solution of the differential equation

$$\frac{dK}{dt} = s\, Y(t) - c\, K(t) \tag{9.6.8}$$

Substituting Eq. (9.6.6) into Eq. (9.6.8) yields the following nonlinear differential equation.

---

[*] Inada, Ken-Ichi. 1963. On a Two-Sector Model of Economic Growth: Comments & Generalization. *The Review of Economic Studies* 30 (2):119–127.

$$\frac{dK}{dt} = s\,K^m\,[A(t)\,L(t)]^{1-m} - c\,K(t)$$

and substituting terms we have

$$\frac{dK}{dt} = s[A(0)\,L(0)]^{1-m}\,\exp\,[(1-m)\,(a+b)t]K^m - c\,K(t)$$

with the following result for our Cobb–Douglas Model system equation

$$\frac{dK}{dt} = K_0\,\exp\,[(1-m)\,(a+b)t]\,K(t)^m - c\,K(t) \tag{9.6.9}$$

where

$$K_0 = s[A\,(t=0)\,L\,(t=0)]^{1-m}$$

We will consider three simple cases for Eq. (9.6.9).

Case I. This case occurs if $K_0 = 0$. The immediate solution for Eq. (9.6.9) is then

$$K = C_0\,\exp\,(-c\,t)$$

where

$$C_0 = K(0)$$

We find for this case that total production output and capital, $K(t)$, is an exponentially declining function of time due to the absence of effective labor input.

Case II. For the case when $m = 1$ and $A(0)\,L(0) \neq 0$, then with $K(0) = s$, Eq. (9.6.9) reduces to the following separable, first-order equation

$$\frac{dK}{dt} = (s-c)K$$

and the solution is given below again with $C_0 = K(0)$

$$K = C_0\,\exp\,(s-c)\,t$$

This solution shows that Capital increases exponentially over time if total production exceeds available capital, $(s-c) > 0$; is a constant if total production equals available capital, C_0; and declines exponentially if available capital exceeds total production, $(s-c) < 0$.

Case III. If $m = 0$ then

$$\frac{dK}{dt} = K_0\,\exp\,[(a+b)\,t - c\,K]$$

And the solution is then

$$K = C_0\,\exp\,(-c\,t) + \frac{K_0}{a+b+c}\,\exp\,[(a+b)\,t]$$

This solution exhibits a variety of time behavior models for total production, effective labor input, and capital depending upon prevailing economic conditions.

For the cases where $m \neq 1$ or $K_0 \neq 0$, the model Eq.(9.6.9) is a nonlinear differential equation with no known analytical solution. However, we know the two solutions for the limiting conditions when $m = 0$ and $m = 1$; we anticipate a mixed temporal behavior for production, labor, and capital. Furthermore, we can bound the solutions for the interval $0 < m < 1$ using perturbation approximations or numerical solutions. A detailed discussion of the Cobb–Douglas Production Function is provided in Wikipedia and the reader may find additional information regarding this important tool for economic analysis.

9.6.3 Model for Cost of Crime to Society

The system to be studied in this example is a simple model that relates the cost of crime to society associated with the threats and damage to life and property and the cost to society for crime prevention and control. Crime and all its various societal costs constitute a very complex system. The interaction and effects of the courts, lawmakers, lawbreakers, law enforcement, penal institutions, and society itself produce a highly coupled, interactive, nonlinear feedback system that is difficult to model accurately. However, one simple system relating crime costs to society and the continuous system block diagram given in Figure 9-15 represents prevention costs. The single input to the system is the differential cost of crime prevention, $d\$_p$, which in general takes the form of societal tax revenues collected by governments (local, state, and federal) for crime prevention and control.

The single output of the system is the effective cost of crime on society, $d\$_c$, which has been expressed in equivalent numerical costs. Although financial losses of property due to crime are generally apparent and readily quantifiable, we may hesitate at assigning a dollar cost for human physical and emotional suffering and loss of life. However, actuaries, insurance adjusters, and cost-benefit studies determine and assign such costs rather routinely. Thus, we will assume that both the costs of crime and crime prevention are measurable and can be expressed in standard monetary values. From Figure 9-15, it can be seen that a feedback loop reflects the response of society to the perceived costs

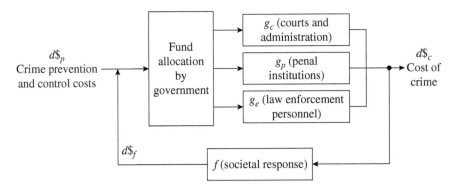

Fig. 9-15 Block diagram for societal crime cost model.

of crime and thus directly affects the funds provided for crime prevention and control. Thus, as the societal cost of crime changes, society provides a feedback response that is reflected in the financial support provided by society to prevent and control subsequent crime.

Let us define the open-loop system kernel (i.e., with no feedback present) as

$$G = g_c + g_p + g_e$$

and the feedback kernel as

$$F = f$$

The individual system equations for the open-loop and feedback kernels are then given by

$$d\$_c = G(d\$_p + d\$_f)$$
$$d\$_f = Fd\$_c$$

Eliminating $d\$_f$, viz., the differential feedback cost, we obtain as the overall system kernel the following standard single-input, single-output system equation with feedback:

$$d\$_c = \frac{G}{1 - GF} d\$_p$$

In the absence of greater knowledge of the system behavior, we will assume for preliminary analysis purposes that the system kernels are all linear in $\$_c$. Specifically, let us assume that

$$G = A\$_c + B$$
$$F = C\$_c + D$$

where A, B, C, and D are arbitrary constants. Then the overall system kernel with feedback can then be written as

$$\left[\frac{1}{A\$_c + B} - C\$_c - D \right] d\$_c = d\$_p$$

which is recognized as a first-order differential equation in separated form and can be immediately integrated to give

$$\frac{1}{A} \text{Ln} (A\$_c + B) - \frac{1}{2} C\$_c^2 - D\$_c = \$_p + K$$

where K is an arbitrary integration constant. As a boundary condition, we will make the reasonable assumption that the cost of crime is a maximum when no funds are spent for crime prevention; so when $\$_p = 0$, then

$$\$_c(\text{maximum}) = \$_{cm}$$

Then we find for the integration constant that

$$K = \frac{1}{A} \mathrm{Ln}(A\$_{cm} + B) - \frac{1}{2} C\$_{cm}^2 - D\$_{cm}$$

So the system solution is expressible as

$$\$_c = \frac{1}{A} \mathrm{Ln} \left(\frac{A\$_c + B}{A\$_{cm} + B} \right) + \frac{1}{2} C \left(\$_{cm}^2 - \$_c^2 \right) + D(\$_{cm} - \$_c) \tag{9.6.10}$$

As an upper limit, assume that there exists a finite maximum cost for crime prevention, $\$_{pm}$, that will force the cost of crime to zero, $\$_c \to 0$, and is given by

$$\$_{pm} = \frac{1}{A} \mathrm{Ln} \left[\frac{B}{A\$_{cm} + B} \right] + \frac{1}{2} C\$_{cm}^2 + D\$_{cm}$$

To eliminate system variable dimensions and normalize and minimize parameters, let us define new dimensionless cause and effect variables as follows:

$$x = \$_c/\$_{cm}, \quad y = \$_p/\$_{pm}$$

where x is the fraction of the maximum crime cost and y is the fraction of the maximum crime prevention cost.

Equation (9.6.10) can then be written as

$$y = a_0 \mathrm{Ln} \left(\frac{x + a_1}{1 + a_1} \right) + b\left(1 - x^2\right) + c(1 - x) \tag{9.6.11}$$

where

$$a_0 = \frac{1}{A\$_{mp}}, \quad a_1 = \frac{B}{A\$_{cm}}, \quad b = C \frac{\$_{cm}^2}{2\$_{pm}}, \quad c = D \frac{\$_{cm}}{\$_{pm}},$$

We observe that when $y = 1$, $x = 0$, so we have that

$$1 = a_0 \mathrm{Ln} \left(\frac{a_1}{1 + a_1} \right) + b + c$$

Equation (9.6.11) can be written then in normalized, reduced form with the parameter c eliminated as follows:

$$y = a_0 \left[\mathrm{Ln} \left(\frac{x + a_1}{1 + a_1} \right) - (1 - x) \mathrm{Ln} \left(\frac{a_1}{1 + a_1} \right) \right] + (bx + 1)\,(1 - x) \tag{9.6.12}$$

It is apparent from Eq. (9.6.12) that there are three independent, essential parameters, a_0, a_1, and b that determine the response characteristics of the system. For Eq. (9.6.12) to exhibit acceptable values, we require that the argument of the natural log functions [i.e., $\mathrm{Ln}(x)$] must be greater than zero. Therefore, we require that $a_1 > 0$ or $a_1 < -1$. Figure 9-16 displays some of the different system responses that can be realized by varying these three system

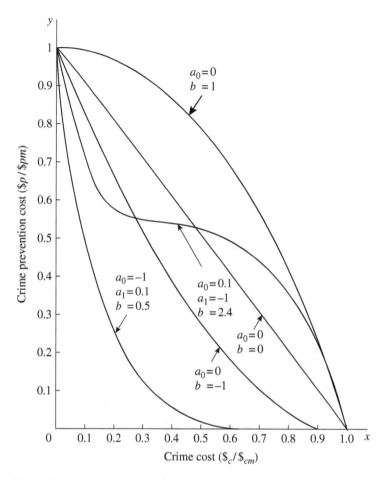

Fig. 9-16 Various responses for societal crime cost system.

parameters and using the range of values for $0 \le y \le 1$ and $0 \le x \le 1$. Even though the original kernels were assumed to be linear, the response of the overall system is obviously nonlinear, and by properly selecting values for the three basic parameters, the system model can be made, hopefully, to adequately describe the real problem.

Equation (9.6.12) represents the general mathematical relationship between the normalized societal cost of crime ($x = \$_c/\$_{cm}$) and the normalized crime prevention cost ($y = \$_p/\$_{pm}$) for the system we have assumed. The three basic parameters, a_0, a, and b, could be determined if three independent operating states, x_1, y_1, x_2, y_2, x_3, y_3, were known for the actual system. Without knowledge of such data, however, it is still possible to impose some general assumptions and thereby draw certain conclusions about the system model. First, we have assumed that the maximum value of x (i.e., $x = 1$) occurs when no funds are spent for prevention (i.e., $y = 0$), and when maximum prevention costs are expended ($y = 1$), there is no crime cost ($x = 0$). If we now assume that every additional dollar spent in crime prevention contributes somewhat to the reduction in crime cost, then we conclude that the system function is monotonic with y (where $1 > y > 0$) and always decreases with increasing x (where $0 < x < 1$).

Thus, for any $\Delta x > 0$,

$$1 \geq y(x) \geq y(x + \Delta x) \geq 0$$

Equation (9.6.12), which is the system response, provides for this inequality that

$$a_0 \left[\text{Ln} \left(\frac{x + a_1}{1 + a_1} \right) - (1 - x) \, \text{Ln} \left(\frac{a_1}{1 + a_1} \right) \right] + (bx + 1)(1 - x)$$

$$\geq a_0 \left[\text{Ln} \left(\frac{x + \Delta x + a_1}{1 + a_1} \right) - (1 - x - \Delta x) \, \text{Ln} \left(\frac{a_1}{1 + a_1} \right) \right] + (bx + b\Delta x + 1)(1 - x - \Delta x)$$

Canceling like terms and combining, we find that

$$a_0 \left[\text{Ln} \left(1 - \frac{\Delta x}{x + \Delta x + a_1} \right) - \Delta x \, \text{Ln} \left(\frac{a_1}{1 + a_1} \right) \right] + \Delta x \left[b \left(2x + \Delta x - 1 \right) + 1 \right] \geq 0$$

$$(9.6.13)$$

Observe that as $\Delta x \to 0$, the left side of the inequality approaches zero. If we assume that Δx is small so that $\Delta x \ll x + \Delta x + a_1$ then

$$\text{Ln} \left(1 - \frac{\Delta x}{x + \Delta x + a_1} \right) \approx \frac{\Delta x}{x + \Delta x + a_1}$$

and inequality (9.6.13) becomes, upon canceling Δx (recall $\Delta x > 0$) and rearranging

$$a_0 \left[\frac{1}{x + \Delta x + a_1} + \text{Ln} \left(\frac{a_1}{1 + a_1} \right) \right] \leq b(2x + \Delta x - 1) + 1 \qquad (9.6.14)$$

If we assume that Δx is sufficiently small that

$$\Delta x < (x + a_1) \quad \text{and} \quad \Delta x < (2x - 1)$$

then inequality (9.6.14) becomes independent of Δx that

$$a_0 \left[\frac{1}{(x + a_1)} + \text{Ln} \left(\frac{a_1}{1 + a_1} \right) \right] \leq b(2x - 1) + 1 \qquad (9.6.15)$$

Thus, in our choices for a_0, a_1, and b, if inequality (9.6.15) is satisfied for all values of x such that $0 \leq x \leq 1$, then Eq. (9.6.12) will be a monotonically decreasing function with x.

Now consider the derivative of y with respect to x in the range $0 \leq x \leq 1$. From Eq. (9.6.12), we find for dy/dx that

$$\frac{dy}{dx} = a_0 \left[\frac{1}{(x + a_1)} + \text{Ln} \left(\frac{a_1}{1 + a_1} \right) \right] + b(1 - 2x) - 1$$

Since $y(x)$ is monotonically decreasing with x, we have that $(dy/dx) \leq 0$ for $0 \leq x \leq 1$, and thus

$$a_0 \left[\frac{1}{(x + a_1)} + \text{Ln} \left(\frac{a_1}{1 + a_1} \right) \right] \leq b(2x - 1) + 1$$

which is identical with inequality (9.6.15). Observe that near $x = 0$ the derivative becomes (recall that either $a_1 > 0$ or $a_1 < -1$)

$$\frac{dy}{dx}\bigg|_{x=0} = a_0 \left[\frac{1}{a_1} + \text{Ln}\left(\frac{a_1}{1 + a_1}\right) \right] + b - 1 \leq 0 \qquad (9.6.16)$$

and

$$\frac{dy}{dx}\bigg|_{x=0} \to 0 \quad \text{if} \quad a_0 \left[\frac{1}{a_1} + \text{Ln}\left(\frac{a_1}{1 + a_1}\right) \right] + b \to 1$$

Also

$$\frac{dy}{dx}\bigg|_{x=0} \to -\infty \quad \text{as} \quad \frac{a_0}{a_1} \to -\infty \quad \text{or} \quad b \to -\infty$$

Furthermore, near $x = 1$ the derivative is

$$\frac{dy}{dx}\bigg|_{x=1} = a_0 \left[\frac{1}{1 + a_1} + \text{Ln}\left(\frac{a_1}{1 + a_1}\right) \right] - b - 1 \leq 0 \qquad (9.6.17)$$

and

$$\frac{dy}{dx}\bigg|_{x=1} \to 0 \quad \text{if} \quad b + 1 \to a_0 \left[\frac{1}{1 + a_1} + \text{Ln}\left(\frac{a_1}{1 + a_1}\right) \right]$$

While

$$\frac{dy}{dx}\bigg|_{x=1} \to -\infty \quad \text{as} \quad \frac{a_0}{a_1} \to -\infty \quad \text{or} \quad b \to \infty$$

These conditions must also satisfy the requirement that $0 \leq y \leq 1$ for $0 \leq x \leq 1$. To select an appropriate set of conditions for the derivative of y with respect to x, consider the following conceptual, graphical response for y and x:

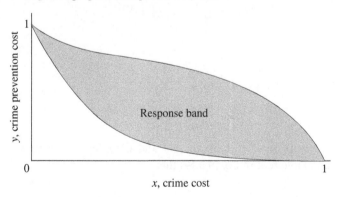

It is reasonable to assume that the slope of y near $x = 0$ should be large and negative, which implies that crime prevention costs increase rapidly when society attempts to eliminate all crime. So from inequality (9.6.16), we conclude that either a_0/a_1 or b is large and negative. Furthermore, the slope of y near $x = 1$ should be small and negative, reflecting the reasonable assumption that in a near-maximum crime situation even small expenditures in crime prevention produce significant reductions in the cost of crime. From inequality (9.6.17), we have that

$$b \approx a_0 \left[\frac{1}{1 + a_1} + \text{Ln} \left(\frac{a_1}{1 + a_1} \right) \right] - 1$$

Substituting this into inequality (9.6.16) for dy/dx ($x = 0$), then

$$a_0 \left[\frac{1 + 2a_1}{a_1(1 + a_1)} + 2\,\text{Ln} \left(\frac{a_1}{1 + a_1} \right) \frac{a_1}{1 + a_1} \right] \leq 2 \qquad (9.6.18)$$

Inequality (9.6.18) is satisfied, for example, for the values $a_0 = -1$, $a_1 = 0.1$, $b = 0.5$. Figure 9-16 gives the response of the system for these particular values.

For another interesting insight into the problem of what are reasonable maximum expenditures that might justifiably be spent by society to prevent crime, recall that

$$\frac{dy}{dx} = \frac{d(\$_p/\$_{pm})}{d(\$_c/\$_{cm})} = \frac{\$_{cm}}{\$_{pm}} \frac{d\$_p}{d\$_c}$$

And if we are willing to spend funds for crime prevention up to a rate such that for each dollar spent for crime prevention, we reduce crime cost by \$1, then we have that

$$\frac{d\$_p}{d\$_c} = -1$$

which implies that each \$1 spent in crime prevention reduces crime cost by \$1. So to find this maximum expenditure for crime prevention, we solve the following equation for x_1:

$$\frac{dy}{dx} = a_0 \left[\frac{1}{x_1 + a_1} + \text{Ln} \left(\frac{a_1}{1 + a_1} \right) \right] + b\,(1 - 2x_1) - 1 = \frac{\$_{pm}}{\$_{cm}}$$

which has a solution for x_1 that

$$x_1 = \frac{1}{4b} \left[a_0\,\text{Ln} \left(\frac{a_1}{1 + a_1} \right) + b\,(1 - 2\,a_1) + \frac{\$_{sm}}{\$_{pm}} - 1 \right] (1 + X)^{1/2}$$

where

$$X = 1 + \frac{8b \left[a_0 + a_1 \left\{ b + \text{Ln} \left(\frac{a_1}{1 + a_1} \right) + \frac{\$_{pm}}{\$_{cm}} - 1 \right\} \right]}{\left[a_0\text{Ln} \left(\frac{a_1}{1 + a_1} \right) + b\,(1 - 2\,a_1) + \frac{\$_{pm}}{\$_{cm}} - 1 \right]^2} \qquad (9.6.19)$$

For our previous example, where $a_0 = -1$, $a_0 = 0.1$, $b = 0.5$, and assuming various values for $\$_{pm}/\$_{cm}$ we find for an acceptable root of Eq. (9.6.19) that

$$x = \begin{cases} 0.28 \ \left(\text{for } \$_{cm} = \$_{pm}\right) \\ 0.63 \ \left(\text{for } \$_{cm} = 0.1\ \$_{pm}\right) \end{cases}$$

So the maximum expenditure that should be devoted to crime prevention is from Figure 9-16,

$$y = \begin{cases} 0.15 \ \left(\text{for } \$_{cm} = \$_{pm}\right) \\ 0.01 \ \left(\text{for } \$_{cm} = 0.1\ \$_{pm}\right) \end{cases}$$

Thus, on the basis of a requirement that every dollar spent for crime prevention should reduce the cost of crime to society by at least $1 or more, the expenditures for crime prevention are

$$\$_p = \begin{cases} 15\% & (\text{if } \$_{cm} = \$_{pm}) \\ 1\% & (\text{if } \$_{cm} = 0.1\,\$_{pm}) \end{cases}$$

9.6.4 Energy Consumption and GNP

A developed nation's total energy consumption (*TEC*) is generally closely related to the nation's economic activity and cost of energy. A standard system equation employed for such national energy modeling has the form

$$d(TEC) = (TEC)\left[a\,\frac{d(GNP)}{GNP} + b\,\frac{d(EIGNP)}{EIGNP}\right] \tag{9.6.20}$$

where

$$TEC = \text{Total national energy consumption in units of } 10^{15} \text{ Btu (quads)}$$
$$GNP = \text{Gross National Product (billions for year 2000 U.S. dollars)}$$
$$EIGNP = \text{Energy Intensity (consumed) per U.S. dollars of GNP}$$

System Eq. (9.6.20) may be integrated to give the model system solution

$$TEC = k\,(GNP)^a(EIGNP)^b$$

where k is the integration constant, and a and b are system constants, all of which are derived by data-fitting (e.g., least squares or regression analysis). Table 9-13 provides the data published by the U.S. Department of Energy required for the model projected to 2025.

Between 1960 and 2010, about 92% of the variation in total energy consumption is attributable to changes in *GNP* only. A sample regression analysis provides as numerical values for k, a, and b the following values

$$TEC = 0.0842\,(GNP)^{0.96}(EIGNP)^{-0.29} \tag{9.6.21}$$

Equation (9.6.21) implies that the *GNP* elasticity for total energy use is 0.96, so a 1% increase in *GNP* leads to a 0.96% increase in energy consumption for constant prices. For energy consumption, the price elasticity is −0.29, so a 1% increase in the *GNP* implies a 0.29% decline in energy use for constant *GNP*.

Table 9-13 U.S. Economic Activity, Total Energy Consumption, and Cost

Year	Gross National Product (GNP)		Total Energy Consumption		Population		Energy Intensity of GNP (EIGNP)	
	Billion $(a)	Annual % Change	Quads Btu(b)	Annual % Change	Millions	Annual % Change	Btu/$	Annual % Change
1960	2502	2.5	45.1	2.3	181	1.7	18,020	−1.0
1965	3191	5.0	54.0	3.7	194	1.4	16,930	−1.2
1970	3772	3.4	67.8	4.7	205	1.1	17,990	1.2
1975	4311	2.7	72.0	1.2	216	1.1	16,700	−1.5

(*Continued*)

Table 9-13   (Continued)

Year	Gross National Product (GNP) Billion $(a)	Annual % Change	Total Energy Consumption Quads Btu(b)	Annual % Change	Population Millions	Annual % Change	Energy Intensity of GNP (EIGNP) Btu/$	Annual % Change
1980	5162	3.7	78.1	1.6	227	1.0	15,130	−3.5
1985	6054	3.2	76.5	-0.4	238	1.0	12,640	−3.5
1990	7113	3.3	84.7	2.1	250	1.0	11,900	−1.2
1995	8032	2.5	91.2	1.5	266	1.2	11,350	−0.9
2000	9817	4.1	99.0	1.7	282	1.2	10,080	−2.3
2005	11,003	2.3	111.5	0.3	296	1.0	8,820	−4.0
2010	11,223	2.0	122.8	0.3	305	1.0	8,476	−3.9
2015	11,447	2.2	134.8	0.3	310	1.0	8,276	−3.9
2020*	11,679	1.9	146.8	0.3	320	1.0	8,176	−3.9
2025*	11,905	2.8						
2030*	12,135	3.1						

(a) $ Chained to year 2000.
(b) Quadrillion Btu or 10^{15} Btu.
Source: Adapted from U.S. Department of Energy Data. (*estimated)

9.6.5   Topical System Problems in Human Society

Agricultural Model. Devise a model for the agricultural output of a specified commodity in terms of the significant factors contributing to that commodity's production

Input:	Land quality and quantity, climate, irrigation and water resources, seeds, plants or animal starts, fertilizer or feed, equipment, technical information and resources, government regulations and control.
Output:	Commodity output in metric tonnage or equivalent cash market value.

Transportation Demand Model. Develop a model that predicts consumer acquisition utilization, and efficiency of automobiles. Include impact of electrification of vehicles (EV).

Input:	Consumer price index, population, average hourly earnings (nonagricultural), disposable income, unemployment rate, new car sales, price of fuel, vehicle miles traveled, and average vehicle efficiency.
Output:	Future fleet efficiency, vehicle miles traveled, new car sales, and fuel and electricity demand.

Dynamic Employment Model. In a plant, business, or economic sector, let the net change in labor employment be ΔE. Then a possible employment model is shown in the following block diagram for the model.

The system inputs are

ΔH = increase from hiring
ΔQ = loss to voluntary quitting
ΔL = loss to layoffs and dismissals
ΔR = loss to retirements and technical improvements)

a) Develop the system equation for the model.
b) Assume that each kernel, g_i, is a constant (establish the proper algebraic sign for each term) and solve the system equation.

Annual Forecast Model for U.S. Economy. Develop a simple mode of the U.S. economy stressing the production side to forecast employment, output, and capital requirements.

Input:	Monetary and fiscal history and policy, resource availability (minerals, fuels, technology), productivity, and population.
Output:	Personal consumption, construction, employment, and prices.

Model for Output of Finished Goods. Develop a model for the essential inputs required to produce a given finished product (e.g., processed food, automobile tires, or cell telephones).

Input:	Capital, labor, equipment, land, raw materials, energy, management, etc.
Output:	Production of the finished goods in appropriate units (e.g., units, kgs, tons, barrels).

Commercial Energy Use Model. Create a model to forecast annual energy use in the commercial sector by end uses and fuel types. Consider as end uses space heating, water heating, cooling, lighting, and others. As the fuel types, consider gas, electricity, oil, and coal.

Input:	Commercial building inventory in units of thousands of square feet area, macroeconomic data, population data, energy requirements per thousand square feet.
Output:	Energy consumption by end-use and fuel type.

Short-Term Electrical Power Demand Model. Provide a model that gives short-term forecasts of hourly load demands by month for specific power supply areas.

Input:	Hourly load demand for area, electrical power costs by time of service and quantity, cooling and heating degree-days, and personal income for area.
Output:	Hourly demand curves and total consumption and growth rate for area.

Age Dependence of Human Values. Common wisdom asserts that one of the important values that accrues with age is wisdom. However, a large sampling of males for each of the age groups considered was asked to rank specified human values by importance, and the results are somewhat surprising. Figure 9-17 ranks 18 important human values by decade age groups from ages 20 through 80 for a large sample of the 1976 U.S. male population. The

Age dependence of human values

Value	Value rank with age							Mean
World peace	1	2	1	1	1	1	1	1.14
Family security	2	1	2	2	2	2	2	1.86
Freedom	3	3	3	3	3	3	11	4.14
Self-respect	7	6	10	5	4	6	6	6.29
Wisdom	5	5	7	7	7	7	7	6.43
Equality	6	4	5	4	9	9	9	6.57
Happiness	4	7	4	10	11	11	3	7.14
Comfortable life	9	14	8	8	12	4	4	8.43
National security	11	10	11	9	8	5	5	8.43
Accomplishments	8	8	9	12	5	10	10	8.86
True friendship	12	9	12	6	14	8	8	9.86
Salvation	10	12	6	11	10	12	12	10.4
Mature love	13	13	14	14	6	14	15	12.7
Inner harmony	14	11	13	13	17	16	16	14.3
World of beauty	15	15	15	15	15	15	14	14.9
Social recognition	17	17	16	16	16	13	13	15.4
Pleasure	16	18	17	17	13	17	17	16.4
Exciting life	18	16	18	18	18	18	18	17.7
Age in years	20	30	40	50	60	70	80	

Fig. 9-17 Ranking of human values by age. *Source:* Adapted from "Future Facts," by Stephen Rosen, 1976.

perceived order of importance or ranking increases vertically upward by a mean value over life span from age 20 to age 80. It is interesting that wisdom, which is ranked fifth at age 20, decreases with age to seventh ranking at age 80, apparently contradicting the common assumption that wisdom (or at least its perceived worth) accrues with age.

a) Treating these 18 human values as model outputs and human male age as the input, develop a system equation (in the form of a tabular set) to predict value ranking with age.
b) We want to develop a scheme that will assign a single numerical score, $s(t)$, for the ranked human values at each age.

The score $s(t)$ at age $t = (20, 30,..., 80)$ is defined as

$$s(t) = \left\{ \frac{\sum_{i=1}^{18} W_i(t_0)\, x_i(t)}{\sum_{i=1}^{18} W_i(t_0)\, x_i(t_0)} \right\}^{1/2}$$

where

$W_i(t_0)$ = weight of each value at age t_0
$x_i(t)$ = vertical rank $(1, 2, ..., 18)$ at age t

Let us assume that the base age t_o is set as 50 so that $s(50) = 1$ and $W_i (50) = 1$ (world peace), 2 (family security),..., 18 (exciting life). Find and plot $s(t)$ for $t = 20, 30,..., 80$. Comment on this numerical result. Is it reasonable? How do human values change with time? Would this function normalized to the age of a given individual (i.e., $t_o =$ your age) provide insight as to how you view the values of other persons?

Human Identification Model. There is a nearly continuous need in our society for individuals to be recognized and uniquely identified by others. At times this identification is vital, such as when cashing a check, appearing in court, or identifying possible terrorists attempting to board an aircraft. Establish a quantitative model that accepts as inputs various human characteristics that might be used to distinguish between individuals. Possible inputs might include the following:

Medical, Actuarial, Criminal, etc. records
Handwritten signature
Facial appearance (e.g., photograph, thermograph, CT scan, eye scans)
Voice characteristics (frequency, pitch, harmonics, timbre, tempo)
Fingerprints (hand, toe, and heel prints)
Combined physiological characteristics
- Height
- Weight
- BMI (body mass index)
- Tattoos, Physical irregularities, Body mass index (BMI)
- Scars and limb irregularities/loss
- Teeth (dental imprints)
- Blood chemistry(blood type/pressure, RH factor, pH, etc.)
- Eye retinal image
- Body odors (dogs can distinguish humans by their unique scent)
- Skeletal x-rays (PET, MRI, CT, etc.)
- Body temperature, pH, etc.
- DNA (test is very definitive and well accepted)
Mental characteristics
- Response to selected questions of past history or identity (polygraph)
- Electroencephalographs
- Language or word tests
- Specific knowledge or physical/mental skill demonstrations

The model output wanted is an identification index that is a quantitative measurement of the value and efficiency of the following factors associated with each identification input (i) per person:

$$\left. \begin{array}{c} \text{Identification} \\ \text{Index(i)} \end{array} \right\} = \frac{(\text{sampling speed}) \times (\text{accuracy probability})}{(\text{sampling cost}) \times (\text{subject discomfort})}$$

Obviously, the value and reliability of the identification index (I/I) will increase with increasing sample speed and the probability for accurate identification. On the other hand, the increasing cost of the sampling input, invasion of privacy, and discomfort to the individual interrogated would reduce the value of the identification index. Establish the model for the identification system and assign quantitative measures to each of the factors contained in the identification index. On the basis of the numerical values obtained for the indexes, rank in order the model inputs for identification. Justify your results. Do these tests meet Legal Requirements for admission in a Court action?

Nuclear War Threat Model. One of the greatest threats imposed on humanity because of the nuclear age is the risk of major terrorist actions or international wars involving nuclear weapons or other Weapons of Mass Destruction (WMD). Obviously, there is great need to reduce the possibility of such a human disaster. However, although desire for peace is nearly universal, a broad recognition and understanding of the major factors contributing to the likelihood of nuclear event (e.g., a nuclear explosion) is not. From the following list of possible factors (or others you believe are more significant), select the four most significant factors that would contribute the most to the threat of nuclear war. Rank these four factors by significance and develop an event tree for the system block diagram (see Section 6.2.5). Assign probability estimates to the kernels and inputs and make an estimate for the output, viz., a nuclear event. See Wikipedia for assessment and data.

Nuclear Event Causal Factors (Possible inputs)
- Terrorist actions
- Level of world tensions
- Scientific developments in nuclear and military field
- World economic conditions
- Arab-Israeli conflict
- World dependence on oil
- Accidental initiation
- Nuclear power infrastructure
- Escalation of conventional and regional conflicts
- Hostilities between North and South Korea and their partners
- Development of nuclear weapons by Iran
- Nuclear arms buildup
- Religious or ideological issues (Taliban, ISIS, Al Qaeda)
- Nuclear weapon use by a country against an adversary (Israel vs. Iran)

Construction Forecasting Model. Estimating infrastructure construction activities (e.g., residential, commercial, highways, railways, etc.) in the economic sector is not only important for the specific industries involved but is an important element in predicting the overall health of the regional and national economy. The output for this construction model will be the change in the economic value dE for construction activities in a given time period (i.e., \$/yr). The basic approach for the development of such forecasting models is to consider the major economic factors (system inputs) that influence the construction field. These factors or inputs are as follows:

I = sector resources (e.g., industrial, commercial, residential, personnel)
P = sector population
E = sector employment or activity
V = vacancy rate or capacity fraction
C = cost of money
T = time

a) Define the selected sector and draw its system model block diagram.
b) Form the system equation for the model.
c) Model each kernel component by dimensional analysis (see Section 6.2.4).
d) Solve the system equation and interpret the result.

Code Breaking Model. As is well known, the frequency of usage of letters in the U.S. English alphabet varies dramatically with the respective letter. Of course, the frequency of letters in another language alphabet is different. For typical U.S. English written material, such as newspapers and magazines, the letter count per 100,000 sequential letters in descending frequency is approximately the following:

English written letter count per 100,000 sequential letters

e	12702	n	6749	l	4025	f	2228	v	978
t	9056	s	6327	c	2782	g	2015	k	772
a	8167	h	6094	u	2758	y	1974	j	153
o	7507	r	5987	m	2406	p	1929	x	150
i	6966	d	4253	w	2360	b	1492	q	95
								z	74

Source: Adapted from – Wikipedia, The Free Encyclopedia (divide by 1000 for percentage).

Now consider an alphabetic cyber code model that accepts words, letter by letter, in a sequential reading fashion for a given cipher letter written in English. Each letter of a word in the cipher to be decoded is counted for its frequency in the cipher. Then with the letters in the cipher ranked in decreasing occurrence, these cipher letters can be compared to the frequency distribution provided in the above table. The first trial decoding of the cipher would replace this frequency distribution with the expected frequency distribution shown above. This process would be repeated with various algorithms for decoding the cipher. As certain frequent words were encountered, they could be parsed from the remaining frequency distribution, and the process repeated until the cipher was adequately decoded.

Decision Analysis Model. A useful quantitative method for assessing decisions and determining those decision policies that should be adopted to satisfy a field of competing groups or requirements is known as Pareto Decision Analysis. To introduce this method, assume that there are only two competing groups or requirements labeled X and Y seeking satisfaction. (Note that the method is easily extended to any number of groups.) Then we must assign a numerical value (x_i, y_i) to each possible alternative or decision D_i of interest that affects both X and Y. The numerical value x_i estimates the value or fractional satisfaction (i.e., $0 \le x_i \le 1$) that decision D_i would provide group X, while y_i reflects the value of that same decision to group Y (i.e., $0 \le y_i \le 1$). Suppose there are n such alternative decisions $(D_1, ..., D_n)$ that are mutually exclusive; that is, only one decision can be implemented, which then excludes the occurrence of any other decision in the set $D_1, ..., D_n$. A plot of the decision set is shown in Figure 9-18.

There are several possible criterions for determining the decision that optimizes the decision choice from the set $(D_1, D_2, ..., D_j)$. Furthermore, as is often the case, the importance of

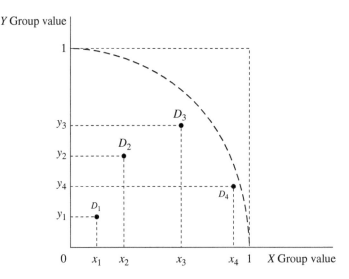

Fig. 9-18 Graphical display for decision analysis.

one group's wishes may exceed the importance of another group, so the decision values x_i are each weighted by W_x and the decision values y_i are each weighted by W_y. Then possible optimization criteria are as follows for determining the "best" decision for a given criterion:

a) Maximize $[W_x x^2_i + W_y y^2_i]^{1/2}$ for all x_i and y_i
    (See example: D_4 in Figure 9-18)

   This choice maximizes the weighted distance from the origin ($x = y = 0$) but fails to maximize the decision value for all groups combined.

b) Maximize $\left\{ \frac{W_x x_i^2 + W_y y_i^2}{(x_i - 1)^2 + (y_i - 1)^2} \right\}^{1/2}$ for all x_i and y_i
    (See example: D_3 in Figure 9-18)

   This choice maximizes the distance from the origin and minimizes the distance to the point (1, 1). Thus, a bias is made to provide some partial satisfaction for both groups. Other policies are possible for optimizing decisions, but these are useful for many decisions resolution cases.

Political Campaign Model. As is well accepted, political action committees (PACs) and other lobbying groups can have a significant impact on the outcome of political elections. Consider the following simple block diagram for an election model, where special interest groups (lobbyists and PACs) serve as feedback systems to the campaign process.

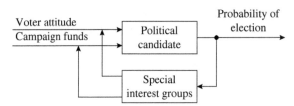

a) Develop a simple quantitative system equation with constant kernels for this model and solve the system equation.
b) What conditions on the feedback kernels (arising from the special interest groups) will maximize the probability of a given election outcome?

Econometric Model. Econometrics is the application of mathematical models for the description of economic behavior. Early studies in econometrics were characterized primarily by attempts to quantify the causal relationship between the price of a commodity (i.e., service or product) and the amount sold at that price. A major goal in such modeling studies is to determine the price elasticity of such commodities and thus what effect a fractional change in cost would have on the fractional change in demand. Accurate assessments of price elasticity are obviously highly desirable as background data for justifying production or employee expansion or contraction for a given commodity. An econometric model for product production developed by Paul Douglas and Charles Cobb at the University of Chicago is as follows:

$$dX = a\,dL + b\,dK$$

where

X = total product output (units produced)
L = labor input (worker-hours)
K = stock of capital equipment (investment units)

Values proposed for the parameters a and b are as follows:

$$2/3 < a < 3/4 \quad \text{and} \quad 1/4 < b\ < 1/3 \qquad \text{with constraint } (a + b) = 1$$

a) Determine the general solution for the system equation.
b) Using the range of values of a and b given previously, assuming the constraint that $(a + b) = 1$ is always satisfied, what are the elasticity's for labor and capital?
c) Assume that labor's full share of output is measured by parameter a, while capital investment is determined by parameter b. How might profits from the product output be equitably distributed?
d) Assume that a and b are linear functions of their respective differential variables. So $a\,(L) = a_0 + a_1\,L$ and $b\,(L) = b_0 + b_1\,K$. Repeat steps (a) through (c) stated above for this system.

Wealth Distribution Model. A well-documented economic phenomenon observed over time for many advanced societies is that the wealth of a society tends to concentrate in a small fraction of the population.

a) Attempt to determine the significant factors or inputs that drive this concentration of wealth.
b) Develop a system equation for the "concentration of wealth" model using the inputs identified in part (a).
c) Obtain a solution (either analytical or computer – e.g., see MATLAB) for the system equation.
d) Introduce reasonable parameter and initial values into the system equation solution and discuss the predicted behavior.

Francis Bacon's Model of Knowledge. In the early 1600s, Francis Bacon embarked on writing the *Magna Instauratio*, which was to be a total reconstruction of the sciences, arts, and indeed, all human knowledge. Although only a small portion of this scholarly work was actually completed, Bacon's classification of knowledge was completed and provides interesting insights regarding the perceived importance of knowledge in the sixteenth century. Table 9-14 provides this classification by Bacon.

Table 9-14   Francis Bacon's Classification of Knowledge

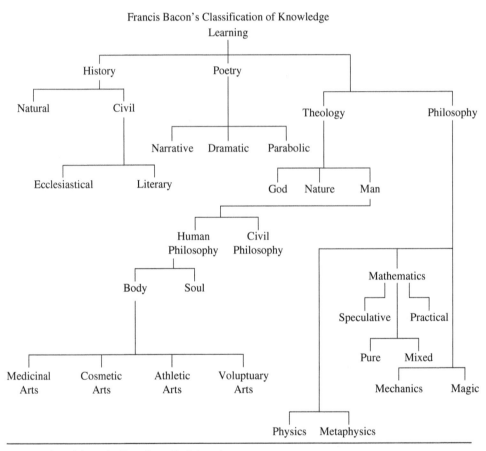

Source: Adapted from the Encyclopaedia Britannica.

a) Compare Bacon's classification of knowledge with that given in Section 9.1 extracted from the *Encyclopaedia Britannica*.

b) What significant categories of knowledge are missing or significantly different between Bacon's classification (Table 9-14) and the *Encyclopaedia Britannica*? (Table 9-1).

c) Establish an organization diagram (see Section 6.1.3) for both knowledge classifications.

Table 9-15 Francis Bacon's Model for National Economies

	System	Input	Output
(a)	Society	Poverty Unemployment Urbanization	Crime rate (crimes/person)
(b)	Business enterprise	Management salary Capital Labor salary	Profits
(c)	Civil litigation	Lawyer selection Pretrial negotiations Jury selection Trial	Winning a suit
(d)	National labor force	White-collar workers Blue-collar workers Service workers Farm workers	Gross national product

(e)	Future economy	Leading economic indicators:
		   Stock prices
		   New orders of durable goods
		   Plant and equipment orders
			Future economic conditions

(f)	Present economy	Coincident indicators:
		   GNP in current dollars
		   Personal income
		   Nonagricultural employment
			Current economic conditions

(g)	Past economy	Lagging indicators:
		   Production inventories
		   Expenditures for plant and equipment
		   Unemployment rate
			Past economic conditions

Source: Wikipedia.

Additional Modeling Problems. For each of the systems given in Table 9-15, define an appropriate system model and the essential elements of the system. Express quantitative variables, and parameters for each of the model inputs and outputs and develop a simple analytical expression for the model's system equation. Determine analytical or other solution forms for the system equation and interpret and evaluate your results.

9.7 Applications of System Science to the Arts

The earliest human art conceived and developed by mankind was language. The broad aspects of language include speaking, listening, counting, measuring, thinking, and reasoning itself. Indeed, it is language and the human skills and intellectual arts it engenders that most distinguishes humans from their animal ancestors. The fine arts are perhaps best described as those human activities whose primary purpose is beauty, pleasure, and recognition of the esthetic aspects of life and the universe. Because of the emotional and spiritual content of the arts, it is often difficult to achieve agreement on the value and content of various art forms and to assess quantitative measurements of the individual or social worth of the arts. Nevertheless, the impact on and importance of the arts to a human society are evident and significant. It is difficult to imagine a world without music, literature, dance, theater, architecture, painting, and the like. Certainly, the arts are a major focus and occupation of humanity, and no individual is truly educated who does not have recognition and appreciation for the human arts. A perspective of the impact of the arts on human enterprise is apparent by considering the brief listing below of the artistic fields associated with human endeavors.

Architecture	Design	Opera
Art & Visual Arts	Drawing	Painting
Calligraphy	Fashion	Photography
Crafts	Film	Poetry
Culinary Art	Language	Sculpture
Dance	Literature	Theatre/Performing Arts
Decorative art	Music	Video

9.7.1 Quantitative Assessment of Language

Written and oral communication is the most important means that humans have of communicating with each other. It is the major means for the transfer of information, observations, feelings, and knowledge itself between humans. The human mind can be considered as the system, with the inputs as "words" in either written or oral form and the output words and actions from the system "the mind" modified and synthesized as the outputs resulting from these inputs into individual

We seek a quantitative model for the human language communication system that provides a measurement of the content and meaning of a phrase of written or oral communication. Such a task is obviously challenging, and our efforts here will be limited to simple, modified subject and verb action statements. To begin, consider the phrase:

."Rather average students can become unusually good learners."

Grammatically, our sequence of words can be diagrammed as follows:

We will limit our quantitative modeling effort of language to the simple, normal subject-verb, adverb-adjective-noun modifier phrase components of communications. Then the question arises. Can we quantitatively measure the communication impact of such phrases? The answer to this question hopefully will be partially answered by our model's success. Norman Cliff established a numerical assignment for common adjectives and adverbs that asserts that the impact on a noun modified by such modifiers can be treated by algebraic multiplication processes. Table 9-16 provides a numerical assignment of the intensifying value of adverbs and the scale value of adjectives. Observe that the intensity values of adverbs are always positive, while adjectives span a range of negative to positive values. Those adjectives with positive scale values are usually associated with favorable, positive, or desirable attributes, while those with negative values are associated with opposite attributes. Observe that the greater the magnitude of the adjective scale value the more intense is the attribute. Obviously, a qualitative database such as language cannot provide an absolute standard of word definitions and meanings in communication, but some level of quantification is possible. The tabulated values presented in Table 9-16 provide a statistical average of common agreement by a selected set of test participants. The utility of Table 9-16 can be demonstrated with an example. Consider the following word phrases, which modify the noun "students." The modifying words are converted to numerical values from Table 9-16, and the multiplier or weight for the noun is then shown as follows:

e.g.,
1. rather average students: $(0.84) \times (-0.79) = -0.66$ Students
2. unusually good learners: $(1.28) \times (3.09) = 3.96$ Learners

We observe that statement (1) is a slightly negative modifier of students with a value of -0.66, while statement (2) is a favorable description of learners with a value of 3.96. Assuming that the students become learners, then the net score or rating for the sentence

Table 9-16 Numerical Weighting for Adverbs and Adjectives

Adverbs	Weight	Adjectives	Weight
slightly	0.54	evil	−2.64
		bad	−2.59
somewhat	0.66	wicked	−2.54
		immoral	−2.48
		inferior	−2.46
rather	0.84	contemptible	−2.20
		disgusting	−2.14
		average	−0.79
fairly	0.88	ordinary	−0.67
(no adverb)	1.00	(no adjective)	+1.00
quite	1.05	charming	+2.39
decidedly	1.16	lovable	+2.43
very	1.25	nice	+2.62
unusually	1.28	pleasant	+2.80
extremely	1.45	good	+3.09
		admirable	+3.12

Source: Adapted: Norman Cliff, Adverbs as Multipliers, Psychological Rev, vol. 66, pp. 27–44, January 1959.

summing (1) and (2) yields 3.29, which is a net favorable statement about average students becoming good learners.

A very negative sentence is "very bad dictators can become extremely evil leaders" and quantitatively from the table, we have the following value:

$$[(1.25) \times (-2.59) + (1.45) \times (-2.64)] = -7.07 \text{ (a very negative assessment)}$$

How might we use these quantitative data about adverbs and adjectives to quantitatively assess communications? An interesting application of the model would be the quantitative assessment of politicians during election campaigns. A numerical judgment could be made for each candidate, regarding policy positions.

A block diagram for a possible model is shown in Figure 9-19.

Observe that a sentence to be parsed enters the first block, where the sentence is sorted into noun phrases (adverb-adjective-noun format) and verb phrases, together with a phrase and verb sequence code to be used for subsequent sentence assembly. The verb phrase is then transmitted to a re-assembler model block awaiting evaluation of the noun phrases. The noun phrases now are processed by a word sorter block, which sorts each phrase into modifying adverb(s), adjective(s), and noun. The adverb and adjective modifiers are then subjected to a table lookup block, where a numerical value is assigned to the modifiers. The noun phrases are then tagged with a numerical value, as we demonstrated earlier. Finally, the complete sentence is assembled and output, together with a numerical quantifier for the nouns in each noun phrase of the sentence. Although our model for quantitative language assessment is crude and limited, it would allow for the comparison and ranking of related statements on a given subject. It would be interesting to use our model for quantitatively assessing the statements that rival politicians make about each other.

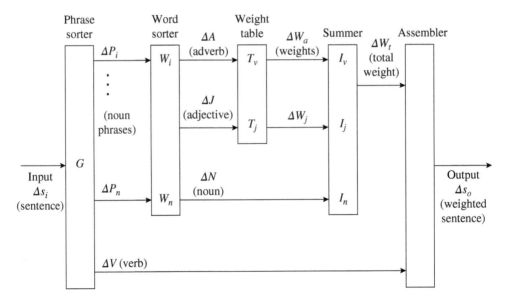

Fig. 9-19 Model for quantitative evaluation of sentences.

9.7.2 Art Awareness Model

One of the major objectives of promoters of the human arts is art awareness appreciation. It is both culturally and financially profitable for those in the arts to spread their talents, performances, and unique cultural skills to the largest audiences that they can. A simple mathematical model for this spread of cultural awareness can be based on a "diffusion-type" equation in which those individuals who are aware of a particular artist or art form share their interest and enlist others to participate. So we propose the system equation

$$dn = D(N - n)^a dt \tag{9.7.1}$$

where

$n =$ number of people at time t aware of particular artist or art form
$N =$ number of people who could become aware of artist or art form
$D =$ spreading or diffusion constant with dimensions of reciprocal time
$t =$ elapsed time for art awareness to spread
$a =$ dimensionless constant that determines rate of spreading

System Eq. (9.7.1) is separable as follows:

$$\frac{dn}{(N-n)^a} = Ddt \tag{9.7.2}$$

Equation (9.7.2) can be directly integrated to give

$$(N - n)^{1-a} = K - (1-a)Dt$$

where K is the arbitrary integration constant. Evaluating this constant, we have that for $t = 0$ and $n(0) = n_o$, then $K = (N - n_o)^{1-a}$. Finally, the system equation solution that results is

$$n(t) = \begin{cases} N\left[(N - n_0)^{1-a} - (1-a)Dt\right]^{1/(1-a)} & \text{for } a \neq 1 \\ N[1 - \exp(-Dt)] + n_0 \exp(-Dt) & \text{for } a = 1 \end{cases}$$

It is interesting to observe that if $a > 1$, then $n(t) \to N$ as $t \to \infty$. However, if $a < 1$, then $n(t) \to N$ as

$$t \to \frac{(N - n_0)^{1-a}}{(1 - a)D}$$

Furthermore, we find that the rate of growth in awareness (i.e., dn/dt) increases as D and N increase. Thus, agents should promote those art forms and artists that can appeal to the greatest potential audiences (N) and spread the most rapidly (D). This observation is clearly evident in the "pop music" field.

9.7.3 Topical System Problems in the Arts

Economic Support of the Fine Arts. Develop a model that predicts the future economic support available for the fine arts in the United States.

Input:	Economic support through gifts, grants, endowments, royalties, ticketing, taxes, and other financial gain inputs.
Output.	Funding for

(1) literature (books, magazines, journals, etc.)
(2) live theater and performances (plays, musicals, opera, concerts, symphonies, etc.)
(3) motion pictures and television (film, DVD's, video tapes, etc.)
(4) fine arts
(5) recorded music (physical and electronic storage).

Obtain data on the annual total funding for each output and predict separately the economic growth for each input and output. What will be the economic status of each of these various art fields for the year 2025?

Model for Value Measurement of the Arts. Art, being a diverse human concept, has diverse intents and meanings for people. Art might be defined as those processes and products of human activity that can be distinguished by their esthetic function. A broader definition might consider art as all forms of human skill and products without respect to esthetic value. Let us attempt to develop a pragmatic, quantitative model for the worth or perceived worth of art to humanity. Consider as a simple model for the worth of art the following block diagram:

Human arts

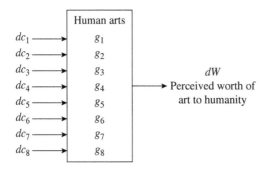

where we have as inputs the following factors:

dc_1 = enhancement of human experience and thought
dc_2 = sublimation, consolation, and escape from reality
de_3 = perception of higher, more ideal, or enhanced reality
dc_4 = source of pleasure and delight
dc_5 = promotion of cultural, humanistic, and historical values
dc_6 = pedagogical and propagandistic tool
dc_7 = purgative or healing factor in human therapy
dc_8 = means for human communication

Obviously, individuals will have strong opinions and biases toward the contribution made by each of these described inputs to the worth of human art. However, the purpose of the model is to attempt to predict and possibly assess the aggregate attitudes toward the worth of art for a distinguishable group, such as a political, ethical, religious, or national population.

If the kernel components for each of the inputs can be quantified, then an analysis and assessment can possibly be made of the ranking of art contributions to a given group and how that group might modify and enhance the system to increase perceived worth from the arts.

As a specific assignment, do the following for this model of art worth:

a) Establish a quantitative basis for measurement of each model inputs, dc_i.
b) Determine a quantitative value (a constant) for each of the model's input kernels, g_i, for a selected group or population.
c) Solve system equation for model and obtain quantitative value for worth of art to given group.
d) What actions can be taken to increase or maximize worth of art to a group?

Photographic Control Model. Photography is a valuable art form that allows the user to preserve a physical record or image of visual events. In practical photography, the important elements that determine the inputs for a satisfactory photographic result are the equipment, the data storage medium, the photographic environment, and of course, the photographer. Determine a quantitative model that provides a satisfactory photograph (e.g., digital image) for the following inputs: p_r (pixel resolution), s shutter speed (in fractions of a second), f stop setting (standard units from 2.5 to 32), and d distance to image (units of meters from 0.1 to infinity).

Distribution Model for Literature. Establish a model that assesses the existing volume (in words) for each of the major literature types. The types for consideration are rhetoric, prosody, drama, lyric, satire, prose, and narrative. Attempt to extend the model to examine the rate (e.g., units of millions of words or megabytes of computer memory required per year) at which new literature is generated for each type and extrapolate these outputs to the year 2030. What conclusions may be drawn?

Model for Musical Composition. Consider the following major components of musical composition: pitch, timing, timbre, harmony, counterpoint, texture, and orchestration. Carefully define each component and produce an organization diagram indicating the ordering and interaction of these components as they relate to the production of a given musical composition (e.g., a symphonic composition). In particular, attempt to identify the coupling and feedback of each component. Also, attempt to rank the components by importance. Subject a standard classical composition (e.g., Beethoven's Fifth Symphony) to an analysis by your model. Assess the model's credibility and acceptability.

Additional Modeling Problems. For each of the systems given in Table 9-17, define an appropriate system model and the essential elements of the system. Express quantitative variables for each of the model inputs and outputs and develop a simple expression for the model's system equation. Determine analytical or other solution forms for the system equation and interpret and evaluate your results.

Table 9-17   Fine Arts Models

System	Input	Output
(a) Work of art	Esthetic criteria Form and content Artistic technique Meaning and significance	Rational evaluation
(b) Fraudulent art	Expert evaluation Scientific tests	Detection and rejection
(c) Dramatic literature	Classical writers	Tragedy Comedy Farce Melodrama Religious drama
(d) Motion pictures	Directing Acting Script Sound Scenery	Quality of production Financial income
(e) Musical compositions	Composers	Sonatas Symphonies Concerti Fugues Chamber music
(f) Dance forms	Dancers	Ballet Modern dance Primitive Folk Popular
(g) Architecture	Renaissance Baroque Neoclassical Romantic Realistic	Assessment

9.8   Applications of System Science to Technology

It is in the field of technology that system science has been most effectively used and very successful in the modeling, analysis, and synthesis of systems. The significant developments in computer technology, manufacturing, communications, electronics, transportation, and so on are evidence of accurate understanding and advanced knowledge acquired through the recognition and exploitation of general systems methods. Undoubtedly, significant progress will continue in the scientific and technical fields, and the results and products of this progress will continue to have a major influence in human endeavors.

9.8.1 Nuclear Reactor Stability with Xenon-135 Dependence

The study of the stability of a nuclear power reactor is primarily concerned with determining whether the power distribution within the reactor core is sufficiently stable to permit operation within the operational limits allowed by the control system. There are numerous aspects of interest relative to the behavior and stability of the power distribution, but we will restrict our attention here to the influence that the important fission product xenon-135, with its enormous thermal neutron absorption cross-section, has on the time behavior and stability of the nuclear power reactor.

For familiarization with xenon-135 and its potential to induce instability of the neutron flux in a nuclear reactor, consider a reactor model in which the neutron flux's energy and spatial dependence is averaged by integration over the reactor core at each point in time. The behavior of this average flux becomes a function of time only, since the energy and spatial dependence are eliminated by the spatial averaging process. The system equations that describe the time behavior of the averaged neutron flux, the xenon-135 concentration, and the iodine 135 concentration are the following:

$$
\begin{bmatrix} d\varphi \\ dX \\ dI \end{bmatrix} = \begin{bmatrix} v\left(v\Sigma_f - \Sigma_a - \sigma_x X\right)\varphi \\ \gamma_{xo}\Sigma_f\varphi + \lambda_i I - \sigma_x X\,\varphi - \lambda_x X \\ \gamma_{io}\Sigma_f\varphi - \lambda_i I \end{bmatrix} dt
\tag{9.8.1}
$$

where

$\varphi\,(t)$ = spatially averaged neutron flux (neutrons/cm^2-s or nts/cm^2-s)
$X\,(t)$ = spatially averaged xenon-135 concentration (Xe atoms/cm^3)
$I\,(t)$ = spatially averaged iodine-135 concentration (I atoms/cm^3)
v = average neutron speed (cm/s)
v = average total neutron yield per fission event (neutrons/fission)
Σ_f = macroscopic cross section for fission (1/cm)
Σ_a = total microscopic cross section for absorption (1/cm)
σ_x = macroscopic absorption cross section for xenon-135
γ_{xo} = average fission yield of xenon-135
γ_{io} = average fission yield of iodine 135 and tellurium-135
λ_x = radioactive decay constant for xenon-135 decay to cesium-135 (1/s)
λ_i = radioactive decay constant for iodine-135 decay to xenon-135 (1/s)

The xenon-135 concentration in the nuclear reactor model results from the production of the fission products tellurium-135 and iodine-135 and the direct fission yield of xenon-135. Since tellurium-135 decays rapidly and cesium-135 is relatively stable, the radioactive decay chain of chief interest is that of iodine-135 decay to xenon-135. It will be assumed that, since the decay time for tellurium is negligible compared to the 6.7-hour half-life of iodine, the fission yield of tellurium appears as a prompt fission yield of iodine. Also, the absorption of neutrons by iodine-135 can be neglected since its thermal neutron absorption cross section is small, and delayed neutrons are not considered except for their effect on the average neutron lifetime.

The nontrivial steady-state or stationary values (ϕ_0, X_0, I_0) occur for the dependent variables of the system Eq. (9.8.1) when $d\phi = dX = dI = 0$, so that

$$\varphi_0 = \frac{\lambda_x \left(v \Sigma_f - \Sigma_a \right)}{\sigma_x \left(\gamma_0 \Sigma_f + \Sigma_a - v \Sigma_f \right)} = \frac{\lambda_x X_0}{\gamma_0 \Sigma_f - \sigma_x X_0} \tag{9.8.2.1}$$

$$X_0 = \frac{v \Sigma_f - \Sigma_a}{\sigma_x} = \frac{\gamma_0 \Sigma_f \varphi_0}{\sigma_x \varphi_0 + \lambda_x} \tag{9.8.2.2}$$

$$I_0 = \frac{\gamma_{i0} \Sigma_f \lambda_x \left(v \Sigma_f - \Sigma_a \right)}{\lambda_i \sigma_x \left(\gamma_0 \Sigma_f + \Sigma_a - v \Sigma_f \right)} = \frac{\gamma_{i0} \Sigma_f \varphi_0}{\lambda_i} \tag{9.8.2.3}$$

where

$$\gamma_0 = \gamma_{i0} + \gamma_{x0}$$

For convenience, as was shown in Section 8.2, the dependent variables will be redefined as follows so that they are dimensionless and assume their steady value when equal to unity.

$$\varphi(t) = \varphi_0 \, \phi\,(t) \tag{9.8.3.1}$$

$$X(t) \;=\; X_0 x(t) \tag{9.8.3.2}$$

$$I(t) \;=\; I_0 \, i(t) \tag{9.8.3.3}$$

Equations (9.8.1) become, under the transformation of variables given in Eqs. (9.8.3), the following:

$$\begin{bmatrix} d\phi \\ dx \\ di \end{bmatrix} = \begin{bmatrix} \rho_0 (1-x)\,\phi \\ (\lambda_0 + \lambda_x)\,(\gamma_x \phi + \gamma_i i) - (\lambda_0 \phi + \lambda_x)\,x \\ \lambda_i (\phi - i) \end{bmatrix} dt \tag{9.8.4}$$

where

$$\rho_0 = v\,\sigma_x X_0 = \frac{1}{\ell}\left(\frac{v \Sigma_f}{\Sigma_a} - 1 \right) = \frac{1}{\ell}\left[\frac{v}{\gamma_0}\left(1 + \frac{\lambda_x}{\lambda_0} \right) - 1 \right]^{-1}$$

$\ell = 1/v \Sigma_a$ (effective reactor neutron lifetime excluding xenon)

$\lambda_0 = \sigma_x \phi_0$

$\gamma_x = \dfrac{\gamma_{x0}}{\gamma_0}$

$\gamma_i = \dfrac{\gamma_{i0}}{\gamma_0}$

$\gamma_i + \gamma_x = 1$

System Eqs. (9.8.4) are nonlinear and difficult to solve analytically. However, a linearized system equation around steady-state operation can be formulated using perturbation methods (see Section 7.2.2) by letting

$$\phi(t) \;= 1 + \delta\phi(t)$$
$$x\,(t) \;= 1 + \delta x(t) \tag{9.8.5}$$
$$i\,(t) \;= 1 + \delta i(t)$$

Substituting Eqs. (9.8.5) into Eqs. (9.8.4) and neglecting products of perturbations gives

$$
\begin{bmatrix} d\delta\,\phi \\ d\delta x \\ d\delta i \end{bmatrix} = \begin{bmatrix} -\rho_0\delta x \\ (\lambda_0 + \lambda_x)\,(\gamma_x\delta\phi + \gamma_i\delta i - \delta x) - \lambda_0\delta\phi \\ \lambda_i(\delta\phi - \delta i) \end{bmatrix} dt
\tag{9.8.6}
$$

Equations (9.8.6) are now linear and amenable to Laplace transformation techniques, which yield in matrix form for the Laplace transform for the dependent variables

$$
\begin{bmatrix} s & \rho_0 & 0 \\ \gamma_i\lambda_0 - \gamma_x\lambda_x & s + \lambda_0 + \lambda_x & -\gamma_i(\lambda_0 + \lambda_x) \\ -\lambda_i & 0 & s + \lambda_i \end{bmatrix} \begin{bmatrix} \delta\langle\phi(s)\rangle \\ \delta\langle x(s)\rangle \\ \delta\langle i(s)\rangle \end{bmatrix} = 0
\tag{9.8.7}
$$

where initial values are assumed to be, at zero time, the following:

$$
\delta\phi(0) = \delta x(0) = \delta i(0)
\tag{9.8.8}
$$

Now matrix Eq. (9.8.8) possesses a nontrivial solution for the dependent variables if and only if the coefficient matrix is singular, and singularity is satisfied only if the determinant of the coefficient matrix vanishes. Expanding the coefficient determinant of Eq. (9.8.8) gives

$$
P(s) = s^3 + A_1s^2 + A_2s + A_3 = 0
\tag{9.8.9}
$$

where

$$
A_1 = \lambda_0 + \lambda_x + \lambda_i
$$

$$
A_2 = \lambda_i\,(\lambda_0 + \lambda_x) - \rho_0(\gamma_i\,\lambda_0 - \gamma_x\,\lambda_x)
$$

$$
A_3 = \rho_0\,\lambda_i\lambda_x
$$

The three roots of the characteristic Eq. (9.8.9), which are generally complex, determine the nature of the time behavior of the perturbed flux, xenon, and iodine. For these solutions to be bounded for all positive values of time, it is necessary that the real part of each of the roots of Eq. (9.8.9) be negative. This condition may be evaluated without explicitly finding the roots by employing the Routh-Hurwitz stability criterion, which states that for any polynomial equation of degree n

$$
P(z) = a_0z^n + a_1z^{n-1} + a_2z^{n-2} + \cdots + a_n = 0
$$

the real part of each root is negative if and only if each coefficient a_i is of the same sign and each determinant D_i is positive, where

$$
D_0 = a_0 > 0
$$

$$
D_1 = a_1 > 0
$$

$$
D_2 = \begin{bmatrix} a_0 & a_1 \\ a_3 & a_2 \end{bmatrix} > 0
$$

$$
D_n = \begin{bmatrix} a_1 & a_0 & \cdots & 0 \\ a_3 & a_2 & \cdots & 0 \\ \vdots & & & \\ a_{2n-1} & a_{2n-2} & \cdots & a_n \end{bmatrix} > 0
$$

Thus, for the bounding of the time solutions of Eqs. (9.8.6), it is necessary that the following four conditions be satisfied:

Condition I : $A_1 > 0$,   so $\lambda_0 + \lambda_x + \lambda_i > 0$

But this is always satisfied for positive or zero (physically real) neutron fluxes.

Condition II : $A_2 > 0$,   so $\lambda_i + \rho_0 \dfrac{\lambda_x}{\lambda_0 + \lambda_x} > \gamma_i \rho_0$

Condition III : $A_3 > 0$,   so $\rho_0 \lambda_i \lambda_i > 0$

But Condition III is always satisfied for $\rho_0 > 0$ (i.e., for $\varphi_0 > 0$).

Condition IV : $A_1 A_2 > A_3$

so we have that

$$
\lambda_i + \rho_0 \frac{\lambda_x}{\lambda_0 + \lambda_x + \lambda_i} > \gamma_i \rho_0
$$

Now the only Conditions that are not trivially satisfied to assure positive parameters are Conditions II and IV. However, it is obvious that if Condition IV is satisfied, then so is Condition II. Thus, the requirement for bounding of the solutions of Eqs. (9.8.6) is only Condition IV. Significantly, ρ_0 is the only parameter, in general, that can be controlled in this reactor model. The decay constants λ_x and λ_i, and the fission yields for a given fuel γ_i, γ_x are natural, physical constants and are not subject to variation or control. Thus, the value of ρ_0, which provides stability in the perturbation sense, is the following

$$
\rho_0 < \frac{\lambda_i}{\gamma_i - \left(\dfrac{\lambda_i}{\lambda_0 + \lambda_x + \lambda_i}\right)} \tag{9.8.10}
$$

A graph for inequality (9.8.10) is given in Figure 9-20 for a set of typical reactor parameters. However, the stability requirement (9.8.10) is extremely restrictive and generally not satisfactory for practical reactor operations. Thus, we may conclude that the model under investigation is inherently unstable in the small perturbation sense, and external control is required to maintain stability. The fact that the fundamental mode of a high flux thermal reactor can be unstable due to the influence of xenon and that external control is required is not surprising, however. The control systems of most reactors are specifically designed to control the fundamental mode of the flux, which is essentially a measure of the power output of the reactor.

Another approximation technique, which is satisfactory at high flux levels ($\phi_0 > 10^{13}$ nts /cm^2-s), is to neglect the decay of xenon compared to xenon burnout (i.e., assume $\lambda_x X \ll \sigma_x X \Phi$). Then, effectively setting $\lambda_x = 0$ and Eqs. (9.8.4) become

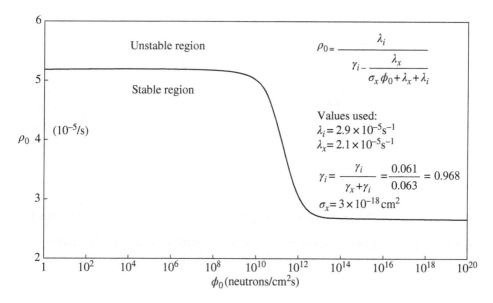

Fig. 9-20 Linear stability criterion for space-independent nuclear reactor kinetics.

$$\begin{bmatrix} d\phi \\ dx \\ di \end{bmatrix} = \begin{bmatrix} \rho_0(1-x) \\ \lambda_0(\gamma_x \phi + \gamma_i\, i - x\, \phi) \\ \lambda_i\, (\phi - i) \end{bmatrix} dt \qquad (9.8.11)$$

where nontrivial stationary values are [see Eqs. (9.8.2)]

$$\phi_0 = \text{arbitrary but require} > 10^{13}\, \text{nts/cm}^2 - \text{s}$$

$$X_0 = \frac{\gamma_0\, \Sigma_f}{\sigma_x}$$

$$I_0 = \frac{\gamma_{i0}\, \Sigma_f\phi_0}{\lambda_i}$$

and

$$\rho_0 = v\, \sigma_x\, X_0 = v\gamma_0\Sigma_f = \frac{1}{\ell}\frac{\gamma_0\Sigma_f}{\Sigma_a} = \frac{1}{\ell}\frac{\gamma_0}{(v-\gamma_0)}$$

To provide a basis for later comparison, first-order perturbation analysis of Eqs. (9.8.11) leads to the following matrix equation for the Laplace transformed dependent variables.

$$\begin{bmatrix} s & \rho_0 & 0 \\ \gamma_{i0}\,\lambda_0 & s+\lambda_0 & -\gamma_i\,\lambda_i \\ -\lambda_i & 0 & s+\lambda_i \end{bmatrix} \begin{bmatrix} \delta\langle\phi\,(s)\rangle \\ \delta\langle x\,(s)\rangle \\ \delta\langle i\,(s)\rangle \end{bmatrix} = 0$$

Now, as before for nontrivial solutions, the determinant of the coefficient matrix must vanish, so

$$s\left[s^2 + (\lambda_0 + \lambda_i)\,s + \lambda_0\,\lambda_i - \gamma_i\,\rho_0\lambda_0\right] = 0$$

and the roots of this polynomial are

$$s_1 = 0$$

$$s_2 = -\frac{\lambda_0 + \lambda_i}{2} + \sqrt{\left(\frac{\lambda_0 - \lambda_i}{2}\right)^2 + \gamma_i\,\rho_0\lambda_0}$$

$$s_3 = -\frac{\lambda_0 + \lambda_i}{2} - \sqrt{\left(\frac{\lambda_0 - \lambda_i}{2}\right)^2 + \gamma_i\,\rho_0\lambda_0}$$

The general solution for $\delta\phi(t)$ in time becomes

$$\delta\phi\,(t) = C_1 + C_2 \exp\,(s_2 t) + C_3 \exp\,(s_3 t) \tag{9.8.12}$$

where the C's are arbitrary integration constants. The perturbed flux in this equation will become unbounded as $t \rightarrow \infty$ unless $C_2 = 0$ or $s_2 < 0$. It is apparent that $s_3 < 0$. Now $s_2 < 0$ if the following is satisfied:

$$\lambda_i > \gamma_i \rho_0 = \frac{1}{\ell}\,\frac{\gamma_0 \Sigma_f}{\Sigma_a} = \frac{1}{\ell}\,\frac{\gamma_0}{(\nu - \gamma_0)}$$

This requirement for stability is the same as that given in expression (9.8.10) if λ_x is set to zero.

A singular solution for $\delta\phi\,(t)$ for the initial conditions that

$$\delta\phi\,(0) = \delta\phi_0$$
$$\delta x\,(0) = \delta x_0$$
$$\delta i(0) = \delta i_0$$

is the following:

$$\begin{aligned}
\delta\phi\,(t) = {} & \frac{1}{s_2\,s_3}[\lambda_0\,\lambda_i\,\delta\phi_0 - \gamma_i\,\rho_0\,\lambda_0\,\delta x_0] \\[4pt]
& + \frac{e^{s_2 t}}{s_2(s_2 - s_3)}[(s_2 + \lambda_0)\,(s_2 + \lambda_i)\,\delta\phi_0 - \gamma_i\,\rho_0\lambda_0\delta i_0 - (s_2 + \lambda_i)\rho_0\delta x_0] \\[4pt]
& + \frac{e^{s_3 t}}{s_3(s_3 - s_2)}[(s_3 + \lambda_0)\,(s_3 + \lambda_i)\,\delta\phi_0 - \gamma_i\,\rho_0\lambda_0\delta i_0 - (s_3 + \lambda_i)\,\rho_0\delta x_0]
\end{aligned}$$

For the condition that

$$\lambda_i = \gamma_i\,\rho_0$$

the roots of the characteristic equation become

$$s_1 = 0(\text{double root})$$
$$s_2 = -(\lambda_0 + \lambda_i)$$

and the general solution for $\delta\phi$ is then

$$\delta\varphi\,(t) = D_1 + D_2\,t + D_3 \exp\,[-(\lambda_0 + \lambda_i)\,t]$$

which is obviously unbounded as $t \to \infty$ unless $D_2 = 0$.

Returning to Eqs. (9.8.11), if the first equation is multiplied by γ_i and then all of Eqs. (9.8.11) are added together, forming the equality

$$\frac{1}{\rho_0} d\phi - (1 - x) \phi \, dt$$
$$= \frac{1}{\lambda_0} dx - (\gamma_x \phi + \gamma_i i - x \phi) dt + \frac{\gamma_i}{\lambda_i} di - (\gamma_i \phi - \gamma_i i) \, dt \tag{9.8.13}$$

and making the obvious cancellations, Eq. (9.8.13) can be integrated to give

$$\frac{1}{\rho_0} (\phi - 1) = \frac{1}{\lambda_0} (x - 1) + \frac{\gamma_i}{\lambda_i} (i - 1) \tag{9.8.14}$$

where initial conditions are assumed to be the following at zero time:

$$\phi(0) = x(0) = i(0) = 1 \tag{9.8.15}$$

Now for convenience, the dependent variables will be defined to reflect their departure from stationary values as follows (this variable transformation is not to be misconstrued with the assumptions of perturbation theory employed earlier, in which the perturbations around stationary values were assumed to be small and products of perturbations were neglected):

$$\phi(t) = 1 + \delta\phi(t) \tag{9.8.16.1}$$
$$x(t) = 1 + \delta x(t) \tag{9.8.16.2}$$
$$i(t) = 1 + \delta i(t) \tag{9.8.16.3}$$

Thus, when $\delta\phi = \delta x = \delta i = 0$, stationary or steady-state conditions exist. Equation (9.8.14) becomes under the transformation of variables in Eqs. (9.8.15) as follows:

$$\frac{1}{\rho_0} \delta\phi(t) = \frac{1}{\lambda_0} \delta x(t) + \frac{\gamma_i}{\lambda_i} \delta i(t) \tag{9.8.17}$$

Now Eq. (9.8.17) plots as a flat plane in three-dimensional space as shown in Figure 9-21. The time solution of Eq. (9.8.17) with initial conditions given by Eqs. (9.8.15) must lie on the plane in Figure 9-21. Any solutions of Eq. (9.8.13) with arbitrary initial conditions will lie on planes parallel to the plane in Figure 9-21.

If the transformations of variables found in Eqs. (9.8.16) are imposed on Eq. (9.8.11) together with Eq. (9.8.17) to eliminate the perturbed iodine concentration, we obtain the following set of coupled equations:

$$\begin{bmatrix} d\delta\phi \\ d\delta x \end{bmatrix} = \begin{bmatrix} -\rho_0 \delta x (1 + \delta\phi) \\ \lambda_i \lambda_0 \left\{ \left(\frac{1}{\rho_0} - \frac{\gamma_i}{\lambda_i} \right) \delta\phi - \left(\frac{1}{\lambda_0} + \frac{1}{\lambda_i} \right) \delta x - \frac{1}{\lambda_i} \delta x \delta\phi \right\} \end{bmatrix} dt \tag{9.8.18.1}$$

Eliminating dt and solving for the perturbed flux, we find

$$d\delta\phi = -\frac{\rho_0}{\lambda_i \lambda_0} \frac{\delta x (1 + \delta\phi)}{\left(\frac{1}{\rho_0} - \frac{\gamma_i}{\lambda_i} \right) \delta\phi - \left(\frac{1}{\lambda_0} + \frac{1}{\lambda_i} \right) \delta x - \frac{1}{\lambda_i} \delta x \delta\phi} \delta x \tag{9.8.18.2}$$

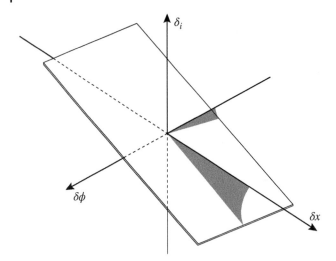

Fig. 9-21 Phase space of space independent kinetic equations at high flux.

If we eliminate the perturbed xenon concentration from the system equations, we have

$$\begin{bmatrix} d\delta\phi \\ d\delta i \end{bmatrix} = \begin{bmatrix} -\rho_0\lambda_0\left(\frac{1}{\rho_0}\delta\phi - \frac{\gamma_i}{\lambda_i}\delta i\right)(1+\delta\phi) \\ \lambda_i(\delta\phi - \delta i) \end{bmatrix} dt \tag{9.8.19.1}$$

Again eliminating dt and solving for the perturbed flux we have

$$d\delta\phi = -\frac{\rho_0\lambda_0}{\lambda_i} \frac{(1+\delta\phi)\left(\frac{1}{\rho_0}\delta\phi - \frac{\gamma_i}{\lambda_i}\delta i\right)}{(\delta\phi - \delta i)} d\delta i \tag{9.8.19.2}$$

Finally, eliminating the perturbed flux we have for the system equation

$$\begin{bmatrix} d\delta x \\ d\delta i \end{bmatrix} = \begin{bmatrix} \lambda_0\rho_0\left(\frac{\gamma_i}{\rho_0} - \frac{\gamma_i^2}{\lambda_i}\right)\delta i - \delta x\left(\frac{\gamma_i}{\lambda_0} + \frac{1}{\rho_0} + \frac{1}{\lambda_0}\delta x + \frac{\gamma_i}{\lambda_i}\delta i\right) \\ \lambda_i\rho_0\left\{\frac{1}{\lambda_0}\delta x + \left(\frac{\gamma_i}{\lambda_i} - \frac{1}{\rho_0}\right)\delta i\right\} \end{bmatrix} dt \tag{9.8.20.1}$$

And solving for the xenon concentration we find that

$$d\delta x = \frac{\lambda_0}{\lambda_i}\left\{\frac{\left(\frac{\gamma_i}{\rho_0} - \frac{\gamma_i^2}{\lambda_i}\right)\delta i - \delta x\left(\frac{\gamma_i}{\lambda_0} + \frac{1}{\rho_0} + \frac{1}{\lambda_0}\delta x + \frac{\gamma_i}{\lambda_i}\delta i\right)}{\frac{1}{\lambda_0}\delta x + \left(\frac{\gamma_i}{\lambda_i} - \frac{1}{\rho_0}\right)\delta i}\right\} d\delta i \tag{9.8.20.2}$$

Generally, Eqs. (9.8.18), (9.8.19), and (9.8.20) each constitute a dynamical system with two degrees of freedom. [Note that Eq. (9.8.17) determines a given dependent variable as a function of the other two dependent variables and thus serves as a constraint imposed on the system.

Furthermore, since equations are autonomous (not explicit functions of time), Eqs. (9.8.18.2), (9.8.19.2), and (9.8.20.2) are viable. Thus, for example, Eq. (9.8.18.2) may be considered as coordinates of a space of two dimensions called a phase space. To each state of the system, there corresponds a point with coordinates in the space, and as time varies, the point produces a path, which describes the time history of the system. The totality of all such paths may be described as the phase portrait of the system. It represents all the possible time histories of the system, any one of which is determined by a single state. Geometrically, this merely means that through every ordinary point of the phase plane passes one, and only one, integral curve defined by the equation.

A special situation arises for a point such that both the numerator and denominator of the phase plane derivatives $(d\delta\phi/d\delta i \to 0/0)$ vanish. Such a point is called a singular point of the system, and it is evident that a singular point is a point of equilibrium.

Typical phase plane solutions of Eqs. (9.8.18) and (9.8.19) are presented in Figures 9-22 through 9-25, with the stability and behavior of the reactor apparent from each figure.

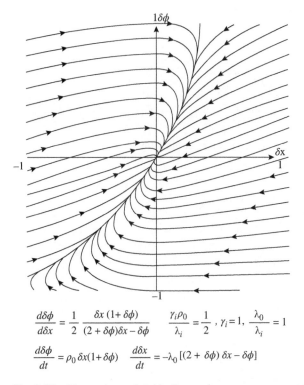

$$\frac{d\delta\phi}{d\delta x} = \frac{1}{2}\frac{\delta x\,(1+\delta\phi)}{(2+\delta\phi)\delta x - \delta\phi} \qquad \frac{\gamma_i\rho_0}{\lambda_i} = \frac{1}{2}, \; \gamma_i = 1, \; \frac{\lambda_0}{\lambda_i} = 1$$

$$\frac{d\delta\phi}{dt} = \rho_0\,\delta x(1+\delta\phi) \qquad \frac{d\delta x}{dt} = -\lambda_0\,[(2+\delta\phi)\,\delta x - \delta\phi]$$

Fig. 9-22 Phase plane of stable flux and xenon response at high flux for $\gamma_i\rho_0 = \lambda_i/2$.

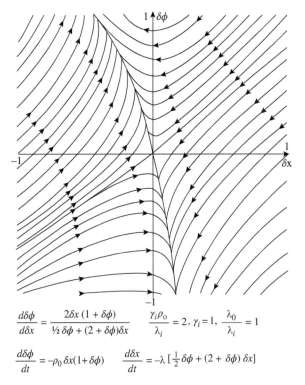

$$\frac{d\delta\phi}{d\delta x} = \frac{2\delta x\,(1 + \delta\phi)}{\tfrac{1}{2}\,\delta\phi + (2 + \delta\phi)\delta x} \qquad \frac{\gamma_i \rho_0}{\lambda_i} = 2,\ \gamma_i = 1,\ \frac{\lambda_0}{\lambda_i} = 1$$

$$\frac{d\delta\phi}{dt} = -\rho_0\,\delta x(1 + \delta\phi) \qquad \frac{d\delta x}{dt} = -\lambda\,[\tfrac{1}{2}\,\delta\phi + (2 + \delta\phi)\,\delta x]$$

Fig. 9-23 Phase plane of unstable flux and xenon response at high flux for $\gamma_i \rho_0 = 2\,\lambda_i$.

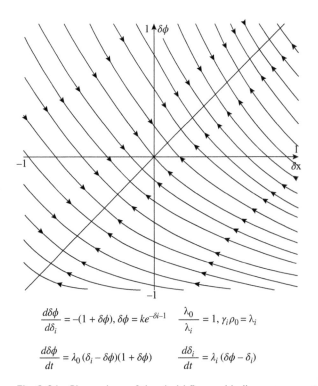

$$\frac{d\delta\phi}{d\delta_i} = -(1 + \delta\phi),\ \delta\phi = ke^{-\delta i-1} \qquad \frac{\lambda_0}{\lambda_i} = 1,\ \gamma_i \rho_0 = \lambda_i$$

$$\frac{d\delta\phi}{dt} = \lambda_0\,(\delta_i - \delta\phi)(1 + \delta\phi) \qquad \frac{d\delta_i}{dt} = \lambda_i\,(\delta\phi - \delta_i)$$

Fig. 9-24 Phase plane of threshold flux and iodine response at high flux for $\gamma_i \rho_0 = \lambda_i$.

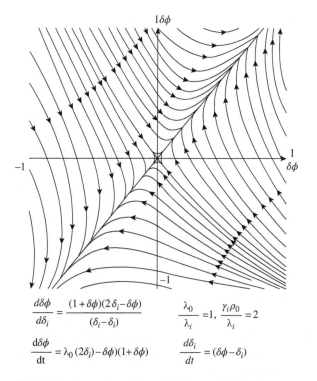

$$\frac{d\delta\phi}{d\delta_i} = \frac{(1+\delta\phi)(2\delta_i - \delta\phi)}{(\delta_i - \delta_i)} \qquad \frac{\lambda_0}{\lambda_i} = 1, \ \frac{\gamma_i p_0}{\lambda_i} = 2$$

$$\frac{d\delta\phi}{dt} = \lambda_0 (2\delta_i) - \delta\phi)(1+\delta\phi) \qquad \frac{d\delta_i}{dt} = (\delta\phi - \delta_i)$$

Fig. 9-25 Phase plane of unstable flux and iodine response at high flux for $\gamma_i p_0 = 2\ \lambda_i$.

9.8.2 Fluid Flow with Friction

The usual system equation employed to model the differential pressure drop for a fluid of density ρ flowing at a speed V along a cylindrical pipe of diameter D and differential length dL is given by the Darcy equation. Observe the negative sign in the Darcy equation, which implies that $dP < 0$ (i.e., the pressure decreases) for $dL > 0$ (i.e., the flow length or travel increases).

$$dP = 4\,\frac{f}{D}\left(\frac{\rho\,V^2}{2}\right) dL \tag{9.8.21}$$

The friction factor f is assumed to be related to the Reynolds number (a dimensionless grouping of fluid flow parameters and variables) as follows:

$$f = K\,Re^n = K\left(\frac{VD\rho}{\mu}\right)^n \tag{9.8.22}$$

The parameter μ is defined as the viscosity of the fluid. We identify Eq. (9.8.21) as a single-input, single-output system with a block diagram representation as follows:

$$dL \longrightarrow \boxed{-\frac{4f}{D}\left(\frac{\rho V^2}{2}\right)} \longrightarrow dP$$

Let us assume that our fluid is incompressible (so that ρ is constant) and flows at constant temperature (so μ is a constant for isothermal flow). We will also assume that we have no loss of fluid mass during flow through the pipe, so the constant mass flow rate m is given by

$$4\,m = \pi\,D^2 V = \pi\,D_0^2 V_0 \tag{9.8.23}$$

where D_0 is a reference pipe diameter with a fluid flow speed V. Solving Eq. (9.8.23) for the speed V and substituting this expression into Eqs. (9.8.21) and (9.8.22), it is possible to express Eq. (9.8.21) in the following form:

$$dP = -2\,K\left(\frac{D_0^2\rho\,V_0}{\mu}\right)^n \rho\,V_0^2\,\frac{D_0^4}{D^{n+5}}\,dL \tag{9.8.24}$$

To solve this system equation, we need only express the pipe diameter D as a function of the fluid pressure P and position L. Let us assume as a specific example that we have a pipe whose diameter changes linearly with length so that

$$D = D_0\,(L/L_0) \tag{9.8.25}$$

Then substituting Eq. (9.8.25) into (9.8.24), we form a separable system equation as follows:

$$dP = -2\,K\left(\frac{V\,D_0^2\rho}{\mu}\right)^n \frac{\rho\,V_0^2}{D_0}\left(\frac{L_0}{L}\right)^{n+5}\,dL$$

This equation can be immediately integrated to give, for the pressure P as a function of position L, the following:

$$P = P_0 - \frac{2K}{n+4}\,Re_0^n\left(\frac{\rho\,V_0^2 L_0}{D_0}\right)\left[1-\left(\frac{L_0}{L}\right)^{n+4}\right] \tag{9.8.26}$$

where we have defined the pressure as P_0 at $L = L_0$ and the Reynolds number Re_0 at position L_0 as

$$Re_0 = \frac{V_0\,D_0\,\rho}{\mu}$$

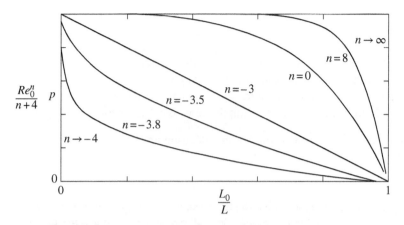

Fig. 9-26 Fluid flow model with friction.

We can express Eq. (9.8.26) in the following dimensionless form for the pressure drop $(p = P/P_0)$:

$$p = \frac{P_0 - P}{2K} \frac{D_0}{\rho V_0^0 L_0} = \frac{1}{n+4} Re_0^n \left[1 - (L_0/L)^{n+4}\right] \tag{9.8.27}$$

and we obtain Figure 9-26 as a graph of Eq. (9.8.27) for various values of n.

9.8.3 Models for Forecasting Electrical Power Demand

Most models for forecasting electrical power demand are usually based on one of the following:

- Estimate of the physical inventory of electrical energy-consuming appliances or equipment that exists and will exist in the future.
- Economic motivations held, and decisions made by the electrical power consumer that will determine future power demands.

The forecasting model based on the anticipated inventory of electrical energy using equipment assumes that a quantitative estimate of the market penetration and acceptance of each class of equipment can be made. The total power consumption for the model is then simply the sum of power consumption for each appliance or equipment class.

The block diagram for our continuous inventory model for electrical power demand for a specific appliance or equipment class is as follows:

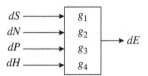

where

dE = change in electrical consumption for appliance/equipment (kWh)
dS = change in the number of appliances per customer (A/C)
dN = change in the number of customers using appliances (C)
dP = change in average power consumed per appliance (kW/A)
dH = change in average time in hours appliance is used (hr. or h)

Note: A is number of appliances, C is the number of customers
And we have for the system equation that

$$dE = g_1 \, dS + g_2 \, dN + g_3 \, dP + g_4 \, dH$$

With model inputs and outputs identified, the next task is to quantify the system kernel. Recalling the mathematical significance for each kernel component, we have for an exact system equation that

$$g_1 = \frac{\partial E}{\partial S}, \quad g_2 = \frac{\partial E}{\partial N}, \quad g_3 = \frac{\partial E}{\partial P}, \quad g_4 = \frac{\partial E}{\partial H}$$

Recalling the development in Section 6.2, we observe that each kernel represents the net output change in E (kW) per unit change in that specific input. As a tentative model, let us assume that

$$g_1 = a\frac{E}{S}, \quad g_2 = b\frac{E}{N}, \quad g_3 = c\frac{E}{P}, \quad g_4 = e\frac{E}{H}$$

which reflects a reasonable power growth model if a, b, c, and e are the change exponents, each assumed to be constant. Then our system equation becomes

$$dE = aE \, \frac{dS}{S} + bE \, \frac{dN}{N} + cE \, \frac{dP}{P} + eE \, \frac{dH}{H}$$

This system equation is separable and can be made exact (as planned) by multiplying by the integrating factor for the system equation, which is $1/E$. So we have upon integration that

$$\int \frac{dE}{E} = a \int \frac{dS}{S} + b \int \frac{dN}{N} + c \int \frac{dP}{P} + e \int \frac{dH}{H} + \text{Ln } K$$

and completing the integration, we have that

$$E = K \, S^a N^b P^c H^e$$

The integration constant K may be evaluated by imposing initial conditions for the system as follows:

$$E_0 = K \, S_0^a N_0^b P_0^c H_0^e$$

So we have as the system solution for the inventory model for electrical power demand for a specific appliance the following.

$$E = E_0 \left(\frac{S}{S_0}\right)^a \left(\frac{N}{N_0}\right)^b \left(\frac{P}{P_0}\right)^c \left(\frac{H}{H_0}\right)^e$$

Assuming that a similar equation applies for each appliance or equipment class, we have that the total electrical demand as predicted by the equipment inventory model is given by

$$E_t = \sum_i E_{0\,i} \left(\frac{S_i}{S_{0i}}\right)^{a_i} \left(\frac{N_i}{N_{0i}}\right)^{b_i} \left(\frac{P_i}{P_{0i}}\right)^{c_i} \left(\frac{H_i}{H_{0i}}\right)^{e_i}$$

where the sum is performed for all classes (i) of electrical appliances.

9.8.4 Topical System Problems in Technology

Overall Availability Model. A complex, interactive system is composed of N essential components, each of which is essential to satisfactory overall operation of the device. Establish a model for the overall availability of the device (i.e., the fraction of the time that the device might be expected to be operational on a continuous duty cycle). Available data on each component are the mean time between failures T_f and the mean time T_r to repair the component.

Detection Model. The probability of detection or discovery of a previously unknown technical innovation is to be modeled. Assume that P_d is the probability that a particular technical development will be made. Let n be the number of independent discovery attempts, each of equal probability p per attempt, for discovering the development.

a) Write the general form of the system equation that expresses the change in the discovery probability dP_d (output) with a change in the number of discovery attempts dn (input).
b) Determine two different quantitative kernels for this system model and solve the system equation.
c) Consider the following system equation for the model:

$$dP_d = p\left(1 - P_d\right) dn$$

Show that the resulting system equation is

$$P_d = 1 - \exp\left(-np\right)$$

d) Assume that the cost of performing each independent discovery attempt is the same, so the total cost for n attempts is proportional to n. Furthermore, if the benefits to be derived from the technical discovery are proportional to P_d^2, show that a maximization of D the benefit to cost ratio occurs when $n_0\, p = 1.2564...$, where n_0 is the number of discovery attempts that results in a maximum benefit-to-cost ratio.
e) If in part (d) the benefit is proportional to P_d show that maximization of the benefit-to-cost ratio results in no attempts at discovery.

Model for Urban Automobile Traffic. The near universal distribution of the automobile provides strong evidence of the profound impact the automobile has had on industry, technology, and human society. With the high vehicular density, particularly in crowded urban areas, efforts to improve vehicle flow have resulted in considerable land area in urban districts dedicated to moving vehicles. Figure 9-27 is a sample of data estimated for vehicle flow as a function of travel time for roads with different speed limits and intersection signals (semaphores and stop signs).

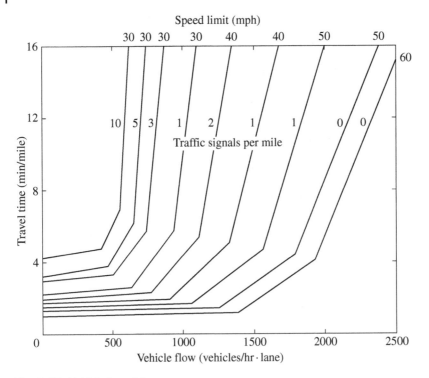

Fig. 9-27 Vehicle flow data.

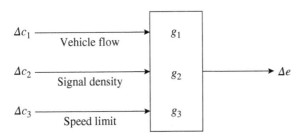

Fig. 9-28 Block diagram for automobile traffic model.

Consider the block diagram in Figure 9-28 for the model where inputs are vehicle flow (vehicle/h – lane), signal density (signal marked intersections/ mile), and local speed limit (mph). The model output is vehicle travel time (miles/minute).

a) On the basis of the graphical data, develop a set of quantitative system kernels for this model that reasonably approximates the data.
b) Attempt to solve the system equations developed in part (a).

Research and Development Task Model. Devise a model that predicts the costs, material, and personnel resources required to accomplish a specified technical task or development. Possible model variables might include personnel, equipment, data resources, financial support, time schedule, and computer facilities. Estimate quantitatively the significance of each model input and output.

Air Traffic Control Model. Develop a model for an air traffic control system associated with a metropolitan airport. Select the significant inputs and outputs for the model as appropriate, which might include aircraft arrivals, departures and docking times, air control space and flight pattern distributions, runway capacity, and meteorological conditions.

Water Reservoir Model. A large water storage reservoir (the system) holds a volume V of water that can vary with time. The seasonal variation of input water (from the surrounding watershed) into the reservoir has a sinusoidal time distribution of the form

$$W_0 \left(1 + \sin 2\pi t/t_0 \right)$$

where t is measured in years. The net loss of water from the reservoir from evaporation is proportional to the surface area of the reservoir, and the surface area is found to be proportional to the reservoir volume. Finally, water is discharged from the reservoir to nearby users at approximately a constant rate. The system output of interest is the time-dependent reservoir volume.

a) Set up the system equation.
b) What is the time-dependent reservoir volume?

Refinery and Petrochemical Model. Establish a model for a single crude oil refinery that can be used to predict petroleum product output capabilities.

Input:	Various crude oil compositions, supply quantities and costs, investment and operating expenses, government regulations.
Output:	Product yields, blending patterns, refinery material balances, production profits.

U.S. National Strategic Petroleum Reserve Model. Formulate a model that forecasts US domestic oil and gas storage for different price-support levels ignoring political factors influencing storage.

Input:	Volume and production capacity of proved reserves, pipeline capacity, receiving refinery characteristics and capacities.
Output:	Oil and gas supply data, minimum acceptable supply prices for oil and gas products.

World Domestic Oil and Gas Supply Model. Develop a model for projecting domestic oil and gas supply based on production, reserves, and exploration activity.

Input:	Data on drilling, reserves, production, and storage capacities.
Output:	Oil and gas production rates.

World Coal Supply Model. Develop a coal supply model that reflects resources, economics, transportation, and environmental constraints.

Input:	Coal deposits, mining capacities, depth and seam size allocation factors, generation plant capacities, demand factors, transmission losses, environmental constraints, transportation rates, extraction and processing costs.
Output:	Coal production and distribution.

National Air Pollution Model. Devise a model that predicts air pollution emissions by type from the major air pollution point sources for a given country.

Input:	Fuel consumption by basic sectors (industrial, commercial, residential, and transportation sectors) for regional sources. Regional economic activity indicators, and environmental standards by year of application, geographical region, and specific pollutant.
Output:	Regional emissions by source type and pollutant type.

Nuclear Fuel Cycle Cost Model. Develop a model that determines the unit price of electricity required to recover expenses incurred by a nuclear power electrical plant for uranium fuel and fuel cycle services.

Input:	Reactor operating parameters such as thermal efficiency, energy extraction from fuel, initial and final fuel assays; fuel cycle parameters such as enrichment plant operations, recycle mode, processing times and efficiencies; and unit prices for fuel cycle services and products.
Output:	Fuel cycle services and products required per unit of generated electricity and nuclear fuel costs per unit of generated electricity.

Radar Power Emanation Model. Consider a model for the power transmission of a radar antenna. The output of the model is the ratio of the power incident on the receiving antenna to the power transmitted (i.e., P/P_0). The inputs to the model are as follows:

A = effective cross section area of receiving antenna (m^2)
R = minimum distance between receiving and transmitting antennas (m)
G = power gain (Watts) of the feeder and radiating antenna (W gain/W)
K = reflection coefficient (m^2)

a) Assume a logarithmic kernel (i.e., $(dy/y) \propto (dx/x)$) and determine a quantitative system equation for the model.
b) Compare your system equation solution with the system equation

$$\frac{P}{P_0} = \frac{AGK}{\left(4\pi R^2\right)^2}$$

Rocket Fuel Consumption Model. Develop a model for the consumption of rocket fuel for spacecraft. Model variables are as follows:

f = fuel consumption in kg (output)
M = initial mass of rocket and fuel in kg (input)
s = rocket flight distance in km (input)

Define the general system equation, develop specific model kernels, and obtain solutions for these system equations.

Ballistic Missile Submarine. A Trident submarine can launch a missile while submerged. The launched missile is the system, and the output of interest is the vertical height or altitude of the missile. Time after launch is the input. Assume that the missile has a constant weight W and the engine produces a constant upward thrust of T. Furthermore, the resistance provided by the water as the missile moves from the submarine to the water surface is proportional to the missile's vertical speed through the water. At the water surface, the missile breaks through into the atmosphere. Neglect air resistance as the missile enters the atmosphere. Using Newton's laws of motion, determine the time-dependent altitude of the missile both in the water and the atmosphere.

Additional Modeling Problems. For each of the systems given in Table 9-18, define an appropriate system model and the essential elements of the system. Express quantitative variables for each of the model inputs and outputs and develop a simple analytical expression for the model's system equation. Determine analytical or other solution forms for the system equation and interpret and evaluate your results.

Table 9-18 Models for Industrial Systems

	System	Input	Output
(a)	Manufacturing	Labor for first unit Time for first unit	Labor for nth unit Time for nth unit
(b)	Influence of Technology	Technology	Economic wealth Education Social institutions
(c)	National energy consumption	Coal Oil Natural gas Uranium Hydroelectric	Transportation Commercial Residential Industrial
(d)	Renewable energy sources	Solar Geothermal Wind Water (hydroelectric dams)	World energy fraction
(e)	Extraction industries	Fossil fuels Salt Stone Metals Sulfur Lithium (electric batteries, etc.) Phosphates	U.S. annual output
(f)	Food industry	Fruits and vegetables Cereals Meats Fish Dairy products	Economic value Human nutrition Production and distribution
(g)	Computer systems	Computer hardware transistorsintegrated circuits LSI VLSI CMOS Quantum computers Bubble memories	Performance speed, storage, cost, size

9.9 Applications of System Science to Religion

Religion with its many facets of beliefs, rituals, sacred writings, and spiritual qualities occupies a prominent role in the thoughts and actions of people. Most moral standards, codes of conduct, ethics, and even many civil laws of local and national governments trace their origins to religious teachings and mores. The impact of religion on humanity may appear subtle and indirect, but through religion's capacity to mold minds and inspire support and allegiance, the true impact of religion is profound and ubiquitous. Much of this influence of religious training is manifest in people through human-based feedback; that is, these religious tenets and practices create a feedback system within the human mind that influences and alters most of the inputs incident on people. Thus, our present actions and thoughts are obviously influenced to a great degree by our past experiences.

So to understand people, we must understand their religions and what these religions represent and teach. The system investigator may balk at modeling religious-based systems, but, in fact, they too are subject to rational study and the principle of causality.

9.9.1 Quality of Life and Belief in God Model

Consider as a system to be modeled a control group of people about whom we seek to estimate and correlate their "quality of life" and "faith in God" as a function of critical events in their lives.

Let us define as continuous input and output variables, together with these variables' respective ranges of values and interpretations, the following:

Q is the quality of life for the control group of persons as measured by an appropriate classification of life qualities. Such qualities might include peace of mind, contentment, morality, happiness, love for others, joy, well-being, and so on. The range of values for Q is as follows:

$$-1 \text{ (lowest quality)} < Q < +1 \text{ (highest quality)}$$

F is the quantitative measure of the control group's belief and faith in a Supreme Being (God, Allah, Elohim, The Eternal, etc.). Evidence of such faith would be measured through such measurable elements as reverence, prayer, piety, holiness, church affiliation, scripture reading, and the like. The range of F is as follows

$$-1 \text{ (disbelief)} < F < +1 \text{ (belief)}$$

T is a quantitative measure (or sum) of significant life events that influence either F or Q or both. Such life events would include illness, employment, marriage, children, death of friends or relatives, etc. The range of T is as follows:

$$0 \text{ (birth)} < T < 1 \text{ (end of life)}$$

We will treat T (*life events*) as the model input and Q (*quality of life*) and F (*faith*) as outputs. Many difficulties and uncertainties associated with our model are probably already apparent but let us withhold criticism of the model and reserve judgment until the results of the model become apparent.

As a graphical representation for this model, consider the following block diagram for which we generalize the model to include full feedback of the outputs.

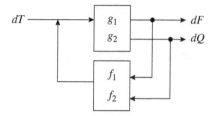

The general system equations for such a single-input, double-output system with feed-back are

$$dF = \frac{g_1}{1 - g_1 f_1 - g_2 f_2} \, dT \tag{9.9.1}$$

$$dQ = \frac{g_2}{1 - g_1 f_1 - g_2 f_2} \, dT \tag{9.9.2}$$

Our justification for including full feedback for the model is based on our anticipation that considerable interaction and feedback among religious faith, quality of life, and life events should exist.

To begin the quantitative phase of modeling the system equation kernels, let us divide Eq. (9.9.1) by Eq. (9.9.2) to eliminate the differential input dT and the denominators of both system equations. Then we have that

$$dF = \frac{g_1}{g_2} \, dQ \tag{9.9.3}$$

The motivation for this step is to eliminate the feedback kernels f_1 and f_2, and if g_1 and g_2 are not explicit functions of T, then it may be possible to more easily model and solve the reduced system equation by assuming that g_1 and g_2 are functions of Q and F only. In particular, for the quantitative kernels considered here, let us assume that both g_1 and g_2 are separable functions of Q and F as follows:

$$g_1 = g_{1f}(F) \, g_{1q}(Q)$$

$$g_2 = g_{2f}(F) \, g_{2q}(Q)$$

With these functions, Eq. (9.9.3) is separable and can be integrated to give

$$\int \frac{g_{2f}(F)}{g_{1f}(F)} \, dF = \int \frac{g_{1q}(Q)}{g_{2q}(Q)} \, dF + K \tag{9.9.4}$$

As a specific model for the kernels g_2 and g_2, let us consider the following linear, separable functions:

$$g_1(F, Q) = (a_1 F + b_1)(c_1 Q + e_1)$$

$$g_2(F, Q) = (a_2 F + b_2)(c_2 Q + e_2)$$

where a's, b's, c's, and e's are arbitrary constants.

Substitution of the particular functions into Eq. (9.9.4) results in the following general system solution:

$$AF + B \operatorname{Ln}(a_1 F + b_1) = CQ + E \operatorname{Ln}[K_0(c_2 Q + e_2)] \tag{9.9.5}$$

where

$$A = \frac{a_2}{a_1}, \qquad C = \frac{c_1}{c_2}$$

$$B = \frac{b_2}{b_1} - \frac{a_2 b_1}{a_1^2}, \qquad E = \frac{e_1}{c_2} - \frac{c_1 e_2}{c_2^2}$$

and K_0 is an arbitrary constant.

Equation (9.9.5) can also be expressed in the alternative form as

$$Q = K(a_1 F + b_1)^{B/E} \exp\left[\frac{1}{F}(AF - CQ)\right] - \frac{e_2}{c_2} \tag{9.9.6}$$

where K is also an arbitrary constant. To normalize the response of the solutions, let us assume that when $Q = 0$ then $F = 0$, so Eq. (9.9.6) becomes

$$Q = \frac{e_2}{c_2}\left\{ \left(1 + \frac{a_1 F}{b_1}\right)^{B/E} \exp\left(\frac{AF - CQ}{E}\right) - 1 \right\} \tag{9.9.7}$$

To provide some insight into the behavior of the solution of the system equation for Q as a function of F as defined by Eq. (9.9.7), let us set $c_1 = C = 0$ and define our system parameters as follows for convenience in graphing:

$$w = \left(\frac{a_2 c_2}{a_1 e_1}\right) F \text{ (faith)}$$

$$x = \left(\frac{c_2}{e_2}\right) Q \text{ (quality of life)}$$

$$y = a_1^2 \frac{e_1}{a_2 c_2 b_1} \text{ (parameter)}$$

$$z = c_2 \frac{(a_1 b_2 - a_2 b_1)}{a_1^2 e_1} \text{ (parameter)}$$

Then Eq. (9.9.7) can be expressed as

$$x = (1 + yw)^z \exp(w) - 1 \tag{9.9.8}$$

A generalized plot of Eq. (9.9.8) is shown in the following figure:

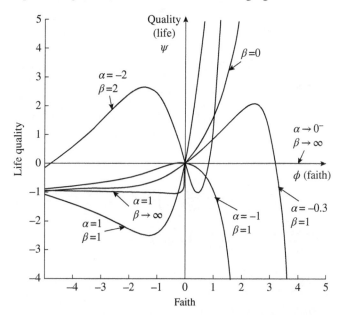

It is apparent from the figure that Eq. (9.9.8) and thus system Eq. (9.9.7) has a broad variety of responses for our model. Suppose the response we believe best fits our anticipated response for the model is that for $y = z = 1$. So in terms of F and Q, we have

$$Q = a_0\left[(1 + F)e^F - 1\right]$$ (9.9.9)

where $a_0 = e_2/c_2$ and $a_1^2 e_1 = a_2 c_2 b_1 = c_2(b_2 a_1 - a_2 b_1)$. To determine what response this equation provides in the presence of feedback, let us express Eq. (9.9.1) as

$$(1/g_1 - f_1 - f_2 g_2/g_1)\, dF = dT$$ (9.9.10)

Now since we can express Q as a known function of F from Eq. (9.9.9), then if the feedback kernels f_1 and f_2 are restricted to be functions of F only, Eq. (9.9.10) is exact and directly integrable. For simplicity, assume that f_1 and f_2 are constants. Then the system equation for F as a function of T as given by Eq. (9.9.10) is found to be

$$\frac{1}{a_1 e_1} \mathrm{Ln}\,(a_1 F + b_1) - f_1 F - \frac{f_2}{e_1}\int X(F)\, dF = T + K$$ (9.9.11)

where

$$X(F) = \left(\frac{a_2 F + b_2}{a_1 F + b_1}\right)[c_2\, a_0(1 + F)\,\exp\,(F) - 1]$$

and K is the integration constant. We observe in Eq. (9.9.11) that the integral of the $X(F)$ will result in exponential integral functions of the first kind, which we prefer to avoid. To avoid these functions, we assume that

$$\frac{(a_2 F + b_2)\,(1 + F)}{a_1 F + b_1} \Rightarrow \frac{a_2}{a_1} F + \frac{b_2}{b_1}$$

and require that

$$a_2 + b_2 = a_1 \frac{b_2}{b_1} + b_1 \frac{a_2}{a_1}$$

Then Eq. (9.9.11) can be expressed as

$$\varphi(F) = T + K \tag{9.9.12}$$

where

$$\varphi(F) = \varphi_1(F) - \varphi_2(F) + \varphi_3(F) + \varphi_4(F)$$

and

$$\varphi_1(F) = \frac{1}{a_1 e_1} \mathrm{Ln}\left(a_1 F + \frac{a_2^2/a_1}{(1 - a_2/a_1)^2}\right)$$

$$\varphi_2(F) = f^2 \frac{a_2 e_1}{a_1 e_2}\left(F + \frac{a_2/a_1}{1 - a_2/a_1}\right) \exp(F)$$

$$\varphi_3(F) = \left(f_2 \frac{a_2 e_2}{a_1 e_1} - f_1\right) F$$

$$\varphi_4(F) = \left(\frac{a_2}{a_1}\right)^2 \frac{e_2}{e_1}\left(\frac{1}{1 - a_2/a_1}\right) \mathrm{Ln}(F + 1)$$

Equation (9.9.12), as was Eq. (9.9.7), also has a broad variety of responses depending on the choices imposed upon the model parameters a_1, a_2, e_1, e_2, f_1, and f_2. Figure 9-29

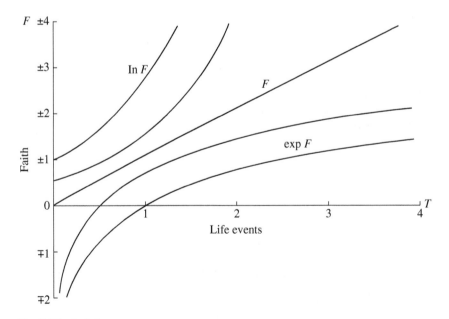

Fig. 9-29 Religion.

indicates several typical generalized responses that allow the model investigator to adapt the model behavior to almost any observation that might be entertained.

9.9.2 Models for the Great Religions

The impact of religion on the thoughts and actions of mankind is profound. Indeed, any attempt to understand the past, present, and future human condition is incomplete without an assessment of the influence that religion, particularly the major faiths exert on the thought and actions of humans. We propose a quantitative model that assesses the effect **e** on mankind arising from the following seven major religions in alphabetical order: Buddhism (B), Christianity (C), Confucianism (Co), Hinduism (H), Islam (I), Judaism (J), and Taoism (T). Suppose then we have as a block diagram with feedback for our continuous system model

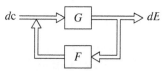

where

$$
dc = \begin{bmatrix} dB & \text{(Buddhism)} \\ dC & \text{(Christianity)} \\ dCo & \text{(Confucianism)} \\ dH & \text{(Hinduism)} \\ dI & \text{(Islam)} \\ dJ & \text{(Judaism)} \\ dT & \text{(Taoism)} \end{bmatrix}, \quad F = \begin{bmatrix} F_b \\ F_c \\ F_{co} \\ F_h \\ F_i \\ F_j \\ F_t \end{bmatrix}
$$

$$
G = \begin{pmatrix} G_b & G_c & G_{co} & G_h & G_i & G_j G_t \end{pmatrix}
$$

Our system equation for this multiple-input system with feedback is given by

$$
de = \frac{G}{1 - GF} dc
$$

$$
de = \frac{G_b\, dB + G_c dC + G_{co} dC_{co} + G_h dH + G_i\, dI + G_j dJ + G_t dT}{1 - G_b F_b - G_c F_c - G_{co} F_{co} - G_h F_h - G_i F_i - G_j F_j - G_t F_t} \tag{9.9.13}
$$

It is undoubtedly true that the output from the model (i.e., the effect on mankind) influences the inputs (i.e., the various religions). We consider this influence through the mechanism of feedback, which is quantitatively modeled by the various F_k functions.

The major challenge imposed by this model is the quantification of the various kernel functions (i.e., F's and G's) for the system. As an initial assessment of the impact of religion on mankind, let us define each G kernel component, G_k, as the anticipated growth (or decline) in adherents that will occur in future for a given religious faith:

$$
G_k\, dk = P_k (1 + I_k)^t dt = P_k \exp \left[t\, \text{Ln} \left(1 + I_k \right) \right] dt \tag{9.9.14}
$$

where

P_k = adherent population for religion k at time t
I_k = mean historical growth rate for religion k
t = time

In general, for small anticipated religious membership growth rates, we have that $I_k << 1$, so we can assume that

$$\text{Ln}\,(1 + I_k) \approx I_k$$

Then Eq. (9.9.14) becomes

$$G_k dk = P_k \exp\,(I_k t)\, dt$$

Furthermore, to make the resulting system equation easily integrable, let us assume for the feedback kernels that

$$F_k = -A/I_k$$

Then system Eq. (9.9.13) provides

$$de = \frac{\sum_{k=1}^{n} P_k \exp\,(I_k t) dt}{1 + A\sum_{k=1}^{n} \frac{P_k}{I_k} \exp\,(I_k t)} = \frac{1}{A} d\left\{ \text{Ln}\left[1 + A \sum_{k=1}^{n} \frac{P_k}{I_k} \exp\,(I_k t)\right] \right\}$$

which can be integrated to give

$$\exp\,(-A\mathbf{e}) = K\left(1 + A\sum_{k} \frac{P_k}{I_k} \exp\,(I_k t)\right) \tag{9.9.15}$$

where K is the integration constant. For $\mathbf{e} = 0$ when $t = 0$, we have for K that

$$K = 1 + A\sum_{k} \frac{P_k}{I_k}$$

We will not pursue this model further, but we observe that as $t \to \infty$ we have from Eq. (9.9.15) that $\exp\,(-A\,\mathbf{e}) \to 0$.

9.9.3 Topical System Problems in Religion

The following models in religion may require some background research by the system investigator, particularly in unfamiliar areas. The standard methods of system science apply, although establishing quantitative measurement bases may provide a challenge. The reader is encouraged to exercise boldness and ingenuity in this relatively "untapped" field for the application of system science.

Ancient Religions. Devise a model that accounts for the significance and contribution made by the various factors that contributed to the existence of ancient religions.

Input:	The sun, moon, fire, climate, diseases, spirits, animals, ancestors, marital status, children, sacrifice, medicines, and drugs.
Output:	The contribution each of these inputs made to ancient religious beliefs and number of followers.

Impact of Religion on a Nation. Develop a model that estimates the impact that religious beliefs and training have on significant factors affecting a given nation.

Input:	Religious beliefs and training.
Output:	Moral conduct, legal system, economics, education, politics, arts, and medicine.

Great Benefactors of Humanity. Devise a model that assigns a quantitative measure of the contribution made by each of the following great benefactors:

- Jesus Christ (Christianity)
- Mohammed (Islam)
- Buddha
- Confucius (The First Teacher)
- Socrates
- Plato
- Aristotle
- Zoroaster
- Lao-tzu

Among the major contributory factors to humanity made by these benefactors consider the following:

- Ethics
- Moral standards
- Bases for judicial systems
- Political systems
- Governments
- Piety
- Social and scientific knowledge

Bases of Religious Functions. Devise a quantitative model that assesses the significance and impact of each of the following aspects of religious practice or thought:

- Religious authority and leadership
- Ritual and sacred rites
- Speculation and prophecy
- Tradition, history, and philosophy
- God's sovereignty, power, and grace
- Mysteries, miracles, the supernatural, and spirituality

Islam. Islam is the religious faith of approximately one-seventh of humans living today. The Islamic creed of the Muslims is erected upon the following five pillars of belief and practice.

1. One God: There is no God but Allah and Mohammed is His Prophet.
2. Prayer: Five times daily: (1) upon rising, (2) at noon, (3) mid-afternoon, (4) after sunset, and (5) before retiring.

3. Charity: Those who are more fortunate than others should ease the burden of those who have less. Those with economic means should distribute annually one-fortieth of the value of all they possess.

4. Ramadan: Able-bodied Muslims should fast from all food and drink each day during the month appointed as Ramadan from daybreak to sunset.

5. Pilgrimage: Once in the lifetime of every Muslim who is physically and economically able, a journey to Mecca is expected.

Using these five pillars as inputs, create a system model for the Muslim faith, Islam. Devise simple quantitative measurement schemes for each input. Select a weighting value for each input and determine an output measure of compliance with these pillars or tests of faith.

Taoism. A basic concept of Taoism (pronounced Dowism) is the relativity of all values and events in life. This idea of "the existence and identity of contraries" is conveyed by the traditional Chinese symbol of "yang and yin" shown here:

The polarity and complete symmetry about any diameter of the outer circle conveys life's basic opposites: good-evil, strong-weak, active-passive, positive-negative, light-dark, summer-winter, male-female, knowledge-ignorance, and so on. Although the principles of yang and yin are those of differences and relative opposition, in fact, the principles are not believed to be exclusive and decoupled but rather complementary and interactive. There is a cross-over, an interaction, and often a strong coupling between opposites. Perhaps this mixing and interchange of quantities presumably independent of one another are somewhat demonstrated in system science through the mechanism of feedback.

Provide several simple models that exhibit feedback and can demonstrate this mixing and altering of inputs and outputs so as to produce markedly different results than anticipated. Discuss the significance of each model and the means or mechanism by which feedback alters the system's anticipated results.

Confucianism. One of the greatest names in Chinese culture is that of Confucius, Kung Fu-Tzu or "The First Teacher." The sayings of Confucius convey great understanding and profound wisdom for the human mind. A particularly profound saying of Confucius that has great significance for human society today and the potential for achievement of peace for mankind is the following simple verse:

- If there be righteousness in the heart, there will be beauty in character.
- If there be beauty in character, there will be harmony in the home.
- If there be harmony in the home, there will be order in the nation.
- If there be order in the nation, there will be peace in the world.

Using a qualitative model approach, develop a graphical (block diagram or signal-flow graph) representation for the sequential inputs and outputs of this verse. Discuss the coupling between cause and effect and the validity of each causal relationship. Assess the model input "righteousness in the heart."

Judaism. It is through the Ten Commandments or Decalogue that Judaism has exerted its greatest impact on humanity. Embracing the Decalogue's impact on Christianity and Islam, the Ten Commandments constitute an essential portion of the moral foundation for half of the world's population today. The Ten Commandments prescribe the minimum standard conduct necessary to achieve social order, and the Decalogue constitutes a model for Hebraic morality and conduct. Vital proscriptions of the Decalogue include the following:

- Thou shalt not kill.
- Thou shalt not commit adultery.
- Thou shalt not steal.
- Thou shalt not bear false witness.
- Thou shalt not covet.

The importance of these commandments resides not in their finality but in their inescapable priority. Serving as the moral foundation for half of mankind, use these commandments to establish a moral system model. Determine a quantitative measure that allows an assessment for compliance with each commandment.

Simple Model for Buddhism. Buddha's approach to religion can be summarized as follows:

- Empiricism: Direct, personal experience is the basic test of truth.
- Scientific: Rational cause and effect sequences are responsible for all existence and events.
- Pragmatism: Good works of mankind are essential to realize worthwhile goals.
- Therapeutic: People are expected to reduce suffering for each other.
- Psychological: Faith should be human centered with the goal of improving mankind.
- Democratic: Religion should be blind to social position and heredity.
- Individualism: The individual is of primary importance.

Devise a model with a quantitative range of measure for each of the preceding inputs for the Buddhist faith.

Karma (Hinduism). The law of karma as recognized and described within Hinduism is the moral law of cause and effect. Therefore, it is interesting and productive to examine this close relationship of the "causality principle," which serves as the basis of system science.

Karma necessitates the application of universal causation to all moral and spiritual issues and consequences. The present condition of each human's inner, spiritual life is the precise consequence of all past thoughts and deeds. The individual human destiny is a direct consequence of the individual's past and future performance, and mankind will reap exactly what it sows' Although karma decrees determinate consequences, the decisions made by individuals that result in these consequences are within the awareness and decision capability of the individual. Each human is a free agent and is therefore wholly responsible for his or her destiny.

The problem assignment here is to carefully analyze and evaluate the basis and implications of karma. What impact do environment, heredity, and random events have on one's destiny? Are individuals fully responsible for their present condition in life? Attempt to diagram a qualitative block diagram for a karma model and select rational inputs and outputs for such a model. Document your assumptions and conclusions.

Model for Christian Ethics. As the World's largest religion, some of the noblest and highest ethical concepts known to mankind are those associated with the gospel or "good news" taught and espoused by Jesus Christ twenty centuries ago. These marvelous concepts of human morality and unselfish love are recognized by Christians as "idealized" codes or ethics for human interactions. Because of the great moral and philosophical worth and impact of Christ's teachings, attempt to quantitatively measure and weigh the value of each of the following ethical inputs:

a) Be a peacemaker
b) Love your enemies
c) Do good to those who hate or despise you
d) Do not judge others
e) Seek righteousness
f) Have a pure heart
g) Do not resist evil directed at you
h) Do your good works secretly
i) Do not seek rewards for your good deeds
j) Be merciful and forgiving of others
k) Do not be overly concerned with your own needs and desires
l) When wronged, forgive the wrongdoer
m) Be meek and humble and serve others without reservation or reward

Establish a range (e.g., 0–10 for human impact) and a measurement basis for each input ethic. Then rank these inputs and assign a numerical weighting for each. Measure your own ethical character and behavior using this Christian model. Is your personal ethical measurement score satisfactory?

Additional Modeling Problems. For each of the systems given in Table 9-19, define an appropriate system model and the essential elements of the system. Express quantitative variables for each of the model inputs and outputs and develop a simple expression for the model's system equation. Determine analytical or other solution forms for the system equation and interpret and evaluate your results.

Table 9-19   Models for Philosophy and Religion

	System	Input	Output
(a)	Religious dogma	Divine love	Religious conviction
		Creation	Unselfish service
		Revelation	Self-sacrifice
		Miracles	Universal love
(b)	Philosophy of religion	Theism	Existence of God
		Atheism	Existence of many Gods
		Agnosticism	Existence of no God
		Pantheism	World distribution
(c)	Religious devotion	Worship	Oblations
		Prayer	Attendance to devotion
		Rituals	Taking the Sacrament
		Symbols	Charitable acts
(d)	Religious practice	Followers	Religious personages:
		Piety	Priests
		Devotion	Prophets
		Virtue	Saints
		Reverence	Monks
			Diviners
			Shamen
			Magicians
			Mystics
(e)	Buddhism	Right view	Aryan eightfold path
		Right aspiration	Actual behavior
		Right speech	
		Right doing	
		Right livelihood	
		Right effort	
		Right mindfulness	
		Right rapture	
(f)	Modern philosophers	Sigmund Freud	Religious views
		Carl Jung	Theist
		William James	Atheist
		Aldus Huxley	Agnostic
		Bertrand Russell	

(*Continued*)

Table 9-19   (Continued)

	System	Input	Output
(g)	Sacred books	Bhagavad-Gita	Religious conviction
		Sutras	Human practice
		Hsin-Hsin Ming	
		Zend-Avesta	
		Koran	
		Talmud	
		Dead Sea Scrolls	
		Holy Bible	
		New Testament	

9.10   Applications of System Science to History

It has been wisely stated that those who are ignorant of human history are destined to repeat it. The common wisdom of this assertion is that we should and must learn from the past, or we will repeat our past mistakes in the future. Important lessons about most human activities can be learned from a careful study (which includes quantitative studies based on system science) of major past events and society's responses to those events. For example, what were the major factors (or inputs) that were causal in bringing about World War I and World War II? How might these factors have been changed, if possible, so as to have avoided or at least minimized the consequences of these wars that eventually engulfed most of the earth's nations and wrought a terrible cost in suffering, loss of life, and property? Are any of these significant inputs (e.g., discontent, revolution, war, terrorism) present today? How should we then assess and deal with these conditions so as to avoid or reduce the risk of crises in the future? Obviously, these are difficult, demanding questions, but they certainly merit our attention.

In this vein, let us briefly examine some historical systems that might be fashioned into useful models of human behavior and motivation and see what we can learn from these models about our society and ourselves.

9.10.1   Expansion Model for Aggressive Societies

History is replete with the occurrence of dominant societal groups who because of their peculiar abilities and resources are able to conquer, subjugate, control, and expand their own society despite a hostile environment or competing societies. We want to model and analyze such a society as a function of time t (yr) and land control r (hectare). Let the quantitative model for the growth of an aggressive, expanding society be described by the following first-order partial differential equation:

$$\frac{\partial n}{\partial t} = \frac{k}{2\pi r}\frac{\partial n}{\partial r} + \lambda n$$

where

n = population density at radial position r and time t (persons/hectare)
k = effective areal expansion constant for population (hectare^2/yr)
λ = effective growth rate in time for the population (1/yr)

Employing the Method of Lagrange for solving first-order partial differential equations outlined in Section 5.2.9, we observe that to solve our partial differential equation we must solve the following coupled set of ordinary first-order differential equations:

$$dt = -2\pi \frac{r}{k} dr = \frac{1}{\lambda n} dn$$

We obtain as coupled solutions that

$$t = -\frac{\pi r^2}{k} + c_1$$

and

$$\lambda t = \text{Ln} n - c_2$$

where c_1 and c_2 are independent integration constants. The general solution for our system model equation then has the form $c_2 = F(c_1)$ or, equivalently,

$$\text{Ln } n = \lambda t + F\left(t + \pi r^2/k\right) \tag{9.10.1}$$

where F is an arbitrary function. Now, if we take antilogarithms of Eq. (9.10.1), then

$$n(r,t) = G\left(t + \pi r^2/k\right) \exp(\lambda t) \tag{9.10.2}$$

where G is our new arbitrary function of both time and position. Suppose for our model that the population $n(0, t)$ is assumed to have the following time behavior:

$$n(0,t) = n_0 \left[a - \exp(-bt)\right]$$

where $a \geq 1$ and $b \geq 0$. Then we have from Eq. (9.10.2) that

$$G(t) = n_0 \left[a - \exp(-bt)\right] \exp(-\lambda t)$$

and replacing t by $(t + \pi r^2/k)$, then

$$G(t + \pi r^2/k) = n_0 \exp\left(-\lambda\left(t + \pi r^2/k\right)\right)\left(a - \exp\left[-b\left(t + \pi r^2/k\right)\right]\right) \tag{9.10.3}$$

So we have for our system model solution from Eqs. (9.10.2) and (9.10.3) that

$$n(r,t) = n_0 \exp\left(-\lambda\left(t + \pi \frac{r^2}{k}\right)\right)\left(a - \exp\left[-b\left(t + \pi \frac{r^2}{k}\right)\right]\right) \tag{9.10.4}$$

To minimize the number of essential system parameters and normalize variables, let

$$x = \left(\lambda \frac{\pi}{k}\right)^{1/2} r, \quad y = \frac{1}{a} \exp(-bt)$$

and

$$N(x,y) = \frac{n(r,t)}{n_0\, a}$$

Then Eq. (9.10.4) can be expressed as a single parameter equation for the input x and output y:

$$N(x,y) = \exp\left(-x^2\right)\left[1 - y\, \exp\left(-cx^2\right)\right] \tag{9.10.5}$$

where

$$c = b/\lambda$$

Figure 9-30 provides a graphical depiction of Eq. (9.10.5) for selected parameter values. The total population n_t of our society model can be determined at any time by integrating Eq. (9.10.4) over all r. The result is found to be

$$n_t(t) = 2\pi \int_0^\infty r\, n(r,t)dr\; = k\, n_0 \left[\frac{a}{\lambda} \; - \; \frac{1}{(\lambda + b)}\, \exp\left(-b\,t\right)\right]$$

and we determine that the model approaches a final population of

$$n_t(t \rightarrow \infty\,) = \frac{a\,k}{\lambda}\, n_0$$

It is interesting to compare this quantitative model with a specific society such as the former Soviet Union under Communism after the revolution of 1917 or modern Israel since 1948. Our model demonstrates the strong tendency of such societies to expand past their original borders.

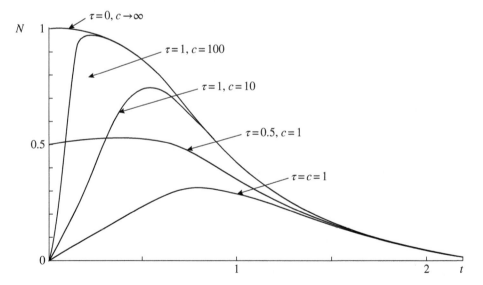

Fig. 9-30 Expansion of aggressive societies.

9.10.2 Historical Growth in Weapons Trade

During the last several decades, the United States and former Soviet Union (USSR) engaged in a competitive world trade in the production and distribution of military weapons to other nations. These weapon systems were sold, exchanged, and given outright to achieve various purposes such as military objectives, political alignments, and other perceived benefits. The impact and history into the twenty-first century fostering civil wars, "police actions," and other military and political events are evident through time.

Assume that the total world demand for military materials is P_t and that a fraction P_{as} of this total can be satisfied by either the United States or the former USSR. Then simple first-order growth equations for the United States and former USSR military trade are

$$dC_a = k_a P_{as}(t)\, dt \tag{9.10.6.1}$$

$$dC_s = k_s P_{as}(t)\, dt \tag{9.10.6.2}$$

where

$$P_{as}(t) = P_t(t) - C_a(t) - C_s(t) \tag{9.10.7}$$

and

$C_a(t)$ = U.S. military trade at time t (\$ equivalents)
$C_s(t)$ = Former USSR military trade at time t (\$ equivalents)
k_a = U.S. growth rate constant (1/yr)
k_s = Former USSR growth rate constant (1/yr)

If we divide Eq. (9.10.6.1) by Eq. (9.10.6.2) and integrate the result, we have that

$$k_s C_a(t) = k_a C_s(t) + K \tag{9.10.8}$$

where K is an arbitrary constant. We may now eliminate $C_s(t)$ from Eq. (9.10.7) using Eq. (9.10.8), and this action entered into Eq. (9.10.6.1) yields

$$\frac{dC_a}{dt} + (k_a + k_s)\, C_a = k_a P_t + K \tag{9.10.9}$$

Equation (9.10.9) is a linear first-order differential equation, which may be solved using the method outlined in Section 4.2.4. The result for growth in U.S. military trade with time is

$$C_a(t) = C_{a0}\, \exp\left[-(k_a + k_s)\, t\right]$$
$$+ k_a \frac{P_{t0}}{(k_a + k_s + k_t)}\, \left[\exp\left(k_t t\right) - \exp\left(-(k_a + k_s)\, t\right)\right] \tag{9.10.10}$$
$$+ \frac{(k_s C_{a0} - k_a C_{s0})}{(k_a + k_s)}\, \left[1 - \exp\left(-(k_a + k_s) t\right)\right]$$

where we have assumed that the total world demand for military materials, P_t increases exponentially in time,

$$P_t = P_{t0}\, \exp\, (kt) \tag{9.10.11}$$

and for $t = 0$ we have $C_a(0)$ and $C_s(0)$ as initial values. The solution of Eq. (9.10.6.2) for USSR military trade is similar to Eq. (9.10.10) and is given by

$$C_s(t) = C_{s0} \exp\left[-(k_a + k_s)\, t\right]$$

$$+ \frac{k_s P_{t0}}{(k_a + k_s + k_t)} \left[\exp\,(k_t t) - \exp\,(-(k_a + k_s)\, t)\right]$$

$$+ \frac{(k_a C_{a0} - k_s C_{s0})}{(k_a + k_s)} \left[1 - \exp\,(-(k_a + k_s)\, t)\right]$$

The major task now confronting the system investigator is the accurate determination of the parameters found in Eqs. (9.10.10) and (9.10.11). However, significant observations can be gleaned from our system model. Observe, for example, that as $t \to t_\infty > > 1/(k_a + k_s)$, from Eq. (9.10.8) we find

$$\frac{C_a(t_\infty) - C_a(0)}{C_s(t_\infty) - C_s(0)} = \frac{k_a}{k_s}$$

Thus, we conclude that the asymptotic ratio of weapons trade achieved by the United States and the Former USSR (Russian Federation) approached the ratio of their respective growth rate constants. The validity of this conclusion is left to the reader to assess and confirm.

9.10.3 Additional Modeling Problems

For each of the systems given in Table 9-20, define an appropriate system model and the essential elements of the system. Express variables for each of the model inputs and outputs and develop a simple (linear) expression for the model's system equation. Determine descriptive and graphical forms for the system model and interpret and evaluate your results.

Table 9-20 Major National and International Events

	System	Input	Output
(a)	Great Depression of 1930s (worldwide)	World War I	Unemployment
		Inflation	Riots
		Stock market collapse	Poverty
		Bank failures	U.S. Government response
(b)	German Third Reich	Adolph Hitler	Rearmament
		Nazism development	War in Europe
(c)	American	English Taxes	Sons of Liberty
	Revolutionary War	Edicts of King George	New Government
		Lack of representation	Impact on England
(d)	French Revolution	Excesses of French Kings	Change in government
		American Revolution	Impact on Europe

Table 9-20   (Continued)

	System	Input	Output
(e)	Battle of Britain (WWII)	German Luftwaffe V1, V2 Rockets	RAF Code breaking (Enigma)
(f)	U.S. labor movement	Labor leaders/Contracts Management Strikes	AFL CIO UAW
(g)	European Common Market	Market members	Market control Economic consequences
(h)	OPEC	Member countries	Crude oil output
(i)	Middle East	UN Israel PLO	Terrorism Retaliatory strikes Arms trade
(j)	Manhattan Project	Hahn & Strassmann (fission discovery)	Trinity Test (first atomic bomb test)
(k)	Civil War in Ireland	Catholics Protestants IRA	Civil strife Terrorism
(l)	Russian Revolution	Tsar World War I	Communism
(m)	U.S. World Trade Center Attack (9/11)	United States Al Qaeda, Taliban	Afghanistan Saudi Government
(n)	World Economic Disruption in 2008	Credit Abuses Mortgages Government Intervention	International Banks Markets Derivatives
(o)	Middle East	Osama bin Laden ISIS, Al Qaeda	Terrorism
(p)	Collapse of USSR	United States Reagan Administration	Russian Federation Mikhail Gorbachev
(q)	US Financial Crisis 2008	Credit Abuses Mortgages Government Intervention	U.S. Banks U.S. Stock Market U.S. Congress response
(r)	Asian Flu Epidemic 1918	World War I	U.S. Spread of Flu Virus
(s)	German attack on Russia (Operation Barbarrosa)	Adolph Hitler Military operations	Joseph Stalin Military operations

(*Continued*)

Table 9-20   (Continued)

	System	Input	Output
(t)	Arab-Israeli Conflict	Edicts of King George	New Government
		Lack of representation	Impact on England
(u)	French Revolution	Excesses of French Kings	Changes in government
(v)	Battle of Britain	German Luftwaffe	RAF
	(WWII)	V1, V2 Rockets	Enigma code breaking
(w)	U.S. labor movement	Labor leaders & Contracts	AFL
(x)	COVID-19 Pandemic	Viral release in Wuhan	Worldwide responses
		China (~2019)	deaths, economic impact
(y)	Middle East	UN (United Nations)	Terrorism
		PLO (Palestine Libration Org)	Arms trade
(z)	Manhattan District	Hahn & Strassmann	J. Robert Oppenheimer
	Project (Atom bomb)	Lisa Meitner	Trinity Test
		Nuclear fission	Allied scientists

General System Science Bibliography

Matter, Energy, Space, and Time

ADLER, L., *Inside the Nucleus*, New American Library, NY, 1963.

ASIMOV, L., *Understanding Physics: The Electron, Proton and Neutron*, New American Library, NY, 1966.

BENNETT, J., et al., *The Cosmic Perspective*, Pearson Addison-Wesley, San Francisco, CA, 2008.

BERTALANFLY, L. VON, *General System Theory*, George Braziller, NY, 1968.

GRIFFITHS, D., *Introduction to Elementary Particles*. Wiley, John & Sons, Inc., 1987

HEMPEL, C., *Philosophy of Natural Science*, Prentice-Hall, Englewood Cliffs, NJ, 1966.

IBERALL, A., *Toward a General Science of Viable Systems*, McGraw-Hill Book Co., NY, 1972.

KANE, G., *Modern Elementary Particle Physics*, Perseus Books, 1987.

KEMENY, J., *A Philosopher Looks at Science*, Van Nostrand Reinhold, NY, 1959.

MERLIN, T., *Astronomy*, John Wiley & Sons, NY, 1959.

PAULING, L., *General Chemistry*, Dover Publications, Inc., NY, 1988

PEACOCK, J., *Cosmological Physics*. Cambridge University Press, 1999.

REICHENBACH, H., *Philosophic Foundations of Quantum Mechanics*, University of California Press, Berkeley, 1944.

ROSSINI, F., *Thermodynamics of Physics and Matter*, Princeton University Press, Princeton, NJ, 1955.

SANDQUIST, G. M., *Introduction to System Science*, Prentice-Hall, Englewood Cliffs, NJ, 1985

SOMMERFIELD, A., *Mechanics, Lectures on Theoretical Physics*, Academic Press, NY, 1964.

SULLIVAN, M. R., *Statistics, Informed Decisions Using Data*, 3rd Edition, Prentice-Hall, 2010.

TOLMAN, R., *Statistical Mechanics*, Oxford University Press, NY, 1938.

WEATHERALL, M., *Scientific Method*, Simon & Schuster, NY, 1968.

WEAVER, J. H., *The World of Physics*, Simon & Schuster, NY, 1987.

WEINBERG, S., *Dreams of a Final Theory*, Pantheon Books, NY, 1992.

WITHROW, G. J., *The Natural Philosophy of Time*, Harper & Row, NY, 1963.

WHITTAKER, E. T., *Analytical Dynamics of Particles and Rigid Bodies*, Cambridge University Press, London, 1927.

Earth Sciences

ALLABY, M., *Dictionary of Earth Sciences*, Oxford University Press, 2008.

BATES, D., *The Planet Earth*, Pergamon Press, Elmsford, NY, 1964.

GATES, D., *Energy Exchange in the Biosphere*, Harper & Row, NY, 1962.

GILLULY, J., *Principles of Geology*, W. H. Freeman, San Francisco, CA 1968.

JACOBS, J., *The Earth's Core and Geomagnetism*, Macmillan, NY, 1963.

KORVIN, G., *Fractal Models in the Earth Sciences*, Elsevier, New York, 1987

Larousse Encyclopedia of the Earth, Prometheus Press, NY, 1961.

MILLAR, C., et al., *Fundamentals of Soil Science*, John Wiley & Sons, NY, 1965.

OLDROYD, D., *Earth Cycles: A Historical Perspective*. Westport, CN, Greenwood Press., 2006.

PETTERSSEN, S., *Introduction to Meteorology*, McGraw-Hill Book Co., NY, 1969.

REHL, H., *Introduction to the Atmosphere*, McGraw-Hill Book Co., NY, 1965.

SMITH, G., *How Does the Earth Work?*, Pearson Prentice Hall, Upper Saddle River, NJ, 2006.

STRAHLER, A. N., *Introduction to Physical Geography*, John Wiley & Sons, NY, 1965.

TARBUCK, E., et al., *Earth Science*, Prentice-Hall, 2002.

THORNBURY, W. D., *Principles of Geomorphology*, John Wiley & Sons, NY, 1969.

TODD, D. K., *Ground Water Hydrology*, John Wiley & Sons, NY, 1959.

TREWARTHA, G. T., *An Introduction to Climate*, McGraw-Hill Book Co., NY, 1968.

YANG, X., *Mathematical Modeling for Earth Sciences*, Dunedin Academic Press, 2008.

Life Sciences

ALBERTS, B. et al., *Essential Cell Biology*, 3rd Edition, Garland Science, Taylor & Francis, New York & London, 2010.

BENNETT, J., et al., *Life in the Universe*, Addison Wesley, CA, MA, 2003.

BUTCHER, J., et al., *Abnormal Psychology*, 14th Edition, Allyn & Bacon, Boston, 2010.

CAMBELL, N. A., J. B. REECE, *Biology*, 6th Edition, Benjamin Cummings, San Francisco, 2002

FREEMAN, S., *Biological Science*, 2nd Edition, Pearson Prentice Hall, Upper Saddle River, NJ, 2005.

FRIEDENBERG, R. M., *Unexplored Model Systems in Modern Biology*, Hafner Publishing Co., NY, 1968.

GELFAND, I. M., Ed., *Models of the Structural-Functional Organization of Certain Biological Systems*, MIT Press, Cambridge, MA, 1971.

GERADIN, L., *Bionics*, World University Library, 1968.

HAMILTON, S., et al., *Biological Science*, 2nd Edition, Prentice-Hall, NJ, 2005.

HARTWELL, L. H. et al., *Genetics from Genes to Genomes*, 2nd Edition, McGraw-Hill, 2004.

JOHNSON, G., *Biology, Visualizing Life*. Holt, Rinehart, and Winston, New York, NY, 2005.

KALMUS, H., *Regulation and Control in Living Systems*, John Wiley & Sons, NJ, 1966.

KLIR, G., M. VALACH, *Cybernetic Modeling*, ILIFFE, London, 1967.

MESAROVIC, M. D., *Systems Theory and Biology*, Springer-Verlag, NY, 1968.

RASHEVSKY, N., *Mathematical Biophysics*, Dover Publications, NY, 1960.

ROSEN, R., *Dynamical System Theory in Biology*, Wiley-Interscience, NY, 1970.

Human Life

BOYD, R., J. SILK, *How Humans Evolved*, Norton & Company, NY, 2003.

BROWN, J., et al., *Biomedical Engineering*, F. A. Davis, Philadelphia, PA, 1971.

Central Intelligence Agency (CIA), The World Factbook, US Government Publishing Office.

Editors of Fortune, *Our Ailing Medical System*, Harper & Row, NY, 1969.

GROVES, C., et al., Eds. *Mammal Species of the World*, 3rd Edition, Johns Hopkins University Press, 2005.

MANKIW, N. G., *Principles of Economics*, 5th Edition, South-Western Cengage Learning, OH, 2009.

NEUMANN, J. VON, O. MORGENSTERN, *Theory of Games and Economic Behavior*, Princeton University Press, Princeton, NJ, 1947.

PITTS, P. M., M. I. POSNER, *Human Performance*, Brooks/Cole Publishing Co., Monterey, CA, 1967.

RASHEVSKY, N., *Mathematical Biophysics*, Dover Publications, NY, 1960.

ROBIN, A., *Biological Perspectives on Human Pigmentation*: Cambridge University Press, 1991.

RUTSTEIN, D. D., M. EDEN, *Engineering and Living Systems*, MIT Press, Cambridge, MA, 1970.

SAGAN, C., *The Dragons of Eden, A Ballantine Book*, Wikipedia, NY, 1978.

SOROKIN, P., *Modern Historical and Social Philosophies*, Dover Publications, NY, 1963.

WEINER, N., *Cybernetics*, MIT Press, Cambridge, MA, 1961.

WETTER, G., *Dialectic Materialism*, Praeger Publishers, NY, 1963.

YAKOWITZ, S. J., *Mathematics of Adaptive Control Processes*, Elsevier, NY, 1969.

Human Society

ALLEN, R. G., *Mathematical Economics*, Macmillan, NY, 1957.

AUDI, R., General Editor, *The Cambridge Dictionary of Philosophy*, 2nd Edition, Cambridge University Press, 2006.

BERRIEU, F. K., *General and Social Systems*, Rutgers University Press, New Brunswick, NJ, 1968.

COLEMAN, J. S., *Introduction to Mathematical Sociology*, Collier-Macmillan, London, 1964.

CRAIG, A. M., et al., *The Heritage of World Civilizations*, 5th Edition, Prentice-Hall, Englewood Cliffs, NJ, 2012.

DE GREENE, K. B., *Systems Psychology*, McGraw-Hill Book Co., NY, 1970.

GORE, W. J., *Administrative Decision-Making, A Heuristic Model*, John Wiley & Sons, NY, 1964.

JENKINS, R., *Foundations of Sociology*. Palgrave MacMillan, London, 2002.

KEMENY, J., J. L. SNELL, *Mathematical Models in the Social Sciences*, Blaisdell Publishing Co., Waltham, MA, 1962.

LAZARSFELD, P. F., N. W. HENRY, eds., *Readings in Mathematical Social Science*, Science Research Associates, Chicago, IL 1966.

LENSKI, G., *Human Societies: An Introduction to Macrosociology*, McGraw- Hill, Inc., NY, 1974.

MCKEAN, R. M., *Efficiency in Government Through Education*, John Wiley & Sons, NY, 1958.

MCKEAN, R. M., *Public Spending*, McGraw-Hill Book Co., NY, 1968.

RAIFFA, H., *Decision Analysis*, Addison-Wesley, Reading, MA, 1970.

RASHEVSKY, N., *Mathematical Biology of Social Behavior*, University of Chicago Press, Chicago, IL, 1959.

SAATY, J. L., *Mathematical Models of Arms Control and Disarmament*, John Wiley & Sons, NY, 1968.

SIMON, H. A., *Models of Man*, John Wiley & Sons, NY, 1957.

STEINER, G., *Strategic Planning*, Free Press, NY, 1979.

STERNBERG, S., et al., *Mathematics and Social Sciences*, Mouton, Paris, 1965.

VICKERS, G., *Value Systems and the Social Process*, Basic Books, NY, 1968.

Arts

CHERRY, C., *On Human Communications*, John Wiley & Sons, NY, 1957.

Encyclopedia of World Art, *15 vols.*, McGraw-Hill Book Co., NY, 1972.

GARDNER, H., Art through the Ages, revised by H. DE LA CROIX and R. G. TANSEY, Harcourt Brace Jovanovich, NY, 1980.

HARRAH, D., *Communications: A Logical Model*, MIT Press, Cambridge, MA, 1963.

MUELLER, R. E., *The Science of Art*, John Day Co., NY, 1967.

TOWER, A., *Future Shock*, Random House, NY, 1972.

Technology and Engineering

AGUILAR, R J., *System Analysis and Design in Engineering, Architecture, Construction and Planning*, Prentice-Hall, Englewood Cliffs, NJ, 1973.

BEDWORTH, D. D., *Industrial Systems, Planning, Analysis, and Control*, Ronald Press, NY, 1973.

BELLMAN, R., *Adaptive Control Processes: A Guided Tour*, Princeton University Press, Princeton, NJ, 1961.

CHESTNUT, H., *Systems Engineering Methods*, John Wiley & Sons, NY, 1967.

CRAMB, A. W., *A Short History of Metals*. Carnegie Mellon University, Pittsburgh, PA, 2007

CRANDELL, S. H., et al., *Analysis of Discrete Physical Systems*, McGraw-Hill, NY, 1967.

CRUMP, T., *A Brief History of Science*, Robinson Publishing, Pittsburg, PA, 2001.

FERRARO, G. P., *Cultural Anthropology: An Applied Perspective*, University of Chicago Press, IL, 2006.

FRANKLIN, U., *Real World of Technology*, House of Anansi Press, 2007.

GOSLING, W., *The Design of Engineering Systems*, John Wiley & Sons, NY, 1962.

GUSTON, D. H., *Between politics and science: Assuring the integrity and productivity of research*, Cambridge University Press, New York, 2000.

HALL, A. D., *A Methodology for Systems Engineering*, Van Nostrand Reinhold, NY, 1962.

HAVILAND, W.A., *Cultural Anthropology: The Human Challenge*. Thomson Reuters, Phoenix, AZ, 2004.

HOBERMAN, C. M., *Engineering Systems Analysis, Charles E*, Merrill, Columbus, OH, 1965.

KOENIG, H. E., et al., *Analysis of Discrete Physical Systems*, McGraw-Hill, NY, 1967.

LOVITT, W., *The Question Concerning Technology*, Harper Torchbooks, New York, NY, 1977.

MONSMA, S. V., *Responsible Technology. Grand Rapids*, W. B. Eerdmans Publishing Co., Grand Rapids, MI, 1986.

OAKLEY, K. P., *Man the Tool-Maker*. University of Chicago Press, Chicago, IL, 1976.

PAYNTER, H. M., *Analysis and Design of Engineering Systems*, MIT Press, Cambridge, MA, 1960.

PORTER, W. A., *Modern Foundations of Systems Engineering*, Macmillan, NY, 1966.

SKOCLUND, V. J., *Similitude: Theory and Applications*, International Textbook Co., Scranton, PA, 1973.

STARK, R. M., R. L. NICHOLLS, *Mathematical Foundations for Design: Civil Engineering Systems*, McGraw-Hill Book Co., NY, 1972.

STIEGLER, B., *Technics and Time, 1: The Fault of Epimetheus*. Stanford University Press, Stanford, CA, 1998.

ZAGAR, N., J. TRIBBIA, eds., *Modal View of Atmospheric Variability: Applications of Normal-Mode Function Decomposition in Weather and Climate Research*, Springer, New York, NY, 2020.

Religion

AUDI, R., Ed., *The Cambridge Dictionary of Philosophy*, 2nd Edition, Cambridge University Press, Cambridge, UK, 2006.

BARZILAI, G., *Law and Religion; The International Library of Essays in Law and Society*, Ashgate, 2007. www.academia.edu.

BRODD, J., *World Religions*. Saint Mary's Press, Winona, MN, 2003.

DESCARTES, R., *Meditations on First Philosophy*, Bobbs-Merrill Co, NY, 1960.

DE GREENE, K. B., ed., *Systems Psychology*, McGraw-Hill Book Co., NY, 1970.

HAISCH, B., *The God Theory: Universes*, San Francisco, CA, 2006.

HENEMANN, *The Origin of Live & Death*, African Creation Myths, 1966.

JAMES, W., *The Varieties of Religious Experiences*, New American Library, NY, 1959.

KRANTZ, D. H., et al., *Foundations of Measurement*, Academic Press, NY, 1971.

LAO TZU, LAO TZU; TAO TE CHING (VICTOR H. Mair translator); Bantam, 1998.

LUCA, R. D., et al., eds., *Handbook of Mathematical Psychology, Volumes 1, II and III*, John Wiley & Sons, NY, 1965.

McMANNERS, J., Ed., *The Oxford Illustrated History of Christianity*, Oxford Press, 2001.

MILES, G. A., *Mathematics and Psychology*, John Wiley & Sons, NY, 1964.

MOURANT, J. A., *Readings in the Philosophy of Religion*, Thomas Y. Crowell Co., NY, 1959.

RADER, M., *The Enduring Questions*, Holt, Rinehart & Winston, NY, 1956.

RUSSELL, B., *Wisdom of the West*, Crown Publishers, NY, 1959.

SALER, B., *Conceptualizing Religion: Immanent Anthropologists, Transcendent Natives, & Unbounded Categories*, Google Books, 1990.

SCHROEDER, G., *The Science of God*, The Free Press, NY, 1997.

SKINNER, B. F., *Beyond Freedom and Dignity*, A. A. Knopf, NY, 1971.

SMITH, H., *The Religions of Man*, New American Library, NY, 1958.

The Holy Bible, King James Version, New American Library, 1974.

The Koran, *Penguin*, 2000.

TRUEBLOOD, D. E., *Philosophy of Religion*, Harper & Row, NY, 1957.

SAINT AUGUSTINE, The Confessions of Saint Augustine, J. K. Ryan translator, Image, 1960.

WINSTON, K., *Encyclopedia of Religion*, 2nd Edition, Macmillan Reference USA, Detroit, 2005.

History

ASIMOV, I., *Asimov's Chronology of the World*, Harper Collins, 1991.

CAMPBELL, B., *Human Evolution*, Aldine Publishing Co., Chicago, IL, 1974.

CARR, E. H. with a new introduction by Richard J. Evans; *What is History?*, Palgrave, Basingstoke, 2001

CLARK, K., *Civilization, A Personal View*, Harper & Row, NY, 1966.

DARLINGTON, C., *The Evolution of Man and Society*, Simon & Schuster, NY, 1969.

DURANT, W., A. DURANT, *Lessons of History*, Simon & Schuster, NY, 1968.

EVANS, R. J., *In Defense of History*, W. W. Norton, 2000.

KAHN, H., *On Thermonuclear War*, Princeton University Press, Princeton, NJ, 1960.

MEADOWS, D. H., et al., *The Limits to Growth*, Universe Books, NY, 1972.

MELKO, M., *The Nature of Civilizations*, Porter Sargent Publishers, Boston, MA, 1969.

MESAROVIC, M., E. PESTEL, *Mankind at a Turning Point*, E. P. Dutton, NY, 1974.

NEVINS, A., *The Gateway to History*, Doubleday & Co., Garden City, NY, 1962.

QUIGLEY, C., *The Evolution of Civilization*, Macmillan, NY, 1961.

RASHEVSKY, N., *Looking at History through Mathematics*, MIT Press, Cambridge, MA, 1968.

SEIDENBERG, R., *Post-Historic Man*, Viking Press, NY, 1974.

TOSH, J., *The Pursuit of History*, Longman, 2006.

VICKERS, G., *Value Systems and Social Process*, Basic Books, NY, 1968.

WALSH, W. H., *Philosophy of History*, Harper & Row, NY, 1960.

WARFIELD, J. H., *Societal Systems*, John Wiley & Sons, NY, 1976.

WELLS, H. G., *The Outline of History*, Doubleday & Co., Garden City, NY, 1961.

10

System Modeling Paradigms

10.1 Background

The modeling steps required for the identification, quantification, and evaluation for a successful system model are fundamental and diverse. To develop an accurate, quantitative system model that can contribute to the existing knowledge, support effective utilization, and bear critical, external, and independent review requires extensive background research and evaluation of the existing knowledge and supporting data regarding the system. Such extensive research effort to access existing data and evaluation is necessary to ensure that the final system model developed will adequately address the system with current information and accurate evaluation of the systems behavior. Then the result is a defensible and useful contribution to the field of System Science.

Wikimedia is the title for the contemporary global collection of a hundred million freely accessible internet data files that provide a growing, diverse data source and virtual data library for system science. The entities that constitute this encyclopedic collection of knowledge are constantly reviewed and updated by users with contemporary information. Thus, the practitioner of system science has immediate access to the current state-of-the-art information and knowledge regarding essentially any field of interest. The worldwide internet has virtually supplanted printed books, encyclopedias, data banks, technical journals, and printed document libraries. Therefore, the successful system science operative must now be conversant with and adept in using these virtual resources.

Obviously, this text's model examples will also suffer to some extent given latest knowledge regarding the system exemplified. Fortunately, the mathematical skills and knowledge presented in this text are more established and stable. However, the system science models portrayed herein may be outdated or even erroneous by the time the information is set in print and distributed and positioned on a bookshelf. The reader must recognize this reality for background information and data for performing current, accurate system science. The reader should become adept and conversant with virtual, internet class data now stored and immediately available in the computer-generated, electromagnetic cloud.

The enlarging Wikimedia movement now encompasses vigorous, advancing internet entities that include Wikipedia (encyclopedia), Wikiversity (learning resources), Wikibooks (textbooks and manuals), Wikidata (knowledge base), Wiktionary (dictionary

Introduction to System Science with MATLAB, Second Edition.
Gary Marlin Sandquist and Zakary Robert Wilde.
© 2023 John Wiley & Sons Ltd. Published 2023 by John Wiley & Sons Ltd.

and thesaurus), and many others which are of significant value for any pursuit in system science. These internet-based entities provide a current (virtually immediate), actively updated and corrected, evolving records, files, books, resources, and data that are of significant value for system science. The Wiktionary is a broad-based translation file that permits immediate translation of a record in over 180 foreign languages to the native language of the system scientist. Thus, a data record in essentially any foreign language unfamiliar to the system investigator can be immediately translated and employed for system science use.

The reader is advised to become well acquainted and practiced with Wikimedia and especially Wikipedia. A free user account with Wikimedia is easily established, and it is advised for the reader to become acquainted with these internet sources. Understand that these sources are active, constantly changing resources for the system scientist. It is not necessary to log in to Wikipedia before viewing, using, or editing pages, although it is useful for storing and accessing visits. To "log in" or "create an account," enter these quoted commands in the Search Wikipedia box and follow the instructions. Observe that all Wiki phrases and commands are subject to change. Currently, if the reader selects "Keep me logged in (for up to 365 days)" the reader does not need to enter a password again if access to Wikipedia is made from the same computer.

10.2 Modeling Paradigm

The following paradigm, as a set of sequential procedures for system modeling, is provided as an example that should be useful to achieve an accurate and satisfactory system model. Recognize that this paradigm is contingent upon the system modeler's sufficient understanding of the system subject and competence with the essential skills for system modeling presented in Chapters 1 through 9 of this text. The proper application of these skills should ensure that the model developed will be credible and can satisfy the sequential steps offered in this chapter for the creation of a new system model.

If the system modeler lacks adequate personal, background knowledge and experience in the particular discipline or subject to be modeled, the modeler should perform focused training in the area associated with the system model subject. Since the range of topics addressed by system science is very broad, it is not unexpected that the system modeler may require specific familiarization and training in the particular subject area selected for model development.

A practical option here is for the modeler to acquire the necessary background in the discipline under study through a formal academic course(s) employing online courses available through the internet. Of particular interest is the broad range of online courses now available from the website edX.org. An extensive field of free online courses are available at edX.org, which provides access to the academic courses offered by Harvard, MIT, Stanford, Princeton, University of California at Berkeley, and other academic institutions. Currently, the total collection of these courses exceeds several thousand and the collection is increasing rapidly. It is highly probable that the specific modeling topic under investigation is also addressed by an online course(s) through edX.org. These online courses include undergraduate and graduate level topics embracing all the disciplines addressed in Chapter 9 of the Text. However, the pursuit of such academic training can be time demanding, so the training effort should be commensurate with the effort required for the particular system model to be developed.

10.3 Essential System Modeling Paradigm Steps

10.3.1 Step-1 Explore and Document

Explore, Review, Record, and Evaluate Background Information to Identify and Support System to be Modeled

This initial step is essential since the system model to be created will depend upon the extent and quality of the background search and the assimilation of adequate information regarding the system. The examination step should be sufficiently comprehensive to satisfy the requirements and expectations for the final system model. Such accuracy requires adequate time and effort to develop a representative system model for the system under examination. The investigator should perform an adequate, comprehensive background investigation of the known information regarding the system selected. This action is essential since the system modeler must ensure that the model developed addresses existing knowledge and is sufficiently comprehensive to address current knowledge regarding the system.

Fortunately, extensive information, resources, facts, and other essential data are now available and readily accessed on the computer-based internet World Wide Web (WWW). We have surpassed earlier times when frequent, physical trips to a major library or repository to access documents, books, encyclopedias, journals, research papers, government reports, etc., were required to access the information and data to perform the study. Considerable time and effort were expended to simply locate and identify cataloged information. Then it was necessary to extract and record pertinent information and identify the source and the material gathered from these sources. Furthermore, these data accessed were often outdated and not current regarding the subject system because of the time delay in publishing and delivering the material to the repository library or other information source.

Now the effective and superior strategy is to search massive and diverse information sources using computer-based data repositories. This is especially true for Wikimedia, addressed earlier for the massive global collection of data internet-based data files, now freely accessible internet data files that provide a growing, diverse data source and virtual library for system science. To effectively manage the enormous quantity of digital data now available, experience with data gathering sources using metadata harvesting* is useful and a very effective choice. Metadata computer resources now replace exhaustive human searching of the vast repository of digital and written data with the enhanced, invisible hand of metadata searching schemes for processing and documenting the data in HTML, XHTML, XML, and RDF formats.

The global, computer-based WWW now accesses most recorded information and provides the system modeler with immense knowledge resources that were unimagined

---

* *Source:* ZENG, M. L. and J. QIN (2008).

and unavailable in the past. The interaction and assessment of web information provide the system modeler with a level of confidence that the web information gained has been reviewed and critiqued by many web users. Web access and effective use greatly reduce the effort and time required to effectively search, gather, process, and record information needed to create a comprehensive system model.

This text will confine the use of metadata harvesting to the universal WWW. The web coupled with extensive, web search engines such as Google Scholar, Microsoft Bing, Yahoo, Wikimedia, Wikipedia, and other computer data sources, are continuously updated and critiqued for their online information. This step establishes the initial scope for the system model under study. Given a directed, extensive background investigation, the investigator assembles and evaluates the information found in Step-1 and then identifies and creates the essential system inputs and the system kernel. Establishing the system inputs (causes) and associated outputs (effects) is an iterative process and provides the background required to accomplish the search process.

Utilization of this new paradigm for gathering of data for system development and modeling requires familiarization and experience with the Web to efficiently access and process this resource for information. We will provide a brief overview of Web nomenclature and data gathering processes using the website Wikipedia to define and demonstrate these Web tools. Useful web tools are "underlined" in the next paragraphs and Table 10-1. "Accessing the Web for System Modelling," the web entry can be copied and inserted into the Wikipedia search bar to gain additional information and access web data regarding the item. Wikipedia and other Wikimedia applications comprise the free internet source that host millions of available articles at innumerable websites and provide a comprehensive and widely used reference work for system science. Wikipedia is now the largest and most read and used reference work in history. A major advantage is Wikipedia's online

Table 10-1 World Wide Web (WWW) Nomenclature Table

World Wide Web or WWW or Web, an information system with documents and web resources identified by a Uniform Resource Locator or URL
Uniform Resource Locator or URL, a web address to a web resource on a computer network that retrieves a data resource.
Resource is written material or data available on the web.
Computer is a machine that performs an arithmetic or logical operation
Web server is computer software that requests a network protocol for a web page via HTTP
Software is a set of instructions and data that operates a computer
Web page or webpage is a hypertext document on a website displayed to a user in a web browser
Hypertext is text displayed by a computer
Website or web site is a web page identified by a domain name on a web server; examples are wikipedia.org, google.com, amazon.com
Web portal is a website that provides information from diverse sources emails, online forums, search engines
Web resource is a resource, digital, written, connected to WWW shown by Uniform Resource Identifiers
Systems Science or Systems Research or System is an interdisciplinary subject in nature, society, applied sciences, etc.
System is a group of interacting elements that obey quantitative rules.
Systems are subjects of study in systems theory

information is under constant review and assessment by a broad, international audience of web users.

A brief overview of the process for the interaction of the system modeler with the WWW is shown below. Extended Web terminology needed by the system modeler is shown in Table 10-1.

Accessing the Web (WWW) for System Modeling

Identification of the essential role that selected <u>search engines</u> and web portals provide for system science development is necessary to fully access and utilize the web for its information content. Although the number and application of websites and associated resources are constantly developing and changing, some current useful existing web resources for system science applications are shown below:

- <u>World Wide Web</u> (<u>WWW</u>) is the information system where web resources are accessed by a <u>user</u> via a web browser, using a software application called a web server to respond to a <u>web query</u> or <u>web search query</u>.
- The <u>web server</u> accepts the <u>web query</u> via a <u>Hypertext Transfer Protocol</u> (<u>HTTPS</u>) or (<u>HTTP</u>) or communication protocol and retrieves the web page with the requested information. HTTPS is the secure version of <u>HTTP</u> that enhances integrity and provides confidentiality for <u>web query</u>.
- A user agent, the system modeler uses the web browser or web crawler to make a request for a specific <u>web</u> resource. <u>HTTP</u> and the <u>web server</u> responds with the <u>search results</u> or an error message if no information is found.
- A <u>web site</u> (or <u>website</u>) is a collection of web pages identified by a common domain name on at least one web server. Notable examples of major websites are wikipedia.org, google.com, yahoo.com, and amazon.com.
- The <u>web portal</u> is a special website that can collect information from many diverse sources.

Accessing Information for System Modeling Using the Web (WWW)

Identification of the essential role that selected search engines and web portals provide for system science development is necessary to fully utilize the web for its vast information. Although the number and application of websites and associated resources is constantly developing and changing, some useful existing web resources for system science applications are shown below:

Google Scholar is a leading search engine that accesses most scholarly literature across a comprehensive array of publishing formats and disciplines. Google Scholar indexes the largest and most complete peer-reviewed online academic journals, books, conferences, papers, dissertations, theses, preprints, abstracts, technical reports, and other scholarly literature, including even court opinions and patents. It is the preferred and obvious search site for initiating data collection for a system science model under development.

Science.gov is the specific web portal and search engine for most U.S. government scientific, technical, and research agencies. This site is the predominant source for access to essentially all U.S. documented scientific and technical information. The web portal, Science.gov is also the U.S. contribution to the international portal WorldWideScience.

WorldWideScience.org enables a system modeler with internet access to perform a single-query search of all national scientific databases and portals in more than 70 countries spanning three-quarters of the world's population.

The site that includes U.S. National Academies of Sciences, Engineering, and Medicine (NASEM or National Academies) is comprised of the following National Academies:

1) National Academy of Sciences (NAS)
2) National Academy of Engineering (NAE)
3) National Academy of Medicine (NAM)

The National Research Council (NRC) produces the reports and other products issued by NASEM and are published by the National Academies Press. The National Institutes of Health (NIH) is the primary agency of the U.S. Government responsible for biomedical research and public health. Biomedicine is the cornerstone of modern health care and laboratory diagnostics and is based on molecular biology and combines all areas of molecular and nuclear medicine. Public health addresses the science and practice of preventing disease, prolonging life, and improving quality of life through research and delivery of services and resources

U.S. Department of Energy (DOE) is the U.S. Government cabinet-level department for policies regarding energy, nuclear programs, energy conservation and research, radioactive waste disposal, and domestic energy production. DOE directs research in genomics and the Human Genome Project. DOE also sponsors most research in the physical sciences, which is conducted at U.S. National Laboratories.

National Science Foundation (NSF) is the independent agency of the U.S. government that supports fundamental research and education in science and engineering. NSF funds over a quarter of all federally supported basic research conducted by U.S. colleges and universities and is a major source of funding for mathematics, computer science, economics, and the social sciences.

Other informational website data for addressing system topics in the following areas include:

- Engineering is the application of scientific principles to analyze, design and build machines, structures, and other items, including bridges, tunnels, roads, vehicles, and buildings. The discipline of engineering encompasses a broad range of more specialized fields of engineering, each with a more specific emphasis on particular areas of applied mathematics, applied science, and types of application. Engineering-related websites address disciplines of bioengineering, environmental systems, civil and mechanical systems, chemical and transport systems, nuclear engineering, electrical and communications systems, design, and manufacturing.
- Computer Science spans a range of topics from theoretical studies of algorithms, computation, and information to the practical issues of implementing computational systems in hardware and software.
- Geosciences address the geological, atmospheric, and ocean sciences
- Mathematical and Physical Sciences viz. (mathematics, astronomy, physics, chemistry, and materials science)

- Social, Behavioral, and Economic Sciences (neuroscience, management, psychology, sociology, anthropology, linguistics, science, and economics).
- Life science including medicine, human anatomy, genetics, population.
- Human life and Human society, including medicine, human culture, anatomy, genetics, population
- The Arts are also addressed along with Religion and History.

10.3.2   Step-2 Define and Contain

Define, Comprehend, and Isolate and Contain the System Model from the External Environment

Given that an adequate review and evaluation of the existing background information regarding the system model has been performed, the modeler should then determine and confirm the actual system features that are to be modeled. What features are significant and are required for developing an accurate system model for the system selected? These features must be identified, collected, and organized from the background information and assembled for their interaction upon the system model. It is very important to isolate the system from its background to realize an effective quantitative model.

10.3.3   Step-3 Select and Develop

Select Essential System Variables (Inputs, Outputs, Parameters, Descriptors) to Quantify the System Model.

The modeler should identify and select the essential variables required to describe the system and the information to be gained from the system model. The modeler must now perform the essential stages to select the system variables, specifically the inputs, outputs, and parameters that control the system and dictates its performance.

A suggested restraint here is to begin with a minimum selection of inputs and outputs. The selection of too many variables initially may result in severe difficulties in achieving a satisfactory analysis of the system. A suggestion is to begin with only a single input, the independent variable and a single output, the dependent variable to initiate the system investigation. The modeler may seek initially to attack the modeling too aggressively with numerous inputs and outputs. Then the set of system equations that results may resist analytical solution and prove to be unsuccessful. The initial modeling effort with a constrained system limited to a single input and output can be very useful. This choice forces the modeler to focus, at least initially, on the most important input(s) and the most important output(s) for the system. Then, given success with this initial effort, this action will prove useful as the modeler adds additional inputs and outputs to improve accuracy and adequate system representation.

Chapter 2 in the Text provides the basic background for establishing the necessary variables and parameters for the system model. That successful effort combined with a simple system kernel provides the basis for developing the essential, quantitative system kernel. Furthermore, a reasonable starting kernel is the use of a constant or linear system model as declared in Section 4.2 and Sections 4.2.3 or 4.2.4 in this text. This promises that the system model can be developed that will successfully deliver the system's desired output(s).

10.3.4 Step-4 Construct and Quantify

Construct and Formulate the Quantitative System Kernel

With the system inputs (independent variables) and the outputs (dependent variables) declared, the next step is to formulate the quantitative system kernel. Consider and initially utilize a simple, linear, constant coefficient kernel if possible. This choice provides experience and confidence in starting the system analysis. Again, it is important to focus on the most important output(s) expected from the system model and the principal input (s) that produces that output. Undoubtedly, it may be necessary then to increase and correct the number of inputs and outputs to correctly address the expected response of the system model.

Assuming that the modeler has commenced a satisfactory preliminary system model effort, the next step is to formulate the quantitative system kernel for the model. Given the disclosed nature of the system from the background investigation performed in Step-1, the system modeler now must focus on the acceptable quantitative system kernel.

The complete set of model kernels is identified and given below. System Kernel Model equations for inputs and outputs are addressed as follows:

1) Single input and single output

   See Text Sections: 3.2.1, 4.2, 6.1, 6.2.1, 6.3, 7.2, 7.3.1, Table 8-1, <u>8.3.3</u>.

2) Single input and multiple outputs: Applicable Sections in 8.
3) Multiple inputs and single output

   See Text Sections: 3.2.2, 4.3, 6.1, 6.2., 6.3, 7.2, 7.3.2, Applicable Sections in 8.

4) Multiple inputs and multiple outputs

   See Text Sections: 3.2.3, 5.2, 6.1, 6.2, 6.3, 7.2, 7.4.1, Applicable Sections in 8.

10.3.5 Step-5 Analyze and Evaluate

Perform Initial System Analysis for Actual Response

Given the preliminary system kernel has been established, the number of inputs and outputs have been identified, and the kernel has been quantified in explicit functional form. This step commences the actual execution of the quantitative model for expected behavior. The modeler can observe the output of the system model and impose the modeler's expectations for the system output from the model.

10.3.6 Step-6 Assess and Re-Evaluate

Evaluate Initial System Model for Expected and Actual Behavior

During this step in the model development process, the modeler repetitively evaluates and corrects the response of the system model for the range of inputs and associated outputs imposed on the system model. This is a repetitive process that the modeler must perform to produce that expected model output.

There are many questions and issues that arise at this point for the model. Should or do the model entities, inputs, outputs, kernel, and modifying parameters need further revision? Are feedback and/or nonlinearities in the system kernel needed or present? Is an increase and expansion of the system variables needed? Are the parameters dependent upon the system inputs and outputs?

This step establishes if the system model requires a modification or extension of the system kernel, does the system require feedback of certain inputs and outputs and should the range of values for the system parameters be altered?

10.3.7   Step-7 Finalize and Confirm

Finalize and Confirm System Model for Response and Fidelity.

Finally, the modeler tests, evaluates, modifies, reassesses, and finalizes the quantitative model developed for the system. The modeler then assembles the final system model with its complete set of variables and parameters and fully analyzes and subjects the system model to a comprehensive evaluation. Upon satisfactory assessment of the system model, the modeler prepares the system model for external (non-modeler) use and evaluation.

10.3.8   Step-8 Resolve and Accept

Perform an External Review of Model Behavior for Acceptance

Given that the system model is deemed to adequately describe the behavior of the system under development, the next step is to disseminate the model to knowledgeable sources for their independent review and assessment. This provides confirmation of the modeler's efforts and provides confidence that the system has potential merit. The modeling issues that may arise and the modelers response and the final system model adopted by the modeler can seek critique and acceptance by competent, informed reviewers of the system model developed.

10.3.9   Step-9 Publish and Disseminate

Assemble Final System Model Results and Disseminate and Publish Work

Given that the system model is satisfactory to the system modeler and external reviews have not produced significant issues or errors, the final step is to disseminate the model to external, knowledgeable sources for their independent review and assessment. This action will provide external confirmation of the modeler's efforts in developing the system model and provides confidence that the system model developed has value and will be utilized.

10.4   Example of Analysis Process after System Identification using MATLAB

This next section describes the process for generating quality images in MATLAB for reporting or publication. The following steps describe the process that will be used:

1) Determine constants, independent, and dependent variables for model
2) Determine ranges of values for model constants, independent, and dependent variables

3) Impose various graphical visualization techniques

4) Prepare for independent review, assessment, and publication

While there are many ways to generate these visualizations, anonymous functions will be used here. This is done to allow for the relevant equation to be defined only once. After that, it can be simplified in any number of ways depending on the desired constants and dependent variables.

1) Determine constants, independent, and dependent variables for model

The first steps when deciding on and which variables are to be varied (independent) and how those variables affect the response (dependent) variable. Any other quantity will need to be assigned some constant value.

Sometimes it may be desirable to vary all possible combinations of variables to observe their effect on the dependent variables. To do this, potentially many more visualizations will be required as the number of variables that can be represented in static graphics is limited.

The number of independent variables for a given graphic is generally determined by the number of axis available. For two-dimensional plots, you can vary one independent variable on the horizontal axis and the dependent variable on the vertical axis. Additional independent variables can be viewed by using multiple lines to the same plot. This can be useful in many circumstances; however, too many lines on a single plot will make it difficult to interpret. Color could also be used here to represent another independent variable.

If color is used as an axis, then contour plots allow for two independent variables to be varied on the horizontal and vertical axis, and the dependent variable is shown with the color contours.

Surface plots use three spatial axes to map two independent variables and one dependent variable. In this section, a surface plot will be used as an example.

For this section, the code required to visualize the normalized population equation from Chapter 8 will be shown. This visualization can be done in a number of ways. For this example, we will plot it as a function. The equation can be written as an anonymous function in MATLAB as follows:

```
pop_norm_eq = @(Tau,Lam,p) p - exp(Tau + Lam.*(p-1).^2); % Implicit
```

The @ symbol is used to define the variables, which in this case are the normalized population, normalized time, and the other normalized parameter. The function is set up so that it is equal to zero. This way MATLAB can solve the implicit function.

2) Determine ranges of values for model constants, independent, and dependent variables

After the constants and independent variables have been defined, they will need values or ranges of values assigned to them. Constant values should be given a relevant value to ensure the physicality of the results. The value ranges for the independent variables are often chosen as well to cover a regime of physical interest (but not always).

It is also important to consider the number of sample points over the range for each variable. Especially if a significant number of independent variables are of .interest, it may be worthwhile to reduce the number of samples over potentially less sensitive independent variables to get a broad macroscopic view of its effects without needing extensive reevaluation.

For the current problem. The following ranges were used to observe the behavior of multiple parameters:

```
interval = [-5, 5, -5, 5 0, 6]; % This is the interval for -/+Tau,
-/+Lam, and -+p
```

The sequence of numbers is important to the visualization function. This could be something like *fplot* or *fsurf* but *fimplicit3* will be used here because the function is implicit. The funtion *fimplicit3* needs to input, the function to be evaluated, in this case *pop_norm_eq*, and the maximum and minimum bounds for each of the inputs. For three variables in *norm*, there are six numbers in *interval*. These values were chosen from trial and error, and the reader is encouraged to do vary the parameters as well.

3) Impose various graphical visualization techniques

After the equation, as well as the variables and ranges have been defined, it is time to visualize the model using multiple different techniques. Types of plots that are common are line plots (2D or 3D), surface plots (3D), contour plots (3D), quiver plots (2D or 3D), scatter plots (2D or 3D), and others. It is recommended to attempt multiple different types of visualizations to see which one is most ideal.

For this work, a surface plot is used to show the three-dimensionality of the problem. It also shows MATLAB's capability to use solve implicit functions that have to be solved numerically.

```
fimplicit3(pop_norm_eq,interval,'MeshDensity',50) %Solve implicit
function
```

This allows for non-analytical solutions to be found. The *MeshDensity* term is added to adjust how the plot is rendered. This can be adjusted or even omitted if desired.

At this stage, the bounds chosen in Step-2 can be adjusted the visualization redone to see how the function responds to different types of inputs. If your function had some variables that were currently being held constant, you could change your constants and independent variables in Step-1 and visualize using a different set of variables.

4) Prepare for independent review, assessment, and publication

After the desired types of graphics and ranges have been identified, then it will be time to prepare the graphics for review and publication where they are needed. The features used to generate the graphic in Chapter 8 are outlined below:

```
% Axis Labels
xlabel('Normalized time (\tau = at)','FontSize',14)
ylabel('Normalized parameters (\lambda = 1/2 abP_0^2)','
FontSize',14) zlabel('Normalized Population
(p=P(t)/P_0','FontSize',14)
```

Label's for the X, Y, and Z axis can be generated. The *FontSize* can be adjusted based on the publication needs.

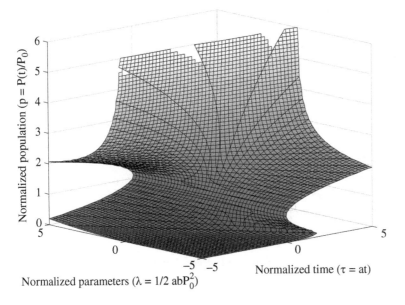

Fig. 10-1 Normalize population change using MATLAB.

```
grid on % Turn on a background grid
colormap bone % Setup colormap to "bone" which is a greyscale
```

It is frequently useful to place a grid on the plot. This is done by calling *grid on*. The color map can also be adjusted by changing the name after the *colormap* function. To generate greyscale images, the *bone* colormap was used.

```
caxis([-6,6]) % Color scale min and max values (for coloring)
view([-42.00 14.34]) % Viewing angle
```

Finally, the color scale axis can be adjusted with the *caxis* function. This sets the plot value for the maximum and minimum of the color scale. These only need to be adjusted if necessary. Finally, the *view* function is called to change the angle at which the 3D plot is viewed to explore the behavior of the system model (Figure 10-1).

10.5 Final Words

The presence of the vast, readily accessible, and continuously expanding resources of information and knowledge now available on the internet has a profound impact upon the practice of system science. These resources include WWW, Wikimedia, Google, and many other information and computerized data resources that provide significant benefits and utility for the field of system science. The ubiquitous personal cell phone is a clear example of our mounting dependence upon the age of the computer.

Truly, a new age of System Science exists and is exhilarating in its potential for effective system study and evaluation. Systems never approached in the past can now be seriously contemplated and examined. Hopefully, some of the great problems of the twenty-first century such as famine, terrorism, war, politics, religion, science, and cosmology are now all candidates for edification and profound inquiry and analysis. The potential understanding and resolution of these historical issues can result in understanding, acceptance, and achievement with a true renaissance for human life, hopefully with peace, justice, reconciliation, and abundant living for all humans. The anguish and fatalism of past philosophers such as Nietzsche and Schopenhauer over the final state and future of humankind may now be put to rest. Furthermore, the wisdom, knowledge, and insight of the great thinkers, the sages, the scientists, and the philosophers of the past available only in print are now open and easily accessible to the system scientist.

Finally, as we close this textbook on System Science, it is sobering, but thrilling to contemplate upon the application of system science on our being as individual humans. Consider the impact and significance of system science for the human family. An individual human being enters this physical existence as the biological combination of a genetic mass consisting of a sperm and ovum that is smaller than a kernel of wheat. Then, given favorable circumstances this zygote eventually develops into a feeling, sentient human being. If circumstances are favorable, this single individual zygote grows into a human being.

In summary, there are two essential, obligatory systems that we each inhabit and function as the dominant role throughout our lives. The two mandatory systems are:

1) our existence as a unique human being;
2) the cosmic home we inhabit.

The fundamental system model that each of us creates is our individual existence as human beings, viz., homo sapiens to which we each belong, witness, and declare as "ourselves." The second essential system is the earth and the cosmos that surrounds our terrestrial home and supports us.

Each of us, as "mortals," and perhaps as "immortals," comprise an open system crudely outlined, subject to change and transformation from the external inputs and feedback of our environment. The environment is the external universe that includes the earth as a celestial home in which the individual exists as a complex, engrossing system that humans only poorly comprehend and operate within.

Bibliography

ZENG, M. L., and J. QIN, *Metadata*, Neal-Schuman Publishers, Inc., NY, 2008.

Index

Introduction to System Science with MATLAB, Second Edition.
Gary Marlin Sandquist and Zakary Robert Wilde.
© 2023 John Wiley & Sons Ltd. Published 2023 by John Wiley & Sons Ltd.